Livestock Diseases and Management

This book series discusses the various infectious diseases affecting the livestock, principle of the disease control, and specific disease management. It discusses the existing strategies to control infectious disease includes animal management programs, vaccination, targeted antimicrobial use, and food hygiene.

Despite public health and veterinary public health improvement within the last century, animal populations remain vulnerable to health threats caused by infectious diseases.

It reviews the current understanding of the zoonotic, emerging, and transboundary animal infections in relation to their transmission, epidemiology, clinical and pathological effects, diagnosis and treatment.

It also examines the recent advancements in the veterinary diagnostics including the existing capabilities, constraints, opportunities, and future potentials. In addition, it elaborates on the conventional and recombinant vaccines that are used in the veterinary medicines and the molecular approaches that have led to the development of new vaccines in recent years. A volume focusing on the various water- and food-borne diseases and its impact on the domestic animals is also a part of this series.

The book series examines the emergence of antimicrobial resistance in livestock, ongoing global surveillance, and monitoring program, its impact on the animal-human interface and strategies for combating resistance.

More information about this series at http://www.springer.com/series/16404

Yashpal Singh Malik • Raj Kumar Singh •
Mahendra Pal Yadav
Editors

Emerging and Transboundary Animal Viruses

Springer

Editors
Yashpal Singh Malik
Biological Standardization
Indian Veterinary Research Institute
Izatnagar, Uttar Pradesh, India

Raj Kumar Singh
ICAR-Indian Veterinary Research Institute
Izatnagar, Uttar Pradesh, India

Mahendra Pal Yadav
ICAR-Indian Veterinary Research Institute
Izatnagar, Uttar Pradesh, India

Sardar Vallabhbhai Patel University of
Agriculture & Technology
Meerut, Uttar Pradesh, India

ISSN 2662-4346 ISSN 2662-4354 (electronic)
Livestock Diseases and Management
ISBN 978-981-15-0401-3 ISBN 978-981-15-0402-0 (eBook)
https://doi.org/10.1007/978-981-15-0402-0

This Springer imprint is published by the registered company Springer Nature Singapore Pte Ltd.
The registered company address is: 152 Beach Road, #21-01/04 Gateway East, Singapore 189721, Singapore

The book is dedicated to all the eminent virologists who own not only inventiveness and creativity but also astonishing compassion. Their resourceful contribution helped in accomplishing this mammoth compilation involving important aspects on animal viral pathogens/diseases having emerging and transboundary description.

Preface

Even if the livestock sector has become a backbone to maintain the economy of many countries, the emerging and transboundary (crosses the national borders) animal viral diseases pose a serious risk to the animal farming sector and food and nutrition security globally. The epidemics and pandemics of a few infectious diseases of animals and/or humans during the past couple of decades have highlighted the significance of emerging infectious diseases (EIDs) due to their direct impact on the economy, welfare and public health. Even though Asia has been recognized as the epicentre of many EIDs and upcoming infections, several new pathogens have also appeared in the recent past in other parts of the world. Additionally, the emergence of new viral diseases/infections, such as Rift Valley fever, West Nile fever, SARS coronavirus, Hendra virus, avian influenza A (H5N1), Nipah virus, Zika virus and swine influenza A (H1N1) virus, from time to time is the flagrant examples threatening adversely both animal and public health globally. Therefore, the emerging, re-emerging and transboundary viral infections have become the prime choice of researchers and public health workers. The leading reason behind the increase in the number of viral diseases is the absence of safe and inexpensive prophylactics and therapeutics. Besides, the control and management procedures adopted for restricting viral diseases remain challenging and require expertise in capacity building for detection and differentiation of the pathogen; development of rapid, sensitive and cost-effective pen-side diagnostic tests/kits; regional and peripheral diagnostic laboratories; clinical and sero-surveillance of the disease in the susceptible and in contact animal populations; border control for transboundary diseases; quarantine facilities, vector control, and restrictions on the movement of animals from the affected areas; and other relevant general health control measures, such as disposal of carcasses, zoo-sanitary measures and management practices.

Currently, numerous research articles/news are available on several animal pathogens. Nevertheless, the book dealing with emerging, re-emerging and transboundary animal viruses is limited. Such a resource is essential for the research community to understand the latest knowledge and trends in this field so that it can be utilized for improving the counteractions. To overcome these issues and fill the gaps, we have come up with a compilation on *Emerging and Transboundary Animal*

Viruses to provide a conversant resource in this area. The significance of viral diseases in animals in the light of the adverse economic impact on the farmers and the livestock industry is also highlighted using appropriate examples. This book provides precise and up-to-date information on animal viral diseases that have emerged in the recent past or are re-emerging due to several complex environmental interplay factors and the ones which are not bounded in restricted national boundaries and have attained the transboundary status. Conclusively, the chapters define present-day information on the existence of emerging, re-emerging and transboundary animal viruses in a global context, with emphasis on the molecular state-of-the-art tools with special reference to the development of diagnostics, prophylactics and therapeutics.

The book covers important viruses/viral diseases of economic/public health concern in animals in various chapters written by more than 50 authors (researchers/academicians/young investigators) from different parts of the world. Additionally, throughout the book, tables and figures provide important clinical data and recommendations, with specific references at the end for readers who want to obtain further details of each topic. In this book, we have included 15 chapters on important animal viruses. The first chapter (Chap. 1) by Dr. Yadav and colleagues provides an overview of emerging and transboundary animal viruses, highlighting the significance of EIDs on public health. This chapter oveviews the experience gained in the control and management of a few important transboundary animal diseases and EIDs along with the successes, constraints, limitations and future research needs for developing better control approaches. Chapter 2 on African Swine Fever Virus by Dr. Alexander Malogolovkin and colleagues from Russia provides virus epidemiology, immunopathobiology and diagnostics with a brief overview of recent advances of virus vaccine development. Similarly, comprehensive information is provided on Classical Swine Fever Virus in Chap. 3 by Dr. Sarma. Coronavirus infections are a serious threat to the swine industry. In Chap. 4, Dr. Vlasova and colleagues discuss the progress on Transmissible Gastroenteritis Virus (TGEV), Porcine Epidemic Diarrhoea Virus (PEDV) and Porcine Deltacoronavirus (PDCoV) of swine. An overview of Torque Teno Virus of swine is given in Chap. 5 by Dr. Ghosh and colleagues. Dr. Malik and colleagues provide an overview of Teschovirus of swine in Chap. 6.

The burden of Flaviviruses has made the whole world vulnerable towards its infection. Animals play a crucial role in Flavivirus life cycle where pigs are the amplifying host for Japanese encephalitis virus, and migratory birds are reservoirs for West Nile virus. Dr. Saxena and team in Chap. 7 provide a detailed account of the Animal Flaviviruses. In Chap. 8, Orbivirus, the largest genus of the family *Reoviridae*, is discussed by Dr. Maan's team. Main emphasis of the chapter is on Bluetongue virus with elucidating taxonomic relationships, epidemiology, replication mechanisms and evolutionary process of these viruses. The next chapter (Chap. 9) by Dr. Virmani's group elaborates the equine influenza virus giving its current situation globally. In the subsequent chapter (Chap. 10), Dr. Sudhakar and team provide an overview of Schmallenberg virus. A tick-borne virus, classified as a BSL4 agent, the Crimean-Congo Haemorrhagic Fever Virus, is explained in Chap.

11 by Dr. Raut and colleagues. In the next chapter (Chap. 12), Dr. Rajkhowa has given a detailed account of an economically important swine disease named Porcine Reproductive and Respiratory Syndrome Virus.

Dr. Balamurugan's team has dealt with Peste-des-petits ruminants virus, a small ruminant morbillivirus (SRMV) belonging to the family *Paramyxoviridae* in Chap. 13. It is highly contagious, OIE notifiable and economically important transboundary animal viral disease of domestic and wild small ruminants, known as 'Plague of Small ruminants'. An overview of Sapelovirus in swine is provided in Chap. 14 by Dr. Malik and team. The last chapter (Chap. 15) is on hepatitis E virus by Dr. Kumar and colleagues. This viral disease is considered highly significant on account of its predominance in both developed and developing nations due to poor sanitation and low-grade drinking water.

We believe that owing to the in-depth knowledge of important animal viruses with high-quality contributions by experts, the present book will be an excellent source of information for the reader. The chapters published could be useful for veterinary professionals, clinicians, public health experts, researchers, students/scholars, animal producers, faculty and students with an interest in virology, viral diseases, epidemiology of viral diseases, viral zoonoses and management of viral diseases and epidemics, the pharmaceutical industry and biomedicine experts and pave the way towards designing and adapting effective and safer therapeutics from clinics to the laboratory for countering important animal viral diseases.

We, the editors, would like to express our gratitude to all the contributors for their support and hard work to make this book compilation a reality. We also extend special thanks to all the peer reviewers whose able expertise and rigorous reviewing of the manuscripts submitted for this book helped the authors to reach publication stages. We are grateful to Springer Nature for accepting our book proposal and extend our special thanks to Dr. Bhavik Sawhney, Associate Editor-Biomedicine, Springer Nature, for providing all the editorial help and high cooperation while processing the manuscripts for its successful publishing.

Izatnagar, India Yashpal Singh Malik
Izatnagar, India Raj Kumar Singh
Izatnagar, India Mahendra Pal Yadav

"Emerging and Transboundary Animal Viruses": A Publication from World Society for Virology

About World Society for Virology

The World Society for Virology (WSV) is a nonprofit organization, 501c3-ID No. 001303257, that was established in 2017 with the mission to strengthen virology research on different viral diseases of humans, animals, plants, and others.

The main objectives of WSV include but not limited to:

1. Gather the virologists worldwide in the main society that does not require a fee for its membership [a great obstacle for many virologists in many countries] and provide help to all whenever possible.
2. Build up a network of scientific collaborations among virologists worldwide.
3. Build international bridges for virology laboratories worldwide.
4. Help virologists worldwide to advance their careers and obtain awards.
5. Provide educational resources free of charge and freely available to all members.
6. Help and facilitate getting scholarship and vacancies for virologists worldwide.
7. Build up databases of virologists based on their field of specialization for remote assistance and guide in case of the existence of any disease outbreak.

For details, visit www.ws-virology.org

About the Book

This book illustrates the prominence and implications of the emerging, re-emerging, and transboundary animal viruses which have become researchers' and public health workers' top precedence to triumph over the last few decades due to the consequential losses in animals as well as the unprecedented threats to the public. Even if the livestock sector has become the backbone to maintain the economy of many countries, the emerging and transboundary (crosses the national boarders) animal viral diseases possess a serious risk to the animal-agriculture sector and food security globally. The research outcomes of twenty-first century are transfiguring our capacity to respond to these animal healths defies swiftly and commendably through advent of urbane diagnostics tools, new generation prophylactic vaccines and therapeutic antimicrobials, and delivery systems. This book is comprised as an integrated approach to encompass comprehensive knowledge by a large team of more than 50 authors (researchers/academicians/young investigators) from different parts of the world. This book describes the precise and up-to-date information on animal viral diseases which have emerged in the recent past or are re-emerging due to several complex environmental interplay factors and the ones which are not bounded in restricted national boundaries and attained the transboundary status. The chapters provide information pertaining to the important viruses of livestock, emphasizing developments in the frontier research areas in studying emerging and transboundary animal viruses. Additionally, throughout the book, tables and figures comprehend the important clinical data and recommendations, with specific references at the end for readers who want to obtain further details of each topic. The significance of viral diseases in animals in the light of economic impact to the farmers and the livestock industry is also highlighted using apposite examples. Conclusively, the chapters define present-day information on existence of emerging, re-emerging, and transboundary animal viruses in global context with emphasis on the molecular state-of-the-art tools with special reference to the development of diagnostics, prophylactics, and therapeutics. As the chapters provided in this compilation are serious concern globally and suggestive of serious consideration at decision-making level, the book also explicitly describes the challenges imposed by the emerging and transboundary viral infections and our preparedness to counter them.

Contents

Editors and Contributors

About the Editors

Yashpal Singh Malik is presently working as "ICAR National Fellow" at the premier Veterinary Institute of the country—Indian Veterinary Research Institute (IVRI), Izatnagar, India. His major research achievements include contributions in viral disease epidemiology, virus-host interactions, microbial biodiversity, characterization, and diagnosis. He acquired advanced training in Molecular Virology from the University of Minnesota, Saint Paul, USA; Division of Virology, Ontario Research Institute, University of Ottawa, Ontario, Canada; and Wuhan Institute of Virology, Wuhan, China. He is a recipient of several prestigious national, state, and academy awards/honors including ICAR Jawaharlal Nehru Award (2001), Young Scientist Award of the Association of Microbiologists of India (2000), and Young Scientist Award from Uttarakhand Council of Science and Technology (2010). He is an active member of noted scientific and professional societies of international and national repute. He has been bestowed with several honors in the form of distinguished Associateships/Membership viz. Associateship of National Academy of Agricultural Sciences (2010); Membership—National Academy of Veterinary Sciences (2010); CSIR—Senior Research Fellowship (1997–2000); ICAR—Junior Research Fellowship (1995–1997); and Academic Merit Scholarship in bachelor's degree (1990–1995). He is an elected Fellow of the Indian Virological Society, Indian Association of

Veterinary Public Health Specialists, Indian Society for Veterinary Immunology and Biotechnology, and National Academy of Biological Sciences. Dr. Malik is a member of the International Committee on Taxonomy of Viruses (ICTV) on Birnaviridae and Picobirnaviridae study group and a managing committee member of the World Society for Virology. He has supervised 3 Ph.D. and 17 M.V.Sc. students. Over the years, he has developed several technologies and diagnostic kits and also has filed two national patents. He has authored 5 books, 25 book chapters, and published 2017 scientific research and review articles in peer-reviewed national/international journals of high impact factor. Dr. Malik has been the editor-in-chief of the *Journal of Immunology Immunopathology* and also edited a special issue of the Springer journal *VirusDisease* on "Enteric Viral Infections in Humans and Animals"; a special issue on "Emerging and Zoonotic Virus Challenges of Developing Nations" in Bentham's *The Open Virology Journal*; a special issue on "Therapeutic Advances and Their Biomedical Perspectives" for the *Journal of Current Drug Metabolism*; and a special issue on "Biomedical Perspectives of Advances in Disease Diagnosis and Therapeutics (BPADDT)" edited for the *Journal of Experimental Biology and Agricultural Sciences*.

Raj Kumar Singh is currently the Director-cum-Vice-Chancellor of the ICAR-Indian Veterinary Research Institute, Izatnagar. Dr. Singh is a noted scientist of high repute with specialization in veterinary microbiology, biotechnology, molecular epidemiology, diagnostics, and vaccinology. He has served as the Head of the Division of Virology, Station-in-Charge at IVRI, Mukteswar campus, Uttarakhand, and later as Director of the NRC on Equines and VTCC, Hisar. Dr. Singh has 10 national patents (granted—2 and filed—8), developed >8 live attenuated vaccines/vaccine candidates, and >26 diagnostic tests/assays/kits. He has authored 2 books, 23 book chapters, and published over 245 scientific research papers, 52 reviews, 15 lead papers, and 24 guest editorials/compendium chapters. Dr. Singh has supervised 8 doctoral and 11 master's students. He received several prominent awards including prestigious ICAR Rafi Ahmad Kidwai Award and Team

Research Award, DBT Tata Innovation Fellowship Award, Agriculture Research Leadership Award, FAO Fellowship for training at the University of California Davis, USA, and many others. He serves as the President of ISVIB and has several distinguished fellowships and life memberships of prestigious professional societies.

Mahendra Pal Yadav former Director, IVRI, Izatnagar, and Vice-Chancellor, Sardar Vallabhbhai Patel University of Agriculture and Technology, Meerut, is among the most renowned scientists working in the field of virology. The major research contributions to his name include the development of indigenous vaccines against equine influenza, infectious laryngotracheitis and colisepticemia in poultry, isolation and characterization of animal viruses, and development of several animal diseases diagnostic kits. Dr. Yadav has served as Professor of Virology (1981–1982) and Professor and Head of the Division of Virology at IVRI Mukteswar (1982–1987), Principal Scientist and In-charge of the Animal Health Unit (1987–1993), and Director of the National Research Centre on Equines, Hisar (1993–2000). He has also served as Director-cum-Vice-Chancellor of IVRI, Izatnagar (2000–2006), and later as Vice-Chancellor of Sardar Vallabhbhai Patel University of Agriculture and Technology, Meerut (2006–2009). As an eminent scientist, he is a recipient of several prestigious awards and fellowships including Fellow—National Academy of Veterinary Science (FNAVS); Fellow—National Academy of Agricultural Sciences (FNAAS), New Delhi, India; Fellow—Indian Association for Advancement of Veterinary Research (FIAAVR); President of Indian Association of Veterinary Microbiologists, Immunologists and Specialists in Infectious Diseases (IAVMI); Fellow—Indian Virological Society (FIVS); Fellow—Indian Society of Veterinary Immunology and Biotechnology (FISVIB); Fellow—Society for Immunology and Immunopathology (FSIIP); and Fellow—Royal Society of Crop Sciences (FRSCP). Dr. Yadav has received many prestigious awards including Chancellor's Medal 1966; Lance Award, 61 Cavalry, India 1996; ICAR Special Award 1998; Major (Mrs.) Malika IAAVR Award 2001; OIE International Meritorious Award (2000); Distinguished

Veterinarian Award 2002 of IAAVR; Dr. P. Richard Masillamony Oration Award 2003; Vocational Education and Development Award 2003; Higher Education and Development International Award 2004; Order of Merit Award in Management 2004, Swadeshi Vigyan Puruskar of Vigyan Bharti, India (2004); Dr. P. Bhattacharya Memorial Award of NAAS (2004); World Intellectual Property Organization (WIPO) Geneva Award (2005); NRDC Meritorious Invention Award (2005); Vigyan Gaurav Award of U.P. Council of Science and Technology (2005); Dr. C.M. Singh Samman 2005; IAVPH Life Time Achievement Award 2006; Society of Plant Research Eminence Award 2007; World Institution Building Programme Life Time Achievement Award, 2008; Lakhi Ram Memorial Award 2008; Lalbahadur Shashtri Honor 2008; Padma Vibhushan Dr. M.S. Swaminathan Agricultural Scientist Award 2009; IAVMI Dr. P. G. Pandey Oration Award 2010; ISACP Life Time Achievement Award, 2011; Dr. A. Lingard Memorial Award, 2012, by the Indian Academy of Environmental Sciences; Dr. M.R. Dhanda Memorial Oration Award, 2012, by IAPH; and Excellence in Virology Award of Indian Virology Society (2013). Dr. Yadav served as President, Indian Virological Society (1996–2003); President, IAVMI (1999 to date); President, IAUA (2006–2007); President NAVS(I) (2011–2013); FAO National Consultant; and Consultant, Haryana Farmers' Commission. He has expertise in areas such as veterinary virology, bacteriology, infectious diseases, vaccinology, biotechnology, and equine infectious diseases.

Contributors

Taruna Anand ICAR-National Research Centre on Equines, Hisar, Haryana, India

Houssam Attoui Institut National de la Recherche Agronomique (INRA), National Institute for Agricultural Research, UMR1161 Virologie, INRA-ANSES-ENVA, Paris, France

Balamurugan Vinayagamurthy ICAR-National Institute of Veterinary Epidemiology and Disease Information (ICAR-NIVEDI), Bengaluru, Karnataka, India

Manjunatha N. Belaganahalli One Health Institute, School of Veterinary Medicine, University of California, Davis, CA, USA

B. C. Bera ICAR-National Research Centre on Equines, Hisar, Haryana, India

Sudipta Bhat ICAR-Indian Veterinary Research Institute, Izatnagar, Uttar Pradesh, India

Kuldeep Dhama ICAR-Indian Veterinary Research Institute, Izatnagar, Uttar Pradesh, India

Pradeep Gandhale ICAR-National Institute of High Security Animal Diseases, Bhopal, Madhya Pradesh, India

Souvik Ghosh Department of Biomedical Sciences, One Health Center for Zoonoses and Tropical Veterinary Medicine, Ross University School of Veterinary Medicine, Basseterre, St. Kitts and Nevis, West Indies

Govindaraj Gurrappa Naidu ICAR-National Institute of Veterinary Epidemiology and Disease Information (ICAR-NIVEDI), Bengaluru, Karnataka, India

Amrita Haikerwal Center for Advanced Research (CFAR), King George's Medical University (KGMU), Lucknow, India

K. Jung Food Animal Health Research Program, Department of Veterinary Preventive Medicine, Ohio Agricultural Research and Development Center, College of Food, Agricultural and Environmental Sciences, The Ohio State University, Wooster, OH, USA

Nassim Kamar Department of Nephrology, Dialysis an Organ Transplantation, CHU Rangueil, INSERM U1043, IFR–BMT, University Paul Sabatier, Toulouse, France

Alyssa Kleymann Department of Biomedical Sciences, Ross University School of Veterinary Medicine, Basseterre, St. Kitts and Nevis, West Indies

Nobumichi Kobayashi Department of Hygiene, Sapporo Medical University, Sapporo, Hokkaido, Japan

Denis Kolbasov Federal Research Center for Virology and Microbiology, Pokrov, Russia

Diwakar D. Kulkarni ICAR-National Institute of High Security Animal Diseases, Bhopal, Madhya Pradesh, India

D. Senthil Kumar ICAR-National Institute of High Security Animal Diseases, Bhopal, Madhya Pradesh, India

Dinesh Kumar School of Bioengineering & Food Technology, Shoolini University of Biotechnology and Management Sciences, Solan, Himachal Pradesh, India

Harsh Kumar School of Bioengineering & Food Technology, Shoolini University of Biotechnology and Management Sciences, Solan, Himachal Pradesh, India

Naveen Kumar ICAR-National Institute of High Security Animal Diseases, Bhopal, Madhya Pradesh, India

Swatantra Kumar Department of Center for Advanced Research (CFAR), King George's Medical University (KGMU), Lucknow, India

S. N. Langel Food Animal Health Research Program, Department of Veterinary Preventive Medicine, Ohio Agricultural Research and Development Center, College of Food, Agricultural and Environmental Sciences, The Ohio State University, Wooster, OH, USA

Narender S. Maan Lala Lajpat Rai University of Veterinary and Animal Sciences, Hisar, Haryana, India

Sushila Maan Lala Lajpat Rai University of Veterinary and Animal Sciences, Hisar, Haryana, India

Yashpal Singh Malik ICAR-Indian Veterinary Research Institute, Izatnagar, Uttar Pradesh, India

Alexander Malogolovkin Federal Research Center for Virology and Microbiology, Pokrov, Russia
London School of Hygiene and Tropical Medicine, London, UK

Peter P. C. Mertens School of Veterinary Medicine and Science, University of Nottingham, Loughborough, UK
The Pirbright Institute, Woking, Surrey, UK

Anamika Mishra ICAR-National Institute of High Security Animal Diseases, Bhopal, Madhya Pradesh, India

Gheyath K. Nasrallah Department of Biomedical Science, Biomedical Research Center, College of Health Sciences, QA Health, Qatar University, Doha, Qatar

Atul Pateriya ICAR-National Institute of High Security Animal Diseases, Bhopal, Madhya Pradesh, India

S. Pavulraj ICAR-National Research Centre on Equines, Hisar, Haryana, India

Tridib Kumar Rajkhowa Department of Veterinary Pathology, College of Veterinary Sciences and Animal Husbandry, Central Agricultural University, Aizawl, Mizoram, India

Ashwin Ashok Raut ICAR-National Institute of High Security Animal Diseases, Bhopal, Madhya Pradesh, India

Parimal Roy Indian Council of Agricultural Research-National Institute of Veterinary Epidemiology and Disease Informatics (ICAR-NIVEDI), Bengaluru, Karnataka, India

Linda J. Saif Food Animal Health Research Program, Department of Veterinary Preventive Medicine, CFAES, Ohio Agricultural Research and Development Center, College of Veterinary Medicine, The Ohio State University, Wooster, OH, USA

Dilip K. Sarma Veterinary Microbiology, Assam Agricultural University, Guwahati, Assam, India

Shailendra K. Saxena Center for Advanced Research (CFAR), King George's Medical University (KGMU), Lucknow, India
CSIR-Centre for Cellular and Molecular Biology, Hyderabad, India

Alexey Sereda Federal Research Center for Virology and Microbiology, Pokrov, Russia

F. Singh ICAR-National Institute of High Security Animal Diseases, Bhopal, Madhya Pradesh, India

Raj Kumar Singh ICAR-Indian Veterinary Research Institute, Izatnagar, Uttar Pradesh, India

V. P. Singh ICAR-National Institute of High Security Animal Diseases, Bhopal, Madhya Pradesh, India

S. B. Sudhakar ICAR-National Institute of High Security Animal Diseases, Bhopal, Madhya Pradesh, India

Nadia Touil Faculté de Médecine et de Pharmacie, Centre de Génomique des Pathologies Humaines (GENOPATH), Université Mohamed V, Rabat, Morocco
Laboratoire de Biosécuritéet de Recherche, Hôpital Militaire d'Instruction Med V de Rabat, Rabat, Morocco

B. N. Tripathi ICAR-National Research Centre on Equines, Hisar, Haryana, India

Nitin Virmani ICAR-National Research Centre on Equines, Hisar, Haryana, India

Anastasia N. Vlasova Food Animal Health Research Program, Department of Veterinary Preventive Medicine, CFAES, Ohio Agricultural Research and Development Center, College of Veterinary Medicine, The Ohio State University, Wooster, OH, USA

Fun-In Wang School of Veterinary Medicine, National Taiwan University, Taipei, Taiwan

Q. Wang Food Animal Health Research Program, Department of Veterinary Preventive Medicine, Ohio Agricultural Research and Development Center, College of Food, Agricultural and Environmental Sciences, The Ohio State University, Wooster, OH, USA

Mahendra Pal Yadav Sardar Vallabh Bhai Patel University of Agriculture and Technology, Meerut, Uttar Pradesh, India
ICAR-Indian Veterinary Research Institute, Izatnagar, Uttar Pradesh, India

Chapter 1
Emerging and Transboundary Animal Viral Diseases: Perspectives and Preparedness

Mahendra Pal Yadav, Raj Kumar Singh, and Yashpal Singh Malik

Abstract The epidemics and pandemics of a few infectious diseases during the past couple of decades have accentuated the significance of emerging infectious diseases (EIDs) due to their influence on public health. Although Asia region has been identified as the epicentre of many EIDs and upcoming infections, several new pathogens have also emerged in the past in other parts of the world. Furthermore, the emergence of new viral diseases/infections, such as Rift Valley fever, West Nile fever, SARS coronavirus, Hendra virus, avian influenza A (H5N1), Nipah virus, Zika virus and swine influenza A (H1N1) virus, from time to time is a glaring example threatening adversely both animal and public health globally. Infectious diseases are dynamic and concerning due to their epidemiology and aetiological agents, which is manifested within a host, pathogen and environment continuum involving domestic animals, wildlife and human populations. The complex relationship among host populations and other environmental factors creates conditions for the emergence of diseases. The factors driving the emergence of different emerging infectious disease (EID) interfaces include global travel, urbanisation and biomedical manipulations for human EIDs; agricultural intensification for domestic animal EIDs; translocation for wildlife EIDs; human encroachment, ex situ contact and ecological manipulation for wildlife–human EIDs; encroachment, new introductions and 'spill-over' and 'spill-back'; and technology and industry for domestic animal–human EIDs. The concepts of sanitary and phytosanitary (SPS) measures and biosecurity have gained recognition globally in almost all the realms of human activities, including livestock health and production management. This chapter provides the experience gained in the control and management of a few important

M. P. Yadav
Sardar Vallabh Bhai Patel University of Agriculture and Technology, Meerut, Uttar Pradesh, India

ICAR-Indian Veterinary Research Institute, Izatnagar, Uttar Pradesh, India

R. K. Singh · Y. S. Malik (✉)
ICAR-Indian Veterinary Research Institute, Izatnagar, Uttar Pradesh, India

Y. S. Malik et al. (eds.), *Emerging and Transboundary Animal Viruses*, Livestock Diseases and Management, https://doi.org/10.1007/978-981-15-0402-0_1

TADs and EIDs along with the successes, constraints, limitations and future research needs for developing better control approaches.

Keywords Transboundary animal disease · Infectious animal diseases · Emerging diseases · Biosecurity and biosafety · Animal virus diseases · Disease management

1.1 Preamble

A 'disease' may be defined as a condition of the living human, animal or plant body or one of its parts that impairs normal physiological functioning and is typically manifested by clinical symptoms and signs. Animal and human diseases are classified in several ways depending on the criteria, such as infectious and non-infectious diseases, contagious and non-contagious diseases, zoonotic and non-zoonotic diseases, and acute and chronic diseases. The infectious diseases are caused by virus, bacteria, mycoplasma, fungi or rickettsia depending on the nature of the etiological agent involved. Among the infectious diseases, viral diseases are most devastating and difficult to control as they usually spread very fast and unlike other infectious diseases have no cost-effective or safe antiviral antibiotics/drugs for the treatment. A new disease occurring in an area/host population spread further and get established in want of awareness, lack of diagnostic facility, expertise or other factors is termed as 'emerging disease'. While disease of past, re-appearing in an area having a susceptible host population(s), is termed as 're-emerging disease' (Lederberg et al. 1992; Daszak et al. 2000). Example of the latter is the re-emergence of human tuberculosis in many countries which earlier had negative status as a result of judicious control measures adopted; it re-appeared subsequently due to the spread of immunosuppressive human AID's virus, namely human immunodeficiency virus (HIV). Introduction of susceptible animal populations/breeds from abroad for cross-breeding to upgrade the local breeds in India and other developing countries has resulted in the re-emergence of protozoon parasitic infections, such as theileria, babesia and anaplasma in the imported stock or their crossbreeds. The disease which is introduced from abroad into a country is termed as 'exotic'.

The infectious diseases are dynamic concerning their epidemiology and etiological agents which is manifested within a host, pathogen and environment continuum involving domestic animals, wildlife and human populations. The complex relationship between these host populations and other environmental factors creates conditions for disease emergence. The spill-over of the new emerging disease may be from 'domestic animals to wild life', 'wildlife to humans' or 'domestic animal to human' or in all categories. The factors driving the emergence of different EID interfaces include global travel, urbanisation and biomedical manipulations for human EIDs; agricultural intensification for domestic animal EIDs; translocation for wildlife EIDs; human encroachment, ex situ contact and ecological manipulation for wildlife–human EIDs; encroachment, new introductions and 'spill-over' and 'spill-back'; and technology and industry for domestic animal–human EIDs.

1.2 Emerging Diseases

The term 'emerging disease' is used to refer to changes in the disease dynamics in the population. Emerging infectious diseases (EIDs) are those which have moved recently into a new host or have enhanced incidences or geographic range or are caused by evolving pathogens (Lederberg et al. 1992; Daszak et al. 2000). This general definition covers a range of infectious diseases of man and animals which pose a significant threat to both medical and veterinary public health. Among the OIE-listed diseases of viral aetiology, major changes have been experienced in the occurrence of rinderpest, peste-des-petits ruminants (PPR), foot-and-mouth disease (FMD), African swine fever (ASF), lumpy skin disease and Rift Valley fever (RVF). Of these, rinderpest presents a success story from the 1990s to 2011 as a result of FAO, OIE, EU and IAEA (International Atomic Energy Agency) guided and co-ordinated programmes including the Pan African Rinderpest Campaign (PARC), NPRE and NREP in India (Yadav 2011), Global Rinderpest Eradication Program (GREP) of the FAO and other national governments where the disease was endemic. These exemplary efforts led to the historic declaration of global rinderpest eradication by the FAO on June 28, 2011.

1.3 Transboundary Diseases

The terms 'exotic disease' and transboundary animal diseases (TADs) are often used interchangeably. Though all transboundary diseases are of exotic origin, all exotic diseases are not included in TAD listing. Many EIDs are also transboundary diseases. The TADs are defined as highly contagious and transmissible epidemic diseases of livestock which have the capability for rapid spread to new areas and regions regardless of national borders and have serious socio-economic and public health consequences. Nearly all diseases affect livestock, poultry, fishes and other animals and adversely impact the quality and quantity of food and other products, such as hides and skins, bones, fibres, wool and animal draft power for tilling, transport and traction. The reduction in animal production, productivity and profitability due to TADs affect the human livelihood. In the present scenario of fast-increasing globalisation, TADs represent a serious threat to the economy and welfare of the public and affected nations as they drastically reduce production and productivity; disrupt trade and travel and local and national economies; and also threaten human health through inferior food quality and zoonotic diseases/infections. As such, consequences of TADs could have a significant detrimental effect on the economy and public health of not only the affected nations but also the whole of the world.

1.3.1 Source of the Pathogens of EIDs and TADs

Possibly the infectious agents which cause emerging and transboundary diseases are already present in the environment and get the opportunity to cause disease under certain altered circumstances. The transmission of the infectious agent could occur between animal and human; between wildlife, human and domestic animals; or between wildlife, domestic animal(s) and human. However, the main source for maintenance and transmission of the infectious agents in nature is determined by the zoonotic pool and spill-over and spill-back mechanisms.

1.3.2 Transboundary Diseases as Potential Threats

TADs have become of great concern due to the risk for national security on account of their economic significance, zoonotic nature and ever-growing threat of newer TADs in future. Among the TADs having zoonotic manifestations, a number of infectious diseases, such as highly pathogenic avian influenza (HPAI), BSE (Mad cow disease caused by prion), West Nile fever, Rift Valley fever, SARS coronavirus, Hendra virus, Nipah virus, Ebola virus, Zika virus and CCHF, to name a few, adversely affecting animal and human health have been in the news in recent times (Malik and Dhama 2015; Munjal et al. 2017; Singh et al. 2017, 2019). The direct and indirect costs due to the FMD outbreak in the UK in 2001 were assessed to be over US$9 billion. Over 150 million chicken died or were destroyed in Southeast Asia in 2004 to control HPAI (H5N1). The Netherlands suffered an economic loss of $2.5 billion due to classical swine fever in 1997–1998. As per the estimates of FAO nearly one-third of the world meat trade was facing import bans on account of BSE, HPAI and other animal diseases. There is evidence to suggest that threats from TADs have increased over the years. The risk of animal disease outbreaks is likely to further grow in future as the higher incomes of people in developing countries will generate more demand for animal protein and products (milk, meat, egg, chicken and fish). The number of animals raised for meat is growing rapidly. During 1990s poultry production in East Asia has increased by about 12% per year to double every 5–6 years. Similar to TADs, new human viral diseases have emerged like Ebola, SARS, Zika, CCHF, Nipah and BSE as well as there is the emergence of new antigenic forms or new biotypes of the existing infectious diseases, such as a hypervirulent strain of IBD in poultry in Europe and highly virulent strain of Newcastle disease in the USA (Riemenschneider 2005; Singh et al. 2017). Vector-borne pathogens, namely, bluetongue, African horse sickness, Rift Valley fever and West Nile fever, have the potential to spread in epidemic forms. Riemenschneider (2005) has deliberated over several issues relevant in the control of TADs as proposed in the Institute of Medicine (IOM) Report (Anonymous 2003). Some of the points which could be responsible for the increased threat of TADs are briefly discussed below.

1.3.2.1 Globalisation and Trade

In the present-day world, higher quantitative levels of animal origin foods, as well as faster trade, new trade routes and air travel, have led to higher risks for contracting new infections and diseases. As it is now possible to reach any part of the world within 24 h which is less than the incubation period of most of the infectious diseases, animals or people carrying the infectious agents go undetected in want of clinical disease/symptoms. Fresh commodities vis-à-vis processed foods that have witnessed an increased trade are more likely to carry the pathogens to distant parts of the world—countries and continents.

1.3.2.2 Intensive Animal Production Systems to Meet the Rising Demand for Animal Protein

Recent decades are witnessing higher demands for animal protein and other nutrients through meat and meat products, milk and milk products, eggs, and fish and fish products as a result of rising incomes in the developing countries and elsewhere which leads to the intensification of production systems and overcrowding of animals. This increased production is often required in peri-urban areas, having large human populations, under suboptimal husbandry practices. In such high-production areas, disease outbreaks affect a greater number of animals at a faster rate and speed, leading to heavy economic losses. Drastic control measures are taken, such as the slaughter of infected and in-contact animals followed by burning or burial is not acceptable to the society at large. For example, the mass slaughter of pigs in the Netherlands in 1997–1998 for the control of CSF virus led to objection from the non-farm population which might influence the application of the stamping-out policy as a disease control approach in future.

1.3.2.3 Impact of Changes in Forest Ecology

Exposure of the domestic animals to forest niches due to deforestation and transformation of tropical rainforests for livestock grazing exposes the domestic livestock to a completely new range of pathogens and vectors which previously circulated in wildlife reservoir niches only. With the domestic livestock being fully susceptible and naïve to these infectious agents, the disease spreads more rapidly and severely in want of lack of diagnostic tests and vaccines against these new pathogens resulting in heavy morbidity, mortality, trade restrictions and economic losses.

1.3.2.4 Influences of Increased Conflicts and Unrest

Nowadays many countries face prolonged civil unrests besides inter- and intra-country conflicts, which may lead to enhanced threat of TADs. Civil disorders are known to disrupt enforcement of quarantine and other control measures due to refugee and army deployments/movements. Breakdown in the institutional support for quarantine and difficulty in gaining access to border area due to landmines make disease surveillance more difficult. Inflows of more food aids for such areas also pose additional risks as the food items may have contaminants.

1.3.2.5 Effect of Climate Change, Global Warming and Microbial Evolution

Climate change and global warming seem to be altering rainfall and weather patterns. Rising temperatures in the northern hemisphere are likely to shift the distribution of insect vectors of bluetongue, African horse sickness, Rift Valley fever and similar vector-borne diseases. The bluetongue virus (BTV) having 27 serotypes occurs in many parts of the world. However, until recently it was never reported from Europe. The sudden incursions of some serotypes into Spain, Italy, Greece, Portugal and the Balkan countries since 1998, followed by Germany, and the recent incursion of BTV serotype 8 in several farms in the Netherlands, Germany and Belgium since August 2006 as well as serotype 1 are also believed to be due to climate change as European weather has become hotter in recent decades. The BTV serotype 8 revealed that this serotype is closest to the Nigerian strain. The incursion is believed to have been caused by the importation of an infected zoo animal or an infected midge. An upsurge of Rift Valley fever was observed in East and West Africa due to climatic changes.

1.3.2.6 TADs as a Serious Threat to National Security

Many factors discussed above make the TADS as a serious threat to national and international security. The developing countries are usually the worst sufferers. Among other factors, veterinary public health services in developing countries are usually much behind than the medical public health services. Moreover, unlike human disease reporting, animal disease reporting systems are usually based on passive reporting rather than active disease surveillance. A few other factors are also responsible for greater threat due to TADs, namely (1) lack of awareness of the farmers about the high-threat epizootic animal diseases; (2) lack of diagnostic facilities for exotic diseases, and under-reporting of animal diseases like HPAI due to the fear of loss of internal and export market till the country gets infection-free status as per OIE-laid-down criteria; and (3) poor and faulty compensation schemes.

1.4 Threat of Bioterror/Biowarfare

In the technological advances made in today's world, there is always a real risk of deliberate misuse of certain infectious agents/pathogens by terrorists as a means of biowarfare between nations to harm the people and/or livestock, poultry and other animals. Potential for pathogenic disease agents not reported previously in a country and being misused or mishandled for bioterrorism is likely to threaten the ecosystem on a large scale. Even new pathogens can be engineered as novel infectious agents. The animal diseases could even be a greater threat than human diseases as these may result in significant economic disruptions, besides causing food poisoning and deterioration, and zoonotic diseases in human beings. As animal diseases get less priority than human infections/diseases in undertaking immediate disease control measures, the threat scenario with the use of animal pathogens for bioterrorism or biowarfare will have many serious consequences. Some of the viruses having significant bioterror potential for humans and or animals include HPAI (H5N1), monkeypox virus, FMD virus, yellow fever virus, Spanish flu virus, poliovirus, AIDS virus (HIV), measles virus, Hendra virus, Nipah virus, SARS coronavirus, BSE prion agent and unknown agent(s) created through biotechnology and gene editing.

1.5 Handling of New Viral Disease Outbreaks

When an exotic viral disease strikes a country for the first time, it may initially affect one animal, few animals or a large number of animals. The strategy to be adopted for containing the outbreak will depend on the nature of the virus, speed of its spread, role of vectors, risk assessment, communication and management, response time and country legislation on disease control and prevention. Thus, there is a need to develop strategic plans for the prevention and control of exotic and TAD on a case-to-case basis. Examples of such viral diseases from Indian perspective include African swine fever (ASF), transmissible gastroenteritis (TGE), and swine vesicular disease in pigs, Rift Valley fever, African horse sickness (AHS), West Nile fever, Eastern equine encephalomyelitis (EEE), Western equine encephalomyelitis (WEE), and Venezuelan equine encephalomyelitis (VEE), FMD virus types 'C', 'SAT I', 'SAT II' and 'SAT III', Nipah virus, Hendra virus, SARS coronavirus, and prion diseases—bovine spongiform encephalopathy (BSE), and scrapie.

1.6 Biosecurity and Biosafety Measures to Combat Viral Infections

Institution of appropriate and timely biosecurity measures is an important instrument for the protection and improvement of animal health. Breach in biosecurity due to ignorance and avoidable lapses in the adoption of timely biosecurity and biosafety measures in the management of livestock, poultry and fish minimise the risks from infectious diseases including EIDs and TADs. Breach in biosecurity in livestock management is often an important reason for the high incidence of zoonotic and other infectious diseases of animals. This is more so in case of the viral diseases of livestock and poultry. Closer contact between wildlife, animals and humans and rearing of livestock and poultry in close association with people promote spread of viral and other infectious diseases which have the potential for threatening health, economies and food security around the world. The emergence of new viral diseases/infections, such as Rift Valley fever, West Nile fever, SARS coronavirus, Hendra virus, avian influenza A (H5N1), Nipah virus, Zika virus and swine influenza A (H1N1) virus, from time to time is a glaring example of zoonotic disease threats adversely affecting both animal health and public health, national economies and food and nutrition security globally. Due to a lacuna in the biosecurity, viral diseases like the FMD had reoccurred in countries where these had not been reported for many decades, including the UK, a developed country.

1.6.1 Biosafety and Biosecurity

Biosafety and biosecurity are interrelated terms but used in different contexts. The guidelines are developed by WHO, FAO and OIE. Biosafety aims at the protection of person(s) at work and the facilities which are dealing with the biological agents, against their exposure to a disease agent, and prevents unintentional exposure to pathogens/toxins or their accidental release. Thus, biosafety is the application of knowledge, techniques and equipment to prevent personal, laboratory and environmental exposure to potentially infectious agents or biohazards. Biosecurity, unlike biosafety, has divergent meanings in different contexts in which it is used. It deals with the protection of microbiological assets from spill-over, theft, loss, diversion or intentional release from laboratories, preventing the import of certain organisms/ toxins. Biosecurity is a set of preventive measures designed to reduce the risk of intentional transmission of infectious diseases to safeguard the facilities containing sensitive biological materials with the potential of a biological weapon. In brief, biosecurity means bio-risk management. Once a disease is eradicated globally, the policy for keeping the wild and vaccine strains of the virus along with vaccine stocks for emergency use and their subsequent destruction is decided by international agencies like FAO, WHO and OIE based on the recommendations of experts in the area.

1.6.2 Biosecurity Policies, Protocols and Action Plans

These include risk assessment; communication and management; quarantine of imported animals at seaports, dry ports and farm; establishment of check posts and vaccination stations at international and interstate borders for clinical surveillance; creation of immune belts at international borders; and planning and conducting structured disease surveillance including clinical surveillance and serosurveillance. Biosafety and biosecurity need to be observed at all levels beginning from farm to national and international levels. For handling the most dangerous transboundary disease pathogens, BSL III and BSL IV laboratories are required to ensure biosafety, biosecurity and biocontainment. Proper zoo sanitary measures, such as quarantine; rodent and vector control; disinfection of animal sheds and premises; proper disposal of dung, urine, feed and fodder wastes; and proper carcass disposal, need to be adopted religiously for effective management of EIDs and TADs.

1.6.2.1 Biosecurity at International Borders

Every country needs strict and foolproof biosecurity mechanism at its international borders as a safeguard against the entry of exotic infectious agents/diseases from abroad along with the import of livestock and other animals and their products. For example, India has contiguous and porous borders with countries like Nepal, Bhutan, Pakistan and Bangladesh, besides free trade with Nepal and Bhutan. Since all these countries are vulnerable to TADs, there is a need for regional biosecurity plan to ensure a biosecure region. It would never be possible to have a biosecure country if the bordering countries do not have effective biosecurity in place.

Different countries are at risk for a number of TADs like anthrax, plague, glanders, Lyme disease, contagious equine metritis, *Salmonella abortus equi*, HPAI virus, FMD virus (SAT 1–3), Lyssavirus, rabies, Hendra and Nipah viruses, West Nile virus, highly pathogenic ND virus, rabbit haemorrhagic disease virus, bovine spongiform encephalopathy (BSE), African horse sickness (AHS), equine encephalomyelitis (EEE, VEE, WEE), equine infectious anaemia, chicken infectious anaemia, equine influenza, vesicular stomatitis, Rift Valley fever, malignant catarrhal fever (MCF) and other TSEs of sheep, goat and deer. Biosecurity measures are required for preventing and containing the ingress of these diseases through international trade. The OIE has facilitated safe trade in animals and animal products by developing effective standards to prevent the spread of animal diseases across the globe. Prevention of transmission of pathogens across intra- and inter-country borders warrants devising of biosecurity measures at par with international standards. Adequate infrastructure comprising check posts and quarantine facilities at seaports, airports and porous international land border are must to check the ingress of viral and other pathogens from across the borders. Diagnostic facilities with trained human resource, and well equipped with instruments and pen-side diagnostic tests/kits, should be in place for ensuring the pathogen-free status of imported livestock and livestock products.

1.6.2.2 Biosecurity Measures at National Level

Biosecurity measures at national level incorporate the components of 'external biosecurity' preventing the ingress of exotic and transboundary animal diseases and 'internal biosecurity' within the country encompassing zonal, compartmental and farm-level biosecurity. Regulations for animal movement through interstate borders in India are in place but need strict implementation. Modern detection systems can be used for identification and tracking of animals and animal products to provide information regarding the origin of the animal, and environmental practices used in production and food safety.

1.6.2.3 Farm Biosecurity

For effective disease prevention and control, integration of biosecurity into every operation at the farm is essential. Farm biosecurity should be inclusive of both 'bio-exclusion' (measures for preventing a pathogen from being introduced to a herd/flock) and 'biocontainment'. The latter addresses the events after the introduction of the pathogen and its ability to spread among susceptible groups of animals at the farm or further spill-over to other farms.

Strict implementation of biosecurity at farm level has played a crucial role in preventing the spread of diseases. A suitable plan addressing important issues, such as location and layout of the farm, animal health practices in place and general management on the farm, needs to be chalked out. It should be flexible to include new knowledge, concepts and technology. A wide range of biosecurity practices have been recommended for different livestock species and production systems, both for specific infection risks or for disease prevention in general. Biosecurity practices have been recommended for cattle, sheep, pig, poultry and fish production systems.

General biosecurity practices and interventions that can be applicable across species and farms include:

1. Maintaining a closed herd procurement/purchase of animals from known sources
2. Minimising the number of animals purchased/transferred/exchanged and the number of herds from which the animals are introduced
3. Avoiding purchases from markets or dealers
4. Appropriate quarantine and testing of animals upon introduction or reintroduction in farm premises
5. Discouraging farming practices such as hiring a bull or stallion and returning it after the breeding season
6. Avoiding the introduction of biological material of uncertain health status
7. Health and vaccination records should be obtained for all the newly introduced animals

Isolation/quarantine of such animals for 2–8 weeks in a separate quarantine facility should be practiced and the animals during this period should be observed for illness/symptoms and screened for important diseases before mixing with other stock at the farm. Laboratory testing of appropriate samples collected during quarantine against important infectious diseases is recommended. The incoming stock can also be given vaccine against the endemic disease prevalent in the area at least 2 weeks before release from quarantine to boost their protective immunity.

Animal diseases can spread from farm to farm resulting in animal sickness, death and economic losses. Visitors to the livestock farm, disease laboratory, birds, rodents, vehicles, feed and fodder and other inanimate objects are often a source of infection. In addition to adverse effects on the economy, there can be negative effects on the environment and human health. The best designs are to implement effective biosecurity practices. Baths by the laboratory workers after and before visiting animal farm or laboratory and putting on gum boots, disposable overall, head gear and gloves should be a mandatory requirement. All effluent from the laboratory should be pre-treated to ensure freedom from pathogens before their disposal to the environment. Disinfectant foot bath for the workers and vehicles entering the farm at the gate, exclusive separate dress and shoes for laboratory and farm workers, minimum movement of the people and animals within the farm during the outbreak period, and personal health and hygiene of the staff are some of the minimum guidelines to strengthen farm biosecurity.

1.7 Early Disease Reporting and Compensation Play a Key Role in Animal Health Management

Timely, rapid and accurate disease reporting based on OIE-approved diagnostic tests is a must for effective detection of the pathogen and instituting early response without giving much time for the disease to spread further. To face the new exotic diseases, it is recommended to have a standard SOP in place along with technical guidelines, decision-taking levels along with adequate provision for funds and legal backup. It has been observed that lack or inadequate compensation for culling the diseased and in-contact animals and poultry and negative effect on the sale, sale price and exports deter the farmers from reporting animal diseases in time which eventually leads to the spread of TADs. For example, due to BSE cattle producers in the USA lost over $1 billion in exports to Japan. Similarly, HPAI resulted in over $1 billion loss in poultry exports for Thailand. Trade concerns also discourage the use of preventive vaccinations for some diseases such as HPAI and FMD. Disease-free countries are generally reluctant to import animals or animal products from the countries practicing preventive vaccinations. With these adverse trade considerations, stamping-out policy was adopted by the UK over the vaccination for FMD in 2001 outbreak. From a public health point of view, a vaccination programme might reduce the viral load circulating in a country and thus reduce the risk of HPAI spreading to humans. However, stamping out rather than vaccination is preferred by most of the countries to declare themselves free from disease/infection at the earliest possible to regain access to exports. To ensure the cooperation of livestock farmers, it is essential to provide adequate and timely compensations to reduce the losses suffered by them on account of culling and closing the units for a few months.

1.8 Failure in Timely Disease Reporting Hastens Spread of TADs

The failure of timely disease reporting hastens the spread of TADs within the country as well as between countries. It is believed that the HPAI (H5N1) avian influenza virus might have been circulating in the poultry for months in the affected region before it was reported to the international authorities in 1996 leading to the wide spread of the disease/infection. An ex ante study of the 2001 FMD outbreak of the UK suggested that the FMD virus was probably introduced 3–4 weeks before it was reported and followed by a ban on livestock movements. Earlier reporting and ban on animal movements would have cut the spread of the disease by about 40%.

1.9 Steps for Control of EIDs and TADs

It is believed that the EIDs and TADs will continue to remain an ever-growing threat to animal and human health, economic sustenance of the world and global environment well-being. However, it is difficult to predict the number of these diseases which could rapidly escalate in a country or region threatening the animal and human life as well as the economy of that region or nation. The rising global demands for meat, particularly in East and South Asia, have put humans and animals together in numbers never seen before in the world. The fact is that the farm biosecurity in these countries where meat production is growing most rapidly is often poor. This scenario creates a great scope for animal diseases to jump species to create human health problems. Some of these issues were thoroughly described in the IOM Report, 2003, and further debated and discussed by Riemenschneider (2005). The steps suggested include early detection and early response, preparedness plans, decentralisation of government structure, international coordination, understanding of ecology, microbial evolution and viral traffic, expanded surveillance system, disease intelligence, preparedness, collaboration and cooperation among government agencies and cross-field partnerships.

1.9.1 Preparedness Plans

In developing countries, the preparedness plans for animal diseases are often unsatisfactory. Incentives such as adequate compensation should be provided to the affected farmers as an impetus for reporting animal diseases. The level of preparedness should be assessed by conducting mock drills. This will help in confidence building for rapid detection and response to both EIDs and TADs that appear suddenly and are capable of spreading to large areas in a short time. Import bans in response to an animal disease outbreak must be based on sound scientific evidence to ensure that the concerned countries also have the incentive to report the disease to international agencies, namely OIE, FAO and WHO. Deficiencies in national veterinary services have been attributed for inability in early detection of the disease and response as investigation, and diagnostic services have deteriorated in many regions. A continuing structural upgradation programme for national veterinary services will have to be taken into account for their transformation from providers of services, such as diagnosis, vaccinations and treatment of sick animals to inspection and quality assurance services. Disease surveillance, early warning and emergency preparedness need to be pursued more vigorously in Africa, the Middle East and Southeast Asia as vital components of national veterinary services.

1.9.2 Decentralisation of Authority

Though public health and national security are under the perspective of national governments, the decentralised government structure and improved international coordination are essential to address the threat of TADs effectively as they do not respect local, regional or national boundaries. Nevertheless, government support at the administrative level is essential to assure sufficient and timely response to avoid the spread of disease through livestock movement controls, closing of live markets, sharing of diagnostic services, expertise, funding, etc.

1.9.3 International Coordination

Technical support and guidance of international agencies, such as FAO, WHO and OIE, are key in the formulation and timely implementation of the plans and modalities for the control and management of EIDs and TADs. The FAO in 1994 established the Global Framework for the control of TADs (GF-TAD) through the Emergency Centre for Transboundary Animal Disease (ECTAD) operations and Emergency Prevention System (EMPRES) for transboundary animal and plant pests and disease initiatives for early warning and response to disease threats, following a collaborative approach to investigation at animal–human–ecosystem interface. These mechanisms have proved to be of immense help and use in the control, prevention and eradication of disease(s).

1.9.4 Understanding of Ecology, Microbial Evolution and Traffic

Microbial evolution, particularly viral evolution, is a continuous process. It is, therefore, necessary to conduct basic research on emerging infectious diseases, both viral and another microbial origin, for providing new insights about the factors responsible for the emergence of new microbes. For understanding the ecology of disease, social factors, viral and microbial traffic and spread, ecological and demographic changes in human and animal populations due to migration and other factors work in tandem leading to precipitation of emerging infections. These signals for viral and microbial traffic should be seen as warning signs. Biodiversity should include microbes and viruses, and environmental impact assessment should include health aspects into account in development planning.

1.9.5 Expanded and Robust Surveillance System

Enhancing surveillance systems by establishing laboratory response network at national, regional and international level is important for which adequate funds should be provided. By linking the laboratories in public and private domains, such networks are expected to enhance the capabilities at all levels to detect and prevent the spread of EIDs, transmitted naturally or intentionally (Anonymous 2003). A network of more than 100 laboratories world over by WHO for a constant survey of influenza viruses is one of the best examples of networking of laboratories for EIDs and TADs. These laboratories should have multidisciplinary teams involving veterinarians, physicians, ecologists, entomologists, vaccinologists, epidemiologists, molecular biologists, immunologists and possibly other specialists.

1.9.6 Disease Intelligence

State-of-the art disease surveillance is required having the capability to forecast when and where a particular disease is likely to occur for more targeted surveillance. Such actionable intelligence may derive from the analysis of changes in climatic conditions, vegetation, wildlife demographics, trade pattern or vector demographics and distribution (Anonymous 2003).

1.9.7 Updating of the Vaccines as per the Current Antigenic Types and Genotypes

The disease-producing microbes, particularly viruses and bacterial agents, often change their antigenic make-up as a result of spontaneous mutations, and immune pressure when the wild strains of the infectious agent persist in the host in the presence of vaccinal antibodies. RNA viruses having segmented genome are more prone to such antigenic changes as a result of recombination, gene deletion, etc. Influenza A viruses of human and animals continuously evolve new virulent variants by exchanging haemagglutinin (H) and neuraminidase (N) genes of various H and N types circulating in human, birds, pigs and other species including equines. With the change in antigenic make-up, the current vaccine strains do not provide protection against the new types of the virus. Similar situations occur in FMD virus having seven types and further subtypes, clades and genotypes: PPR virus and Newcastle disease virus, to name a few. New antigenic types of a virus or pathogen may also be introduced from abroad through imported livestock and poultry. Hence, there is a need to have a plan in place to upgrade the vaccine by incorporating the current strains of the pathogen which induce strong and lasting immunity. This will require the setting of repositories of field strains isolated from disease outbreaks, particularly the ones from vaccine failure cases. Such updates of vaccines are routinely followed for influenza vaccines for poultry, equines and human, and FMD and CSF vaccine for livestock.

1.10 Lessons Learnt from TADs

1.10.1 Benefits of Vaccination

Vaccination is a valuable and well-tested method in preventive veterinary medicine for promoting animal health and welfare and reducing the risk of human exposure to several zoonotic pathogens. Prophylactic immunisation practices, principles and vaccination protocols have helped in significantly reducing the prevalence of many life-threatening viral and bacterial diseases. The risks of not vaccinating their stock on account of lack of awareness among the stakeholders, non-availability of cost-effective diagnostics and vaccines, and poor delivery of veterinary services to the livestock farmers can have serious consequences on livelihoods of rural livestock producers. Effective vaccination programmes if implemented properly with a broader perspective are likely to reduce the need for antimicrobials, which in turn can help reduce the risk of emergent antimicrobial resistance. The World Veterinary Association (WVA) and Health for Animals believe that it is essential for the global veterinary profession to educate the public, particularly animal keepers and producers, about the benefits of vaccination for animals and humans.

The major objectives and motive of veterinary vaccines are to protect, improve and promote the health and welfare of companion and food animals; increase the production of livestock in a cost-effective manner; and prevent animal-to-human transmission of infectious diseases from domestic animals and wildlife to humans through animal-origin food, close contact and other mechanisms. These diverse aims have led to different approaches to the development of veterinary vaccines from crude but effective whole-pathogen preparations to molecularly defined subunit vaccines, genetically engineered organisms or chimeras, vectored antigen formulations and naked DNA injections for immunisation of animals.

1.10.2 Research on Vaccines and Vaccinology

The final successful outcome of vaccine research and development is the generation of a product that will be available in the marketplace on demand and suitable to be used in the field to achieve desired outcomes. Successful veterinary vaccines have been produced against major bacterial, viral, protozoan and multicellular pathogens, which led to successful field application and adaptation of novel technologies. These veterinary vaccines have had, and continue to have, a major impact not only on animal health and production but also on human health through increasing safe food supplies, namely milk, meat, eggs and fish, and preventing animal-to-human transmission of infectious diseases. The continued interaction between the researchers from veterinary and medical streams and health professionals will be a major impetus for adapting new technologies, providing animal models of human diseases and confronting new and emerging infectious diseases. Over 100 different veterinary vaccines are currently commercially available (Meeusen et al. 2007).

1.10.2.1 Multivalent Vaccines

Multivalent (bivalent, trivalent and polyvalent) vaccines should be given preference over monovalent vaccines to cover more than one disease prevalent during control programmes to save money, time and other expenses and also to reduce the burden on implementing agencies, such as veterinarians and para-health livestock workers.

1.10.2.2 Monitoring of Vaccine Response

There should be a system in place to conduct post-vaccination sero-monitoring in the field by appropriate agencies for finding evidence for adequate seroconversion in the randomly collected samples as per standard procedure preferably using DIVA tests to differentiate between vaccine-induced immune response and the one induced by the virulent virus.

1.10.2.3 Import Risk Analysis

The application of risk analysis concerning the spread of disease on account of international trade in live animals and their products, namely, import risk analysis (IRA), has been largely driven by the Sanitary and Phytosanitary (SPS) Agreement of the World Trade Organization (WTO). The IRA standard established by the World Organisation for Animal Health (OIE), and associated guidance, meets the needs of the SPS agreement. The use of scenario trees is the core modelling approach adopted to represent the steps necessary for the hazard to occur. There is scope to elaborate scenario trees for commodity IRA so that the quantity of hazard at each step is assessed (Peeler et al. 2015).

The dependence between exposure and establishment of the hazard suggests that they should fall within the same subcomponent. IRA undertaken for trade reasons must include an assessment of consequences to meet SPS criteria. The integration of epidemiological and economic modelling may open a path for better methods. Matrices have been used in qualitative IRA to combine estimates of entry and exposure, and consequences with likelihood, but this approach has flaws, and better methods are needed. IRA standards and guidance provided by OIE indicate that the volume of trade should be taken into account. Some published qualitative IRAs have assumed current levels and patterns of trade without specifying the volume of trade, which constrains the use of IRA to determine mitigation measures (to reduce risk to an acceptable level) and whether the principle of equivalence, fundamental to the SPS agreement, has been observed. It is questionable whether qualitative IRA can meet all the criteria set out in the SPS agreement. Nevertheless, scope exists to elaborate the current standards and guidance, so that they better serve the principle of science-based decision-making.

1.11 Disease-Free Zones

Options for trade from disease-free zones and disease-free compartments and trading in safe commodities are now available to have a positive mechanism for facilitating international trade. In India, FMD-Control Program (FMD-CP) is already in operation intending to create FMD-free zones. Similar zones can be created for other diseases like HS, bluetongue, sheep pox, goat pox, PPR and other important diseases. Compartmental biosecurity is the new concept for the management of biosecurity in a compartment through a single set of biosecurity measures. Creation of zones/compartments will ensure a boost in international trade of livestock and poultry products. In India, legislation regarding the movement of animals across these zones and compartments are required by the central and state governments.

1.12 Financing Is Key for the Control and Management of TADs

1.12.1 Economic Burden from Transboundary Diseases

The TADs are a threat to animal health and production and cause huge losses to the economy of nations. Recent outbreaks of bovine spongiform encephalopathy (BSE), foot-and-mouth disease (FMD) and highly pathogenic avian influenza (HPAI) have unfolded the real and growing global threat that animal diseases pose to livestock systems and to human health and welfare. The TADs adversely affect the trade in live animals and their products. The detection of one BSE-positive animal in 2003 in the USA led to an 80% drop in beef exports during 2003–2004. Similarly, the losses in the UK were estimated to be over US$9 billion during the ill-fated 2001 FMD outbreak. The economic losses due to HPAI (H5N1) avian influenza have been estimated from 0.5% to 1.5% of GDP in Thailand and Vietnam by Rushton et al. (2005). The 2003 outbreak of avian influenza due to H7N7 strain in the Netherlands destroyed as many as 30 million birds. Direct losses due to FMD in India have been estimated to the extent of INR 230,000 million per annum (Anonymous 2017–18). PPR has been estimated to cause global losses between US$1.45 billion and $2.1 billion per year. The disease causes economic losses in India to the tune of INR 11,070 million per year (Tripathi et al. 2018).

The FAO of the United Nations and the World Organisation for Animal Health (OIE) have opined that on an average one newly emergent animal disease has been detected per year over the last few decades, while three-fourths of these diseases being zoonotic are transmissible to humans. An outbreak of FMD in the UK in 2001, bluetongue in 2007, equine influenza in Australia in 2007 and India in 2008–2009, and highly pathogenic avian influenza (HPAI) in Maharashtra and adjoining parts of India in 2006 caused severe damage and economic losses to the farmers and industry.

1.12.2 Availability of Adequate Financial Support

Availability of adequate financial support for animal health R&D, especially in developing countries, is not always readily ensured. As the livestock keepers in these countries are mostly socio-economically poor, the local and national governments should come forward to support these programmes, particularly for the landless and marginal farmers keeping pigs, sheep, goats, backyard poultry and low-producing bovine stocks by providing incentives or subsidies for diagnostics and vaccines. Raising venture fund for emergency disease control through public and private partnership could be considered to meet the urgent requirements, besides farmer-friendly insurance policies for livestock health protection.

1.13 'One World–One Health' Approach

For important TADs, such as avian influenza, PPR and FMD, multinational, regional or global programmes under the supervision of FAO, WHO and OIE under 'One Health' concept are suggested for better coordination and results. The 'One World–One Health' (OWOH) concept steered by FAO, WHO and OIE has its roots in the interaction between living beings including humans, animals and pathogens, and the environment is considered as a unique dynamic system in which the health of each component is interconnected and dependent with other components. Nowadays, a newly integrated 'One Health One Medicine' approach reflects this interdependence with a holistic view of the ecological system. The OWOH can be defined as a collaborative and a multidisciplinary effort at the local, national and global level to guarantee an optimal healthy status for humans, animals and environment. The control of infectious diseases, which have influenced the course of human history, is to be considered strictly related to the One Health concept.

1.14 Experiences Gained with Some TADS

1.14.1 Avian Influenza

After its first occurrence in 1996 in China, the highly pathogenic avian influenza (HPAI) A virus (H5N1) has affected more than 60 countries in Asia, Europe, Africa and North America. The virus affected wild birds as well as domestic poultry. Sporadic cases of transmission to humans in close contact of infected birds with sizeable mortality raised the pandemic concern of 'bird flu'. After the first report of the H5N1 virus from India and Bangladesh in 2006 and 2007, respectively, both these countries are experiencing outbreaks almost every year. Between February–March 2006 and February 2019, India incurred an expenditure of more than INR 1098.7 million, including INR 263.7 million for compensation and INR 835 million on the culling of 8.349 million birds (Anonymous 2018–19). Avian influenza viruses (AIVs) have become a continued threat to global health and economy. After its first outbreak in 2014, the H5N8 HPAI serotype disseminated very fast from Korea to other parts of Asia, Europe and North America, a feature not observed in case of other highly pathogenic AIVs. However, the pathobiological features of the virus that favoured its global translocation are not known. Results of simulation studies undertaken in migratory birds to identify pathobiological features supporting AIV intercontinental dissemination risk suggest that characteristic differences exist among H5N8 and other AIV subtypes, e.g. H5N6 and H5N1 that have not spread as rapidly. Lower infection recovery and mortality rates and migration recovery rates also favour translocation in migratory bird populations. Although India has been reporting H5N1 AIV since 2006, the H5N8 virus was first time reported in 2017 from migratory birds and poultry in the states of Delhi, Madhya Pradesh, Kerala, Karnataka, Punjab and Haryana. Studies undertaken on comparative epidemiology of influenza viruses H5N1 and H7N9 among human and bird populations to find out similarities and differences between the two viruses in their genetic characteristics, distribution patterns in human and bird populations and postulated mechanisms of global spread (Bui et al. 2014) indicated that H7N9 viruses are diversifying at a much greater rate than H5N1 viruses. Analyses of certain H7N9 strains demonstrated similarities with engineered transmissible H5N1 viruses, which make it more adaptable to the human respiratory tract. These differences in the epidemiology of H5N1 and H7N9 viruses in human and birds raise further questions as to how H7N9 has spread at a greater rate than the H5N1 virus.

1.14.2 African Swine Fever

African swine fever (ASF) is a highly contagious, deadly emerging disease of pigs in many countries. Although first described in 1921 and it affected more than 50 countries in Africa, Europe and South America, several key issues about its pathogenesis, immune evasion and epidemiology remain uncertain (Arias et al. 2017). In the absence of a vaccine, the disease causes greater sanitary, social and economic impacts on swine herds compared to many other swine diseases. Currently, ASF is present in sub-Saharan Africa, Sardinia, the Trans-Caucasus, the Russian Federation and Central and Eastern states of the European Union. The disease continues to spread, with first reports in China (August 2018), Bulgaria (August 2018), Belgium (September 2018) and Vietnam (February 2019) highlighting the increasing threat of ASF to the global pig industry (Netherton et al. 2019). Ongoing outbreaks have also been reported in Hungary, Latvia, Moldova, Poland, Romania, Russia, South Africa, Ukraine, Cambodia, North Korea, Vietnam and Laos. The disease was rampant in China during 2019, and about half of China's breeding pigs died or were slaughtered. The threat of ASF looms large as presently no licensed vaccine is available against this disease, and further research is desired in this area for the development of live attenuated vaccines for ASFV. It has been possible to generate pigs resistant to classical swine fever virus and PRRS virus (Burkard et al. 2018) by using genetic modification of the host species. Genetic modification can be attempted as a viable solution to increase the host resistance to ASFV. Wild suids, namely warthog or bush pig, sequences could be engineered into the domestic pig genome to produce animals in which replication of ASF virus and/or disease burden after ASFV infection is reduced. However, to generate pigs fully resistant to ASFV infection, a more effective strategy such as targeting the virus receptors on the host cell to block the entry of virus and viral replication may be attempted.

Different clinical courses of ASFV infection in pigs have been described based on the virulence of the virus isolates, and sequencing the genomes of isolates of reduced virulence has identified virus genes associated with this phenotype. Targeted gene modifications and deletions and testing of the genetically modified viruses in macrophages and pigs have contributed to an understanding of virulence factors and how the virus modulates host responses. In the absence of a vaccine and rapid spread of ASF in Europe and Asia, the main emphasis should be on strict customs and border protection to keep the negative countries free from ASF virus infection/disease.

Research is required on priority to explore the virulence genes and genes related to host protection and immune evasion, role of multigene families in antigenic variability, mechanism of evasion of the immune response, factors determining viral persistence and infection outcomes, and interactions between ASFV and wild African suids, which are tolerant to ASFV infection. Such studies will provide a complete understanding of the pathogenesis of ASF. The specific role of different hosts including wild suids, vectors and environmental factors in disease propagation needs to be elucidated for understanding different epidemiological scenarios. In this regard, the northern European scenario in which infected wild boars drive disease transmission and maintenance needs to be investigated further. Presently, ASF has become of great significance in China and a real threat to the pig and pork production. The affected countries are planning to compensate for the losses in pork production by increasing broiler poultry production.

Gaps in sanitary control of wild boar populations make ASF control difficult. Raising awareness among veterinarians, hunters and farmers should be the priorities for ASF control. Advances in non-invasive sampling are required to facilitate surveillance in affected areas. Current and future tests need to be optimised for non-invasive matrices. The availability of a confirmatory serological test and cell lines for replacing primary cell cultures should be the priorities for future work. Availability of safe and potent vaccine against ASF could benefit disease control and prevention substantially, but despite some advances such vaccine is still lacking (Arias et al. 2017).

1.14.3 Foot-and-Mouth Disease

After the successful eradication of rinderpest from the globe in 2011, foot-and-mouth disease (FMD) of cloven-footed animals is another OIE-listed important viral disease inflicting heavy economic losses and adversely affecting the trade of livestock and livestock products from endemic countries to FMD-free nations/regions. Knight-Jones et al. (2016) have given a detailed account of global FMD research update along with gaps and an overview of global status and research needs. The conclusions are drawn to highlight that currently available vaccines and control tools have enabled FMD eradication from many countries of the developed world. However, in many developing countries, FMD remains uncontrolled. The main reason given is that biosecurity measures that have been fundamental to successful FMD control in the developed country are difficult to be implemented effectively in developing countries due to obvious reasons. In the present scenario, improved vaccines, with longer lasting protection against a wider range of FMDV strains and lower production costs, could be the single most important development to enhance our ability to control FMD. Although encouraging progress has been made with several novel vaccine candidates, addressing key limitations of the current inactivated vaccines, a commercial vaccine is yet awaited. While new discoveries are crucial, current vaccines have been used to effectively control FMD on numerous occasions. However, for imparting better immunity, FMD vaccines should be subjected to adequate quality assurance and be made available in sufficient quantity to provide desired coverage following appropriate strategy. There is also a need for better training and support in the design and execution of vaccine-based FMD control programmes. Another area of research is genetic and molecular studies on the virus to elucidate host-virus interactions. More powerful tools and analyses are increasing our understanding of various aspects of FMDV evolution, ecology and epidemiology. This, in turn, should benefit many areas of FMD research, from basic virology to the vaccine and diagnostic development. Furthermore, improved genetic technologies have the potential to reveal information crucial for control, such as transmission chains, vaccine match and level of virus circulation.

FMD control has been prioritised by many governments around the world. Besides traditional bastions of established research institutes in Europe and North and South America, notable work is being conducted in China, India and Africa. Experiences in South America and Europe have shown that through decades of sustained investment FMD can be controlled, even in regions where once it was rampant and control was seemingly impossible. However, if improved and more widespread FMD control is to be achieved, continued investment in FMD research at the local and international level is a must. Improved DIVA diagnostics increase our ability to detect infected animals in vaccinated populations. Greater confidence in the ascertainment of FMD status of animals and products has, in turn, opened the way for international standards for trade and disease control that are more efficient and less restrictive. Rigorous licensing procedures increase the time taken for new technologies for diagnostic kits and vaccines to reach the market. However, if authorisation is less rigorous, substandard products may be released onto the market. Hence, there is a need to balance these two requirements. Relaxation should be provided for necessary changes to the existing technologies, such as changing vaccine strains, particularly when the need is urgent (Knight-Jones et al. 2016).

1.14.4 Rinderpest: Now a Disease of Past

Rinderpest, also known as 'cattle plague', was once a deadly serious threat to the livestock industry and agriculture economy in several regions of the globe, particularly in Asia, Africa, Europe and the Americas. It periodically swept through Old World, resulting in devastating epizootics and huge economic losses. The disease could be successfully eliminated from the globe with mass vaccination programmes, zoo sanitary measures, policy support, international cooperation and political will.

The morbidity and mortality rates in newly exposed naïve populations could be as high as 95–100% leading to enormous economic losses. In India mortality rate of about 200,000 animals were recorded among the affected bovine population of 400,000 per annum during the first half of the 1950s, indicating average mortality of 50%. Throughout the history of humankind, the social, economic and ecological consequences due to rinderpest had been more catastrophic, even changing the history of nations and empires. In India, the presence of rinderpest was confirmed by the Cattle Plague Commission (Hallen et al. 1871). This disease has been conquered successfully by following mass vaccination along with zoo sanitary measures. The FAO declared the global eradication of rinderpest on 28 June 2011, marking it the first ever viral disease of animals eradicated globally about three decades after the eradication of smallpox, a viral disease of humans in 1980 (Yadav et al. 2016). Constraints of availability of quality vaccine in sufficient quality, freeze-drying of vaccines and maintenance of cold chain for a vaccine in tropical countries, lack of infrastructure for structured clinical surveillance and sero-surveillance were some of the limitations in executing the mass vaccination programmes. In India, dividing the country into four zones based on the epidemiological picture of the disease and adopting strategic and focused vaccinations at interstate and international borders and migration routes of bovines and caprine for creating immune belts, coupled with rigorous clinical and sero-surveillance, were of great help in achieving freedom from the infection. The financial support and/or technical guidance from FAO, OIE, EU and IAEA were the driving forces in achieving infection-free status for India in the year 2006. With the successful eradication of rinderpest, the livestock sector across the globe became safer, and consequently the living standard of livestock farmers improved.

The success of rinderpest control and eradication proved a rewarding experience and landmark for the veterinary services in India, providing capacity building and confidence among field veterinarians, researchers, policy planners and donor agencies and other stakeholders to undertake a successful control programme of livestock diseases at the national level. The freedom of the country from rinderpest not only enabled the growth of the dairy industry in India but has also boosted the export of meat and other dairy products in the recent decade. Today India tops not only in milk production in the world but is also the largest exporter of buffalo meat. Cost–benefit analyses indicated that every dollar spent on rinderpest control programme gained about $20 to the Indian dairy industry through more milk, meat and draft power for better agricultural productivity (Uppal 2011).

1.15 Conclusion and Prospects

In the fate of rampant threat due to EIDs and TADs, a diverse, dynamic and well-planned structured disease surveillance and monitoring approach would be the key for the sustainability and welfare of healthy livestock production systems of any country. Preparedness for combating the prevailing, emerging, re-emerging EIDs and TADs requires robust monitoring and precision detection systems that are flexible, feasible and adaptable under field conditions. In this regard, pen-side diagnostic tests/lab-on-the chip tools are the need of the hour. The hurdles of sampling need to be curtailed opting non-invasive methods for sample collection from different animal species and wildlife. Transparency in disease reporting needs to be adhered to and reported to OIE. Because of trading in animals and animal products, the international obligation for OIE reportable diseases of high importance must be followed by all member countries of WTO. It is high time to apply developed diagnostics and molecular detection tools in the field to ensure fast detection and confirmation of pathogens capable of causing diseases in humans and animals. This must be accompanied by national-level disease surveillance, monitoring and networking to enable an early warning system for infectious diseases based on forecasting (Saminathan et al. 2016). Due priority is also required for development and application of new potent, safe and affordable vaccines and vaccine delivery systems and adopting innovative vaccination programmes and immunomodulatory and effective therapeutic modalities, which would help in devising timely prevention and control strategies against viral and other infectious diseases. Besides these, good management and standard biosecurity and biosafety measures/practices and appropriate hygienic and zoo sanitary and quarantine measures should be observed.

Moreover, on-the-spot control and checking of the spread of pathogens and adequate trade restrictions as envisaged under the SPS agreement of WTO also need to be followed. A holistic vision and approaches are required for timely implementation of these concepts and strategies along with the strengthening of various multidimensional research and development programmes supported by appropriate funding resources. These measures will greatly help to minimise disease incidences and outbreaks, and lessen economic burdens due to infectious animal diseases and boost livestock and poultry health, reproduction and production to strengthen sustainable growth of livestock and poultry industry. Reduction in pandemic threats and public health concerns eventually lead to an improvement in the socioeconomic status and welfare of the society at large under 'One Health' umbrella. Application of artificial intelligence (AI), GPS, remote sensing and traceability in disease detection and management needs priority attention in developing countries. Similarly, the latest techniques of gene editing, base editing, nanotechnologies, electronic nose, etc. should be applied for efficient disease diagnosis and drug delivery. While planning the breeding policies for livestock and poultry, both higher production performance and health of the progeny should be given equal weightage. Modern techniques should be used for developing disease resistance (absolute or partial) in livestock and poultry using indigenous germplasm.

Acknowledgements All the authors of the manuscript thank and acknowledge their respective universities and institutes.

Conflict of interest: There is no conflict of interest.

References

Anon (2003) Institute of Medicine (IOM) report

Anonymous (2017–18) Annual report, ICAR-Directorate of foot-and-mouth disease, Mukteswar, India, pp 1–88

Anonymous (2018–19) Annual report, Department of Animal Husbandry, Dairying and Fisheries, Ministry of Agriculture and Farmers Welfare, Govt. of India, New Delhi, pp 61–64. http://dahd.nic.in

Arias M, Jurado C, Gallardo C, Fernandez-Pinero J, Sanchez-Vizcaıno JM (2017) Gaps in African swine fever: analysis and priorities. Transbound Emerg Dis 65(S1):235–247. https://doi.org/10.1111/tbed.12695

Bui C, Bethmont A, Chughtai AA, Gardner L, Sarkar S, Hassan S, Seale H, MacIntyre CR (2014) A systematic review of the comparative epidemiology of avian and human influenza A H5N1 and H7N9—lessons and unanswered questions. Transbound Emerg Dis 63(6):602–620. https://doi.org/10.1111/tbed.12327

Burkard C, Opriessnig T, Mileham AJ, Stadejek T, Ait-Ali T, Lillico SG et al (2018) Pigs lacking the scavenger receptor cysteine-rich domain 5 of CD163 are resistant to porcine reproductive and respiratory syndrome virus 1 infection. J Virol 92:e00415-18. https://doi.org/10.1128/JVI.00415-1

Daszak P, Cunningham AA, Hyatt AD (2000) Emerging infectious diseases of wildlife—threats to biodiversity and human health. Science 287:443–449. https://doi.org/10.1111/tbed.12180

Hallen JHB, Mcleod K, Charles JG, Keer HC, Allijan MM (1871) The cattle plague commission report to government of India. Calcutta Publication, Calcutta, pp 1–999
Knight-Jones TJD, Robinson L, Charleston B, Rodriguez LL, Gay CG, Sumption KJ, Vosloo W (2016) Global foot-and-mouth disease research update and gap analysis: 1—overview of global status and research needs. Transbound Emerg Dis 63(S1):3–13. https://doi.org/10.1111/tbed.12528
Lederberg J, Shope RE, Oaks SC Jr (eds) (1992) Emerging infections: microbial threats to health in the United States. National Academy of Sciences, Washington, DC
Malik YS, Dhama K (2015) Zika virus—an imminent risk to the world. J Immunol Immunopathol 17(2):57–59. https://doi.org/10.5958/0973-9149.2015.00019.2
Meeusen ENT, Walker J, Peters A, Pastoret P-P, Jungersen G (2007) Current status of veterinary vaccines. Clin Microbiol Rev 20(3):489–510. https://doi.org/10.1128/CMR.00005-07
Munjal A, Khandia R, Dhama K, Sachan S, Karthik K, Tiwari R, Malik YS, Kumar D, Singh RK, Iqbal HMN, Joshi S (2017) Advances in developing therapies to Combat Zika virus: current knowledge and future perspectives. Front Microbiol 8:2677. https://doi.org/10.3389/fmicb.2017.01469
Netherton CL, Cornell S, CTO B, Dixon LK (2019) The genetics of life and death: virus-host interactions underpinning resistance to African swine fever, a viral hemorrhagic disease. Front Genet 10:402. https://doi.org/10.3389/fgene.2019.00402
Peeler EL, Reese RA, Thrush MA (2015) Animal disease import risk analysis—a review of current methods and practice: open access article. Transbound Emerg Dis. 62:480–490. https://doi.org/10.1111/tbed.12180
Riemenschneider CH (2005) Avian influenza and other transboundary animal diseases, Director, Liaison Office for North America, Food and Agriculture Organization of the United Nations. Presentation at "Health in Foreign Policy Forum 2005", Washington, DC, 4 Feb 2005
Rushton J, Viscarra R, Guerne Bleich E, McLeod A (2005) Impact of avian influenza outbreaks in the poultry sectors of five South East Asian countries (Cambodia, Indonesia, Lao PDR, Thailand, Viet Nam) outbreak costs, responses and potential long term control. World's Poultry Sci. J. 61(1):491–514
Saminathan M, Rana R, Ramakrishnan MA, Karthik K, Malik YS, Dhama K (2016) Prevalence, diagnosis, management and control of important diseases of ruminants with special reference to Indian scenario. J Exp Biol Agric Sci 4(3S):3338–3367. https://doi.org/10.18006/2016.4(3s).338.367
Singh RK, Dhama K, Malik YS, Ramakrishnan MA, Karthik K, Tiwari R, Khandia R, Munjal A, Saminathan M, Sachan S, Desingu PA, Kattoor JJ, Iqbal HMN, Joshi SK (2017) Ebola virus—epidemiology, diagnosis and control: threat to humans, lessons learnt and preparedness plans—an update on its 40 year's journey. Vet Quart 37(1):98–135. https://doi.org/10.1080/01652176.2017.1309474
Singh RK, Dhama K, Chakraborty S, Tiwari R, Natesan S, Khandia R, Munjal A, Vora KS, Latheef SK, Karthikh K, Malik YS, Singh R, Chaicumpaj W, Mourya DT (2019) Nipah virus: epidemiology, pathology, immunobiology and advances in diagnosis, vaccine designing and control strategies—a comprehensive review. Vet Quart 39(1):26–55. https://doi.org/10.1080/01652176.2019.1580827
Tripathi BN, Kumar N, Barua S (2018) Peste des Petits ruminants: sheep and goat plague. Today and Tomorrow's Printers and Publishers, New Delhi, pp 1–180
Uppal PK (2011) FAO sponsored final project report "National testimonies" under the Global Rinderpest Eradication Programme (GREP)-(GCP/GLO/302/EC), pp 1–134
Yadav MP (2011) FAO sponsored final project report on "Laboratory contributions for rinderpest eradication in India" under the Global Rinderpest Eradication Programme (GREP)-(GCP/GLO/302/EC), pp 1–58
Yadav MP, Uppal PK, Rao JR (2016) Animal sciences. In: Singh RB (ed) 100 Years of agricultural sciences in India. NAAS, New Delhi, pp 158–258

Chapter 2
African Swine Fever Virus

Alexander Malogolovkin, Alexey Sereda, and Denis Kolbasov

Abstract African swine fever virus introduction to naïve swine population leads to high mortality and losses among susceptible animals. ASF epidemic in Russia (2007–to date) and lately in Eastern Europe highlights severe socio-economic consequences of this disease. The disease epidemiology is rather complex in endemic territories since many factors are involved in virus transmission. The disease control is only based on stamping-out policy and rapid virus diagnostics, since no effective and safe vaccine is available. This chapter focuses on African swine fever epidemiology, immunopathobiology and diagnostics with a brief overview of recent advances of ASF vaccine development.

Keywords African swine fever · Epidemiology · Prevention · Diagnostics and control

2.1 Prologue

African swine fever (ASF) is arguably the most dangerous swine disease, which threatens wild boar and domestic population worldwide. The mortality rate is approaching 100%, once the disease is introduced into the new territory. Many African countries, the Caucasus Republics, the Russian Federation and lately Eastern European countries are experiencing ASF outbreaks in swine farms and wildlife. Recent reports on the epidemiological situation with ASF in South-East Asia worryingly suggest that this disease may have reached the pandemic range. In 2019, devastating ASF outbreaks were documented in China, Cambodia, Myanmar,

A. Malogolovkin (✉)
Federal Research Center for Virology and Microbiology, Pokrov, Russia

London School of Hygiene and Tropical Medicine, London, UK
e-mail: alex.malogolovkin@lshtm.ac.uk

A. Sereda · D. Kolbasov
Federal Research Center for Virology and Microbiology, Pokrov, Russia

Y. S. Malik et al. (eds.), *Emerging and Transboundary Animal Viruses*, Livestock Diseases and Management, https://doi.org/10.1007/978-981-15-0402-0_2

Vietnam, and Laos, where millions of pigs were culled with the desperate attempt to stop the disease transmission. In China, 440 million pigs (50% of the world's pigs) have either died from AFSV or been killed to stamp out the virus (Gogin et al. 2013; Nurmoja et al. 2017a; Oganesyan et al. 2013; Okoth et al. 2013; Owolodun et al. 2010; Pejsak et al. 2014).

African swine fever has been first observed in Africa by Montgomery (De Kock et al. 1940; Edgar et al. 1952). The disease has been identified in domestic pigs, which were demonstrating the clinical signs similar to hog cholera (classical swine fever). Several following experiments carried out by Hess, Hay, DeTray, Plowright and Malmquist have described virus isolation and the main fundamental concepts of ASF virus biology, transmission and pathogenesis (Anderson 1986; Bool et al. 1970; Hammond and Detray 1955; Pan and Hess 1985; Pan et al. 1980; Parker et al. 1969).

African swine fever is an emerging transboundary disease. ASF outbreaks were registered in many countries around the world outside Africa: Portugal, Spain, France, the Netherlands, Italy, USSR, Brazil, Cuba and Haiti (Boinas et al. 2011; Caporale et al. 1988; Costard et al. 2013; Korennoy et al. 2017; Lyra 2006; Terpstra and Wensvoort 1986). ASF outbreaks can be registered very far from endemic territories, a 1000 km away from the outbreaks. In Spain ASF lasted for more than 30 years that seriously affected national swine production industry, but the disease was successfully eradicated due to strict control policy, effective surveillance programme and thorough research of virus diagnostics and epidemiology (Arias et al. 2001; Pastor et al. 1989; Sanchez-Vizcaino et al. 1981).

Since 1978, ASF has been registered in Sardinia (Italy) and remains an issue for local pig producers and veterinary authorities (Jurado et al. 2018; Mur et al. 2018). The modern history of ASF has been started in 2007 (Fig. 2.1) when the disease outbreaks have been notified in Georgia (Costard et al. 2009; Onashvili et al. 2012). Since then, ASF rapidly affected the Caucasus republics and the Russian Federation. The virus was introduced in the wild boar population and then subsequently transmitted to domestic pigs. In Russia, in 2017, the Federal Service for Veterinary and Phytosanitary Surveillance (Rosselkhoznadzor) reported that during 2007–2017, >1000 ASF outbreaks resulted in deaths of >800,000 pigs in 46 regions across Russia. Production of backyard swine industry decreased by almost half, from 1119 tons of pork in 2007 to 608 tons of pork in 2017 (Kovalev 2017). However, highly industrialised pig farms showed increased production every year during this same period, despite the ASF epidemic.

The disease epidemiology in Russia and the lessons learned from 10 years of ASF endemicity will be presented in the respective section. African swine fever virus (ASFV) is an aetiological agent of the disease. ASFV is the only DNA arbovirus that can infect and replicate in both soft ticks and pigs (Alonso et al. 2018). Such extreme host range together with complex virus genome organisation makes ASFV the unique and sole member of the *Asfarviridae* family so far. In the following sections, we discuss the peculiarity of ASFV transmission and immunopathogenesis in different hosts.

ASFV in infected hosts replicates in mononuclear cells (monocytes/macrophages) and has a sophisticated and multifunctional system of immune evasion

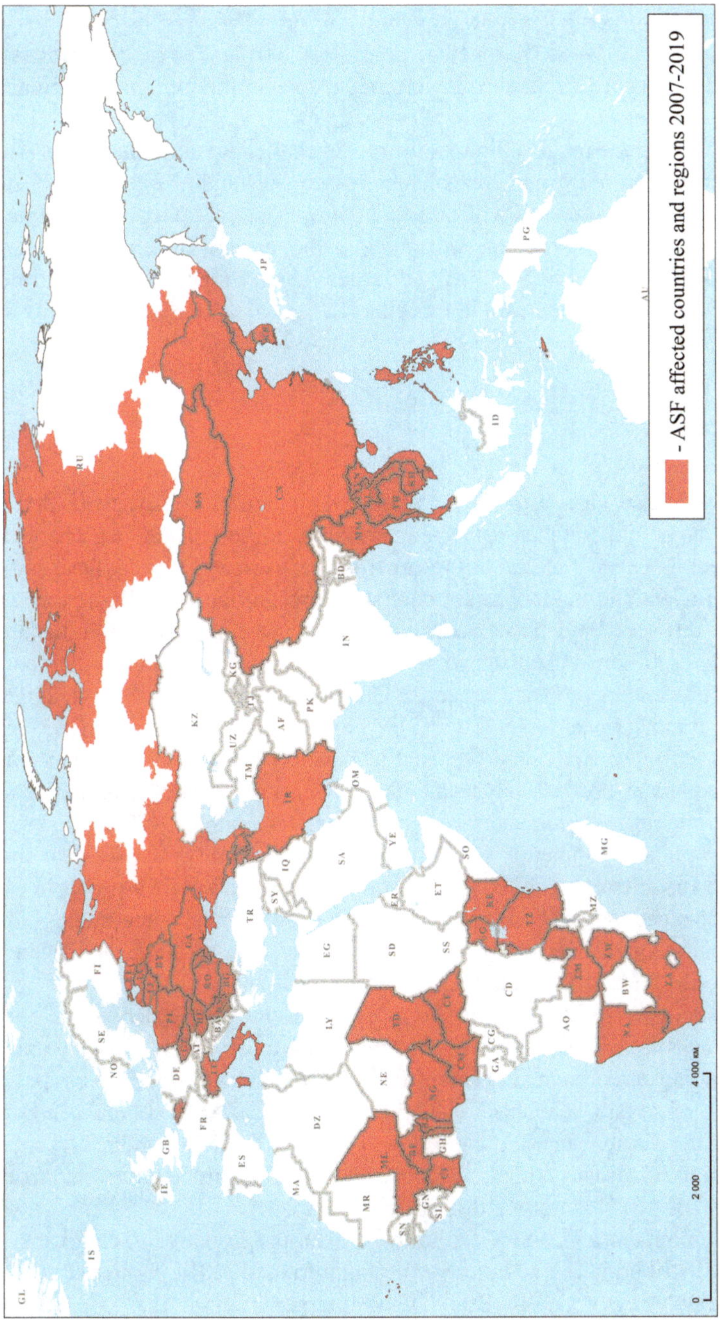

Fig. 2.1 African swine fever-affected countries and regions from 2007 to 2019

(Reis et al. 2017a), which makes the virus a "perfect killer" and still undefeated pathogen. The effective and safe vaccine is not available against ASFV, but some research groups presented encouraging and very promising results of future ASFV vaccine (Arias et al. 2018; Dixon et al. 2013; Rock 2017). The recent advances and knowledge gaps of ASFV vaccine development are summarised in the vaccine section of this chapter.

Here, we guide the readers through the recent challenges and solutions in African swine fever epidemiology and control, and discuss outstanding questions for ASFV vaccine research. The readers will also find the updated references for specific topics of ASFV biology, prevention, control and pathogenesis. The reference list is not complete, and we would like to thank all researchers for their valuable contribution to the ASF research summarised in this chapter.

2.2 Virus

African swine fever virus is a sole member of *Asfarviridae* family, genus *Asfivirus*. ASFV is a large and complex dsDNA arbovirus (Alonso et al. 2018). The genome length varies between 165 and 194 kbp from the isolates. Virions have multiple membrane layers and nucleoprotein core structure. Surface membrane (envelope) consists of different lipid forms and glycosylated proteins. The virion diameter is around 170–190 nm. ASFV genome structure is like the other members of nucleocytoplasmic large DNA viruses (NCLDV) and consists of a single molecule of double-stranded linear DNA. The ASFV genome on two termini spanned by terminal inverted repeats covalently closed in flip-flop form. Only 21 ASFV whole-genome sequences are publicly available in GenBank. More information about ASFV genome organisation and replication is available on the ICTV website.

ASFV replicates efficiently in mononuclear-phagocytic cells, resident macrophages and specific reticular cells of natural hosts. In vitro, ASFV grows in monocyte/macrophages and can be adapted to endothelial cell lines. Some studies indicate that adaptation of ASFV to endothelial cells may lead to attenuated phenotype (Carlson et al. 2016; O'Donnell et al. 2016).

Antigenic diversity of ASFV is the most represented in Eastern Africa, where different transmission cycles are involved in disease transmission. Based on nucleotide sequencing of core capsid protein P72 (B646L) of ASFV, 23 genotypes have been identified so far (Achenbach et al. 2017). Historical ASF outbreaks in Europe, USSR and the Caribbean were caused by genotype I ASFV strains. Recent ASF epidemic in the Caucasus republics, Russia and Eastern Europe has started from the introduction of ASFV genotype II into Georgia in 2007.

Additional markers of ASFV typing have been proposed for tracing back virus origin and distribution in endemic areas (Gallardo et al. 2014; Goller et al. 2015). Central variable region (CVR, B602L) and intergenic region (I73R-I329R) allow complement ASFV genotyping. Several ASFV IGR variants have been identified among the ASFV isolates isolated in Russia and Eastern Europe from 2012 to 2018.

Alternatively, ASFV isolates have been divided into serotypes based on haemadsorption inhibition assay (HAI) and cross-protection in vivo experiments. So far, eight serotypes have been identified, but more likely exist (Malogolovkin et al. 2015a; Sereda et al. 1994; Sereda and Balyshev 2011). Recently, genetic signatures of serotype specificity have been identified in CD2v (EP420R, haemagglutinin) and C-type lectin-like proteins (EP153R) (Malogolovkin et al. 2015b). This approach may fill the knowledge gap between ASFV genetic and antigenic diversity.

ASFV strains may cause acute, moderate and chronic disease forms. Several virulence factors have been identified in ASFV genome. The ASFV isolates may lose some members of multigene families MGF360/530 (Borca et al. 2018; O'Donnell et al. 2016) or MGF110 that lead to an attenuated phenotype (Zani et al. 2018). The recombinant ASFV strains with deleted interferon inhibitor genes or CD2v (Abrams et al. 2013; Monteagudo et al. 2017; Neilan et al. 2002) also had decreased virulence for domestic pigs. Some controversial results have been obtained about the role of CD2v protein in virulence and protection using different virus models (Burmakina et al. 2016; Monteagudo et al. 2017).

ASFV has unique characteristic by haemadsorbing red blood cells around infected macrophages (Fig. 2.2). Initially, this phenomenon was used for differential diagnostics of CSF and ASF. Later, ASFV CD2v protein was identified as a virus haemagglutinin (Galindo et al. 2000; Rodríguez et al. 1993). Interestingly, some ASFV have truncated or interrupted CD2v (EP402R) and as a result have not demonstrated haemadsorbing ability. Some non-haemadsorbing ASFV strains have

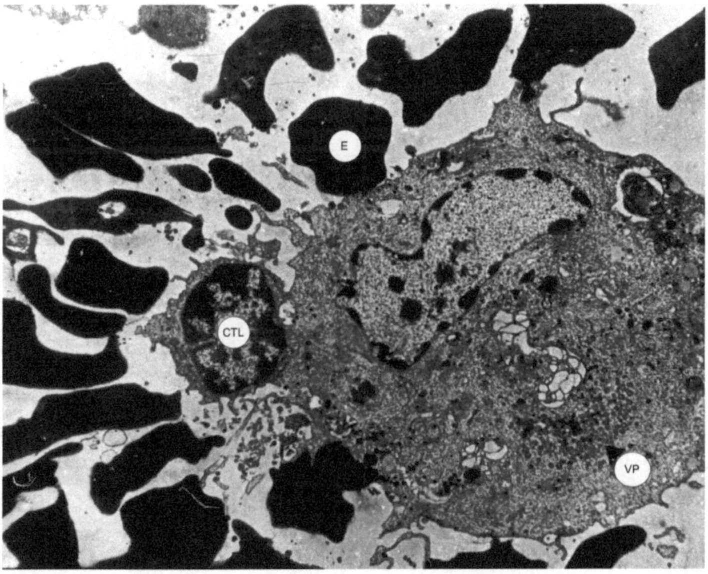

Fig. 2.2 Macrophage infected by ASFV surrounded by red blood cells (E) and attached by cytotoxic T-lymphocyte, VP - viroplasm (electron microscopy of an ultrathin section)

attenuated phenotype and have been used as a model for vaccine research (King et al. 2011; Sanchez-Cordon et al. 2017).

2.3 Epidemiology of Disease

The thorough and updated reviews about ASF epidemiology in Europe and Africa have been published recently (Bosch et al. 2017; Brown et al. 2018; Cisek et al. 2016; Gogin et al. 2013; Mur et al. 2012). We would highlight some additional aspects of ASF epidemiology in Russia and emphasise on the main risk factors that have been identified over 10 years of the epidemic.

ASF is present in Russia since December 2007 when it was first introduced in the North Caucasus regions, and over the past 10 years it spread from Russia throughout Eastern Europe, affecting domestic and wild boar, reached the Baltic countries and became endemic (Gogin et al. 2013). The epidemic was caused by the genotype II ASFV virus which caused up to 100% mortality in domestic and wild boar and could be considered as a self-limiting disease (Malogolovkin et al. 2012). Earliest data with ASFV Armenia/2008 strain revealed high mortality among domestic pigs and wild boar, but moderate contagiousness of the virus (Gabriel et al. 2011; Pietschmann et al. 2015). Nevertheless, recent reports from the Baltic states demonstrate the increased number of survived seropositive wild boar in some areas (Nurmoja et al. 2017a, b).

The Russian Federation consists of 85 federal subjects, and by veterinary legislation every federal subject is responsible for African swine fever control and prevention. The situation with the disease depends on the capacity and resources of regional authorities. Every federal subject has a different structure of pig production sector, and if the proportion of backyard production is high it puts the region into the group of high risks in terms of ASF introduction and makes it much more difficult to control it.

During the 10 years period, since the first ASFV introduction into the territory of the Russian Federation (from December 2007 to 2017), 1274 outbreaks of ASF have been reported. More than 50% of them were observed among domestic pigs in small private holdings or backyard farms, 7% from a total number of outbreaks originated from industrial pig farms and around 40% of cases in wild boars (Fig. 2.3). Despite the increasing number of ASF cases almost every year, in the last 12 years pig census has raised on 7 million heads in Russia (Karaulov et al. 2018).

After 10 years of disease circulation in the territory of Russia, it is still complicated to collect accurate and up-to-date information about the pig population in backyard farms. Due to this limitation, uncontrolled animal movements play a crucial role in disease distribution (Sánchez-Vizcaíno et al. 2012). The unknown number of animals in backyards does not allow veterinary services to control pig health on backyard farms. Some cases of ASF were detected only after reporting about disease suspicion by pig owners; in many times it happened too late after the first virus introduction and did not allow to define index case and apply control measures on time. The weakness of veterinary service in one region and late application to the outbreaks lead to disease spreading to the neighbour territories.

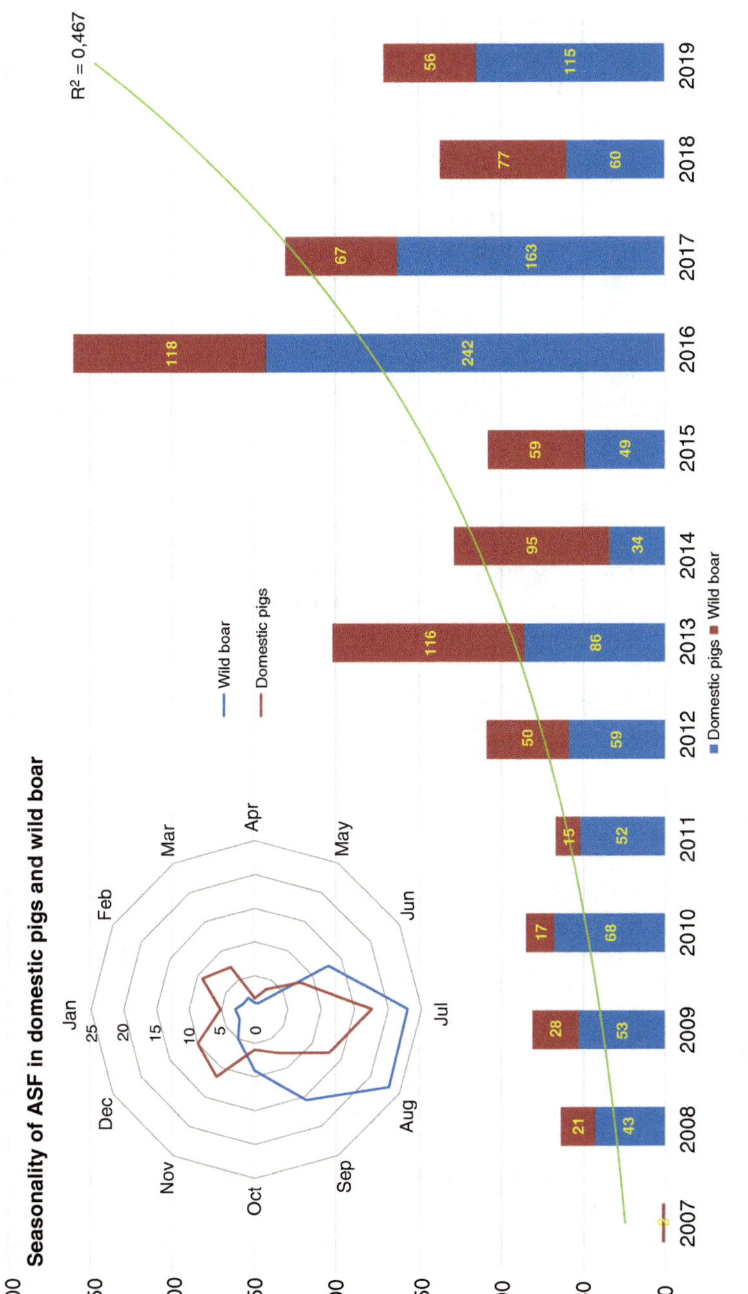

Fig. 2.3 The cumulative number of ASF outbreaks in Russia between 2007 and 2017 and its seasonality

By acting national regulation, after ASF notification, all movements of pigs must be banned on the suspected farm. If ASF is confirmed by laboratory tests, stamping-out must be applied by regional authorities as soon as possible in radius from 5 up to 100 km. Regional authorities define control and surveillance zones. The weakness of this eradication strategy is a high possibility of the late report by the owner and low motivation for regional authorities to expand the control zone because of high expenses for the stamping-out compensations.

The main risk factors of disease spreading are still the same after 10 years: movement of the infected/sick pigs to the new territory and pork products contaminated by ASFV, late reports about the disease and weak cooperation between pig owners, veterinarians and hunters (Kolbasov et al. 2018b).

The role of wild boars in disease introduction to the new regions is controversial. It's clear that white boar is responsible for short-distance spread of the disease among wild boar, but "ASF jumps" on thousands of kilometres in a very short period cannot be explained by this mechanism. The wild boar population can be divided into two parts—wild boar in native habitats and wild boar in-game grounds for hunting purposes. Unfortunately, hunters and managers of gaming grounds have no responsibility for animal health (Kolbasov et al. 2018b). In case ASF is detected in dead wild boar, carcasses of the dead animals are incinerated in place, under the supervision of regional veterinary service, and passive surveillance is applied in this territory. From the very beginning, there are two main driving forces of ASFV epidemic: the socio-economical aspects and human behaviour, whereas the role of wild boar is still not completely understood. There are many examples demonstrating the spatial pattern of the disease characterised by "jumping" spread caused by the illegal movement of pigs and pork products (Kolbasov et al. 2018a).

One of the main factors for ASFV introduction into domestic pig herds is low or unappropriate biosecurity: about 80% of ASF outbreaks have been registered in the backyard sector. Most of them are linked to illegal trade and uncontrolled movements of infected pigs. It is known that wild boar plays a critical role in the introduction of the virus into the new territories through administrative borders. Meanwhile, the involvement of ASF-affected wild boar in the virus distribution into the commercial farms has never been confirmed. Moreover, there are many pieces of evidence of illegal disposal of domestic pigs' carcasses in the forest, following by detections of ASFV in the wild boar population. ASF can be relatively easily controlled in domestic pigs. It is also worth noting that the ticks are not involved in the current ASF epidemic in Eastern Europe and Russia.

It's interesting to compare the results of the epidemiological investigation of African swine fever outbreaks in 1977 in USSR with the data collected from current ASF outbreaks in Russia (Korennoy et al. 2017). The main risk factors remain the same after almost 40 years.

Here are some risk factors that have been identified in 1977 (Jurkov et al. 2014):

- Infected food waste
- Sales of infected meat products
- Trade of infected animals
- Economic ties of farms and enterprises located in non-affected areas to ASF-affected zones

2.4 Transmission

2.4.1 Sylvatic Cycle

ASF is endemic in most African countries from the Sahara desert to the south. In Europe, the island of Sardinia has the longest history of ASF epidemic so far (Mur et al. 2018; Sánchez-Vizcaíno et al. 2012). The major route of ASFV transmission is a contact of susceptible animals (domestic pigs or wild boar) with either infected animals or fomites or contaminated pig products or via tick bites.

Historically, warthogs (*Phacochoerus africanus*) and indigenous African pigs (*Phacochoerus africanus, Potamochoerus porcus, Potamochoerus larvatus, Hylochoerus meinertzhageni*) were considered as the main reservoirs of ASFV in nature in Africa. Later studies have demonstrated no horizontal or vertical transmission of ASFV in wild African pigs. Several unsuccessful attempts have been made to confirm the direct transmission of ASFV from seropositive warthogs to domestic pigs (Anderson et al. 1998). Therefore, soft ticks were considered as a potential player of the sylvatic cycle of ASFV transmission. The warthogs live in the barrows, where soft ticks are frequent neighbours (*Ornithodoros porcinus porcinus, O. porcinus moubata, O. moubata*).

ASFV-infected ticks bite young warthogs, which easily recover the disease, but ASFV titre in the blood may reach 2–3 lg HAU/mL, which is enough to initiate a new round of sylvatic cycle in the soft ticks (Burrage 2013; Plowright et al. 2002).

In the Iberian Peninsula, another species of soft ticks (*Ornithodoros erraticus*) as an ASFV competent vector was found (Bastos et al. 2006a). There is no doubt that *O. erraticus* was involved in ASF epidemic in Spain. ASFV-infected soft ticks may survive up to 5 years that may lead to serious concerns in ASF endemic countries (Boinas et al. 2004). Recent studies have shown that ASFV strain Georgia 2007/1, which is currently circulating in Europe, replicates efficiently in *O. erraticus*, collected in Southern Portugal (Diaz et al. 2012).

Several other soft tick species may also be competent for ASFV replication. Thus, in the USA *O. puertoricensis, O. turicata, O. talaje, O. dugesi* and *O. coriaceus* (Hess et al. 1987) are potential ASFV vectors as well (Hess et al. 1987). It has been proven that in *O. coriaceus*, ASFV may persist after 4 months. However, a transovarial transmission has not been demonstrated (Sánchez-Vizcaíno et al. 2009).

In African tick *O. moubata*, ASFV may be transmitted transovarially and transstadially and by sexual contact (Hess et al. 1989; Plowright et al. 1970; Rennie et al. 2001). These data support the hypothesis of a long-term ASFV persistence in tick populations without pig's involvement. It is most likely that evolutionarily ASFV is an arthropod virus taking into account the taxonomic relationship and sylvatic cycle (Makarov et al. 2016). Some data suggest that ASFV and soft ticks have coherent evolution, and several virus host-range genes are involved in this process. However domestic pigs are unnatural hosts, and the virus may lose some host-range genes during replication in domestic pigs (Afonso et al. 2004; Burrage et al. 2004; Dixon and Wilkinson 1988). Hopefully, new genetic data of the soft tick's genome sequencing will help to dissect the ASFV evolution and origin.

Wild African pigs and soft tick form a sylvatic cycle of ASFV transmission (Parker et al. 1969). However, the soft ticks transmit ASFV to domestic pigs as well. The efficiency of virus transmission correlates with virus titre in ticks' salivary glands and coxal gland that may reach 4–6 lg HAU/mg (Bastos et al. 2006b).

Around 100 species of *Ornithodoros* are known, and 7 species have been identified in the territory of former USSR. Among them, *O. papillipes* in middle Asia and *O. verrucosus* in the Caucasus republics are vectors of the relapsing fever as well as *O. moubata* in Africa and *O. erraticus* in Iberian Peninsula (Fillipova 1966). Therefore, these species of ticks are of interest for ASFV epidemiology.

In eastern European countries, several swine farms with high biosecurity level have been affected by ASFV. To estimate the role of other haematophagy, several studies have been carried out. In experimental settings, the stable fly (*Stomoxys calcitrans*) fed with blood from ASFV-infected animal caused ASF in domestic pigs. These data suggest that other blood-sucking insects may play a role in ASFV transmission at least within a herd (Olesen et al. 2018). In another study, blood from rodents and birds was tested for ASFV without positive results (EFSA AHAW Panel (EFSA Panel on Animal Health and Welfare) 2014).

2.4.2 Fomite Transmission

One of the most important questions of virus transmission in natural settings is its tenacity in different environments. This is a crucial issue for studying disease distribution, modelling and risk assessment. The minimum infectious dose for ASFV Georgia 2007/1 for the oronasal route of inoculation is 10 HAU/mL according to Kovalenko and Sidorov (1973).

ASFV in infected and sick animals is shed via saliva, nasal excretes, faeces, urine, genital excretes and blood. All these ASFV-contained excretes may contaminate ground, feed and water. ASFV shedding in faeces coincides with fever (Greig and Plowright 1970). After primary fever onset in 2–3 days, ASFV titre in nasal swabs may approach 4–5 lg HAU_{50}/mL. In some studies, ASFV was isolated in faeces and oral swabs after 70 days of infection (de Carvalho Ferreira et al. 2012).

At low temperature (4–6 °C) in faeces and urine from sick animals, ASFV was isolated in 159–253 days and 60–87 days, respectively (EFSA AHAW Panel (EFSA Panel on Animal Health and Welfare) 2014). In hot climate environment, the stables where sick animals have been housed, ASFV has been found from 5 to 14 days, and in Spain—up to 3 months (Kovalenko and Sidorov 1973). In mild and cold climate environment of central Russia in contaminated forest sandy ground with pH 4.5–4.6, ASFV survived for 112 days (Smirnov and Butko 2011).

Based on the previous data where the minimal infectious dose of 10 HAU/mL was identified, ASFV remains infective in faeces and urine up to 8–15 days at 4 °C and up to 3–4 days at 37 °C (Davies et al. 2017).

In wintertime, frozen excretes from sick wild boar and virus-contaminated environments may be a potential source of infection after thawing. Several approaches have been developed for ASFV surveillance in wild boar population. One of the most promising is a rope with feed attractants (Chichikin et al. 2012). An alternative approach for ASFV surveillance in wild boar is a faeces sampling, since ASFV may survive in faeces for quite a long time. Faeces samples might be used for ASFV genome detection using PCR (de Carvalho Ferreira et al. 2012). According to de Carvalho Ferreira et al. (2014), ASFV DNA has been identified in faeces till 98 days at 4–12 °C and 35 days at 37 °C, in urine—126 days at 4–37 °C and in saliva—35 days at 4 °C and 14 days at 12–21 °C.

2.5 Immunopathobiology

The pathogenesis and immunopathology of ASF are similar to the most human and animal haemorrhagic fevers. ASFV infects predominantly mononuclear cells (monocytes/macrophages)—the most prominent component of T-cell-mediated immunity. Infected monocytes/macrophages release a wide range of cytokines, which severely affect different cell types (lymphocytes, endothelial cells) leading to apoptosis and cell damage (Penrith 2009; Penrith et al. 2004). An increase of production of proinflammatory cytokines such as IL-1α, IL-6α and TNF-α coincides with haemorrhagic fever symptoms (e.g. fever, vascular damage) (Salguero et al. 2002; Sánchez-Cordón et al. 2005). ASFV replication in monocytes/macrophages leads to its damage and apoptosis. The components of damaged and disrupted mononuclear cells also activate endothelial cells and slow down the coagulation system (Salguero et al. 2008).

Initially entered to mononuclear cells, ASFV is able to infect several other cell types (i.e. neutrophils, megakaryocytes, tonsillar epithelial cells, hepatocytes, kidney cells, granulocytes, presumable dendritic cells) especially in the later disease state (Greig et al. 1967; Sierra et al. 1990). In acute ASF, severe pathomorphological changes and haemorrhages are found in lymphoid organs (spleen, lymph nodes, thymus) and kidney (Kleiboeker 2002; Ramiro-Ibáñez et al. 1997).

The clinical course of ASF varies from unapparent to chronic forms and mostly depends on the ASFV isolate (EFSA AHAW Panel (EFSA Panel on Animal Health and Welfare 2014). First haemorrhagic lesions may appear at 3 days postinfection with virulent ASFV and coincide with monocyte/macrophage destruction. The average incubation period following infection with ASFV virulent strain lasts for 2–7 days, and rarely longer. The mortality rate of ASF may approach 100% but ranges between ASFV isolates (Mebus 1988). Typical clinical signs of acute ASF may include high fever, bloody diarrhoea, respiratory discharges, cyanosis and haemorrhagic lesions. Often, central nervous system symptoms, such as ataxia and convulsion, may be observed in the late stage of infection. ASFV causes the same clinical symptoms in wild boar regardless of their age and sex (Blome et al. 2013; Gabriel et al. 2011). The readers may find some pathological pictures of ASF on the EFSA photo portal (https://efsa.maps.arcgis.com/apps/PublicGallery/index.html?appid=dfbeac92aea944599ed1eb754aa5e6d1).

ASF epidemic in Russia and Europe initially was caused by highly virulent ASFV strain (genotype II). In experimental settings, the animals died within 10 days postinfection with no survivors. However, since 2012 more reports notify moderate ASFV variant in Eastern Europe and Russia with an increasing number of survivors, especially among wild boar (Arias et al. 2018; Gallardo et al. 2018). Several ASFV genetic variants have been identified in Russia and Eastern Europe. Nevertheless, the link between genetic changes and disease course is not fully understood.

ASFV strictly impairs innate immune system (e.g. IFN response, TLR, MHC). ASFV modulates different stages of host immune response and has sophisticated mechanisms of immune evasion. Several ASFV genes have been identified as virus virulence factors and IFN inhibitors (Afonso et al. 2004; Reis et al. 2017b). Recent advances in gene-editing approaches may help to design and produce safe and immunogenic recombinant ASFV strain, lacking virulence factors that might be a very promising vaccine candidate in future.

2.6 Diagnostics

Since no vaccine is available to prevent ASF infection, quarantine measures for liquidation and prevention of the disease are carried out. It is necessary to carry out a laboratory diagnostic for obtaining the information for surveillance and eradication programmes. Positive diagnosis means the identification of animals, which are or earlier were infected with ASFV. It includes detection and identification of the infectious virus, DNA, specific antigens and antibodies (Agüero et al. 2004; Oura et al. 2013).

For ASF diagnostic, a wide range of methods are used. Infectious virus is determined with a biological assay on pigs and using haemadsorption test (HAD) in swine macrophages (Orfei et al. 1968). Virus-specific antigens are identified using primary (direct) immunofluorescence, immunoperoxidase method and ELISA. ASFV DNA is identified using PCR (Fernández-Pinero et al. 2013; James et al. 2010; King et al. 2003). For virus-specific antibody detection, OIE recommends indirect ELISA and for confirmation indirect immunofluorescence or Western blot (World Organisation for Animal Health (OIE) 2012).

From the end of 2015, epizootic, serological and genetic research showed a prominent increase of seropositive animal incidence, which is especially visible in wild boar population in unfavourable EU countries. These results suggest that some animals may have a chronic form of the disease (Olsevskis et al. 2016; Smietanka et al. 2016; Wozniakowski et al. 2016).

Gallardo et al. (2015) evaluated methods of ASF diagnostic of viral DNA detection, antigens and antibodies for experimental and field samples (Gallardo et al. 2015). The authors in parallel investigated 785 field and experimental samples, obtained from pigs, infected with genotype II using three PCR assays for ASF virus genome detection (Agüero et al. 2003; Fernández-Pinero et al. 2013). It was shown that several DNA-positive samples were more for 3.3% in UPL (Universal Probe

Library)–PCR than in PCR assays, recommended by OIE. DNA was easily detected using both PCR assays when high virus levels were found in blood and tissues in the clinical phase of infection.

The results of diagnostic efficacy comparison of ELISA (ELISA Ingezim K2; Ingenasa, Madrid, Spain) and three PCR assays are summarised in Gallardo et al. (2013) and Oura et al. (2013). The authors notify that field samples with bad sampling and storage conditions can dramatically decrease ELISA sensitivity.

It may be denoted by the antigen-antibody complex formation in seropositive animals' tissues and blocking of the interaction between antigen ASFV and specific conjugate to it. In most cases in EU countries, ASF has an acute form and causes deaths with a high level of virus accumulation in all the tissues (Gabriel et al. 2011; Gimenez-Lirola et al. 2016; Guinat et al. 2016). That is because it is rational to use other diagnostic methods in parallel with ELISA.

Immunoperoxidase method (IPT) showed greater sensitivity than indirect ELISA in studying serum samples, obtained in dynamics from experimentally infected 30 domestic pigs by ASFV genotype II. Immunoperoxidase method allowed detecting ASFV antibodies at an earlier stage of infection than using indirect ELISA. Diagnostic sensitivity of indirect ELISA varied from 22% to 50% in comparison of the IPT depending on the assays. Low sensitivity of indirect ELISA can be associated with the fact that samples were selected from the animals with an acute form of ASF before the accumulation of antibodies, which was determined using indirect ELISA.

Nevertheless, it is necessary to search for antibodies from the trophy and fallen animals for understanding a complete epizootic situation. Positive results using indirect ELISA should always be confirmed by alternative methods as secondary (indirect) immunofluorescence and Western blot, according to OIE recommendations (World Organisation for Animal Health 2012). In the authors' opinion, even though there are a lot of good verified methods of ASF diagnostic, obtained results show that UPL-PCR in combination with IPT is the most reliable method for early ASF virus genome and antibody detection.

Interesting results were obtained by evaluating an epidemic situation of ASF and the warthogs in Serengeti nature park in Tanzania (Misinzo 2012). The authors investigated serum samples using indirect ELISA (OIE-ELISA) and Western blot (OIE-IB) for ASFV antibody detection, and blood and organ samples for genome virus detection. According to indirect ELISA results, 100% (34/34) of warthogs were seropositive. Analysis of organ samples using PCR showed that only 8.8% (3/34) were weakly positive. However, the authors could not isolate the infectious virus in any samples. Results correlated with other information that most of the warthogs were seropositive (Heuschele and Coggins 1969).

It should be noted that during the ASF epidemic in Russia in 2007–2012, both PCR and direct immunofluorescence were used that guaranteed 100% of diagnostic accuracy. Specific antibodies in organ tissues from infected pigs were not prominent as they were detected only from 45% of animals (33% from wild boars and 49% from domestic pigs). The levels of specific antibodies varied from 4.3 to 9.0 $\log_2$ in domestic pigs and wild boar. It was noticed that antibodies, even in high concentrations, did not block intracellular antigens and did not prevent from detecting them using direct immunofluorescence and ELISA (Strizhakova et al. 2016).

While choosing a diagnostic test, it is important to consider the phase of the infection and what form of the disease is caused by circulated ASFV. In acute and subacute forms, infectious virus and DNA can be detected before the manifestation of clinical signs. Serological conversion intervenes from the 7th to the 11th day postinfection and antibodies can be detected in the course in the rest of an animal's life. Positive test on the infectious virus (or antigens) presence and DNA indicated that during sampling tested animals were already infected. On the other hand, a positive test for antibodies to ASFV pointed out that animals were infected more than a week ago and/or survived after ASFV infection.

2.7 Prevention and Control

Control over African swine fever (ASF) is complicated due to the lack of specific prevention measures. Only rapid diagnostics and strict stamping-out of the strategy of infected animals may help to eradicate the disease and stop virus transmission. The safe and efficacious vaccine against ASF is not available now, but recent advances of vaccine research demonstrate very encouraging results. Below the readers may find the short historical overview and current approaches about ASF vaccine research.

2.7.1 Inactivated Vaccines

Many attempts have been made to produce traditional inactivated vaccines from ASF using infected macrophages fixed with glutaraldehyde, lysates of primary and passage cell culture treated with ultraviolet radiation, freon, and ionic and nonionic detergents, inactivated with β-propiolactone of purified virions fixed on bovine erythrocytes, mycobacteria and γ-globulin (Blome et al. 2014; Kovalenko and Sidorov 1973; Makarov et al. 2016; Mebus 1988; Petrov et al. 2018). Others and we have not observed any protective effect of inactivated virus formulations. On the contrary, in several cases, the enhancement of disease of immunised pigs is compared with the control group after their infection with virulent isolates of the ASFV (Hess 1981; Mebus 1988; Stone et al. 1968).

2.7.2 Subunit Vaccines

Studies on the development of subunit vaccines have facilitated the search for potentially protective proteins. The effects of pig immunisation made by purified infected cells or recombinant proteins p30, p54, p72 and CD2v were considered to be potentially protective (Barderas et al. 2001; Gómez-Puertas et al. 1996; Gutiérrez-Castañeda et al. 2008; Kollnberger et al. 2002).

Pig immunisation by recombinant proteins p30 and p54 expressed in baculovirus led to a delay in the onset of clinical symptoms of the disease after animal challenge with a virulent isolate (Gómez-Puertas et al. 1998). Pigs' immunisation with a pool of glycoproteins of the ASFV in liposomes induced the antibody titre formation but led to an acceleration of the animal deaths after challenge. As a result of pig immunisation with serotype-specific major glycoprotein CD2v (gp 110–140) of ASFV purified from macrophages infected with ASF, in liposomes, 67% of the animals were protected after infection from death, but not from the reinfection (Sereda et al. 1994). ASFV CD2v is directly involved in the process of haemadsorption in the infection of sensitive cells with ASFV (Rodríguez et al. 1993). It is determined that the results of genotyping at the locus encoding CD2v correspond to the distribution of isolates and strains of ASFV over seroimmunotypes (Malogolovkin et al. 2015a). Immunisation with recombinant baculovirus carrying the CD2v gene of the ASFV protected from subsequent challenge with a virulent strain (Argilaguet et al. 2013). Presumably, this protein can be the main inducer of CTL.

So, most researchers consider the proteins p30, p54 and CD2v required for the immunological defence induction against ASF, but none of them correlates with protection.

2.7.3 Live Vaccines

Live attenuated vaccines against ASFV are a promising tool to dissect the mechanism of protection and find hidden signatures of immune correlates. Several elegant and thorough reviews have been published recently with an emphasis on live attenuated vaccine characteristics (Arias et al. 2018; Rock 2017; Souto et al. 2016).

In the Pokrov Institute, we attenuated, created, selected and isolated a number of strains and ASFV variants that do not cause the death of domestic pigs and are able to form protection from subsequent infection with homologous virulent isolates (Vishnjakov et al. 1991).

The research on the development of protection against ASF leads to the discovery of attenuated strains of eight serotypes. The attenuated candidate vaccine strain was able to protect 75–90% of the animals after 10–14 days post-challenge with virulent ASFV homologous in the serotype (Sereda and Balyshev 2011). Despite the fact that the selected attenuated strains met the established requirements for protection and harmlessness, they had some differences in several biological characteristics: the duration and level of viremia and the timing of the formation of virus-specific protection. Thus, an attenuated FK-135 strain (serotype 4) at a dose of 104.0 HAU50 creates protection on the 7th–10th days in 92–100% of pigs, and the strain of MK-200 (serotype 3) in 50% of animals, and only on the 14th day—in 82–92%.

The disadvantages of live vaccines are the attenuated virus carriage to different extents, the probability of subsequent partial restoration of its virulence, the development of subclinical infection sometimes changing into a chronic form, and insufficient protection in immunocompromised animals, for example pregnant sows.

It is known about the isolation of naturally attenuated strains of the ASFV, for example OURT88/3 or NH/P68. Immunised pigs were protected from infection with homologous virulent strains (Boinas et al. 2004; Leitão et al. 2001; Mulumba-Mfumu et al. 2017). Protection levels varied from 66% to 100%, depending on the pigs and the way and dose of injection. Attempts to use natural attenuated strains of the ASFV as vaccines have shown side effects. Some of the vaccinated pigs developed unfavourable reactions, including pneumonia, locomotor disorders, necrotic foci, abortions and death (Gallardo et al. 2015).

Recombinant ASF viruses with gene deletions involved in immune evasion, such as thymidine kinase (TK), 9GL (B119L) gene, NL (DP71L) gene and several members of the 360 and 505 multigene families (MGF360/505), cause total attenuation of initially virulent isolates and induce the development of protection against homologous virulent isolates (Afonso et al. 1998; Neilan et al. 2002; O'Donnell et al. 2017).

Multiple deletions of the six members of MGF360 and 505 in combination with the 9GL gene resulted in the virulence strain loss of the ASFV Georgia07/01, but did not protect animals after subsequent infection with a virulent homologous virus (O'Donnell et al. 2016). In contrast, the Georgia07/01 virulent isolate, modified by removing the virulence factors 9GL and UK, in contrast to the 9GL modified only by removal, acquired protective properties. These results indicate that the successive removal of the second virulence factor can make recombinant live attenuated ASF viruses much safer. So far, gene-edited recombinant ASFV strains are the most promising candidates for a vaccine against ASF.

2.7.4 DNA Vaccines

It is conceptually important that DNA vaccines are potentially safe for animals and induce antigen-specific cellular immunity. A significant problem of candidate DNA vaccines is the relatively low acquisition of DNA by the cells in vivo, especially in large mammals. Several approaches have been proposed to overcome this problem (Leifert et al. 2004; van Drunen Littel-van den Hurk et al. 2004).

The similarity between the HA haemagglutinin of HA virus (or CD2v) and the leukocyte CD2 molecule suggested that it can target lymphocytes expressing CD2 receptors (CD48 and CD58) to viral antigens in antigen-presenting cells (Borca et al. 1994; Rodríguez et al. 1993). The addition of HA enhanced both the humoral and cellular responses against the chimeric PQ protein (p54 and p30), after three intramuscular injections of the corresponding DNA construct. The enhancement of the immune response to sHA injection may also be due to the presence of T-helper cell epitopes in this molecule.

Another strategy is based on targeting the encoded viral antigens to the places of antigen presentation using the variable fragments of single-chain (ScFv) antibodies that specifically recognise cell antigens on the surface of antigen-presenting cells (Grossmann et al. 2009). The efficacy of the invariant epitope of class II hypoglycaemia (APCH1) as a genetic adjuvant in vivo was confirmed by pig immunisation with plasmid pCMV-APCH1PQ in which the APCH1 gene is fused to the chimera open reading frame for PQ. DNA constructs encoding only PQ did not induce the formation of antibodies in pigs, while immunisation with pCMVAPCH1PQ caused both the synthesis of PQ-specific antibodies and the activity of T-helpers targeted for class II histocompatibility antigen, indicating an adjuvant effect of the APCH1 molecule. However, this candidate DNA vaccine did not protect pigs from subsequent infection with ASFV (Barderas et al. 2001).

Despite the notified humoral response to the immunisation of pCMV-sHAPQ, the pigs were not protected from challenge. To avoid unfavourable induction of antibodies and to enhance specific CD8+ T-cell responses, a pCMV-UbsHAPQ construct was developed that codes for the antigenic determinants p30, p54 and sHA fused to cellular ubiquitin. As expected, immunisation with pMVV-UbsHAPQ did not induce a humoral response in pigs, but provided partial protection against ASFV challenge, confirming the importance of the T-cell response in protecting against this virus. The achieved protection was not enhanced by an increase in the multiplicity of administration of the DNA vaccine, which may reflect a lack of boost effect for the T-cell response induced after the first administration. With twofold immunisation of pCMV-UbsHAPQ, 2 out of 6 pigs survived, while only 4 of them survived. Presumably, boost strategy negatively affects in terms of providing protection. According to the authors, fourfold immunisation with pCMV-UbsHAPQ could lead to weak induction of antibodies exacerbating the disease, which in turn can suppress the protective effect of induced CD8+ T cells (Argilaguet et al. 2012).

Immunisation with DNA expression library is considered as a promising trend in the development of protection against emerging diseases (Talaat and Stemke-Hale 2005). The protection of the ASFVUblib DNA library, represented by short fragments of the ASFV genome combined with the ubiquitin gene in the plasmid pCMV-Ub, was studied to enhance the induction of specific CTL (Lacasta et al. 2014). The obtained 4029 clones (total 130,000 bp) covered about 76% of the viral genome.

Vaccination capabilities based on the use of BacMam viruses, which are baculovirus vectors encoding virus proteins under the control of vertebrate promoters, are established, which provides the high expression of the transgene in mammalian cells (Argilaguet et al. 2013).

Immunogenicity of BacMam-sHAPQ was determined after a threefold administration of 10^7 pfu with a 15-day interval. Then, all animals were infected with a homologous isolate E75 at a dose of 102 HAU50. As expected, there were no such responses before infection in a control group, and in four out of six pigs immunised with BacMam-sHAPQ specific T-cell responses appeared. Therefore, with the vaccination of BacMam-sHAPQ, the pigs' protection against sublethal homologous infection with ASFV is possible in the absence of antibody induction. Besides, the induced defence and stimulation of T cells are directly related.

2.7.5 Antiviral Formulations

Antiviral drugs and approach are of interest to study ASFV reproduction and potentially might be used in vivo to inhibit virus replication. A "specific" target for several antiviral agents is viral DNA polymerase. With the ASFV, phosphonoacetic acid (PTC) was tested, which is an effective inhibitor of the activity of this enzyme. PTC was equally effective both in experiments on cell cultures and in animals.

Based on the data analysis of the synthesis efficiency, reproducibility and therapeutic activity with ASF, three compounds were selected: PTC, PTC complex with 7-amino-1,3,5-triazaadamantane (A-14) and potassium pyridine salt of PTC-230, which prevented the death of more than 80% of infected animals at 100% death in the control group. The use of PTC in combination with metisazone, an inhibitor of the synthesis of "late" virus-specific proteins, under conditions of micro-epizootic, prevented the death of all gilts that were in contact with patients, with 100% death of animals to which the compound was not administered. The possibility (doses, terms and multiplicity of administration of chemotherapy drugs) of reducing the virus level carrying in ASF was shown. The possibility of obtaining type-specific sera is established bypassing the attenuation of virulent strains. This allows reducing the time of obtaining serum 3–12 times, which is important for serotyping the virus (Zubairov et al. 2017).

It has been shown that fluoroquinolones are capable of inhibition of ASFV replication in vitro (Freitas et al. 2016). Particularly, genistein may block ASFV infection in Vero cells and swine macrophage (Arabyan et al. 2018).

Small interfering RNA, targeted ASFV topoisomerase II, decreases virus yield (up to 99.7%) and several infected cells (75.5%) (Freitas et al. 2016).

Promising results have been demonstrated using CRISPR-Cas9 technology-targeted ASFV DNA. Cas-9 and guided RNA aimed at 71–78 codons of p30 virus protein (CP204L) were able to decrease virus titre up to fourfold in wild boar lung cell (WSL). This alternative approach may help to design and create naturally resistant swine in future (Hubner et al. 2018).

2.7.6 Control of ASF

The main risk factors of ASF in Russia are uncontrolled animal movement and contaminated pig products (raw meat, ham, fat and skin); waste and transport; corpse of infected animals (wild boar either domestic pigs); improper herd vaccination using "one needle"; and infected wild boar. The range of sanitary measures in response to outbreaks is exhaustive and listed in national contingency plan according to OIE recommendations. In case of ASF outbreak, based on national legislation and our experience, three steps of disinfection are recommended: cleaning and washing using alkaline solutions, regular disinfection using bactericidal foam and final disinfection with fine aerosols of disinfectants (Sereda et al. 2015).

To shorten the herd replenishment after ASF outbreak, the farmers may initiate biological control of disinfection. Sentinel pigs may be introduced into the farm (10% out of total herd population) for controlling the efficiency of disinfection. The testing period lasts for 60 days; after that, blood and serum samples from sentinel pigs are tested by PCR and ELISA, respectively. Moreover, ground and sewage samples from farm-associated territories are collected. The samples are sent to ASF reference laboratories for biological assay. In case of negative results in PCR and ELISA, and negative biological assay, the disinfection should be well completed, and herd should be ready for replenishment (Sereda et al. 2016). If some positive results are received from sentinel pigs, either ground samples, the disinfection should be repeated.

ASF epidemic in the wild boar population is more complicated. ASFV is transmitted from one animal to another many times that may lead to severe changes in virus virulence. Also, the corpse of dead wild boar is the source of ASF in wild fauna, since ASFV may survive for several months in autumn-winter seasons.

Therefore, despite the fact that there is no vaccine available against ASF, history knows many positive examples of the disease eradication in several continents (Lyra 2006; Peritz 1981; Sánchez-Vizcaíno et al. 2009; Wilkinson 1986). Nowadays, understanding of the disease transmission, pathogenesis and rational biosecurity measurements is crucial for risk mitigation of ASF introduction.

Acknowledgements All the authors of the manuscript thank and acknowledge their respective universities and institutes.

Conflict of interest: There is no conflict of interest.

References

Abrams CC, Goatley L, Fishbourne E, Chapman D, Cooke L, Oura CA et al (2013) Deletion of virulence associated genes from attenuated African swine fever virus isolate OUR T88/3 decreases its ability to protect against challenge with virulent virus. Virology 443:99–105. https://doi.org/10.1016/j.virol.2013.04.028

Achenbach JE, Gallardo C, Nieto-Pelegrin E, Rivera-Arroyo B, Degefa-Negi T, Arias M et al (2017) Identification of a new genotype of African swine fever virus in domestic pigs from Ethiopia. Transbound Emerg Dis 64(5):1393–1404. https://doi.org/10.1111/tbed.12511

Afonso CL, Zsak L, Carrillo C, Borca MV, Rock DL (1998) African swine fever virus NL gene is not required for virus virulence. J Gen Virol 79:2543–2547

Afonso CL, Piccone ME, Zaffuto KM, Neilan J, Kutish GF, Lu Z et al (2004) African swine fever virus multigene family 360 and 530 genes affect host interferon response. J Virol 78:1858–1864. https://doi.org/10.1128/JVI.78.4.1858-1864.2004

Agüero M, Fernández J, Romero L, Mascaraque CS, Arias M, Sánchez-Vizcaíno JM (2003) Highly sensitive PCR assay for routine diagnosis of African swine fever virus in clinical samples. J Clin Microbiol 41:4431–4434. https://doi.org/10.1128/JCM.41.9.4431-4434.2003

Agüero M, Fernández J, Romero LJ, Zamora MJ, Sánchez C, Belák S et al (2004) A highly sensitive and specific gel-based multiplex RT-PCR assay for the simultaneous and differential diagnosis of African swine fever and Classical swine fever in clinical samples. Vet Res 35(5):551–563. https://doi.org/10.1051/vetres:2004031

Alonso C, Borca M, Dixon L, Revilla Y, Rodriguez F, Escribano JM, Consortium IR (2018) ICTV virus taxonomy profile: Asfarviridae. J Gen Virol 99(5):613–614. https://doi.org/10.1099/jgv.0.001049

Anderson EC (1986) African swine fever: current concepts on its pathogenesis and immunology. Revue Scientifique et Technique, Office International Des Epizooties 5:477–486

Anderson EC, Hutchings GH, Mukarati N, Wilkinson PJ (1998) African swine fever virus infection of the bushpig (Potamochoerus porcus) and its significance in the epidemiology of the disease. Vet Microbiol 62:1–15. https://doi.org/10.1016/S0378-1135(98)00187-4

Arabyan E, Hakobyan A, Kotsinyan A, Karalyan Z, Arakelov V, Arakelov G et al (2018) Genistein inhibits African swine fever virus replication in vitro by disrupting viral DNA synthesis. Antivir Res 156:128–137. https://doi.org/10.1016/j.antiviral.2018.06.014

Argilaguet JM, Pérez-Martín E, Nofrarías M, Gallardo C, Accensi F, Lacasta A et al (2012) DNA vaccination partially protects against African swine fever virus lethal challenge in the absence of antibodies. PLoS One 7:e40942. https://doi.org/10.1371/journal.pone.0040942

Argilaguet JM, Pérez-Martín E, López S, Goethe M, Escribano JM, Giesow K et al (2013) BacMam immunization partially protects pigs against sublethal challenge with African swine fever virus. Antivir Res 98:61–65. https://doi.org/10.1016/j.antiviral.2013.02.005

Arias M, Romero L, Agüero M, Canals A, Zamora MJ, Sánchez-Vizcaíno JM (2001) Eradication strategies of infectious diseases: African swine fever and porcine reproductive and respiratory syndrome (PRRS). Magyar Allatorvosok Lapja 123:40–46

Arias M, Jurado C, Gallardo C, Fernandez-Pinero J, Sanchez-Vizcaino JM (2018) Gaps in African swine fever: analysis and priorities. Transbound Emerg Dis 65(Suppl 1):235–247. https://doi.org/10.1111/tbed.12695

Barderas MG, Rodríguez F, Gómez-Puertas P, Avilés M, Beitia F, Alonso C, Escribano JM (2001) Antigenic and immunogenic properties of a chimera of two immunodominant African swine fever virus proteins. Arch Virol 146:1681–1691. https://doi.org/10.1007/s007050170056

Bastos AP, Nix RJ, Boinas F, Mendes S, Silva MJ, Cartaxeiro C et al (2006a) Kinetics of African swine fever virus infection in Ornithodoros erraticus ticks. J Gen Virol 87:1863–1871. https://doi.org/10.1099/vir.0.81765-0

Bastos AP, Portugal RS, Nix RJ, Cartaxeiro C, Boinas F, Dixon LK et al (2006b) Development of a nested PCR and its internal control for the detection of African swine fever virus (ASFV) in Ornithodoros erraticus. Arch Virol 151:819–826. https://doi.org/10.1007/s00705-005-0654-2

Blome S, Gabriel C, Beer M (2013) Pathogenesis of African swine fever in domestic pigs and European wild boar. Virus Res 173:122–130. pii: S0168-1702(12)00415-7. https://doi.org/10.1016/j.virusres.2012.10.026

Blome S, Gabriel C, Beer M (2014) Modern adjuvants do not enhance the efficacy of an inactivated African swine fever virus vaccine preparation. Vaccine 32(31):3879–3882. https://doi.org/10.1016/J.VACCINE.2014.05.051

Boinas FS, Hutchings GH, Dixon LK, Wilkinson PJ (2004) Characterization of pathogenic and non-pathogenic African swine fever virus isolates from Ornithodoros erraticus inhabiting pig premises in Portugal. J Gen Virol 85:2177–2187. https://doi.org/10.1099/vir.0.80058-0

Boinas FS, Wilson AJ, Hutchings GH, Martins C, Dixon LJ (2011) The persistence of African swine fever virus in field-infected Ornithodoros erraticus during the ASF endemic period in Portugal. PLoS One 6:e20383. https://doi.org/10.1371/journal.pone.0020383

Bool PH, OrdaS A, SaNchez Botija C (1970) Fluorescent antibody test for African swine fever. Revista Del Patronato de Biologia Animal 14:115–132

Borca MV, Kutish GF, Afonso CL, Irusta P, Carrillo C, Brun A et al (1994) An African swine fever virus gene with similarity to the T-lymphocyte surface antigen CD2 mediates hemadsorption. Virology 199:463–468. https://doi.org/10.1006/viro.1994.1146

Borca MV, O'Donnell V, Holinka LG, Ramirez-Medina E, Clark BA, Vuono EA et al (2018) The L83L ORF of African swine fever virus strain Georgia encodes for a non-essential gene that interacts with the host protein IL-1beta. Virus Res 249:116–123. https://doi.org/10.1016/j.virusres.2018.03.017

Bosch J, Rodriguez A, Iglesias I, Munoz MJ, Jurado C, Sanchez-Vizcaino JM, de la Torre A (2017) Update on the risk of introduction of African swine fever by wild boar into disease-free European union countries. Transbound Emerg Dis 64(5):1424–1432. https://doi.org/10.1111/tbed.12527

Brown A-A, Penrith ML, Fasina FO, Beltran-Alcrudo D (2018) The African swine fever epidemic in West Africa, 1996-2002. Transbound Emerg Dis 65(1):64–76. https://doi.org/10.1111/tbed.12673

Burmakina G, Malogolovkin A, Tulman ER, Zsak L, Delhon G, Diel DG et al (2016) African swine fever virus serotype-specific proteins are significant protective antigens for African swine fever. J Gen Virol 97(7):1670–1675. https://doi.org/10.1099/jgv.0.000490

Burrage TG (2013) African swine fever virus infection in Ornithodoros ticks. Virus Res 173:131–139. https://doi.org/10.1016/j.virusres.2012.10.010

Burrage TG, Lu Z, Neilan JG, Rock DL, Zsak L (2004) African swine fever virus multigene family 360 genes affect virus replication and generalization of infection in Ornithodoros porcinus ticks. J Virol 78:2445–2453. https://doi.org/10.1128/JVI.78.5.2445-2453.2004

Caporale V, Rutili D, Nannini D, di Francesco C, Ghinato C (1988) Epidemiology of classical swine fever in Italy from 1970 to 1985. Revue Scientifique et Technique, Office International Des Epizooties 7:599–617

Carlson J, O'Donnell V, Alfano M, Velazquez Salinas L, Holinka LG, Krug PW et al (2016) Association of the host immune response with protection using a live attenuated African swine fever virus model. Viruses 8(10):291. https://doi.org/10.3390/v8100291

Chichikin AY, Gazaev IK, Tsybanov SZ, Kolvasov D (2012) A non-contact method for selecting saliva from a wild boar in African swine fever. Veterinariya 6:26–28

Cisek AA, Dabrowska I, Gregorczyk KP, Wyzewski Z (2016) African swine fever virus: a new old enemy of Europe. Ann Parasitol 62(3):161–167

Costard S, Wieland B, de Glanville W, Jori F, Rowlands R, Vosloo W et al (2009) African swine fever: how can global spread be prevented? Philos Trans R Soc Lond B Biol Sci 364:2683–2696. Retrieved from 19687038%5CnPM

Costard S, Mur L, Lubroth J, Sanchez-Vizcaino JM, Pfeiffer DU (2013) Epidemiology of African swine fever virus. Virus Res 173:191–197. https://doi.org/10.1016/j.virusres.2012.10.030

Davies K, Goatley LC, Guinat C, Netherton CL, Gubbins S, Dixon LK, Reis AL (2017) Survival of African swine fever virus in excretions from pigs experimentally infected with the Georgia 2007/1 isolate. Transbound Emerg Dis 64(2):425–431. https://doi.org/10.1111/tbed.12381

de Carvalho Ferreira HC, Weesendorp E, Elbers ARW, Bouma A, Quak S, Stegeman JA, Loeffen WLA (2012) African swine fever virus excretion patterns in persistently infected animals: a quantitative approach. Vet Microbiol 160:327–340. https://doi.org/10.1016/j.vetmic.2012.06.025

de Carvalho Ferreira HC, Weesendorp E, Quak S, Stegeman JA, Loeffen WLA (2014) Suitability of faeces and tissue samples as a basis for non-invasive sampling for African swine fever in wild boar. Vet Microbiol 172:449–454. https://doi.org/10.1016/j.vetmic.2014.06.016

De Kock G, Robinson EM, Keppel JJG (1940) Swine fever in South Africa. Onderstepoort J Vet Sci 14:31–93

Diaz AV, Netherton CL, Dixon LK, Wilson AJ (2012) African swine fever virus strain Georgia 2007/1 in Ornithodoros erraticus ticks. Emerg Infect Dis 18:1026–1028. https://doi.org/10.3201/eid1806.111728

Dixon LK, Wilkinson PJ (1988) Genetic diversity of African swine fever virus isolates from soft ticks (Ornithodoros moubata) inhabiting warthog burrows in Zambia. J Gen Virol 69:2981–2993. https://doi.org/10.1099/0022-1317-69-12-2981

Dixon LK, Abrams CC, Chapman DDG, Goatley LC, Netherton CL, Taylor G, Takamatsu HH (2013) Prospects for development of African swine fever virus vaccines. Dev Biol 135:147–157. https://doi.org/10.1159/000170936

Edgar G, Hart L, Hayston JT (1952) Studies on the viability of the virus of swine fever. Report 14th International Veterinary Congress, vol 2, pp 387–391

EFSA AHAW Panel (EFSA Panel on Animal Health and Welfare) (2014) Scientific opinion on African swine fever. EFSA J 12(4):3628. https://doi.org/10.2903/j.efsa.2014.3628

Fernández-Pinero J, Gallardo C, Elizalde M, Robles A, Gómez C, Bishop R et al (2013) Molecular diagnosis of African swine fever by a new real-time PCR using universal probe library. Transbound Emerg Dis 60:1–11. https://doi.org/10.1111/j.1865-1682.2012.01317.x

Fillipova NA (1966) Arachnids. In: Fauna of the USSR

Freitas FB, Frouco G, Martins C, Leitao A, Ferreira F (2016) In vitro inhibition of African swine fever virus-topoisomerase II disrupts viral replication. Antivir Res 134:34–41. https://doi.org/10.1016/j.antiviral.2016.08.021

Gabriel C, Blome S, Malogolovkin A, Parilov S, Kolbasov D, Teifke JP, Beer M (2011) Characterization of African swine fever virus Caucasus isolate in European wild boars. Emerg Infect Dis 17(12):2342–2345. https://doi.org/10.3201/eid1712.110430

Galindo I, Almazán F, Bustos MJ, Viñuela E, Carrascosa AL (2000) African swine fever virus EP153R open reading frame encodes a glycoprotein involved in the hemadsorption of infected cells. Virology 266:340–351. https://doi.org/10.1006/viro.1999.0080

Gallardo C, Soler A, Nieto R, Carrascosa AL, De Mia GM, Bishop RP et al (2013) Comparative evaluation of novel African swine fever virus (ASF) antibody detection techniques derived from specific ASF viral genotypes with the OIE internationally prescribed serological tests. Vet Microbiol 162:32–43. https://doi.org/10.1016/j.vetmic.2012.08.011

Gallardo C, Fernandez-Pinero J, Pelayo V, Gazaev I, Markowska-Daniel I, Pridotkas G et al (2014) Genetic variation among African swine fever genotype II viruses, eastern and central Europe. Emerg Infect Dis 20(9):1544–1547. https://doi.org/10.3201/eid2009.140554

Gallardo C, Soler A, Nieto R, Sanchez MA, Martins C, Pelayo V et al (2015) Experimental transmission of African swine fever (ASF) low virulent isolate NH/P68 by surviving pigs. Transbound Emerg Dis 62(6):612–622. https://doi.org/10.1111/tbed.12431

Gallardo C, Nurmoja I, Soler A, Delicado V, Simón A, Martin E et al (2018) Evolution in Europe of African swine fever genotype II viruses from highly to moderately virulent. Vet Microbiol 219:70–79. https://doi.org/10.1016/j.vetmic.2018.04.001

Gimenez-Lirola LG, Mur L, Rivera B, Mogler M, Sun Y, Lizano S et al (2016) Detection of African swine fever virus antibodies in serum and oral fluid specimens using a recombinant protein 30 (p30) dual matrix indirect ELISA. PLoS One 11(9):e0161230. https://doi.org/10.1371/journal.pone.0161230

Gogin A, Gerasimov V, Malogolovkin A, Kolbasov D (2013) African swine fever in the North Caucasus region and the Russian Federation in years 2007-2012. Virus Res 173:198–203. https://doi.org/10.1016/j.virusres.2012.12.007

Goller KV, Malogolovkin AS, Katorkin S, Kolbasov D, Titov I, Höper D et al (2015) Tandem repeat insertion in African swine fever virus, Russia, 2012. Emerg Infect Dis 21:731–732. https://doi.org/10.3201/eid2104.141792

Gómez-Puertas P, Rodríguez F, Oviedo JM, Ramiro-Ibáñez F, Ruiz-Gonzalvo F, Alonso C, Escribano JM (1996) Neutralizing antibodies to different proteins of African swine fever virus inhibit both virus attachment and internalization. J Virol 70:5689–5694

Gómez-Puertas P, Rodríguez F, Oviedo JM, Brun A, Alonso C, Escribano JM (1998) The African swine fever virus proteins p54 and p30 are involved in two distinct steps of virus attachment and both contribute to the antibody-mediated protective immune response. Virology 243:461–471. https://doi.org/10.1006/viro.1998.9068

Greig A, Plowright W (1970) The excretion of two virulent strains of African swine fever virus by domestic pigs. J Hyg 68:673–682. https://doi.org/10.1017/S0022172400042613

Greig AS, Boulanger P, Bannister GL (1967) African swine fever. V. Cultivation of the virus in primary pig kidney cells. Can J Comp Med Vet Sci 31:24–31

Grossmann C, Tenbusch M, Nchinda G, Temchura V, Nabi G, Stone GW et al (2009) Enhancement of the priming efficacy of DNA vaccines encoding dendritic cell-targeted antigens by synergistic toll-like receptor ligands. BMC Immunol 10:43. https://doi.org/10.1186/1471-2172-10-43

Guinat C, Gubbins S, Vergne T, Gonzales JL, Dixon L, Pfeiffer DU (2016) Experimental pig-to-pig transmission dynamics for African swine fever virus, Georgia 2007/1 strain-CORRIGENDUM. Epidemiol Infect 144:3564–3566. https://doi.org/10.1017/S0950268816001667

Gutiérrez-Castañeda B, Reis AL, Corteyn A, Parkhouse RME, Kollnberger S (2008) Expression, cellular localization and antibody responses of the African swine fever virus genes B602L and K205R. Arch Virol 153:2303–2306. https://doi.org/10.1007/s00705-008-0246-z

Hammond RA, Detray DE (1955) A recent case of African swine fever in Kenya, East Africa. Am Vet Med Assoc 126:389–391

Hess WR (1981) African swine fever: a reassessment. Adv Vet Sci Comp Med 25:39–69

Hess WR, Endris RG, Haslett TM, Monahan MJ, McCoy JP (1987) Potential arthropod vectors of African swine fever virus in North America and the Caribbean basin. Vet Parasitol 26:145–155. https://doi.org/10.1016/0304-4017(87)90084-7

Hess WR, Endris RG, Lousa A, Caiado JM (1989) Clearance of African swine fever virus from infected tick (Acari) colonies. J Med Entomol 26:314–317

Heuschele WP, Coggins L (1969) Epizootiology of African swine fever virus in warthogs. Bull Epizoot Dis Afr 17:179–183

Hubner A, Petersen B, Keil GM, Niemann H, Mettenleiter TC, Fuchs W (2018) Efficient inhibition of African swine fever virus replication by CRISPR/Cas9 targeting of the viral p30 gene (CP204L). Sci Rep 8(1):1449. https://doi.org/10.1038/s41598-018-19626-1

James HE, Ebert K, McGonigle R, Reid SM, Boonham N, Tomlinson JA et al (2010) Detection of African swine fever virus by loop-mediated isothermal amplification. J Virol Methods 164:68–74. https://doi.org/10.1016/j.jviromet.2009.11.034

Jurado C, Fernandez-Carrion E, Mur L, Rolesu S, Laddomada A, Sanchez-Vizcaino JM (2018) Why is African swine fever still present in Sardinia? Transbound Emerg Dis 65(2):557–566. https://doi.org/10.1111/tbed.12740

Jurkov GG, Peskovatskov AP, Shpackov AK, Mackarevich VG, Cherevatenko BN, Balabanov VA (2014) Report on the results of studying the routes of African swine fever (ASF) entry and spreading over Odessa region in 1977

Karaulov AK, Shevtsov AA, Petrova ON, Korennoy FI, Vadopolas TV (2018) The forecast of African swine fever spread in Russia until 2025. Veterinaria i Kormlenie 3:12–14

King DP, Reid SM, Hutchings GH, Grierson SS, Wilkinson PJ, Dixon LK et al (2003) Development of a TaqMan PCR assay with internal amplification control for the detection of African swine fever virus. J Virol Methods 107:53–61. https://doi.org/10.1016/S0166-0934(02)00189-1

King K, Chapman D, Argilaguet JM, Fishbourne E, Hutet E, Cariolet R et al (2011) Protection of European domestic pigs from virulent African isolates of African swine fever virus by experimental immunisation. Vaccine 29:4593–4600. https://doi.org/10.1016/j.vaccine.2011.04.052

Kleiboeker SB (2002) Swine fever: classical swine fever and African swine fever. Vet Clin N Am Food Anim Pract 18:431–451

Kolbasov D, Titov I, Tsybanov S, Gogin A, Malogolovkin A (2018a) African swine fever virus, Siberia, Russia, 2017. Emerg Infect Dis 24(4):796–798. https://doi.org/10.3201/eid2404.171238

Kolbasov DV, Gogin A, Malogolovkin A (2018b) Ten years with African swine fever—lessons learned. In: 25th International Pig Veterinary Society congress, p 37

Kollnberger SD, Gutierrez-Castañeda B, Foster-Cuevas M, Corteyn A, Parkhouse RME (2002) Identification of the principal serological immunodeterminants of African swine fever virus by screening a virus cDNA library with antibody. J Gen Virol 83:1331–1342

Korennoy FI, Gulenkin VM, Gogin AE, Vergne T, Karaulov AK (2017) Estimating the basic reproductive number for African swine fever using the Ukrainian historical epidemic of 1977. Transbound Emerg Dis 64:1858–1866. https://doi.org/10.1111/tbed.12583

Kovalenko Y, Sidorov MA (1973) [Reservoirs and mode of circulation of African swine fever virus in nature]. Sel'skokhozyaistvennaya Biologiya 8:598–606

Kovalev YI (2017) Veterinariy v svinovodstve. In: Swine production in Russia 2015-2020: current challenges, risks and solutions. Novosibirsk, 18–19 May

Lacasta A, Ballester M, Monteagudo PL, Rodríguez JM, Salas ML, Accensi F et al (2014) Expression library immunization can confer protection against lethal challenge with African swine fever virus. J Virol 88(22):13322–13332. https://doi.org/10.1128/JVI.01893-14

Leifert JA, Rodriguez-Carreno MP, Rodriguez F, Whitton JL (2004) Targeting plasmid-encoded proteins to the antigen presentation pathways. Immunol Rev 199:40–53. https://doi.org/10.1111/j.0105-2896.2004.0135.x

Leitão A, Cartaxeiro C, Coelho R, Cruz B, Parkhouse RME, Portugal FC et al (2001) The non-haemadsorbing African swine fever virus isolate ASFV/NH/P68 provides a model for defining the protective anti-virus immune response. J Gen Virol 82:513–523

Lyra TMP (2006) The eradication of African swine fever in Brazil, 1978-1984. Revue Scientifique et Technique (International Office of Epizootics) 25:93–103

Makarov V, Nedosekov V, Sereda A, Matvienko N (2016) Immunological conception of African swine fever. Zool Ecol 26(3):236–243. https://doi.org/10.1080/21658005.2016.1182822

Malogolovkin A, Yelsukova A, Gallardo C, Tsybanov S, Kolbasov D (2012) Molecular characterization of African swine fever virus isolates originating from outbreaks in the Russian Federation between 2007 and 2011. Vet Microbiol 158(3–4):415–419. https://doi.org/10.1016/j.vetmic.2012.03.002

Malogolovkin A, Burmakina G, Titov I, Sereda A, Gogin A, Baryshnikova E, Kolbasov D (2015a) Comparative analysis of African swine fever virus genotypes and serogroups. Emerg Infect Dis 21(2):312–315. https://doi.org/10.3201/eid2102.140649

Malogolovkin A, Burmakina G, Tulman ER, Delhon G, Diel DG, Salnikov N et al (2015b) African swine fever virus CD2v and C-type lectin gene loci mediate serological specificity. J Gen Virol 96(4):866–873. https://doi.org/10.1099/jgv.0.000024

Mebus CA (1988) African swine fever. Adv Virus Res 35:251–269

Misinzo G (2012) African swine fever virus, Tanzania, 2010–2012. Emerg Infect Dis 193:319–328. https://doi.org/10.3201/eid1812.121083

Monteagudo PL, Lacasta A, Lopez E, Bosch L, Collado J, Pina-Pedrero S et al (2017) BA71DeltaCD2: a new recombinant live attenuated African swine fever virus with cross-protective capabilities. J Virol 91(21):e01058-17. https://doi.org/10.1128/JVI.01058-17

Mulumba-Mfumu LK, Achenbach JE, Mauldin MR, Dixon LK, Tshilenge CG, Thiry E et al (2017) Genetic assessment of African swine fever isolates involved in outbreaks in the democratic Republic of Congo between 2005 and 2012 reveals co-circulation of p72 genotypes I, IX and XIV, including 19 variants. Viruses 9(2):E31. https://doi.org/10.3390/v9020031

Mur L, Martínez-López B, Sánchez-Vizcaíno J (2012) Risk of African swine fever introduction into the European Union through transport-associated routes: returning trucks and waste from international ships and planes. BMC Vet Res 8:149. https://doi.org/10.1186/1746-6148-8-149

Mur L, Sanchez-Vizcaino JM, Fernandez-Carrion E, Jurado C, Rolesu S, Feliziani F et al (2018) Understanding African Swine Fever infection dynamics in Sardinia using a spatially explicit transmission model in domestic pig farms. Transbound Emerg Dis 65(1):123–134. https://doi.org/10.1111/tbed.12636

Neilan JG, Zsak L, Lu Z, Kutish GF, Afonso CL, Rock DL (2002) Novel swine virulence determinant in the left variable region of the African swine fever virus genome. J Virol 76:3095–3104. https://doi.org/10.1128/JVI.76.7.3095-3104.2002

Nurmoja I, Petrov A, Breidenstein C, Zani L, Forth JH, Beer M et al (2017a) Biological characterization of African swine fever virus genotype II strains from north-eastern Estonia in European wild boar. Transbound Emerg Dis 64(6):2034–2041. https://doi.org/10.1111/tbed.12614

Nurmoja I, Schulz K, Staubach C, Sauter-Louis C, Depner K, Conraths FJ, Viltrop A (2017b) Development of African swine fever epidemic among wild boar in Estonia—two different areas in the epidemiological focus. Sci Rep 7(1):12562. https://doi.org/10.1038/s41598-017-12952-w

O'Donnell V, Holinka LG, Sanford B, Krug PW, Carlson J, Pacheco JM et al (2016) African swine fever virus Georgia isolate harboring deletions of 9GL and MGF360/505 genes is highly attenuated in swine but does not confer protection against parental virus challenge. Virus Res 221:8–14. https://doi.org/10.1016/j.virusres.2016.05.014

O'Donnell V, Risatti GR, Holinka LG, Krug PW, Carlson J, Velazquez-Salinas L et al (2017) Simultaneous deletion of the 9GL and UK Genes from the African swine fever virus Georgia 2007 isolate offers increased safety and protection against homologous challenge. J Virol 91(1):e01760-16. https://doi.org/10.1128/JVI.01760-16

Oganesyan AS, Petrova ON, Korennoy FI, Bardina NS, Gogin AE, Dudnikov SA (2013) African swine fever in the Russian Federation: spatio-temporal analysis and epidemiological overview. Virus Res 173:204–211. https://doi.org/10.1016/j.virusres.2012.12.009

Okoth E, Gallardo C, Macharia JM, Omore A, Pelayo V, Bulimo DW et al (2013) Comparison of African swine fever virus prevalence and risk in two contrasting pig-farming systems in South-west and Central Kenya. Prev Vet Med 110:198–205. https://doi.org/10.1016/j.prevetmed.2012.11.012

Olesen AS, Lohse L, Hansen MF, Boklund A, Halasa T, Belsham GJ et al (2018) Infection of pigs with African swine fever virus via ingestion of stable flies (Stomoxys calcitrans). Transbound Emerg Dis 65:1152–1157. https://doi.org/10.1111/tbed.12918

Olsevskis E, Guberti V, Serzants M, Westergaard J, Gallardo C, Rodze I, Depner K (2016) African swine fever virus introduction into the EU in 2014: experience of Latvia. Res Vet Sci 105:28–30. https://doi.org/10.1016/j.rvsc.2016.01.006

Onashvili T, Donduashvili M, Borca M, Mamisashvili E, Goginashvili K, Tighilauri T et al (2012) Countermeasures for the control of African Swine Fever in Georgia. Int J Infect Dis 16:e268. https://doi.org/10.1016/j.ijid.2012.05.920

Orfei Z, Persechino A, Lupini PM, Cassone A (1968) Haemadsorption test in the diagnosis of African swine fever in Italy. Atti Soc Ital Sci Vet 21:850–854

Oura CAL, Edwards L, Batten CA (2013) Virological diagnosis of African swine fever—comparative study of available tests. Virus Res 173:150–158. https://doi.org/10.1016/j.virusres.2012.10.022

Owolodun OA, Yakubu B, Antiabong JF, Ogedengbe ME, Luka PD, John Audu B et al (2010) Temporal dynamics of African swine fever outbreaks in Nigeria, 2002-2007. Transbound Emerg Dis 57:330–339. https://doi.org/10.1111/j.1865-1682.2010.01153.x

Pan IC, Hess WR (1985) Diversity of African swine fever virus. Am J Vet Res 46:314–320

Pan IC, Shimizu M, Hess WR (1980) Replication of African swine fever virus in cell cultures. Am J Vet Res 41:1357–1367

Parker J, Plowright W, Pierce MA (1969) The epizootiology of African swine fever in Africa. Vet Rec 85:668–674

Pastor MJ, Laviada MD, Sanchez-Vizcaino JM, Escribano JM (1989) Detection of African swine fever virus antibodies by immunoblotting assay. Can J Vet Res = Revue Canadienne de Recherche Veterinaire 53:105–107. https://doi.org/10.4314/nvj.v27i2.3518

Pejsak Z, Truszczyński M, Kozak E, Markowska-Daniel I (2014) [Epidemiological analysis of two first cases of African swine fever in wild boar in Poland] Analiza epidemiologiczna dwóch pierwszych przypadków afrykańskiego pomoru świń u dzików w Polsce. Medycyna Weterynaryjna 70:369–372. Retrieved from http://www.scopus.com/inward/record.url?eid=2-s2.0-84903171078&partnerID=tZOtx3y1

Penrith ML (2009) African swine fever. Onderstepoort J Vet Res 76:91–95

Penrith ML, Thomson GR, Bastos ADS, Phiri OC, Lubisi BA, du Plessis EC et al (2004) An investigation into natural resistance to African swine fever in domestic pigs from an endemic area in southern Africa. Revue Scientifique et Technique, Office International Des Épizooties 23:965–977. Retrieved from http://www.ncbi.nlm.nih.gov/pubmed/15861893%5Cnhttp://www.cabdirect.org/abstracts/20053080761.html?resultNumber=18&start=0&q=("'hog+cholera"+OR+"'classical+swine+fever")+AND+yr:[2000+TO+2012]+AND+(cattle+OR+sheep+OR+goats+OR+pigs+OR+poultry)+%5Cn%5Cn%5Cn&fq=gl_

Peritz FJ (1981) The evolution of African swine fever in Latin America and F.A.O.'s corresponding action programme. Bulletin de l'Office International Des Epizooties 93(469):499

Petrov A, Forth JH, Zani L, Beer M, Blome S (2018) No evidence for long-term carrier status of pigs after African swine fever virus infection. Transbound Emerg Dis 65:1318–1328. https://doi.org/10.1111/tbed.12881

Pietschmann J, Guinat C, Beer M, Pronin V, Tauscher K, Petrov A et al (2015) Course and transmission characteristics of oral low-dose infection of domestic pigs and European wild boar with a Caucasian African swine fever virus isolate. Arch Virol 160(7):1657–1667. https://doi.org/10.1007/s00705-015-2430-2

Plowright W, Perry CT, Peirce MA (1970) Transovarial infection with African swine fever virus in the argasid tick, Ornithodoros moubata porcinus, Walton. Res Vet Sci 11:582–584

Plowright W, Thomson GR, Neser JA (2002) African swine fever. In: Infectious diseases of livestock with special reference to Southern Africa, vol 2, pp 567–599. Retrieved from http://www.oie.int/eng/normes/mmanual/2008/pdf/2.08.01_ASF.pdf

Ramiro-Ibáñez F, Ortega A, Ruiz-Gonzalvo F, Escribano JM, Alonso C (1997) Modulation of immune cell populations and activation markers in the pathogenesis of African swine fever virus infection. Virus Res 47:31–40. https://doi.org/10.1016/S0168-1702(96)01403-7

Reis AL, Goatley LC, Jabbar T, Sanchez-Cordon PJ, Netherton CL, Chapman DAG, Dixon LK (2017a) Deletion of the African swine fever virus gene DP148R does not reduce virus replication in culture but reduces virus virulence in pigs and induces high levels of protection against challenge. J Virol 91(24):e01428-17. https://doi.org/10.1128/JVI.01428-17

Reis AL, Netherton C, Dixon LK (2017b) Unraveling the armor of a killer: evasion of host defenses by African swine fever virus. J Virol 91(6):e02338-16. https://doi.org/10.1128/JVI.02338-16

Rennie L, Wilkinson PJ, Mellor PS (2001) Transovarial transmission of African swine fever virus in the argasid tick Ornithodoros moubata. Med Vet Entomol 15:140–146. https://doi.org/10.1046/j.1365-2915.2001.00282.x

Rock DL (2017) Challenges for African swine fever vaccine development—"... perhaps the end of the beginning". Vet Microbiol 206:52–58. https://doi.org/10.1016/j.vetmic.2016.10.003

Rodríguez JM, Yáñez RJ, Almazán F, Viñuela E, Rodriguez JF (1993) African swine fever virus encodes a CD2 homolog responsible for the adhesion of erythrocytes to infected cells. J Virol 67:5312–5320

Salguero FJ, Ruiz-Villamor E, Bautista MJ, Sánchez-Cordón PJ, Carrasco L, Gómez-Villamandos JC (2002) Changes in macrophages in spleen and lymph nodes during acute African swine fever: expression of cytokines. Vet Immunol Immunopathol 90:11–22. https://doi.org/10.1016/S0165-2427(02)00225-8

Salguero FJ, Gil S, Revilla Y, Gallardo C, Arias M, Martins C (2008) Cytokine mRNA expression and pathological findings in pigs inoculated with African swine fever virus (E-70) deleted on A238L. Vet Immunol Immunopathol 124:107–119. https://doi.org/10.1016/j.vetimm.2008.02.012

Sánchez-Cordón PJ, Núñez A, Salguero FJ, Carrasco L, Gómez-Villamandos JC (2005) Evolution of T lymphocytes and cytokine expression in classical swine fever (CSF) virus infection. J Comp Pathol 132(4):249–260. https://doi.org/10.1016/j.jcpa.2004.10.002

Sanchez-Cordon PJ, Chapman D, Jabbar T, Reis AL, Goatley L, Netherton CL et al (2017) Different routes and doses influence protection in pigs immunised with the naturally attenuated African swine fever virus isolate OURT88/3. Antivir Res 138:1–8. https://doi.org/10.1016/j.antiviral.2016.11.021

Sanchez-Vizcaino JM, Slauson DO, Ruiz-Gonzalvo F, Valero F (1981) Lymphocyte function and cell-mediated immunity in pigs with experimentally induced African swine fever. Am J Vet Res 42:1335–1341

Sánchez-Vizcaíno JM, Martínez-Lópeza B, Martínez-Avilés M, Martins C, Boinas F, Vial L et al (2009) Scientific review on African Swine Fever. CFP/EFSA/AHAW/2007/2, pp 1–141. https://doi.org/10.2903/sp.efsa.2009.EN-5

Sánchez-Vizcaíno JM, Mur L, Martínez-López B (2012) African swine fever: an epidemiological update. Transbound Emerg Dis 59:27–35. https://doi.org/10.1111/j.1865-1682.2011.01293.x

Sereda AD, Balyshev VM (2011) [Antigenic diversity of African swine fever viruses]. Vopr Virusol 56(4):38–42. Retrieved from http://www.ncbi.nlm.nih.gov/pubmed/21899069
Sereda AD, Anokhina EG, Makarov VV (1994) [Glycoproteins from the African swine fever virus]. Vopr Virusol 39:278–281. Retrieved from http://www.ncbi.nlm.nih.gov/entrez/query.fcgi?cmd=Retrieve&db=PubMed&dopt=Citation&list_uids=7716925
Sereda A, Selyaninov Y, Egorova I, Balyshev V, Bureev I, Kushnir A et al (2015) Testing means and methods of disinfection for the African swine fever virus. Veterinariya 7:51–55
Sereda A, Zhivodeorov S, Baluishev V, Strizhackova O, Sindryackova I, Lunitsin A, Kolbasov D (2016) Experience of determining disinfection level in carrying out African swine fever eradication campaigns in pig farms using a sentinel method. Veterinariya 6:44–48
Sierra MA, Carrasco L, Gómez-Villamandos JC, Martin de las Mulas J, Méndez A, Jover A (1990) Pulmonary intravascular macrophages in lungs of pigs inoculated with African swine fever virus of differing virulence. J Comp Pathol 102:323–334. https://doi.org/10.1016/S0021-9975(08)80021-7
Smietanka K, Wozniakowski G, Kozak E, Niemczuk K, Fraczyk M, Bocian L et al (2016) African swine fever epidemic, Poland, 2014–2015. Emerg Infect Dis 22(7):1201–1207. https://doi.org/10.3201/eid2207.151708
Smirnov AM, Butko MP (2011) Stability of the pathogen and measures to combat African swine fever. Veterinarian 6:2–7
Souto R, Mutowembwa P, van Heerden J, Fosgate GT, Heath L, Vosloo W (2016) Vaccine potential of two previously uncharacterized African swine fever virus isolates from southern Africa and heterologous cross protection of an avirulent European isolate. Transbound Emerg Dis 63:224–231. https://doi.org/10.1111/tbed.12250
Stone SS, DeLay PD, Sharman EC (1968) The antibody response in pigs inoculated with attenuated African swine fever virus. Can J Comp Med (Gardenvale, Quebec) 32:455–460
Strizhakova OM, Lyska VM, Malogolovkin AS, Novikova MB, Sidlik MV, Nogina IV et al (2016) Validation of an ELISA kit for detection of antibodies against ASF virus in blood or spleen of domestic pigs and wild boars. Sel'skokhozyaistvennaya Biologiya 51(6):845–852. https://doi.org/10.15389/agrobiology.2016.6.845eng
Talaat AM, Stemke-Hale K (2005) Expression library immunization: a road map for discovery of vaccines against infectious diseases. Infect Immun 73(11):7089–7098. https://doi.org/10.1128/IAI.73.11.7089-7098.2005
Terpstra C, Wensvoort G (1986) African swine fever in the Netherlands. Tijdschr Diergeneeskd 111:389–392
van Drunen Littel-van den Hurk S, Babiuk SL, Babiuk LA (2004) Strategies for improved formulation and delivery of DNA vaccines to veterinary target species. Immunol Rev 199:113–125. https://doi.org/10.1111/j.0105-2896.2004.00140.x
Vishnjakov I, Mitin N, Karpov G, Kurinnov VJ (1991) Differentiation African and classical swine fever viruses. Veterinariya 4:28–31
Wilkinson PJ (1986) African swine fever in Belgium. State Vet J 40:123–129; 12 ref.
World Organisation for Animal Health (OIE) (2012) Manual of diagnostic tests and vaccines for terrestrial animal
Wozniakowski G, Kozak E, Kowalczyk A, Lyjak M, Pomorska-Mol M, Niemczuk K, Pejsak Z (2016) Current status of African swine fever virus in a population of wild boar in eastern Poland (2014-2015). Arch Virol 161(1):189–195. https://doi.org/10.1007/s00705-015-2650-5
Zani L, Forth JH, Forth L, Nurmoja I, Leidenberger S, Henke J et al (2018) Deletion at the 5′-end of Estonian ASFV strains associated with an attenuated phenotype. Sci Rep 8(1):6510. https://doi.org/10.1038/s41598-018-24740-1
Zubairov M, Selyaninov Y, Roshchin A, Khokhlov P (2017) Antiviral activity and therapeutic and preventive effect of phosphonoacetic acid and its derivatives. Electr J Chem Saf 1:146–157. https://doi.org/10.25514/CHS2017111440

Chapter 3
Classical Swine Fever Virus

Dilip K. Sarma

Abstract Classical swine fever virus (CSFV), belonging to the genus *Pestivirus* of the family *Flaviviridae* and species *Pestivirus C*, is the aetiological agent of a disease called classical swine fever (CSF) or hog cholera, which is a devastating transboundary disease of pigs across the globe. Besides domestic pigs, the disease has been reported from wild pigs, and pygmy hog (*Porcula salvania*) and wild boar are also reservoir hosts. The CSFV size ranges from 40 to 60 nm, and the genome of the virus consists of a single-stranded, positive-sense RNA. Based on sequence analysis, 3 genotypes and 14 subgenotypes of CSFV have so far been reported, and a shift in the historical genotypes 1–2 has been observed in the recent past. Although many countries are now free from CSF, the disease is still widely prevalent in most of the South American, Asian and South East Asian countries possibly due to high pig population, low vaccination coverage and poor biosecurity. The virus spreads in different ways, and after entry into the susceptible hosts produces various clinical manifestations. The peracute form of CSF has disappeared, while acute form occurs commonly in younger animals and chronic form is more common with reports of absence of clear symptoms in adult animals. Confirmatory diagnosis of the disease is to be made at the early stage to prevent its spread and control of CSF is important to sustain pig production to meet the growing demand of pork in different parts of the world.

Keywords Classical swine fever virus · CSF · CSFV · Genotype · Subgenotype · Epidemiology · Diagnosis · Vaccine and control

3.1 Preamble

Classical swine fever virus (CSFV) is the causative agent of classical swine fever (CSF) or hog cholera, which is a serious, economically damaging disease of pigs, and it can spread in an epizootic form to domestic and wild pig population. It is a listed disease of the World Organisation for Animal Health (OIE), and because of its high economic impact on pig production, it is to be reported to the OIE (Edwards

D. K. Sarma (✉)
Veterinary Microbiology, Assam Agricultural University, Guwahati, Assam, India

Y. S. Malik et al. (eds.), *Emerging and Transboundary Animal Viruses*, Livestock Diseases and Management, https://doi.org/10.1007/978-981-15-0402-0_3

et al. 2000). Although many countries have implemented control and eradication programmes successfully, CSF is a global problem. Due to globalisation, there is more movement of people and intensification of pig trade and increase in domestic pigs, the most susceptible host, and wild boars which act as reservoirs of CSFV, it may continue as an important health problem of pigs worldwide (Moennig and Becher 2015). The disease causes enormous economic loss due to high mortality and morbidity in the affected animals. Economic loss due to CSF in India was estimated to be 4.29 billion INR (Singh et al. 2016), out of which mortality alone accounted for 74.07% (3.184 billion INR), and the loss due to morbidity accounted for 25.93% (1.114 billion INR).

Role of pig farming in livelihood and socio-economic upliftment of the rural people in many parts of the world needs no further emphasis. Therefore, a significant change in terms of pork production, particularly on quality and hygienic pork production, is required. The two major challenges in the piggery sector are poor biosecurity and increasing trends of viral diseases. It has been reported that about 40% of the world's pigs are raised under extensive and low biosecurity conditions (Postel et al. 2017). More than 60% of the world pigs are in Asia and backyard pig farming is largely practised. Although backyard pig farms have less number of pigs per unit, due to low biosecurity standards, little interaction with the veterinary services, poor management and disease prevention, they facilitate the easy spread of the disease. Therefore, scientific piggery with improved biosecurity, proper management and adequate prevention and control programme is the need of the hour to minimise the disease burden and economic loss due to the disease.

3.2 Taxonomy and Structure of CSFV

The CSFV belongs to the genus *Pestivirus* of the family *Flaviviridae* along with the other genera such as *Flavivirus*, *Hepacivirus* and *Pegivirus* (Beer et al. 2015). The genus *Pestivirus* also includes bovine viral diarrhoea viruses 1 and 2 (BVDV-1 and -2), border disease virus (BDV) and several unclassified atypical pestiviruses, from giraffe, pronghorn antelope, Bungowannah virus and atypical porcine pestivirus, which have some genetic and antigenic relationship with CSFV. Two new pestiviruses named as Aydin/04 and Burdur/05 within the border disease virus group and reported from Turkey showed a close genetic and antigenic relationship with CSFV (Postel et al. 2015).

Based on the whole-genome sequence studies and antigenic differences, pestiviruses have been assigned to 11 different species (Smith et al. 2017), and CSFV belongs to *Pestivirus C* species.

The virion of CSFV has icosahedral symmetry, and the viral RNA wrapped by a core protein. The outer layer of the virion is a lipid envelope, which has three envelope glycoproteins, and the virion size ranges between 40 and 60 nm (Moennig et al. 2003). A schematic diagram of CSFV is shown in Fig. 3.1.

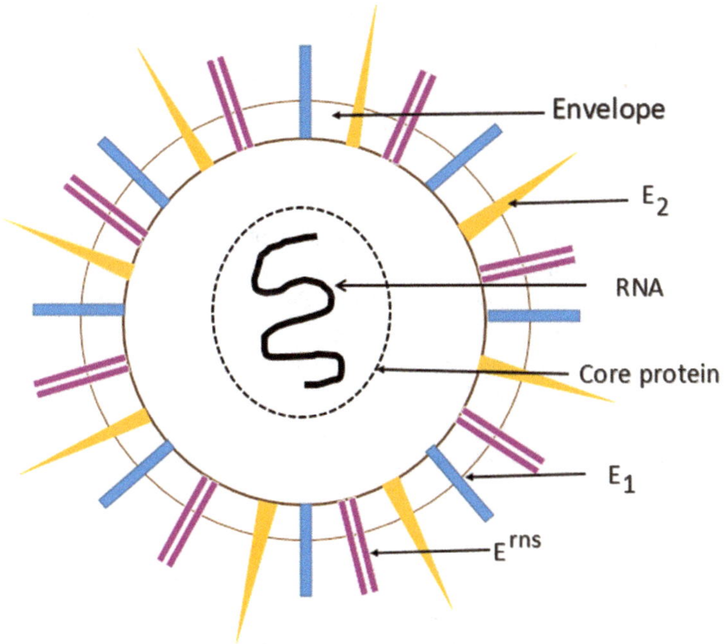

Fig. 3.1 Schematic diagram of classical swine fever virus showing the major structural components such as RNA, core protein, envelope and envelope proteins

3.3 The Genome

The genome of CSFV contains a single-stranded, positive-sense RNA of 12.3 kb in size (Wengler et al. 1995). The genome carries a single large open reading frame (ORF), which encodes a polyprotein of approximately 3898 amino acids and two non-translated regions (NTR) at 5′ and 3′ of the genome. The NTR at the 5′ end carries an internal ribosome entry site (Kolupaeva et al. 2000). The viral genome encodes for anon-structural autoprotease (N^{pro}) located at the N-terminus followed by the four structural proteins such as Core (C), E^{rns}, E1 and E2 and eight non-structural proteins, viz. N^{pro}, p7, NS2, NS3, NS4A, NS4B, NS5A and NS5B (Meyers and Thiel 1996). NS5B represents the RNA-dependent RNA polymerase (Lackner et al. 2006) and NS3 acts as protease (Tautz et al. 1997). The core protein C is a small protein; it plays a role in the formation of the nucleocapsid and also acts as a regulator of gene expression (Liu et al. 1998). E^{rns} (E0) glycoprotein was termed as gp44, which has an approximate size of 44 kDa and is composed of 227 amino acids and loosely associated with the viral envelope (van Gennip et al. 2000). E^{rns} glycoprotein lacks the transmembrane domain and is secreted from infected cells (Lin et al. 2000) along with the virions. It has been reported that this protein plays a role in the attachment and entry of the virus into the host cell (Hulst et al. 2000). E1 glycoprotein has a molecular weight of 33 kDa and is the smallest envelope protein, which plays a role in viral adsorption to host cells (Fernandez-Sainz et al. 2009). E2 is the major envelope glycoprotein; it contains 373 amino acids and has a molecular weight of approximately 55 kDa. Being the major immunogenic protein E2 plays an important role in the induction of the immune responses during the virus infection and also induces the neutralising antibody response. E2 protein of CSFV contains four antigenic domains (A, B, C and D), which are located within the N-terminal half of the protein (Qi et al. 2008). The genes encoding the structural proteins are present towards 5′ end of the genome and most of the genes encoding the non-structural proteins are located mainly in the 3′ end of the genome (Moennig et al. 2003). One of the non-structural proteins of CSFV is p7. It is a small hydrophobic polypeptide with a molecular mass of 7 kDa. The p7 plays a role in pore formation activity and this pore-forming protein of CSFV is involved in the process of the virus virulence in swine (Gladue et al. 2012). The other non-structural proteins NS2–3 to NS5B play a role mostly in the viral replication. A schematic diagram showing the genome organisation of CSFV is shown in Fig. 3.2.

3.4 Attachment, Entry and Replication of CSFV

The two envelope glycoproteins E^{rns} and E2 of CSFV play an important role in the virus infection. They are responsible for the attachment and entry of the virus into host cells. Heparan sulphate (HS), which is ubiquitously present on the surface of many cell types, is used by the viruses for attachment. Besides, another cellular receptor termed as laminin receptor (LamR) has been identified as a receptor for attachment of CSFV to the host cell (Chen et al. 2015). After attachment, entry of the virus occurs through clathrin-mediated endocytosis. The virus entry is dependent on dynamin (dynamin is a 100 kDa GTPase that plays an essential role in cellular membrane fusion during vesicle formation), pH and cholesterol and it requires Rab5 and Rab7 (Shi et al. 2016). Rab proteins are peripheral membrane proteins and possess GTPase (guanosine triphosphatase) fold, which regulates many steps of membrane trafficking, including vesicle formation and membrane fusion. The two envelope glycoproteins of CSFV, viz. E1 and E2, form heterodimers via disulphide

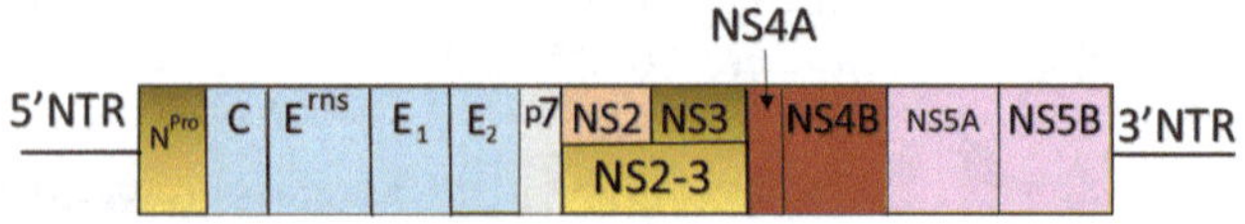

Fig. 3.2 Schematic diagram (not in scale) showing genome organisation of CSFV. Genes which encode the structural proteins (C, E^{rns}, E1 and E2) are towards the 5′NTR. The gene encoding non-structural protein N^{pro} is also located towards 5′NTR and the genes encoding other non-structural proteins (P7, NS2, NS3, NS4A, NS4B, NS5A and NS5B) are located towards 3′NTR. Cleavage of NS2–3 protein into NS2 and NS3 is continuous with cytopathogenic strain of CSFV and free NS3 protein can be detected in cells infected only with cytopathogenic strains of CSFV (Meyer 2011)

bridges between cysteine residue present in the virion and the heterodimers mediate the process of viral entry. The fusion of the viral envelope with the cellular membrane is pH dependent and is triggered by the acidification of the endosome. It has been demonstrated that two peptides of E2 protein mediate fusion between the viral envelope and host cell membrane. After uncoating, the viral genome is released and translated into viral proteins. Besides the structural proteins, non-structural proteins such as NS3, NS4A, NS4B, NS5A and NS5B are required for CSFV replication (Risager et al. 2013). The non-structural protein NS2 harbours an auto-protease activity that is responsible for the cleavage of NS2–3 (Lackner et al. 2006). Previous studies have shown that the uncleaved NS2–3 is crucial for the generation of infectious viral particles for CSFV (Moulin et al. 2007) and cleavage of NS2–NS3 is continuous in case of cytopathogenic strains of CSFV.

3.5 Genetic Diversity

To determine the genetic diversity a 150-nucleotide (nt) fragment of the 5′NTR and a 190-nt fragment of the E2 are mostly used for phylogenetic studies of CSFV (Greiser-Wilke et al. 2006). Besides, a 409-nt fragment of the region coding for the polymerase gene NS5B was employed (Paton et al. 2000a). But, the European Union (EU) Reference Laboratory for CSF has recommended full-length E2-encoding sequences for reliable phylogenetic analysis of CSFV (Postel et al. 2012). The full-length sequencing is useful for quasi-species analyses, high-resolution molecular epidemiology study and also investigation of the virulence determinants of CSFV (Töpfer et al. 2013).

The CSFV isolates across the globes were placed into three genotypes, viz. 1, 2 and 3 (Lowings et al. 1996), and each of the genotypes has different subgenotypes such as 1.1, 1.2, 1.3, 2.1, 2.2, 2.3, 3.1, 3.2, 3.3 and 3.4 (Paton et al. 2000a). Another subgenotype 1.4 has been reported from Cuba (Postel et al. 2013). Two new subgenotypes (1.5 and 1.6) circulating in Brazil are recently described (Silva et al. 2017). Besides, a new subgenotype designated as CSFV subgenotype 1.7 circulating in Ecuador has been reported (Rios et al. 2018).

Recently, a new genotyping scheme of CSFV has been proposed based on complete E2 gene sequences available in the GenBank. In this scheme, 5 genotypes (1–5) and 14 subgenotypes (7 subtypes each for genotypes 1 and 2) of CSFV have been proposed (Rios et al. 2018).

The phylogenetic analyses of CSFV strains have shown a link between genotype and geographical location (Bartak and Greiser-Wilke 2000). For example, the subgenotypes 1.1, 1.2 and 1.3 are found in Asia, South America and Russia; subgenotypes 2.1, 2.2 and 2.3 occur in Europe and parts of Asia; and subgenotypes 3.1, 3.2, 3.3 and 3.4 occur only in Asia. Most of the historical strains, including the vaccine strains of CSFV, belong to genotype 1, and genotype 2 contains most of the current strains and genotype 3 includes most of the strains distributed in Asia (Paton et al. 2000a). Most of the CSFV isolated from outbreaks in Western Europe in the 1990s

Table 3.1 CSFV subgenotypes reported from different countries of some continents

	CSFV subgenotypes/genotype										
Continents	1.1	1.2	1.3	1.4	1.5	1.6	1.7	2.1	2.2	2.3	3
Asia	+	+	+	–	–	–	–	+	+	+	+
North America (Cuba)	–	+	–	+	–	–	–	–	–	–	–
South America	+	–	+	–	+	+	+	–	–	–	–
Europe	–	–	–	–	–	–	–	+	+	+	–
Africa (Madagascar)	–	–	–	–	–	–	–	+	–	–	–
Australia	–	–	–	–	–	–	–	–	–	–	–
Antarctica	–	–	–	–	–	–	–	–	–	–	–

+ Reported, – Not reported

belonged to Group 2. In Europe and Asia, switching of field viruses from genotypes 1 and 3 to that of genotype 2 could be observed in recent years (Cha et al. 2007). The global distributions of the CSFV subgenotypes in different continents are shown in Table 3.1.

The predominant subgenotype reported from India was 1.1 (Sarma et al. 2011). Subsequently, several workers from different parts of India have undertaken phylogenetic characterisation of CSFV isolates and reported the presence of other subgenotypes like 1.2, 2.1 and 2.2 (Patil et al. 2010; Roychoudhury et al. 2014). Complete genome sequences of CSFV subgenotypes 1.1, 2.2 and 2.1 from India have been reported (Gupta et al. 2011; Kumar et al. 2014; Ahuja et al. 2015). In a recent study, the genetic diversity of subgenotype 2.1 isolates revealed ten sub-subgenotypes (2.1a–2.1j), out of which sub-subgenotype 2.1d isolates were circulated only in India, and the rest nine sub-subgenotypes were from China (Gong et al. 2016).

It has been reported that the CSFV genotype 1 has remained relatively constant over time without dynamic changes, whereas CSFV genotype 2 experienced gradual expansion since 1980 with slight fluctuations from 2000 to 2010 (Kwon et al. 2015). The level of genetic diversity of genotypes 2 has increased manyfold compared to the genotype 1. The evolutionary rates estimated for the E2 gene of CSFV genotypes 1 and 2 were 5.76×10^{-4} and 17.29×10^{-4} substitutions/site/year, respectively. Compared to genotype 1, genotype 2 circulated in a more wider population in response to lower immune pressures for a shorter period and with a faster evolutionary rate (Kwon et al. 2015) and the higher evolutionary rate of the genotype 2 is possibly influenced by a lower immune pressures. However, the genetic diversity of CSFV does not result in antigenic diversity.

3.6 Epidemiology

3.6.1 Geographic Distribution

A disease similar to CSF was first reported in France in 1822 (Hanson 1957) followed by the record of the disease in Ohio, USA, in 1833. Subsequently, the disease was widespread in Europe and America, and by the end of the twentieth century the dis-

Table 3.2 Member countries recognised as free from CSF according to the provisions of Chap. 15.2 of the Terrestrial Code as on May 2018 (source: OIE 2018)

Australia	Denmark	Luxembourg	Slovakia
Austria	Finland	Mexico	Slovenia
Belgium	France	New Caledonia	Spain
Bulgaria	Germany	New Zealand	Sweden
Canada	Hungary	Norway	Switzerland
Chile	Ireland	Paraguay	The Netherlands
Costa Rica	Italy	Poland	The United Kingdom
The Czech Republic	Japan	Portugal	USA
	Liechtenstein	Romania	

ease could spread to many parts of the globe. Countries like North America (except Cuba), Australasia and parts of Northern Europe have achieved the eradication of CSF, and many countries have successfully maintained freedom in the absence of vaccination (Edwards et al. 2000). Canada has been free of CSF since 1963, and the official eradication scheme in the USA started in 1961; the last case of CSF was recorded in 1976 (Wise 1986). Brazil is partly free from the disease. No report of CSF has been recorded in Uruguay since 1991, and the country is officially free of the disease. The last case of CSF in Chile was recorded in August 1996 and vaccination against CSF was prohibited throughout the country from October 1997, and Chile has also been declared free of CSF (without vaccination) from April 1998 (Edwards et al. 2000).

As on May 2018, the member countries which are recognised as free from CSF according to the provisions of Terrestrial Code (Chap. 15.2) of the World Organisation for Animal Health (OIE 2018) are given in Table 3.2. Though Japan was declared free from CSF, the disease re-emerged in Japan on September 2018 after 26 years, affecting domestic pigs and wild boars. The CSFV involved in the disease belongs to an emerging clade within the 2.1 subgenotype (Postel et al. 2019). Outbreaks of CSF occur almost regularly in Asia and South East Asia, and the largest viral diversity has been reported in these regions (Paton and Greiser-Wilke 2003). The disease has not been confirmed on the African continent apart from Madagascar, where the disease is endemic (Penrith et al. 2011). CSF entered into Madagascar following the introduction of pigs from Europe in 1965, and the disease became endemic ever since (Penrith et al. 2011). In India, the disease is enzootic and has been reported from all the pig-producing states of the country (Sarma et al. 2008a).

3.6.2 Host Susceptibility

Members of the family *Suidae* are susceptible to CSFV, and the natural hosts of CSFV include domestic and wild pigs. Collared peccaries are also thought to be naturally susceptible to CSFV (OIE 2007). Human beings are not susceptible to the virus. Isolation and detection of CSF virus in domestic and wild pigs have been reported by several workers in India (Sarma et al. 2008b; Barman et al. 2014). Susceptibility of warthogs (*Phacochoerus africanus*) and bush pigs (*Potamochoerus larvatus*) was also demonstrated (Everett et al. 2011). Detection of CSFV in pygmy hog (*Porcula salvania*), an endangered species, has also been reported from Assam, India (Barman et al. 2012). Although CSFV does not replicate in ruminants, ruminant pestiviruses may be found in pigs. Laboratory animals are not usually susceptible, but CSFV has been successfully adapted in rabbits leading to attenuation of the virus for pigs.

3.6.3 Survivability of the Virus

Survivability of CSFV has been determined under physical and chemical conditions. Due to the lipid envelope of the virus, detergents and lipid solvents inactivate the virus easily (McKissick and Gustafson 1967). Inactivation of the virus can be made with 1% formalin, 2% hypochlorite, 6% cresol, 5% phenol and 2% sodium hydroxide (Moennig and Plagemann 1992). The average half-life for the virus is between 2 and 4 days at 5 °C, but only 1 to 3 h at 30 °C and less than 1 min at a temperature above 100 °C (Downing et al. 1977). Significant differences have been observed in the survival of the virus in faecal matters kept at different temperatures (Weesendorp et al. 2008). Different time and temperatures were used to determine the inactivation, and the virus gets inactivated within 1 min at 90 °C and 5 min at 70 °C (Rehman 1987). The virus is comparatively stable at lower temperatures, due to which handling of the samples in the laboratory and shipment of diagnostic samples should be undertaken at a lower temperature. In general, diagnostic samples should be kept at 4 °C, although keeping the samples for short periods at room temperature (20–25 °C) may not damage the virus (Edwards et al. 2000). In refrigerated meat, the virus can survive for months and in frozen meat for years. The virus can withstand pH in the range of above 5 to below 10, but inactivation of the virus occurs rapidly at pH 3 or below, and above pH 10 (Terpstra 1991). In contaminated pens, the virus can survive up to 4 weeks during winter.

3.6.4 Transmission

Classical swine fever virus can spread in different ways and routes. Healthy pigs can get the infection through the direct contact with infected pigs as the virus is shed in the saliva, urine and faeces of infected animals (Laeven et al. 1999). Infected wild boars can spread the virus to domestic pigs through direct contact or indirectly through the meat of infected wild pigs. The virus can excrete in semen, and female pigs are infected during natural service. Spread of the virus through artificial insemination using contaminated semen has also been reported (Floegel et al. 2000). The virus can be transmitted from infected sows to their offspring in utero as the virus can cross the placental barrier. Fresh pork from diseased pigs has been documented as a risk factor for transmission of CSFV. There is also evidence that birds can transmit CSFV (FAO 2010). Fomites such as vehicles, equipment, clothing and contaminated needle may spread the virus. Biting insects can mechanically spread the virus (Van Oirschot 2004).

Besides the above, other factors may increase the risk of introduction and spread of CSFV. Difficulties in early detection of CSF based on clinical symptoms in domestic pigs; variability of symptoms or lack of typical clinical signs; confusion with symptoms of other diseases such as porcine dermatitis nephropathy syndrome, PRRS and congenital tremor; and failure to diagnose the disease properly at the field level facilitate spread of the disease. Late reporting, inadequate control measures, poor hygiene and bad management also favour transmission of the disease. Increase in domestic pig population in some geographical regions, an increase of wild boar population and close contact of domestic pigs with wild boars represent additional risks of transmission and spread of CSFV. Due to the absence of organised pig slaughterhouses in many countries like India, ante-mortem screening of the pigs brought for slaughter is not carried out. Transport of diseased pigs to the wet markets and slaughter of such pigs increase the chance of spread of the virus.

3.6.5 Pathogenesis

The entry of the virus under natural conditions mostly occurs through oral or nasal routes. After entry into the host, the CSFV first infects the epithelial cells of tonsils and then spreads to adjacent lymph nodes. Viraemia develops within 24 h and virus spreads from the initial site of the virus replication to other lymphoid tissues (spleen, Peyer's patches, lymph nodes and thymus), endothelial cells, bone marrow and circulating leukocytes. The virus has a particular affinity for the cells of the reticuloendothelial cell system affecting macrophages and dendritic cells. Spread of the virus to pancreas, brain, heart, gall and urinary bladders, mandibular salivary and adrenal glands, thyroid, liver and kidney is also observed (Trautwein and Leiss 1988). Within 3–4 days of infection, many epithelial type cells are also affected, and the virus is present in excretions and secretions of the infected pigs. In the later part of the infection, the spread of the virus can take place in keratinocytes, hair follicle epithelial cells and mesenchymal cells in the dermis. The virus causes lymphoid tissue depletion leading to immunosuppression, which makes the pigs more susceptible to other infections. Due to the involvement of bone marrow its damage leads to leukopenia and thrombocytopenia. The virus causes damage to the endothelial cell and this results in haemorrhages at many sites. In chronic cases of CSF, there may be antigen-antibody complex deposition in the kidney, and this leads to glomerulonephritis. In pregnant sows and gilts, CSFV may cross the placenta and infect some or all of the foetuses. The effect of in utero infection depends on the stage of pregnancy, and there may be foetal mummification, abortion, delivery of stillborn piglets or persistently infected live piglets.

Lymphocytopenia is one of the most typical symptoms of CSFV infection. After infection with virulent CSFV, the number of the platelets, NK cells, Th cells, Tc cells, cdT cells and CD3− CD4− CD8− and CD3+ CD4+ CD8+ lymphocytes are significantly decreased (Zhou et al. 2009). Many studies have proven that medium or highly virulent CSFV can cause lymphocyte apoptosis, which leads to lymphocytopenia and thrombocytopenia (Sanchez-Cordon et al. 2005). Tumour necrosis factor (TNF) released from CSFV-infected macrophages may induce apoptosis in both infected and uninfected bystander cells (Choi et al. 2004). IL-1 is another important inflammatory cytokine against infection, which causes fever, hyperalgesia, vasodilation and hypotension (Contassot et al. 2012). Several studies showed that IFN-alpha is detectable in blood 2 dpi with highly virulent CSFV infection, with maximum levels at 3–5 dpi. Severe lymphopenia in CSFV infection is associated with the IFN-alpha response, and the infected animals may display depleted B and T lymphocytes. These facts indicate that the apoptosis of lymphocytes is closely related to the untimely overexpression of IFN-alpha (Summerfield et al. 2006), and this is common with highly virulent CSFV infection. But infection with low-virulent strains of CSFV induces no or lower levels of IFN-alpha and pro-inflammatory cytokines (Summerfield et al. 2006). The CSFV exhibits a high affinity for phagocytes of reticuloendothelial cell system, and infection of these cells in the endothelium leads to an increase in vascular permeability, lymphopenia, thrombocytopenia, coagulation disorders and atrophy of the thymus and bone marrow (Gomez-Villamandos et al. 2003). Further studies have revealed that CSFV induces apoptosis in lymphocytes and neutrophil-lineage cells of the bone marrow (Sato et al. 2000).

3.6.6 Clinical Symptoms and Lesions

Depending on the virulence of CSFV strains and host response, the clinical manifestation of the disease can be divided into different forms like peracute, acute, chronic and persistent. However, under field conditions change in the clinical forms of the disease is usually observed. Virulent CSFV strains normally caused peracute to acute disease with a short incubation period, but this has changed as the peracute form has almost disappeared (Meyers and Thiel 1996). Therefore, the clinical forms of CSF can be divided into acute (transient or lethal), chronic, congenital and mild forms. The incubation period for CSF is generally between 3 and 10 days (Moennig and Greiser-Wilke 2008). Under field conditions, symptoms of the disease are sometimes evident even after 4 weeks of the virus infection, or even later (Laeven et al. 1999). The clinical picture of the acute form is more commonly seen in piglets up to 12 weeks of age. A constant feature in the acute form of the disease is pyrexia, usually higher than 41 °C, but in adults the temperature may not exceed 40 °C. Besides, affected animals show anorexia, huddling, lethargy, conjunctivitis, respiratory signs and transient constipation followed by diarrhoea. There may be neurological signs such as staggering gait with weakness of hind legs, incoordination of movement and convulsions. Hyperaemia or haemorrhages on the lower part of the abdomen, ear, tail and inner side of the limbs are also present (Moennig et al. 2003).

Post-mortem examination of dead pigs revealed subcutaneous ecchymotic haemorrhages and swollen, oedematous, haemorrhagic superficial lymph nodes. Spleen shows congestion with small raised hemorrhagic areas and typical infarction. Severely congested mesenteric blood vessels and petechial haemorrhages on the epicardial surface of the heart, non-collapsing haemorrhagic lungs with pneumonic areas, and enlarged and congested liver are also observed. Pinpoint haemorrhages in the subcapsular region of the kidneys resembling 'turkey egg' appearance are a characteristic feature in acute infection. Petechial and ecchymotic haemorrhages may be seen in the epiglottis, gallbladder mucosa and ileocaecal junction.

There are no pathognomonic histopathological changes, although typical changes are present in most of the lymphoid tissues. The lymph nodes and tonsils often show focal to diffuse areas of haemorrhages and depletion of lymphoid follicles along with necrotic cellular debris. Haemorrhages in the red pulp and depletion of the lymphocyte in the splenic corpuscles are present. Kidneys show areas of haemorrhage in the cortico-medullary regions and interstitial nephritis characterised by infiltration of mononuclear cells in the interstitial spaces. Lungs show lesions of interstitial pneumonia with extensive haemorrhages in alveoli and sero-fibrinous exudates in the air spaces. Thickened inter-alveolar septa along with infiltration of mononuclear cells and congestion of alveolar capillaries can be seen. Peribronchiolar areas show haemorrhages, and hyperplasia of lymphoid tissue with the formation of lymphoid aggregates. Changes in the liver are congestion of central vein and sinusoidal spaces and centrilobular degenerative changes with focal areas of necrosis in the hepatocytes. Lymphoid cell necrosis and depletion of lymphocytes in the lymphoid follicles of the Peyer's patches may be seen. The cerebral cortex may show perivascular cuffing along with focal or diffuse gliosis.

In the acute form of the disease, the death of the affected animals usually occurs 2–4 weeks after the virus infection. Mortality can reach up to 100% depending on the age of the animal and the virulence of the virus strain (Moennig et al. 2003). However, the virulence of a strain is not always determined (Mittelholzer et al. 2000). Due to the immunosuppressive condition after CSFV infection, secondary infections can complicate the disease course and symptoms.

The initial symptoms in the chronic form of CSF are similar to acute infection, but with the progress of the disease non-specific signs like intermittent fever, diarrhoea and growth retardation are observed. Death of chronically infected pigs may occur within 2–3 months. Pigs with a chronic form of the disease usually shed CSFV constantly from the onset of clinical signs until death. Antibodies against CSFV may be temporarily detected in serum samples, as the immune system starts to produce antibodies although such antibodies are not able to eliminate the virus from the host (Depner et al. 1996). In the chronic form, pathological changes are less typical, especially the lack of haemorrhages on organs and serosae. Pigs with chronic diarrhoea show necrotic and ulcerative lesions on the ileum, the ileocaecal valve and the rectum. In chronic cases, ulcers with raised edges called button ulcers are often present in the caecum and colon of the affected pigs.

The congenital form of CSF develops when infection occurs during pregnancy. The CSFV can infect the foetus in utero as the virus can pass the placental barrier. The outcome of such in utero infection depends on the stage of gestation. Infection during the first trimester of gestation leads to repeat breeding and foetal mummification, whereas infection during the last trimester mostly results in abortion, malformation or birth of weak or dead piglets. Piglets born under such conditions are often immunotolerant, and such animals not only act as carriers of the virus but also can shed the virus for transmission. Occasionally these piglets show wasting and tremor. The course of congenital infection is termed as late-onset CSF (Kaden et al. 2005; Moennig et al. 2003). Due to non-cytopathogenic (ncp) CSFV infection during 50 and 70 days of pregnancy persistently infected (PI) piglets are born, and such piglets usually develop immunotolerance. These PI piglets remain healthy, but the death of such piglets occurs, and during the period of survival they can shed the virus.

The mild form of CSF is usually observed in older animals and is characterised by transient fever, anorexia and recovery. Recovered animals developed long-lasting immunity.

3.6.7 Immune Response

The CSFV interacts with the host immune system, interaction is complex, and it includes the induction of innate and adaptive immune responses. The envelope glycoproteins, E^{rns} and E2, can elicit neutralising antibodies and can induce protective immunity independently (Konig et al. 1995). However, E2 is mainly responsible for the production of neutralising antibodies and E2 alone is capable of inducing protective antibodies in animals. Also, animals infected with CSFV develop antibodies against non-structural protein NS3, but these antibodies have no neutralising capacity. The other proteins of CSFV are less immunogenic. Although T-cell epitopes were detected in the viral structural and non-structural proteins (Armengol et al. 2002), the role of cellular immunity in CSFV infection is poorly understood. Expression of IFN-stimulated genes, as well as other immune response genes after CSFV infection, has been reported (Summerfield and Ruggli 2015). After CSFV infections, the neutralising antibodies are usually detectable within 2 weeks, but in the chronic form of CSF neutralising antibodies are detectable at the end of the first month of infection. Passive immunity can protect piglets during the first 5 weeks of life, but this cannot stop virus replication and shed completely.

3.7 Laboratory Diagnosis

3.7.1 Samples of Choice

The samples suitable for laboratory diagnosis of CSFV are tonsillar tissue, lymph nodes, spleen and distal ileum from the necropsied animals. The nictitating membrane (translucent third eyelid) also provides a useful source for detection of CSFV in pigs, since this membrane does not undergo autolysis rapidly than the internal organs (Teifke et al. 2005). Whole-blood samples during the viraemic stage may be taken from live pigs. The samples should be collected from several animals if possible (Van Oirschot 2004). Besides, serum samples should be collected from suspected pigs and convalescent pigs after >3 weeks of the disease for laboratory detection of virus-specific antibodies.

3.7.2 Isolation of the Virus

Isolation of CSFV using cell culture is still considered the gold standard in the diagnosis of CSF. Several cells are used for isolation of CSFV from tissues, whole blood or blood components. Detection of CSFV from whole blood or plasma is considered more sensitive in the early phase of infection as compared with the leukocyte fraction (Gisler et al. 1999). The virus can be isolated in primary cells derived from pig kidney or in cell lines like PK-15, SK6, PS and STE (swine testicular epithelioid) cells. Due to the non-cytopathic nature of CSFV, growth of the virus in cell culture is required to detect by the use of immunological and molecular techniques.

3.7.3 Detection of CSFV Antigen

A wide range of immunological techniques have been used for laboratory diagnosis of the disease, and there is rapid progress over the last decades. The sensitivity of the fluorescence antibody test (FAT) for the detection of CSFV antigen under field conditions (Bouma et al. 2001) has been evaluated. Tonsil samples collected from CSF-suspected pigs and experimentally infected pigs have been tested by FAT where the sensitivity of the test for detecting the viral antigen in the tonsil tissues from suspected cases was estimated to be 75%, and the sensitivity of the test from the samples originating from experimentally infected animals was estimated to be 99%. Using the direct FAT, CSFV antigen can be detected in frozen tissue sections within 2 h (Van Oirschot 2004). In the tonsil of infected animals, the CSFV antigen can be detected even at the second day of infections. Besides tonsil, the virus antigen can be detected in the lymph node, spleen and ileum of infected pigs. However, FAT cannot distinguish between CSFV and other pestiviruses unless specific monoclonal antibodies are used in the test (Van Oirschot 2004). A double-antibody sandwich ELISA (enzyme-linked immunosorbent assay) was standardised for detection of CSF virus antigen in clinical samples (Sarma and Sarma 1995). Comparison of sandwich and dot ELISA for detection of CSFV antigen in tissues of naturally infected pigs and pigs slaughtered for human consumption has been studied, and statistical analysis showed excellent agreement between the two tests (Sarma and Meshram 2008). Antigen ELISA was reported to be promising to detect early infection of CSFV in experimentally infected piglets well before clinical manifestations (Raut et al. 2015). ELISA also offers a rapid way of screening large numbers of samples but is less sensitive than the RT-PCR (Clavijo et al. 1998; Van Oirschot 2004). Immuno-chromatographic strip tests with polyclonal/monoclonal colloidal gold conjugation have been reported as a rapid pen-side test for CSFV antigen detection (Zhang et al. 2007). Development of a rapid immunomagnetic bead assay for detection of CSFV antigen has also been reported (Conlan et al. 2009).

3.7.4 Detection of CSFV Nucleic Acid

Molecular diagnostics particularly PCR-based techniques are widely used for detection of the viral nucleic acid and these are comparable or even higher than virus isolation (Paton et al. 2000b; Handel et al. 2004). Furthermore PCR-based technique can detect CSFV infections significantly earlier than conventional virus isolation technique. For the detection of viral RNA in infected tissues, a fluorescent in situ hybridisation (FISH)-based method has been reported (Nagarajan and Saikumar 2012). Using PCR, the viral RNA can be detected even in formalin-fixed tissues (Singh et al. 2005). Detection of CSFV RNA in frozen tissue by reverse transcriptase (RT)-PCR has also been reported (Chopade et al. 2010). Besides the conventional RT-PCR, real-time PCR systems (Zhao et al. 2008; Wen et al. 2010) have been developed and these offer higher specificity and sensitivity for detection of CSFV nucleic acid.

Use of a reverse transcription loop-mediated isothermal amplification (RT-LAMP) assay for rapid detection of CSFV has been reported (Chen et al. 2010). The test showed 100-fold higher sensitivity than the standard RT-PCR method and the samples that were negative in RT-PCR could be found positive in the RT-LAMP. Besides, the RT-LAMP used for CSFV failed to detect other viruses like bovine viral diarrhoea virus (BVDV) and porcine reproductive and respiratory syndrome virus (PRRSV). Since RT-LAMP involves low cost and produces rapid results, it can be used as a tool for CSFV surveillance in the field, especially in developing countries (Chen et al. 2010). Development of RT-LAMP assay as a sensitive and specific method for the detection of CSFV RNA and use of hydroxynaphthol blue (HNB) dye in the assay have also been reported (Wongsawat et al. 2011). Besides diagnosis, molecular tools are now useful to find out the relationships of CSFV isolated from different outbreaks and to understand the epidemiology and virus evolution.

3.7.5 Detection of CSFV Antibodies

Several diagnostic techniques have been developed for detecting the CSFV-specific antibodies in serum of infected animals. Virus neutralisation is the most commonly used assay for detection of CSFV antibody. Besides, blocking ELISA, indirect ELISA and neutralisation peroxidase-linked assay (NPLA) have been developed and used for antibody detection of CSFV (Terpstra et al. 1984; Clavijo et al. 2001; Langedijk et al. 2001; Lin et al. 2005). Use of immuno-chromatographic strip test for rapid detection of CSFV antibodies was also reported (Li et al. 2012). However, these tests cannot differentiate the infected from vaccinated animals, and due to the cross-reaction of CSFV with BVDV, false-positive results may occur while detecting the virus antibodies. The neutralisation peroxidase-linked assay (NPLA), FAT, virus neutralisation and ELISA may be used for serological diagnosis or surveillance and these tests are also prescribed by the OIE for screening serum samples from pigs for international trade.

Despite the progress made, the cross-reactions of CSFV with ruminant pestiviruses pose further challenges in the diagnosis of CSF. The two ruminant pestiviruses, Aydin/04 and Burdur/05 detected in a sheep, and a goat in Turkey showed cross-reactions with CSFV. Sequencing of the full genomes of the two viruses revealed that they form a distinct group located between CSFV and border disease virus with a close relation to CSFV both genetically and antigenically (Becher et al. 2012; Postel et al. 2015). In contrast and fortunately, it was recently shown that atypical porcine pestivirus (APPV) in pigs does not interfere with CSF diagnosis (Postel et al. 2017).

3.7.6 Prevention and Control

The two main strategies used globally for the prevention and control of CSF are systematic vaccination and a stamping-out policy. In CSF endemic countries, prevention and control depend primarily on vaccination programmes. Modified live vaccines (MLVs) or live attenuated vaccines are routinely used to prevent CSF. Subunit vaccines based on E2 are also being used (Huang et al. 2014) by some countries. Although MLVs provide better protection than E2 subunit vaccines, MLVs do not allow differentiation between infected and vaccinated animals (DIVA). Even then, MLVs are being widely used because CSF is endemic in most countries and there is a limited market for the more advanced CSF marker (DIVA) vaccines and companion improved discriminatory diagnostic tests (Greiser-Wilke and Moennig 2004). The live attenuated vaccines using C-strain, developed in China in the mid-1950s, have been used as a major control strategy in China. Attenuation of the C-strain was initially made by passaging in rabbit (lapinised vaccine); later on cell culture attenuated vaccines are developed. The C-strain of the CSFV was considered as the gold standard vaccine for the control of CSF (Dewulf et al. 2004) in many parts of the world. Early protection even at 6 days of post-vaccination by the

C-strain has been reported due to the production of CSFV-specific IFN gamma. The C-strain vaccine can induce detectable neutralising antibodies at 2–3 weeks post-vaccination. The C-strain vaccine can elicit a protective immune response against all CSFV genotypes (Suradhat and Damrongwatanapokin 2003). Different CSFV strains have been used in the production of live attenuated vaccines. Besides C-strain, other effective vaccine strains are GPE, Thiverval, Brescia, PAV-250, etc. These strains are also found safe for prophylactic vaccination and usually provide a long duration of immunity (de Smit et al. 2001). Due to the problems of differentiation of infected and vaccinated animals, live attenuated whole-virus vaccine used in animals and their products needs to be banned from international trade for about 18 months (Greiser-Wilke and Moennig 2004; Moennig et al. 2003).

To prevent the spread of CSFV infection originating from wild boar some field trials were conducted using C-strain as a live vaccine in wild boar (Kaden et al. 2000). Since the handling of wild boars for intramuscular vaccination is difficult, oral vaccination using live attenuated vaccine virus given in the form of baits has been attempted. One of the disadvantages of this oral vaccine is the improper immunisation of all categories of animals; the baits are mainly taken up by adult animals, which develop a good immunity, but young animals are deprived of getting the baits. Another drawback of this vaccine was also the difficulty in differentiating vaccinated and infected animals. Several possibilities have been explored for the development of marker or so-called DIVA vaccines. For the DIVA vaccines one could use small parts of the virus, e.g. single expressed proteins (Moormann et al. 2000), or only parts of proteins, like single peptides (Dong and Chen 2005), or deletion of gene coding envelope protein-like E^{rns}. E2 subunit marker (DIVA) vaccines are available commercially in some countries. Besides, the European Medicines Agency (EMA) in 2014 licensed one of the chimeric marker vaccine candidates, 'CP7_E2alf' (Renson et al. 2013). Although the vaccine is still under investigation, it is expected that this could be a powerful tool for emergency vaccination of domestic and wild pigs. DNA vaccines using the viral DNA incorporated in a plasmid vector are also useful to induce immunity (Hammond et al. 2001). Protection against CSFV infections can also be achieved with deletion mutants of the whole virus (van Gennip et al. 2002) or with chimeric viruses (van Gennip et al. 2000; de Smit et al. 2001). Chimeric viruses are based on infectious clones of the CSFV C-strain in which one of the major antigenic proteins, E^{rns} or E2, or only parts of them, are replaced with the corresponding parts of BVDV (de Smit et al. 2001) or vice versa (Reimann et al. 2004).

It has been reported that CSF vaccines using genotype 1 strains can provide an advantageous immune environment for the survival of genotype 2 CSFV strains, which can continuously evolve to escape the immune mechanism. Studies have shown that inadequate usage of the CSF MLVs in endemic areas results in evolutionary changes in CSFV and this emphasised the need to develop mitigation strategies to minimise the risk associated with the emergence of vaccine-escaping mutants (Yoo et al. 2018). To mitigate the evolution of CSFV, proper and effective vaccination strategies are required, in addition to continuous monitoring and molecular characterisation of circulating CSFV. Besides, there is a need to search for a more potent and stable vaccine virus strain to eradicate CSF and eliminate the CSFV evolution.

Considering the problems of repeated uses and difficulty in differentiating vaccinated from infected animals with the conventional attenuated CSF virus vaccine, researches have been undertaken for the development of antiviral drugs and a number of drugs like prostaglandin and tunicamycin are found to inhibit the multiplication and spread of CSFV in PK-15 and SK-6 cells (Freitas et al. 1998; Tyborowska et al. 2007). Recently, newly designed thioglycosyl analogues of glycosyl-transferase substrates were tested for antiviral activity against CSFV and two of the glycoconjugates not only inhibited the viral multiplication of CSFV but also efficiently suppressed the production of viral proteins in SK-6 cell (Gawolek̦ et al. 2017). Use of small interfering RNA (siRNA) to control CSFV genome replication and viral particle production has been reported. The results showed that the siRNAs caused a 4–12-fold reduction in viral genome copy number and suppressed the production of infectious virus by up to 467-fold for 72–84 h (Xu et al. 2008). Therefore, the application of siRNA strategy for controlling CFSV could become a promising alternative to conventional methods of prevention.

Control of CSF in wild pig population is challenging, although vaccines have been used to combat the disease. Live attenuated C-strain of CSFV has already been tried as baits for oral delivery in wild boar. But for a successful mass oral vaccination programme in wild boar, it is crucial to develop oral bait that is easily detectable, palatable and effectively ingested and this is quite difficult (Rossi et al. 2015).

3.8 Conclusion and Perspectives

Considering the impact of CSF on pig production, control and eradication of the disease are important to ensure food security and rural livelihood and to aid economic growth. Despite the significant achievements made by many countries, the progress towards global eradication of the disease is slow. Economic and social factors of many Asian countries adversely affected the control. Availability of quality vaccines and their large-scale use to provide adequate protection to a maximum number of the pig population in some of the Asian countries are important issues. Approximately 65% of the world's pig population is in Asia, and with the increasing demand for pork and pork products there is a priority to address the diseases particularly like CSF to sustain the pig production. In many developed countries, the main challenge is the control of the disease in wild boar. Differentiation of infected from vaccinated animals with the available diagnostics is another important issue which needs to be addressed. Robust, pen-side multiplex diagnostics for rapid and specific diagnosis of the disease under field condition is a need more particularly for the developing countries due to the emergence of pestiviruses and their cross-reaction with CSFV. Eradication of CSF globally should be the main thrust in the future, and ideal vaccine virus strains and novel antiviral drugs along with continuous surveillance are essential to complete this challenging task.

Acknowledgements All the authors of the manuscript thank and acknowledge their respective universities and institutes.

Conflict of interest: There is no conflict of interest.

References

Ahuja A, Bhattacharjee U, Chakraborty AK et al (2015) Complete genome sequence of classical swine fever virus subgenogroup 2.1 from Assam, India. Genome Announc 3:e01437-14

Armengol E, Wiesmuller KH, Wienhold D et al (2002) Identification of T-cell epitopes in the structural and non-structural proteins of classical swine fever virus. J Gen Virol 83:551–560

Barman NN, Bora DP, Tiwari AK, Kataria RS et al (2012) Classical swine fever in the pygmy hog. Rev Sci Tech Off Int Epiz 31:919–930

Barman NN, Bora DP, Khatoon E et al (2014) Classical swine fever in wild hog: report of its prevalence in northeast India. Transbound Emerg Dis 63:540–547

Bartak P, Greiser-Wilke I (2000) Genetic typing of classical swine fever virus isolates from the territory of the Czech Republic. Vet Microbiol 77:59–70

Becher P, Schmeiser S, Oguzoglu TC et al (2012) Complete genome sequence of a novel pestivirus from sheep. J Virol 86:11412. https://doi.org/10.1128/JVI.01994-12

Beer M, Goller KV, Staubach C et al (2015) Genetic variability and distribution of classical swine fever virus. Anim Health Res Rev 16:33–39

Bouma A, Stegeman JA, Engel B et al (2001) Evaluation of diagnostic tests for the detection of classical swine fever in the field without a gold standard. J Vet Diagn Investig 13:383–388

Cha SH, Choi EJ, Park JH et al (2007) Phylogenetic characterization of classical swine fever viruses isolated in Korea between 1988 and 2003. Virus Res 126:256–261

Chen L, Fan X-Z, Wang Q et al (2010) A novel RT-LAMP assay for rapid and simple detection of classical swine fever virus. Virol Sin 25:59–64

Chen J, He WR, Shen L et al (2015) The laminin receptor is a cellular attachment receptor for classical swine fever virus. J Virol 89:4894–4906

Choi C, Hwang KK, Chae C (2004) Classical swine fever virus induces tumor necrosis factor-alpha and lymphocyte apoptosis. Arch Virol 149:875–889

Chopade NA, Deshmukh VV, Rautmare SS et al (2010) Detection of classical swine fever virus from frozen tissue by RT-PCR. Anim Sci Rep 4:56–59

Clavijo A, Zhou EM, Vydelingum S et al (1998) Development and evaluation of a novel antigen capture assay for the detection of classical swine fever virus antigen. Vet Microbiol 60:155–168

Clavijo A, Lin M, Riva J et al (2001) Application of competitive enzyme-linked immunosorbent assay for the serologic diagnosis of classical swine fever virus infection. J Vet Diagn Investig 13:357–360

Conlan JV, Khounsy S, Blacksell SD et al (2009) Development and evaluation of a rapid immunomagnetic bead assay for the detection of classical swine fever virus antigen. Trop Anim Health Prod 41:913–920

Contassot E, Beer HD, French LE (2012) Interleukin-1, inflammasomes, autoinflammation and the skin. Swiss Med Wkly 142:w1359

de Smit AJ, Bouma A, Deekluijver EP et al (2001) Duration of the protection of an E2 subunit marker vaccine against classical swine fever after a single vaccination. Vet Microbiol 78:307–317

Depner EKR, Rodriguez OA, Pohlenz O et al (1996) Persistent classical swine fever virus infection in pigs infected after weaning with a virus isolated during the 1995 epidemic in Germany: clinical, virological, serological and pathological findings. Eur J Vet Pathol 2:61–66

Dewulf J, Laevens H, Koenen F et al (2004) Efficacy of E2-sub-unit marker and C-strain vaccines in reducing horizontal transmission of classical swine fever virus in weaner pigs. Prev Vet Med 65:121–133
Dong XN, Chen YH (2005) Candidate peptide-vaccines induced immunity against CSFV and identified sequential neutralizing determinants in antigenic domain A of glycoprotein E2. Vaccine 24:1906–1913
Downing DR, Carbrey EA, Stewart WC (1977) Preliminary findings on a thermal inactivation curve for Hog Cholera virus. In: Agri. Res. Seminar on Hog Cholera/Classical Swine Fever and African Swine Fever. USDA Veterinary Laboratory, Ames, Iowa
Edwards S, Fukusho A, Lefevre PC et al (2000) Classical swine fever: the global situation. Vet Microbiol 73:103–119
Everett H, Crooke H, Gurrala R et al (2011) Experimental infection of common warthogs (Phacochoerus africanus) and bushpigs (Potamochoerus larvatus) with classical swine fever virus. I: susceptibility and transmission. Transbound Emerg Dis 58:128–134
FAO (2010) Good practices for biosecurity in the pig sector—issues and options in developing and transition countries. FAO animal production and health paper 169, Rome
Fernandez-Sainz I, Holinka LG, Gavrilov BK et al (2009) Alteration of the N-inked glycosylation condition in El glycoprotein of classical swine fever virus strain Brescia alters virulence in swine. Virology 386:210–216
Floegel G, Wehrend A, Depner KR et al (2000) Detection of classical swine fever virus in semen of infected boars. Vet Microbiol 77:109–116
Freitas TR, Caldas LA, Rebello MA (1998) Prostaglandin A1 inhibits replication of classical swine fever virus. Mem Inst Oswaldo Cruz 93:815–818
Gawolek GP, Chaubey B, Szewczyk B et al (2017) Novel thioglycosyl analogs of glycosyltransferase substrates as antiviral compounds against classical swine fever virus and Hepatitis C virus. Eur J Med Chem 137:247–262
Gisler ACF, Nardi NB, Nonnig RB et al (1999) Classical swine fever virus in plasma and peripheral blood mononuclear cells of acutely infected swine. J Vet Med B 46:585–593
Gladue DP, Holinka LG, Largo E et al (2012) Classical swine fever virus p7 protein is a viroporin involved in virulence in swine. J Virol 86:6778–6791. https://doi.org/10.1128/JVI.00560-12
Gomez-Villamandos JC, Salguero FJ, Ruiz-Villamor E et al (2003) Classical swine fever: pathology of bone marrow. Vet Pathol 40:157–163. https://doi.org/10.1354/vp.40-2-157
Gong W, Wu J, Lu Z, Zhang L et al (2016) Genetic diversity of subgenotype 2.1 isolates of classical swine fever virus. Infect Genet Evol 41:218–226
Greiser-Wilke I, Moennig V (2004) Vaccination against classical swine fever virus: limitations and new strategies. Anim Health Res Rev 5:223–226
Greiser-Wilke I, Dreier S, Haas L et al (2006) Genetic typing of classical swine fever viruses—a review. Dtsch Tierarztl Wochenschr 113:134–138
Gupta PK, Saini M, Dahiya SS et al (2011) Molecular characterization of lapinized classical Swine Fever vaccine strain by full-length genome sequencing and analysis. Anim Biotechnol 22:111–117
Hammond JM, Jansen ES, Morrissy CJ et al (2001) A prime-boost vaccination strategy using naked DNA followed by recombinant porcine adenovirus protects pigs from classical swine fever. Vet Microbiol 80:101–119
Handel K, Kehler H, Hils K et al (2004) Comparison of reverse transcriptase–polymerase chain reaction, virus isolation, and immunoperoxidase assays for detecting pigs infected with low, moderate, and high virulent strain of classical swine fever virus. J Vet Diagn Investig 16:132–138
Hanson RP (1957) Origin of hog cholera. J Am Vet Med Assoc 131:211–218
Huang YL, Deng MC, Wang FI et al (2014) The challenges of classical swine fever control; modified live and E2 subunit vaccines. Virus Res 179:1–11

Hulst MM, van Gennip HGP, Moormann RJM (2000) Passage of classical swine fever virus in cultured swine kidney cells selects virus variants that bind to heparin sulfate due to a single amino acid change in envelope protein E-rns. J Virol 74:9553–9561

Kaden V, Lange E, Fischer U et al (2000) Oral immunisation of wild boar against classical swine fever: evaluation of the first field study in Germany. Vet Microbiol 73:239–252

Kaden V, Steyer H, Schnabel J et al (2005) Classical swine fever (CSF) in wild boar: the role of the transplacental infection in the perpetuation of CSF. J Vet Med B Infect Dis Vet Public Health 52:161–164

Kolupaeva VG, Pestova TV, Hellen CU (2000) Ribosomal binding to the internal ribosomal entry site of classical swine fever virus. RNA 6:1791–1807

Konig M, Lengsfeld T, Pauly T et al (1995) Classical swine fever virus: Independent induction of protective immunity by two structural glycoproteins. J Virol 69:6479–6486

Kumar R, Rajak KK, Chandra T et al (2014) Whole-genome sequence of a classical swine fever virus isolated from the Uttarakhand State of India. Genome A 2:e00371-14

Kwon T, Yoon SH, Kim KW et al (2015) Time-calibrated phylogenomics of the classical swine fever viruses: genome-wide Bayesian coalescent. PLoS 10:e0121578. https://doi.org/10.1371/journal.pone.0121578

Lackner T, Thiel HJ, Tautz N (2006) Dissection of a viral autoprotease elucidates a function of a cellular chaperone in proteolysis. Proc Natl Acad Sci USA 103:1510–1515

Laeven AH, Koenen OF, Deluyker EH et al (1999) Experimental infection of slaughter pigs with classical swine fever virus: transmission of the virus, course of the disease and antibody response. Vet Rec 145:243–248

Langedijk JPM, Middel WG, Meloen RH et al (2001) Enzyme-linked immunosorbent assay using a virus type specific peptide based on a subdomain of envelope protein E(rns) for serologic diagnosis of pestivirus infections in swine. J Clin Microbiol 39:906–912

Li X, Li W, Shi X et al (2012) Development of an immunochromatographic strip for rapid detection of antibodies against classical swine fever virus. J Virol Meth 180:32–37

Lin M, Lin F, Mallory M, Clavijo A (2000) Deletions of structural glycoprotein E2 of classical swine fever virus strain Alfort/187 resolve a linear epitope of monoclonal antibody WH303 and the minimal N-terminal domain essential for binding immunoglobulin G antibodies of a pig hyperimmune serum. J Virol 74:11619–11625

Lin M, Trottier E, Mallory M (2005) Enzyme-linked immunosorbent assay based on a chimeric antigen bearing antigenic regions of structural proteins Erns and E2 for serodiagnosis of classical swine fever virus infection. Clin Diagn Lab Immunol 12:877–881

Liu JJ, Wong ML, Chang TJ (1998) The recombinant nucleocapsid protein of classical swine fever virus can act as a transcriptional regulator. Virus Res 53:75–80

Lowings P, Ibata G, Needham J et al (1996) Classical swine fever diversity and evolution. J Gen Virol 77:1311–1321

McKissick GE, Gustafson DP (1967) In vivo demonstration of lability of hog cholera virus to lipolytic agents. Am J Vet Res 28:909–914

Meyer D (2011) Epitope mapping of the structural protein Erns of classical swine fever virus. Thesis submitted in partial fulfilment of the requirement for the degree Doctor Rerurm Naturatium. Institute of Virology, Hannover

Meyers G, Thiel HJ (1996) Molecular characterization of pestiviruses. Adv Virus Res 47:53–118

Mittelholzer C, Moser C, Tratschin J et al (2000) Analysis of classical swine fever virus replication kinetics allows differentiation of highly virulent from avirulent strains. Vet Microbiol 74:293–308

Moennig V, Becher P (2015) Pestivirus control programs: how far have we come and where are we going? Anim Health Res Rev 16:83–87

Moennig V, Greiser-Wilke I (2008) Classical swine fever virus. In: Mahy BWJ, van Regenmortel MHV (eds) Encyclopedia of virology. Elsevier, Amsterdam, pp 525–532

Moennig V, Plagemann PGW (1992) The pestiviruses. Adv Virus Res 41:53–98

Moennig V, Floegel NG, Greiser-Wilke I (2003) Clinical signs and epidemiology of classical swine fever: a review of new knowledge. Vet J 165:11–20

Moormann RJ, Bouma A, Kramps JA et al (2000) Development of a classical swine fever subunit marker vaccine and companion diagnostic test. Vet Microbiol 73:209–219

Moulin HR, Seuberlich T, Bauhofer O et al (2007) Nonstructural proteins NS2-3 and NS4A of classical swine fever virus: Essential features for infectious particle formation. Virology 365:376–389

Nagarajan K, Saikumar G (2012) Fluorescent in-situ hybridization technique for the detection and localization of classical swine fever virus in infected tissues. Veterinarski Arhiv 82:495–504

OIE (2007) Manual of diagnostic tests and vaccines for terrestrial animals. http://www.oie.int/eng/norms/manual/asummary.htm

OIE (2018) Terrestrial animal health code, 27th edn. World Organization for animal health, Paris, France

Patil SS, Hemadri D, Shankar BP et al (2010) Genetic typing of recent classical swine fever isolates from India. Vet Microbiol 141:367–373

Paton DJ, Greiser-Wilke I (2003) Classical swine fever—an update. Res Vet Sci 75:169–178. https://doi.org/10.1016/S0034-5288(03)00076-6

Paton DJ, McGoldrick A, Greiser-Wilke I et al (2000a) Genetic typing of classical swine fever virus. Vet Microbiol 73:137–157

Paton DJ, McGoldrick A, Belak S (2000b) Classical swine fever virus: a ring test to evaluate RT-PCR detection methods. Vet Microbiol 73:159–174

Penrith ML, Vosloo W, Mather C (2011) Classical swine fever (hog cholera): Review of aspects relevant to control. Transbound Emerg Dis 58(3):187–196

Postel A, Schmeiser S, Bernau J et al (2012) Improved strategy for phylogenetic analysis of classical swine fever virus based on full-length E2 encoding sequences. Vet Res 43:50

Postel A, Schmeiser S, Perera CL et al (2013) Classical swine fever virus isolates from Cuba a new subgenotype 1.4. Vet. Microbiol 161:334–338

Postel A, Schmeiser S, Oguzoglu TC et al (2015) Close relationship of ruminant pestiviruses and classical swine fever virus. Emerg Infect Dis 21(4):668–672

Postel A, Busch SA, Petrov A et al (2017) Epidemiology, diagnosis and control of classical swine fever: recent developments and future challenges. Transbound Emerg Dis 64:1–14

Postel A, Nishi T, Kameyama K et al (2019) Re-emergence of classical swine fever, Japan, 2018. Emerg Infect Dis 25(6):1228–1231. https://doi.org/10.3201/eid2506.181578

Qi Y, Liu LC, Zhang BQ et al (2008) Characterization of antibody responses against a neutralizing epitope on the glycoprotein E2 of classical swine fever virus. Arch Virol 153:1593–1598

Raut S, Dattatraya RK, Kishore KR et al (2015) Detection of classical swine fever virus antigen and nucleic acid on blood of experimentally infected piglets. Adv Anim Vet Sci 3:1–6

Rehman S (1987) Virucidal effect of the heat treatment of waste food for swine. Tierarztliche Umschau 42(11):892–896

Reimann I, Depner K, Trapp S et al (2004) An avirulent chimeric Pestivirus with altered cell tropism protects pigs against lethal infection with classical swine fever virus. Virology 322:143

Renson P, Dimna ML, Keranflech A et al (2013) CP7-E2alf oral vaccination confers partial protection against early classical swine fever virus challenge and interferes with pathogeny-related cytokine responses. Vet Res 44:9. https://doi.org/10.1186/1297-9716-44-9

Rios L, Nunez JI, de Arce HD et al (2018) Revisiting the genetic diversity of classical swine fever virus: a proposal for new genotyping and sub genotyping schemes of classification. Transbound Emerg Dis 65(4):963–971

Risager PC, Fahnøe U, Gullberg M et al (2013) Analysis of classical swine fever virus RNA replication determinants using replicons. J Gen Virol 94:1739–1748

Rossi S, Staubach C, Blome S et al (2015) Controlling of CSFV in European wild boar using oral vaccination: a review. Front Microbiol 6:1–11. https://doi.org/10.3389/fmicb.2015.01141

Roychoudhury P, Sarma DK, Rajkhowa S et al (2014) Predominance of genotype 1.1 and emergence of genotype 2.2 classical swine fever viruses in north-eastern region of India. Transbound Emerg Dis 61(Suppl. 1):69–77
Sanchez-Cordon PJ, Nunez A, Salguero FJ (2005) Lymphocyte apoptosis and thrombocytopenia in spleen during classical swine fever: role of macrophages and cytokines. Vet Pathol 42:477–488
Sarma DK, Meshram DJ (2008) Comparison of sandwich and dot ELISA for detection of CSF virus antigen in pigs. Ind Vet J 85:915–918
Sarma DK, Sarma PC (1995) ELISA for detection of hog cholera virus antigen. Ind J Anim Sci 65:650–651
Sarma DK, Krishna L, Misri J (2008a) Classical swine fever in pigs and its status in India: a review. Ind J Anim Sci 78:1311–1317
Sarma DK, Mishra N, Rajukumar K et al (2008b) Isolation and characterization of classical swine fever virus from pigs in Assam. Ind J Anim Sci 78:37–39
Sarma DK, Mishra N, Vilcek S et al (2011) Phylogenetic analysis of recent classical swine fever virus (CSFV) isolates from Assam, India. Comp Immunol Microbiol Infect Dis 34:11–15
Sato M, Mikami O, Kobayashi M et al (2000) Apoptosis in the lymphatic organs of piglets inoculated with classical swine fever virus. Vet Microbiol 75:1–9
Shi BJ, Liu CC, Zhou J et al (2016) Entry of classical swine fever virus into PK-15 cells via a pH-, dynamin-, and cholesterol-dependent, clathrin-mediated endocytic pathway that requires Rab5 and Rab7. J Virol 90:9194–9208
Silva MN, Silva DM, Leite AS et al (2017) Identification and genetic characterization of classical swine fever virus isolates in Brazil: a new subgenotype. Archiv Virol 162:817–822
Singh VK, Kumar GS, Paliwal OP (2005) Detection of classical swine fever virus in archival formalin-fixed tissues by reverse transcription-polymerase chain reaction. Res Vet Sci 79:81–84
Singh B, Bardhan D, Verma MR et al (2016) Incidence of classical swine fever in pigs in India and its economic evaluation with a simple mathematical model. Anim Sci Rep 10:3–9
Smith DB, Meyers G, Bukh J et al (2017) Proposed revision to the taxonomy of the genus Pestivirus, family *Flaviviridae*. J Gen Virol 98:2106–2112
Summerfield A, Ruggli N (2015) Immune responses against classical swine fever virus: between ignorance and lunacy. Front Vet Sci 2:1–9. https://doi.org/10.3389/fvets.2015.00010
Summerfield A, Alves M, Ruggli N et al (2006) High IFN-alpha responses associated with depletion of lymphocytes and natural IFN-producing cells during classical swine fever. J Interf Cytokine Res 26:248–255
Suradhat S, Damrongwatanapokin S (2003) The influence of maternal immunity on efficacy of a classical swine fever vaccine against classical swine fever virus, genogroup 2.2, infection. Vet Microbiol 92:187–194
Tautz N, Elbers K, Stoll D, Meyers G, Thiel HJ (1997) Serine protease of pestiviruses: determination of cleavage sites. J Virol 71:5415–5422
Teifke JP, Lange E, Klopfleisch R et al (2005) Nictitating membrane as a potentially useful postmortem diagnostic specimen for classical swine fever. J Vet Diagn Investig 17:341–345
Terpstra C, Bloemraad M, Gielkens AL (1984) The neutralizing peroxidase- linked assay for detection of antibody against swine fever virus. Vet Microbiol 9:113–120
Terpstra C (1991) Hog cholera: an update of present knowledge. Brit Vet J 147:397–406
Töpfer A, Höper D, Blome S et al (2013) Sequencing approach to analyze the role of quasispecies for classical swine fever. Virology 438:14–19
Trautwein G, Leiss B (1988) Pathology and pathogenesis of the disease, classical swine fever and related infections. Martinus Nijhoff Publishing, Boston, MA, pp 27–54
Tyborowska J, Zdunek E, Szewczyk B (2007) Effect of N-glycosylation inhibition on the synthesis and processing of classical swine fever virus glycoproteins. Acta Biochim Pol 54:813–819
van Gennip HGP, van Rijn PA, Widjojoatmodjo MN et al (2000) Chimeric classical swine fever viruses containing envelope protein E-RNS or E2 of bovine viral diarrhoea virus protect pigs against challenge with CSFV and induce a distinguishable antibody response. Vaccine 19:447–459

van Gennip HGP, Bouma A, van Rijn PA et al (2002) Experimental non-transmissible marker vaccines for classical swine fever (CSF) by trans-complementation of E(rns) or E2 of CSFV. Vaccine 20:1544–1556

Van Oirschot VJT (2004) Hog cholera. In: JAW C, Tustin RC (eds) Infectious diseases of livestock, 2nd edn. Oxford University Press, Oxford, pp 975–986

Weesendorp E, Stegeman A, Willie LA et al (2008) Survival of classical swine fever virus at various temperatures in faeces and urine derived from experimentally infected pigs. Vet Microbiol 132:249–259

Wen G, Yang J, Luo Q et al (2010) A one-step real-time reverse transcription-polymerase chain reaction detection of classical swine fever virus using a minor groove binding probe. Vet Res Commun 34:359–369

Wengler G, Bradley DW, Collett MS et al (1995) Family *Flaviviridae*. Virus taxonomy. In: Murphy FA, Fauquet CM et al (eds) Sixth report of the international committee on taxonomy of viruses. Springer, New York, pp 415–427

Wise GH (1986) Eradication of hog cholera from the United States. In: Woods GT (ed) Practices in veterinary public health and preventive medicine in the United States. Iowa State University Press, Ames, IA, pp 199–223

Wongsawat K, Dharakul T, Narat P et al (2011) Detection of nucleic acid of classical swine fever virus by reverse transcription loop-mediated isothermal amplification (RT-LAMP). Sci Res 3:447–452

Xu X, Guo H, Xiao C et al (2008) In vitro inhibition of classical swine fever virus replication by siRNAs targeting Npro and NS5B genes. Antivir Res 78:188–193

Yoo SJ, Kwon T, Kang K et al (2018) Genetic evolution of classical swine fever virus under immune environments conditioned by genotype 1-based modified live virus vaccine. Transbound Emerg Dis 65:735–745. https://doi.org/10.1111/tbed.12798

Zhang C, Huang Y, Zheng H et al (2007) Study on colloidal gold strip in detecting classical swine fever virus. Fujian J Anim Hus Vet Med 6:13

Zhao JJ, Cheng D, Li N et al (2008) Evaluation of a multiplex real-time RT-PCR for quantitative and differential detection of wild-type viruses and C-strain vaccine of classical swine fever virus. Vet Microbiol 126:1–10

Zhou YC, Wang Q, Fan XZ et al (2009) The changes of peripheral blood leucocytes subpopulation after challenge with CSFV virulent strain Shimen. Chin J Virol 25:303–308

Chapter 4
Porcine Coronaviruses

A. N. Vlasova, Q. Wang, K. Jung, S. N. Langel, Yashpal Singh Malik, and L. J. Saif

Abstract Transmissible gastroenteritis virus (TGEV), porcine epidemic diarrhoea virus (PEDV), and porcine deltacoronavirus (PDCoV) are enteropathogenic coronaviruses (CoVs) of swine. TGEV appearance in 1946 preceded identification of PEDV (1971) and PDCoV (2009) that are considered as emerging CoVs. A spike deletion mutant of TGEV associated with respiratory tract infection in piglets appeared in 1984 in pigs in Belgium and was designated porcine respiratory coronavirus (PRCV). PRCV is considered non-pathogenic because the infection is very mild or subclinical. Since PRCV emergence and rapid spread, most pigs have become immune to both PRCV and TGEV, which has significantly reduced the clinical and economic importance of TGEV. In contrast, PDCoV and PEDV are currently expanding their geographic distribution, and there are reports on the circulation of TGEV-PEDV recombinants that cause a disease clinically indistinguishable from that associated with the parent viruses. TGEV, PEDV and PDCoV cause acute gastroenteritis in pigs (most severe in neonatal piglets) and matches in their clinical signs and pathogenesis. Necrosis of the infected intestinal epithelial cells causes villous atrophy and malabsorptive diarrhoea. Profuse diarrhoea frequently combined with vomiting results in dehydration, which can lead to the death of piglets. Strong immune responses following natural infection protect against subsequent homologous challenge; however, these viruses display no cross-protection. Adoption of advance biosecurity measures and effective vaccines control and prevent the occurrence of diseases due to these porcine-associated CoVs. Recombination and reversion to virulence are the risks associated with generally highly effective attenuated vaccines necessitating further research on alternative vaccines to ensure their safe application in the field.

A. N. Vlasova (✉) · Q. Wang · K. Jung · S. N. Langel · L. J. Saif
Department of Veterinary Preventive Medicine, Food Animal Health Research Program, Ohio Agricultural Research and Development Center, College of Food, Agricultural and Environmental Sciences, The Ohio State University, Wooster, OH, USA
e-mail: vlasova.1@osu.edu

Y. S. Malik
Division of Biological Standardization, ICAR-Indian Veterinary Research Institute, Izatnagar, Bareilly, Uttar Pradesh, India

Y. S. Malik et al. (eds.), *Emerging and Transboundary Animal Viruses*, Livestock Diseases and Management, https://doi.org/10.1007/978-981-15-0402-0_4

Keywords Porcine coronaviruses · Transmissible gastroenteritis virus · Porcine epidemic diarrhoea virus · Porcine deltacoronavirus · Enteropathogens · Porcine respiratory coronavirus · Pathology · Detection · Vaccine

4.1 Prologue

All known porcine coronaviruses (CoVs) belong to the genera *Alphacoronavirus*, *Betacoronavirus* and *Deltacoronavirus* of the subfamily *Coronavirinae*, in the family *Coronaviridae*, of the order *Nidovirales* (de Groot et al. 2008) [https://data.ictvonline.org/taxonomy-search.asp?msl_id=30 (Fig. 4.1)]. Affections of gastrointestinal, respiratory, peripheral and central nervous systems are usually visualised. Five swine CoVs are recognised: (1) the transmissible gastroenteritis virus (TGEV), first defined in 1946; (2) the porcine respiratory coronavirus (PRCV), a mutant of TGEV, isolated in 1984; (3) the porcine epidemic diarrhoea virus (PEDV), isolated in 1977; (4) the PHEV (porcine haemagglutinating encephalomyelitis virus) isolated in 1962; and (5) the PDCoV (porcine deltacoronavirus) described in 2012. The first two, TGEV and PRCV, belong to the *Alphacoronavirus* 1 species together with closely associated CoVs of cats and dogs, and PEDV and human CoVs (229E and NL63) form distinct species in the *Alphacoronavirus* genus. PHEV and PDCoV belong to the *Beta-* (*Betacoronavirus* 1 species) *and Deltacoronavirus* genera, respectively. PDCoV is closely related to the deltacoronaviruses from Asian leopard cats and Chinese ferret badgers (Ma et al. 2005). While PRCV induces primarily subclinical infections in pigs, enteropathogenic swine alphacoronaviruses (TGEV, PEDV, SeCoV, porcine enteric alphacoronavirus) and PDCoV are allied with a severe enteric disease of variable severity depending on the animal age and immune status. One serotype is recognised for each swine CoV species.

TGEV and PEDV have been reportedly co-circulating in Eurasia and the USA. Recently in Europe, a pathogenic recombinant TGEV/PEDV variant (swine enteric coronavirus, SeCoV) was recognised and described (Akimkin et al. 2016; Belsham et al. 2016; Boniotti et al. 2016). SeCoV that contains PEDV S protein on a TGEV backbone apparently leads to disease clinically indistinct from the TGEV- and PEDV-produced ones (Table 4.1). Additionally, a novel bat-HKU2-like porcine coronavirus [porcine enteric alphacoronavirus (PEAV), GDS04 strain] associated with severe diarrheal disease in suckling piglets was identified in Southern China in 2017 (Gong et al. 2017) (Table 4.1). However, its prevalence and adaptation status to the swine host are unknown.

Currently, PHEV, the only porcine betacoronavirus, has a worldwide prevalence (Li et al. 2016). In neonatal piglets devoid of maternal antibodies (generally in those purchased from infection-free herds), PHEV causes either encephalomyelitis or a condition characterised by vomiting and wasting. Generally, maternal immunity

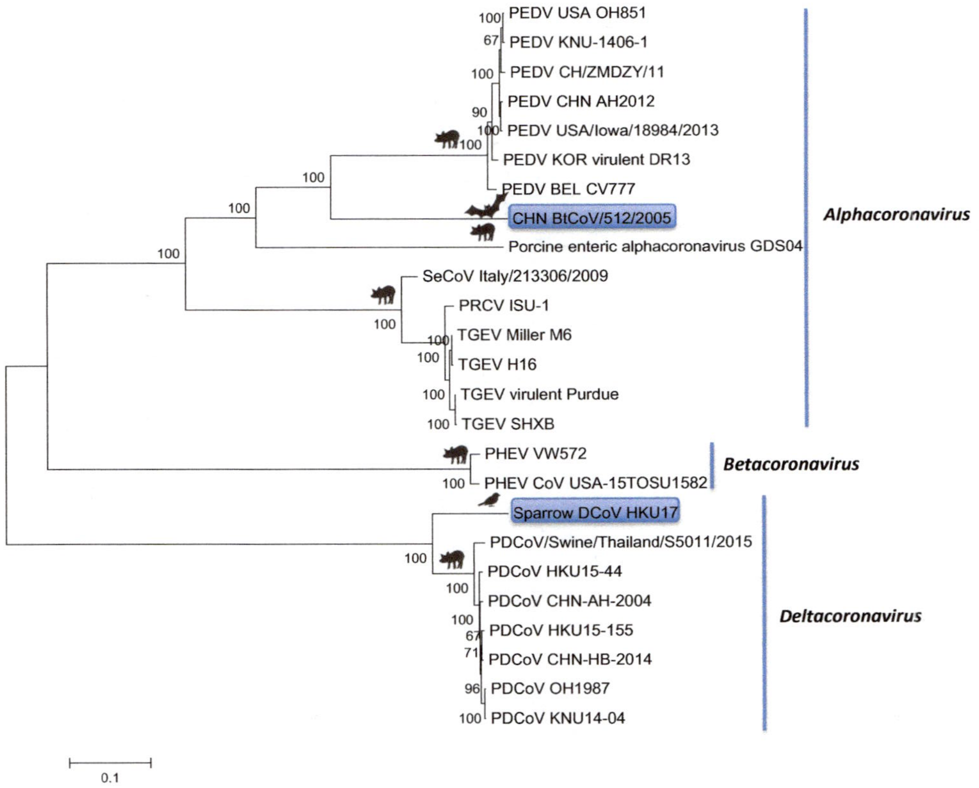

Fig. 4.1 Phylogenetic tree of porcine coronaviruses of the *Alpha-*, *Beta-* and *Deltacoronavirus* genera. Closed circles indicate potential ancestral non-porcine coronaviruses. Bootstrapping with 1000 replicates was used to determine the reliability of each node

Table 4.1 Comparative pathogenesis of porcine enteric CoVs

	Symptoms	Clinical disease onset	Lesions	Morbidity	Mortality	Most affected age group
TGEV	Diarrhoea, vomiting, dehydration	24 h	Jejunum, ileum	100%	Up to 100%	Neonatal piglets up to 3 weeks of age
PEDV	Diarrhoea, vomiting, dehydration	24–36 h	Jejunum, ileum	100%	Up to 100%	Neonatal piglets up to 3 weeks of age
SeCoV	Diarrhoea, vomiting, dehydration	24–36 h	Jejunum, ileum	100%	Up to 100%	Neonatal piglets up to 3 weeks of age
PEAV	Diarrhoea, dehydration	3–4 days	?	100%	?	Neonatal piglets up to 3 weeks of age
PDCoV	Diarrhoea, vomiting, dehydration	1–3 days	Jejunum, ileum, colon	Up to 100%	40–80%	Neonatal piglets up to 3 weeks of age

protects piglets which have taken colostrum for up to 15 weeks, while in pigs older than 3–4 weeks and adult swine the infection is mostly subclinical. Therefore, it is seldom considered to be of economic importance. However, a recent report of an uncommon respiratory (influenza-like) presentation and increasing prevalence of PHEV in adult exhibition swine in the USA (Lorbach et al. 2017) may indicate a potential tropism shift that could lead to a substantial change in its epidemiology. To clear this complex epidemiological position, continuing monitoring and development of state-of-the-art rapid and reliable tools and techniques are needed to confirm and provide a clear differential diagnosis (Kim et al. 2001; Masuda et al. 2016).

CoVs are enveloped and pleomorphic, 60–160 nm in diameter. Swine CoV shaves single-stranded, polyadenylated, large genomic RNA (25–30 kDa) of positive-sense polarity that is infectious. The genome profile, replication strategies as well as protein expression match to other human and animal CoVs (Enjuanes and Van der Zeijst 1995; Gonzalez et al. 2003; Laude et al. 1993). Most porcine CoVs have four basic structural proteins: a large surface glycoprotein (S, spike protein that forms a monolayer of club-shaped spikes defined as the corona); a small membrane protein (E); an integral membrane glycoprotein (M); and a nucleocapsid protein (N). However, PHEV contains a haemagglutinin-esterase (HE) protein that forms a second shorter layer of surface spikes (de Groot et al. 2008). TGEV, PEDV and PDCoV also transcribe 1–2 accessory proteins encoded by open reading frame (ORF)3 (TGEV and PEDV), ORF6 (PDCoV), and ORF7 (TGEV and PDCoV). The complete genome organisation is 5′UTR-ORF1ab, S, ORF3, E, M, ORF6, N and ORF7-3′UTR.

An overall nt and amino acid sequence similarity of 96–98% among TGEV and PRCV proposes that PRCV evolved from TGEV. Two characteristic features of the PRCV genome that may account for its altered tissue tropism include a large

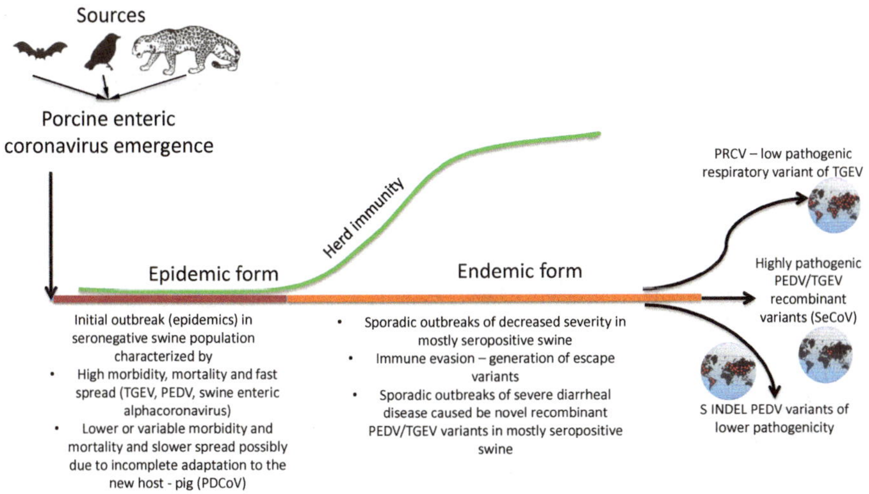

Fig. 4.2 Different stages of evolution of swine enteric coronaviruses

omission (621–681 nt) in the N-termini of the S gene resulting in a reduced S protein size and variable sequence changes in the ORF3 (Ballesteros et al. 1997; Sanchez et al. 1999).

While there is no evidence of the existence of different PEDV serotypes (Lin et al. 2015a), genetically, PEDV strains are classified into two groups: (1) classical (isolates from Eurasia that are genetically similar to the prototype CV777 strain) and (2) emerging PEDV strains (Lin et al. 2016; Vlasova et al. 2014). All classical PEDV strains contain inserts and omissions in the spike gene (S INDEL) that are not present in the majority of the highly virulent emerging PEDV strains (Vlasova et al. 2014). Thus, these highly virulent strains that originally emerged in China in 2010 and transmitted to the USA, Europe and other Asian parts are referred to as emerging non-S INDEL PEDV strains. Recombinants between these two major groups of PEDV contain a set of deletions-insertions in their spike gene identical to those of the classical strains. They are called S INDEL strains and circulate in Asia, Europe and the USA. Additionally, a few reports described other uncommon PEDV variants that bear large deletions (194–216 aa) in the N-terminal domain (NTD) of the S protein and designated as S1 NTD-del type of PEDV (Diep et al. 2017; Oka et al. 2014; Suzuki et al. 2015). Unlike the altered tissue tropism from enteric TGEV to respiratory PRCV, these (S INDEL and S1 NTF have been reported-del) PEDV strains have kept their enteric predilection, but with lower virulence (Suzuki et al. 2016; Hou et al. 2017).

Swine enteric CoVs (TGEV, PEDV and PDCoV) are highly contagious and are associated with severe disease forms such as diarrhoea and vomition, and increased mortality in young ones (often 100%). They can cause sporadic outbreaks (endemic) or large-scale epidemics in swine-producing countries (Fig. 4.2). No specific treatments are available for any of the swine enteric CoVs that so far have resisted eradication efforts in different geographic regions. In this chapter, we have reviewed the diseases due to CoVs that continue evolving in domestic and wild swine, as well as another possible reservoir (avian or bat species) or secondary hosts including carnivores, or via the interspecies spread, recombination and generation of deletion escape variants. We also review PRCV that has lost its enteric tropism but is capable of inducing protective immune responses against TGEV that altered its global epidemiology.

4.2 Pathogenesis and Clinical Signs

4.2.1 TGEV

Extensive necrosis of mature enterocytes of jejunum and ileum within 24 h after infection results in reduced enzymatic activity (alkaline phosphatase, lactase, etc.), disrupted digestion, and cellular electrolyte (including sodium) balance. These changes primarily lead to the deposition of fluid in the intestinal lumen, acute malabsorptive diarrhoea (Moon 1978). The loss of extravascular protein and copious dehydration in piglets can be fatal (Butler et al. 1974). The latter can also lead to metabolic acidosis and hyperkalaemia, causing abnormal cardiac function.

TGE gross lesions are limited to the gastrointestinal tract. The distension of the stomach and the small intestine are seen to be filled up with curdled milk and sometimes petechial haemorrhages are visualised (Hooper and Haelterman 1966a). The small intestinal wall is thin and transparent. The villous atrophy in the jejunum and lesser in the ileum regions are the major TGE lesions and are more pronounced in neonatal pigs than in ≥3-week-old piglets (Moon 1978; Hooper and Haelterman 1966b). The increased severity of TGEV infection results in higher mortality (often 100%) in piglets less than 2 weeks of age that decreases in older pigs (Table 4.1). Although swine of any age is susceptible to TGEV, the mortality in TGEV seropositive groups and swine more than 5 weeks of age is usually low. Mechanisms that represent age-dependent susceptibility to clinical ailment comprise the slower substitution of tainted villous epithelial cells with newly differentiated enterocytes migrating from crypts in newborn pigs (Moon 1978). These lesions are similar to PEDV/PDCoV (Debouck et al. 1981; Jung et al. 2015a) lesions, but more severe than those caused by rotavirus (RV) (Bohl et al. 1978). Pathologic observations and degree of villous atrophy are exceptionally variable in pigs from endemic herds (Pritchard 1987).

Lungs (alveolar macrophages) and mammary gland tissues are recognised extra-intestinal sites for TGEV replication (Kemeny et al. 1975). Hitherto report shows pneumonia due to oronasal infection of pigs with TGEV (Underdahl et al. 1975), and the clinical significance of mammary gland infection is imprecise. However, agalactia is frequently observed in TGEV-affected sows, and TGEV spreads quickly among the population.

4.2.2 PRCV

PRCV replicates efficiently in porcine type 1 and 2 pneumocytes and is seen in epithelial cells of the nares, trachea, bronchi and bronchioles, and alveoli, and on occasion in alveolar macrophages (Atanasova et al. 2008; Pensaert et al. 1986; O'Toole et al. 1989). It can be noticed in blood and tracheobronchial lymph nodes. After experimental infection, nasal PRCV shedding usually lasts for 4–6 days. Pulmonary lesions and clinical signs subside consequently with an increase in the virus-neutralising (VN) antibody titres (Atanasova et al. 2008). Although PRCV is sometimes found in enterocytes, it does not spread efficiently to adjacent epithelial cells (Cox et al. 1990), and faecal shedding is low or undetectable.

PRCV predominantly causes upper and lower respiratory tract disease. The lesions appear to include lung and bronchointerstitial pneumonia, with regular peribronchiolar and perivascular lymphohistiocytic handcuffing (Atanasova et al. 2008; Cox et al. 1990; Halbur et al. 1993; Jung et al. 2007). The PRCV-induced bronchointerstitial pneumonia results in (1) thickened alveolar septa due to macrophage and lymphocyte infiltration, (2) hypertrophy and hyperplasia of type 2 pneumocytes, (3) aggregation of cell debris and inflammatory leukocytes in alveolar and bronchiolar lumina because of airway epithelial necrosis and (4) peribronchiolar or perivascular lymphohistiocytic inflammation.

4.2.3 PEDV

Clinical signs are evident between 22 and 36 h postinfection and match with the peak of viral replication (Table 4.1). The clinical presentation (watery malabsorptive diarrhoea, vomiting, depression and anorexia) and pathological lesions of PEDV are clinically indistinguishable from those of TGEV (Debouck et al. 1981; Coussement et al. 1982).

Morbidity is nearly 100% in piglets and variable in sows. Neonates below 1 week of age often die due to severe dehydration, and mortality touches 50–100%, whereas mortality is low in older pigs and they recover within a week. In sows, severity of diarrhoea is constant and frequently shows only depression and anorexia. Similarly, fattening pigs may develop watery faeces and can become anorexic and depressed within a week. As with TGEV, slower enterocyte turnover and immature innate immune system may add to the more extreme clinical signs, higher mortality and slower recuperation in PEDV-tainted piglets in contrast to weaned pigs (Jung et al. 2015a; Moon et al. 1975; Annamalai et al. 2015).

Each outbreak generally lasts for ~3–4 weeks; however, it might be longer on large breeding farms with multiple, isolated units and variable levels of lactogenic immunity in gilts/sows. PEDV-exposed pregnant sows can provide sufficient lactogenic immunity to protect their piglets, and PED outbreaks stop. After the passage of acute outbreak, diarrhoea may persist and is recurrent in weaned pigs, resembling endemic TGE form (Martelli et al. 2008).

The severity of lesions and the virus replication levels in naturally and experimentally infected suckling piglets vary for classical PEDV, emerging non-S INDEL and S INDEL PEDV strains (Jung et al. 2015a; Coussement et al. 1982; Kim and Chae 2003; Pospischil et al. 1981; Sueyoshi et al. 1995; Lin et al. 2015b; Madson et al. 2014). Lesions remain localised to the small intestine that is swollen and filled up with watery, yellowish liquid. Microscopic examination shows syncytia, vacuolation and shedding of small intestinal enterocytes primarily on the proximal villi. Similar to TGEV, PEDV infection results in degeneration of enterocytes that reduces the villous height:crypt depth (VH: CD) ratios and the enzymatic activity. Although PEDV antigens were detected in colonic epithelial cells, no associated histopathologic changes have been observed (Debouck et al. 1981).

Viral RNA has been confirmed in the serum, and different tissues (including lung, spleen, liver and muscle) of pigs euthanised during PEDV infection (Suzuki et al. 2016; Jung et al. 2014, 2015a; Lohse et al. 2017; Chen et al. 2016a; Park and Shin 2014) with high RNA titres in the serum of 7–8 $\log_{10}$ GE/mL coinciding with peak RNA titres in faeces (11–12 $\log_{10}$ GE/mL) (Jung et al. 2015a). Additionally, PEDV RNA is identified in 40.8% (20/49) of sow milk samples during the epidemics caused by emerging PEDV strains (Sun et al. 2012).

4.2.4 PDCoV

The clinical signs are observed within 1–3 days after PDCoV infection in suckling and older pigs. Although clinical symptoms are similar (Table 4.1), they are less pronounced compared to PEDV and TGEV infections (Chen et al. 2015; Hu et al. 2016; Jung et al. 2015b; Ma et al. 2015). They include acute, watery diarrhoea due to malabsorption induced by the massive loss of absorptive enterocytes. Additional signs may include vomiting, dehydration, weight loss, lethargy and death. Vacuolation of the infected colonic epithelial cells may inhibit water and electrolyte reabsorption. The seronegative pigs are susceptible to PDCoV infection at any age, with high morbidity that can reach 100% in piglets. Evaluation of filed cases in the USA, China and Thailand in 2014 shows that PDCoV infection is associated with up to 40–80% mortality among suckling pigs (Anon 2014). The infection on breeding establishments remains self-limiting and stops when pregnant sows develop lactogenic immunity adequate to secure their offspring.

Gross lesions include thinned and transparent intestinal walls (jejunum to the colon) with a collection of a lot of yellow liquid with gas. Often stomach is found bloated with curdled milk.

PDCoV replicates in the epithelial cell of the large and small intestine. Lesions look like those seen in TGEV and PEDV infections but are mild (Chen et al. 2015; Hu et al. 2016; Jung et al. 2015b; Ma et al. 2015). Histological findings are intense, multifocal to diffuse, mild to extreme atrophic enteritis of jejunum and ileum, at some point joined by mild vacuolation of caecal and colonic epithelial cells (Jung et al. 2015b). Amid acute infection, PDCoV antigens are available in the villous epithelium of the mid-jejunum to the ileum and a lesser degree, in the duodenum, and caecum/colon (Jung et al. 2016a). PDCoV antigens may also be noticed in immune cells of the intestinal lamina propria, Peyer's patches and mesenteric lymph nodes (Hu et al. 2016). Inflammatory cell (macrophage, lymphocyte and neutrophil) infiltration can be observed in the lamina propria. Acute necrosis of PDCoV-infected enterocytes (Jung et al. 2016a) results in marked villous atrophy in jejunum and ileum, but not duodenum or large intestine, which coincides with fewer PDCoV antigen-positive duodenal, caecal or colonic epithelial cells (Chen et al. 2015; Jung et al. 2015b). Acute-phase viremia with low PDCoV RNA titres in serum is observed (Chen et al. 2015; Hu et al. 2016). After recovery of pigs from clinical disease, huge amount of PDCoV antigens are found in the gut lymphatic tissues (Hu et al. 2016). Additionally, low or moderate quantities of PDCoV RNA, but not antigens, are detected in multiple organs, feasibly as of viremia (Chen et al. 2015; Ma et al. 2015; Jung et al. 2016b). Decreased levels of PDCoV shedding (compared with PEDV and TGEV) in the faeces may be indicative of its incomplete adaptation to pigs and can contribute to its slower spread among swine herds and the lower mortality of nursing pigs (Jung et al. 2015b).

4.3 Incidence and Prevalence of the Disease

4.3.1 TGEV

TGEV was first detected in the USA in 1946 from outbreaks of acute diarrhoea with high mortality in piglets (Doyle and Hutchings 1946). Since then the disease has been reported in several pig-rearing countries practicing intensive pig farming system, including Europe, Asia (Japan, Korea, Malaysia and Taiwan), the Americas (North, Central and South) and Africa (Zaïre, Ghana). Despite the widespread application of vaccines, TGEV infections were a prime reason for enteric disease and mortality in piglets in the USA and globally in the 1960s–1980s. The presence and extensive prevalence of PRCV, a deletion mutant of TGEV, narrowed the clinical impact of TGE (Laude et al. 1993; Pensaert et al. 1986, 1993; Brown and Cartwright 1986; Pensaert 1989; Yaeger et al. 2002). Currently, sporadic outbreaks of profuse diarrhoea in piglets due to TGEV in TGEV/PRCV seronegative herds are yet to be confirmed in North America, Europe and Asia. However, careful differentiation between TGEV and emerging TGEV/ PEDV recombinants may be needed.

Two epidemiologic forms of TGE are apparent: epidemic and endemic. Epidemic TGE noticed transcendently in seronegative flocks. After entry, the illness transmits quickly to swine of any age, particularly in winters, with inappetence, vomition or diarrhoea in affected animals. Suckling pigs exhibit prominent clinical signs and get quickly dehydrated. Lactating sows usually show anorexia and agalactia, with reduced milk production, which further adds to piglet mortality.

Endemic TGE indicates the persistence of the virus and disease in a group perpetuated by the continuous influx of susceptible swine. It is a classic sequel to a primary outbreak and occurs in seropositive animals that regularly have farrowing (Stepanek et al. 1979), additions of the herd or mixing of susceptible pig population. In endemic groups, TGEV spreads slowly among grown-up pigs (Pritchard 1987). Sows are most of the time resistant and asymptomatic and will transfer a variable level of passive lactogenic immunity to their offspring. In these groups, mild TGEV diarrhoea is seen with mortality under 10–20% in pigs from ~6 days of age until ~2 weeks post-weaning.

4.3.2 PRCV

PRCV infects the respiratory tract with limited or no shedding in faeces (Pensaert 1989). The first isolation of PRCV was from Belgium in 1984 (Pensaert et al. 1986) and 1989. PRCV was detected in the USA in the herds without a prior history of TGEV infection or vaccination (Hill et al. 1990; Wesley et al. 1990). Antibodies produced in PRCV-infected pigs neutralise the TGEV.

Since the first report, the virus has been introduced rapidly in Europe (Laude et al. 1993; Brown and Cartwright 1986; Have 1990; van Nieuwstadt et al. 1989) and attained endemic status worldwide, including entering TGEV-free countries (Laude et al. 1993; Pensaert 1989; Pensaert et al. 1993). A serological survey from the USA in 1995 demonstrated that clinically healthy pigs from different herds were found to be seropositive for PRCV (Wesley et al. 1997) in Iowa State.

4.3.3 PEDV

The classical PEDV strains were the cause of several epidemics with heavy mortality in Europe from 1971 until the late 1980s. However, after 2000, reports are very rare. In Italy, an epidemic involving 63 herds occurred in 2005 and 2006 where pigs of all ages were found affected, but mortality was mainly limited to suckling piglets (Martelli et al. 2008). Because of the low clinical importance of the disease, no surveillance studies were conducted until the emergence of new PEDV variants in Europe in 2014. Note that the historical prevalence of classical PEDV in the European swine population is unknown. An emerging non-S INDEL strain led to an outbreak in Ukraine in 2014; while outbreaks in European countries (Germany, Belgium, France, the Netherlands and Slovenia) were confirmed as of S INDEL strains (Lin et al. 2016).

Infections associated with classical PEDV strains were originally reported in China in the late 1970s. Since then, PED has spread among swine farms and became leading cause of viral diarrhoea, despite the use of vaccines (targeting the prototype PEDV strain CV777) (Wang et al. 2016a; Xuan et al. 1984). In Japan PEDV was first detected in 1982 (Takahashi et al. 1983) and outbreaks continued during the 1990s with mortality between 30% and 100% in suckling pigs (Sueyoshi et al. 1995; Kuwahara et al. 1988). PEDV was first recognised in Korea in 1993 (Kweon et al. 1993) and became prevalent accounting for more than 50% of the enteric viral infections diagnosed in suckling piglets in the 1990s (Chae et al. 2000; Hwang et al. 1994). In India, 21.2% of 528 serum samples from pigs (2–6 months old) were confirmed positive for PEDV antibodies (Barman et al. 2003). In Thailand, between 2007 and 2008, PED occurred in eight provinces (Puranaveja et al. 2009) affecting all age-group pigs but mortality reached 100% in newborn piglets.

In October 2010, a massive series of PEDV outbreaks occurred in China, resulting in tremendous economic losses (Sun et al. 2012, 2016). These outbreaks were caused by the emerging extremely virulent non-S INDEL strains, and the mortality in neonatal piglets reached 50–100% (Wang et al. 2016b). Subsequently, emerging S INDEL strains were also identified in China (Wang et al. 2016b), demonstrating that classical and emerging (non-S INDEL and S-INDEL PEDV) strains are co-circulating in China.

In 2013, the emerging PEDV strains started spreading to other countries outside China. At the same period, outbreak due to highly virulent non-S INDEL PEDV strains started in US swine farms (Stevenson et al. 2013), which was followed by outbreaks associated with milder S-INDEL PEDV strains in January 2014 (Wang et al. 2014a). However, S INDEL strains were retrospectively found in pig samples collected as early as June 2013 (Vlasova et al. 2014). In 2013–2014, PEDV led to the loss of nearly 10% of the US swine stock (7 million pigs) and the associated profits. By 2018, PEDV had reportedly spread to 39 US states and Puerto Rico (www.aphis.usda.gov/animal-health/secd). In the USA, the last big outbreak of PEDV occurred in the spring of 2014, and the emerging non-S INDEL PEDV strains have reportedly spread to the neighbouring countries, including Canada and Mexico (Lin et al. 2016).

In 2013, the emerging non-S INDEL PEDV strains were identified in Japan (Masuda et al. 2015) and then reportedly spread to South Korea (Kim et al. 2015), Vietnam (Vui et al. 2014), Thailand (Cheun-Arom et al. 2015), Taiwan (Lin et al. 2014) and the Philippines (Kim et al. 2016). The S-INDEL PEDV strains were detected in Japanese (Suzuki et al. 2015) and Korean (Lee et al. 2014) farms in 2013 and 2014, respectively. To date, other PEDV variants that carry a large deletion in the NTD of the S protein were detected exclusively in pigs in Japan. Besides the avirulent TTR-2 strain (Suzuki et al. 2016), 15 novel field PEDV mutants with large deletions in the NTD of S gene ranging from 582 to 648 nt were detected from diarrhoeic pig faecal or intestinal samples collected between December 2013 and June 2015 (Diep et al. 2017). Interestingly, all of these samples contained at least two PEDV strains with distinct large genomic deletions, and the majority of these PEDV strains were confirmed to contain an intact S gene. These variants with large deletions in the S gene were found in both primary and recurring PED outbreaks. In summary, classical, emerging non-S INDEL and S-INDEL PEDV strains exist in Asia and Europe, but only emerging non-S INDEL and S-INDEL PEDV strains are currently confirmed to circulate in the Americas. No PEDV infections have been reported in the African and Australian continents so far.

4.3.4 PDCoV

The initial surveys in China and Hong Kong during 2007–2011 identified DCoVs in pigs and wild birds (Woo et al. 2012). Nonetheless, the DCoV presence was previously confirmed at Chinese live-animal markets in 2005–2006 in small mammals, including Chinese ferret badgers and Asian leopard cats (Dong et al. 2007). Because their helicase and S genes are closely related with PDCoV, interspecies transmission of DCoVs among Asian carnivores, porcine and avian species is suggested. The origin of PDCoV remains unclear, but considering its recent emergence, its adaptation to pigs may be incomplete.

In early 2014, the first outbreak of PDCoV-associated diarrhoea was documented in swine, in Ohio. Among intestinal or faecal samples obtained from diarrhoeic pigs from 5 Ohio farms, 92.9% were found to be positive for PDCoV by RT-PCR (Wang et al. 2014b). The PDCoV sequences shared high nucleotide similarity with the two Chinese prototype PDCoV strains, HKU15-44 and HKU15-155, identified in 2012. Other two genetically similar PDCoV strains, USA/IA/2014/8734 and SDCV/USA/Illinois121/2014, were detected in US states during the same time period (Li et al. 2014; Marthaler et al. 2014a). After this PDCoV has been confirmed in 19 US states (www.aphis.usda.gov/animal-health/secd) but is still less widespread than PEDV. PDCoV origin in US swine is unknown; however, serologic evidence suggests that it was circulating in US swine before its recognition in 2014 (Sinha et al. 2015; Thachil et al. 2015).

PDCoV has also been documented in Canada (March 2014), Korea (April 2014), mainland China (2015), Thailand (2015) and Vietnam and Lao PDR. A Korean study described that of 691 faecal samples from 59 farms collected from diarrhoeic pigs from 2014 to 2015, only 2 samples originating from a single farm were PDCoV RNA positive (Lee et al. 2016a). The two Korean PDCoV strains, SL2 and SL5, were genetically closely related to the US PDCoV strains but differed from the older Korean strain KNU14-04. Following the first PDCoV identification in Hong Kong in 2012 (Woo et al. 2012), it was confirmed in diarrhoeic pigs in mainland China (Dong et al. 2015; Song et al. 2015). PDCoV high prevalence (>30%) and frequent co-infections with PEDV (51%) were reported. All Chinese PDCoV strains shared high nt identities (≥98.9%) with the global PDCoV strains. The Thai PDCoV strains shared the highest nt identities (≥98.4%) with the Chinese strain CHN-AH-2004 (Janetanakit et al. 2016), forming, however, a cluster separate from Chinese and US strains (Zhang 2016). Strains genetically close to the Thai PDCoV lineage were found in Lao PDR, whereas strains of the US PDCoV lineage were detected in Vietnam (Saeng-Chuto et al. 2017).

4.4 Immunobiology

4.4.1 TGEV/PRCV

Infection of adult swine results in noticeable serum antibodies that could persist for 6 months to several years (Stepanek et al. 1979). Serum antibody presence confirms serological evidence for TGEV or PRCV infection; their association with protection against TGEV is not established. Swine that recover from TGEV infection show protection from subsequent short-term challenge, as of intestinal mucosal immunity (Brim et al. 1995; Saif et al. 1994; VanCott et al. 1993, 1994).

Protective immunity depends on the secretory IgA (sIgA) antibodies produced by intestinal plasma cells (Saif et al. 1994; VanCott et al. 1993, 1994). Intestinal and serum IgA TGEV antibodies and antibody-secreting cells (ASCs) have been detected in pigs after oral, but not after parenteral, administration with TGEV (Saif et al. 1994; VanCott et al. 1993, 1994; Kodama et al. 1980). Presence of IgA antibody (likely of intestinal origin) in the serum is considered to be an indicator of active immunity to TGE (Kodama et al. 1980; Saif 1999). Besides local antibody responses, cell-mediated immunity (CMI) is also important in protection against TGEV infections (Brim et al. 1995; Frederick et al. 1976; Shimizu and Shimizu 1979). Retinoic acid may enhance CMI after immunisation of piglets with an inactivated TGEV vaccine by increasing the trafficking of $CD8^{+}$ T cells to lymph nodes and small intestine (Chen et al. 2016b). It was conceived that a low natural killer (NK) cell activity might associate with the higher susceptibility of piglets and parturient sows to TGEV infection (Cepica and Derbyshire 1984). A virulent (SHXB) however not attenuated (STC3) TGEV strain impeded the capacity of porcine intestinal DCs or monocyte-derived DCs to sample antigen, migrate and incite T-cell expansion in vivo and in vitro (Zhao et al. 2014), suggestive of the immune-suppressive potential of TGEV.

PRCV-inoculated pigs developed potent systemic and bronchus-associated but not gut-associated antibody, ASC and T-cell responses (VanCott et al. 1993; Brim et al. 1994). Other than quantitative differences in the count of IgA antibodies in the sow milk induced by TGEV or PRCV exposure, putative divergences in virus epitopes observed by the milk IgA antibodies were recommended (De Diego et al. 1992).

Subsequently to extensive decrease in epidemic outbreaks of TGE in Europe following the widespread distribution of PRCV, the research suggested that PRCV infections induce partial immunity against TGEV, as proven by a reduced length and duration of virus shedding and diarrhoea in most of the pigs examined (Brim et al. 1995; VanCott et al. 1994; Cox et al. 1993; Wesley and Woods 1996). This partial immunity is associated with rapid increases in TGEV-VN antibodies (Cox et al. 1993; Wesley and Woods 1996) and numbers of IgG and IgA ASCs in the intestines (Saif et al. 1994; VanCott et al. 1994). Movement of PRCV IgG and IgA ASCs from the bronchus-associated lymphoid tissues (BALT) to the gut of the PRCV-exposed pigs after TGEV challenge may clarify the quick anamnestic response and induces the partial protection (VanCott et al. 1994). However, neonatal pigs needed a week after PRCV encounter to develop partial immunity to subsequent TGEV challenge (Wesley and Woods 1996).

Circulating passive antibodies (primarily IgG), acquired with colostrum, protect the neonate against systemic, however not the intestinal, infection (Hooper and Haelterman 1966a; Saif and Sestak 2006). Between the first 7 days of lactation, IgA prevails in milk and IgG reduces. Researchers have reviewed the mechanisms of passive immunity to TGEV infection (Saif and Sestak 2006; Chattha et al. 2015; Saif and Jackwood 1990; Saif and Bohl 1979). Swine recovered from TGE spread passive immunity to their suckling pigs with colostrum or milk (lactogenic immunity) containing TGEV virus-neutralising antibodies (Hooper and Haelterman 1966a) that nullify the ingested TGEV in the intestinal lumen. This is accomplished naturally when piglets feed on immune mother regularly or experimentally by constant feeding of antiserum.

IgA TGEV antibodies in milk are stable in the gut and give the best protection, but IgG antibodies are likewise protective if high titres are restored in milk after parenteral or systemic vaccination (Bohl and Saif 1975) or by ingestion of colostral IgG antibodies (Stone et al. 1977). Following TGEV infection and antigenic stimulation in the gut, IgA immunocytes move to the mammary gland where they confine and produce IgA antibodies into colostrum and milk that play a major role in passive protection of suckling pigs (Saif and Sestak 2006; Saif and Jackwood 1990; Saif and Bohl 1979; Bohl and Saif 1975). This "gut-mammary" immunologic axis was first proposed concerning TGEV infections in swine (Bohl et al. 1972; Saif et al. 1972), providing the first concept for a common mucosal immune system.

4.4.2 PEDV

Even though all age pigs are susceptible to PED, highest mortality is seen in 1-week piglets and younger, and their survival is dependent on the transfer of maternal antibodies, especially VN and sIgA (via colostrum and milk from immunised or previously exposed sows). The components of lactogenic security depicted for TGEV infection apply to PEDV also (described previously in TGEV section) (Chattha et al. 2015; Langel et al. 2016). Pigs lose lactogenic protection at weaning becoming susceptible to PEDV infection. Humoral immune response to PEDV infection is quite the same as seen for TGEV (described in TGEV section) (Saif and Sestak 2006; Chattha et al. 2015). VN antibodies appear in the serum but do not play any significant role in protection against the clinical disease since protection basically relies on the presence of sIgA antibodies in the intestinal mucosa (Chattha et al. 2015; Langel et al. 2016). Immunity may not persist long, yet a fast anamnestic response upon re-exposure may decrease the severity of reoccurring disease or even prevent it.

At least 11 proteins [ORF1ab-encoded NS proteins (nsp1, nsp3, nsp5, nsp7, nsp14–16), structural proteins (E, M, N) and accessory protein ORF3s] have been recognised as IFN antagonists allowing PEDV to evade host interferon (IFN) responses (Ding et al. 2014; Wang et al. 2015; Zhang et al. 2016a). Decreased innate immune responses (specifically frequencies and function of NK cells) (Annamalai et al. 2015) likely contribute to the increased severity of PEDV infection in suckling vs. older (weaned, finisher, adult) pigs, as observed for TGEV infections (Derbyshire et al. 1969).

4.4.3 PDCoV

The immune response to PDCoV infection in pigs is still unclear but presumed to be similar to that mentioned for TGEV and PEDV. Gnotobiotic pigs orally inoculated with the original or tissue-culture-developed PDCoV strain (OH-FD22) exhibited serum IgG, IgA and VN antibodies by 14 dpi that shows a peak at 24 dpi when the pigs had recouped from the clinical form and faecal virus shedding (Hu et al. 2016). Similar to TGEV and PEDV, the supply of maternal antibodies with colostrum and milk from immune sows, especially IgA and VN antibodies, should neutralise PDCoV in the gut protecting young piglets (Bohl et al. 1972; Saif et al. 1972).

4.5 Diagnosis

4.5.1 TGEV/PRCV

Since clinical signs and atrophic enteritis due to TGEV are also seen in other enteric infections (RV, PEDV, PDCoV and coccidia), lab-based diagnosis of TGE must be followed utilising the following tests: identification of viral antigen or nucleic acids in faeces, virus seclusion from samples or detection of TGEV antibodies.

Similar methods are used for the diagnosis of PRCV, but with an emphasis on respiratory specimens (nasal swabs or lung homogenates). Evaluation of clinical signs, histologic lesions and viral antigen distribution in tissues might yield a presumptive diagnosis.

Probing of TGEV antigen in small intestinal enterocytes by immunofluorescence (IF) (Pensaert et al. 1970) or immunohistochemical (IHC) (Shoup et al. 1996) techniques using monoclonal antibodies (MAb) against the highly conserved TGEV N protein may be conducted for formalin-fixed or frozen tissues harvested in the early stage of infection.

A mono- or polyclonal antibody-based enzyme-linked immunosorbent assay (ELISA) is utilised to recognise TGEV antigens in cell cultures, intestinal contents and faeces (Lanza et al. 1995; Sestak et al. 1996, 1999a; van Nieuwstadt et al. 1988) or PRCV antigen in nasal swabs or lung homogenates (Lanza et al. 1995); however, the sensitivity of available ELISA tests is generally lower than that of RT-PCR assays. RT-PCR or real-time qPCR is now more commonly used for detection of TGEV and distinguishing TGEV, PRCV, PDCoV and PEDV (Kim et al. 2000, 2001; Masuda et al. 2016; Costantini et al. 2004; Ogawa et al. 2009). PRCV/TGEV distinction is usually performed through PCR assay based on the S gene deletion region in PRCV strains. Multiplex RT-PCR and real-time qPCR have been used for the concurrent identification of major porcine viruses related to diarrhoea including RV, TGEV, PDCoV and PEDV (Masuda et al. 2016; Ogawa et al. 2009). Also, multiplex microarray hybridisation assay was used for quick differential identification of eight CoVs, including TGEV (Chen et al. 2005).

Transmission electron microscopy (TEM) is used to show TGEV in the faecal or intestinal contents of infected pigs. Further, immune EM (IEM) possesses many advantages over TEM and is more sensitive and better in differentiating TGEV from PEDV and PDCoV (Saif et al. 1977).

Primary and secondary pig kidney cells (Bohl and Kumagai 1965) or cell lines (Laude et al. 1981), porcine thyroid cells (Witte 1971) and McClurkin swine testicle (ST) cell line (McClurkin and Norman 1966) are advocated for TGEV isolation from infected pig faeces or intestinal contents. Typical cytopathic effects (CPE) might be negligible upon primary isolation of field strains and require additional blind passages. The CPE observed comprises swollen, round cells with a balloon-like appearance (Bohl and Kumagai 1965). For observing viral CPE or plaques, the addition of pancreatin or trypsin to the ST cell culture media as well as using older cells can improve detection of viral CPE and plaques (Bohl 1979).

ST cells and pig kidney cells are best for isolation of PRCV from a nasal swab or lung tissue homogenates. PRCV- and TGEV-induced CPE are comparable with syncytia usually seen that is also observed in PEDV grown in Vero cells (Hofmann and Wyler 1988; Ksiazek et al. 2003). Detection of virus in the cell culture is confirmed by VN, IF or IEM using specific TGEV antiserum or differential monoclonal antibodies (Garwes et al. 1988) and virus-specific RT-PCR (Enjuanes and Van der Zeijst 1995; Laude et al. 1993; Kim et al. 2000).

TGEV antibodies can be identified by several serological tests. However, TGEV serology is complex, as both TGEV and PRCV prompt VN antibodies that are quantitatively and qualitatively similar (Pensaert 1989). A blocking ELISA that uses MAbs to TGEV differentiates easily from PRCV (Garwes et al. 1988; Bernard et al. 1989; Callebaut et al. 1989). However, blocking ELISA works better on a group basis since certain pigs with low antibody titres to TGEV or PRCV may not show positivity (Callebaut et al. 1989; Sestak et al. 1999b; Simkins et al. 1993). Further, the accuracy of commercial ELISAs for distinguishing US strains of PRCV and TGEV is low (Sestak et al. 1999a).

4.5.2 PEDV

As for TGEV, diagnosis of PEDV must be based on clinical signs and lab identification of viral RNA, viral antigens or increased PEDV-species antibodies. For the location of PEDV RNA, the most broadly utilised lab technique is conventional PCR (Kim et al. 2001; Ishikawa et al. 1997) or real-time RT-PCR (Kim et al. 2007), which is sensitive, specific and rapid for the detection of viral RNA in different clinical samples. Although loop-mediated isothermal amplification (LAMP) assays (Ren and Li 2011; Yu et al. 2015) are recognised more recently as highly sensitive in detecting PEDV RNA, they have still not reached diagnostic labs. In situ hybridisation can be utilised to recognise PEDV RNA in fixed tissues (Stadler et al. 2015).

Direct display of PEDV and additionally its antigens is done utilising IF or IHC tests on the small intestinal tissues of piglets euthanised close to the onset of diarrhoea and before the desquamation of enterocytes. The virus particles can be shown in direct EM or IEM on diarrhoeic pig faeces. However, IEM must be used to distinguish PEDV from other CoVs, viz. TGEV and PDCoV, as all CoVs possess indistinguishable morphology.

Successful isolation of PEDV in Vero cells is increased when using intestinal contents/homogenates compared to faeces (Oka et al. 2014; Chen et al. 2014). ELISA tests based on polyclonal antibodies and MAbs are available for detection of PEDV antigens in faeces (Callebaut et al. 1982; Carvajal et al. 1995).

Paired serum samples are requisite for serologic diagnosis of endemic PEDV. Recently, IgG and IgA antibodies to PEDV are observed in oral fluids, suggesting that they may be appropriate to monitor prior herd exposure to PEDV (Bjustrom-Kraft et al. 2016). PEDV antibodies have been demonstrated using indirect ELISAs based on the whole cell-culture-adapted virus antigens (Carvajal et al. 1995; Hofmann and Wyler 1990; Kweon et al. 1994; Thomas et al. 2015), S/N viral proteins extracted from infected Vero cells (Knuchel et al. 1992; Oh et al. 2005) or bacteria or mammalian expression systems (Wang et al. 2015; Gerber et al. 2014; Gerber and Opriessnig 2015; Okda et al. 2015; Paudel et al. 2014). Blocking and competitive ELISAs have additionally been used for the detection of PEDV antibodies utilising MAbs or polyclonal antibodies (Carvajal et al. 1995; Okda et al. 2015; van Nieuwstadt and Zetstra 1991). Serum IgG antibodies against the N proteins of PEDV can be detected by 9–14 dpi; they peak around 21 dpi and then decline gradually (Okda et al. 2015). The VN test in Vero cell culture is used to assess VN antibodies to PEDV (Thomas et al. 2015; Okda et al. 2015; Paudel et al. 2014). These serological assays have generally been employed to screen prior exposure to the virus and to assess the viability of vaccines.

Overall, use of a laboratory test is suggested which must differentiate PEDV infections from TGEV, SeCoV and PDCoV. Especially for SeCoV that is a recombinant of TGEV (backbone) and PEDV (S protein), a selective assay based on detecting TGEV (any gene aside from S gene) and PEDV (S gene) would be preferred.

4.5.3 PDCoV

Selective laboratory assays must be opted to distinguish PDCoV infection from related PEDV, TGEV and RV infections. The methodologies discussed for TGEV and PEDV apply to PDCoV diagnosis. The confirmatory finding of PDCoV infection incorporates detection of PDCoV RNA or antigens in the faeces or intestinal substance/tissues. Diagnosis can also be made utilising RT-PCR assays that target a conserved region of PDCoV M or N genes (Marthaler et al. 2014b; Wang et al. 2014c), IF or IHC using virus-specific MAbs or polyclonal antibodies (Chen et al. 2015; Jung et al. 2015b; Ma et al. 2015), and in situ hybridisation (Jung et al. 2015b). A duplex real-time RT-PCR assay for detection of PDCoV and/or differentiation from PEDV in intestines and faeces was developed (Zhang et al. 2016b).

Direct EM can be used to display PDCoV viral particles in faeces from diarrhoeic pigs. However, IEM use must distinguish PDCoV from PEDV or TGEV (Jung et al. 2015a) using hyperimmune or convalescent sera. However isolation of PDCoV has been attempted with limited success in LLC-PK or ST cells, except in few strains (OH-FD22) (Hu et al. 2015). The other serologic assays in use for the diagnosis of PDCoV are IF, VN and ELISA. Serum and milk PDCoV antibodies of different isotypes have been quantified using ELISA based on cell culture virus antigen (Ma et al. 2016) or S1 and N viral proteins (Thachil et al. 2015; Okda et al. 2016; Su et al. 2016).

4.6 Transmission, Risk Factors and Stability

4.6.1 TGEV

TGEV is more stable in frozen conditions, but becomes fragile at room or higher temperatures. In an experimental study, the virus infectivity was maintained for more than 8 weeks at 5 °C, 2 weeks at 20 °C and 24 h at 35 °C in liquid manure slurry (Haas et al. 1995). Further, the virus is highly photosensitive and gets inactivated less than 6 h in the exposure of sunlight or ultraviolet light (Cartwright et al. 1965; Haelterman 1962). It also gets inactivated to 1% Lysovet (phenol and aldehyde), 0.03% formalin, 0.01% beta-propiolactone, 1 mM binary ethylenamine, sodium hypochlorite, NaOH, iodines, quaternary ammonium compounds, ether and chloroform (VanCott et al. 1993; Brown 1981). TGEV field strains are resistant to trypsin and relatively stable in pig bile, and at pH 3 (Laude et al. 1981), letting the virus survive in the stomach and small intestine.

In temperate climates, TGE is a seasonal disease with most outbreaks occurring in the winter months. It is assumed that this could be due to the virus stability in frozen form, and frailty to heat or sunlight (Haelterman 1962), permitting easy virus spread among herds in winter on fomites or animals. Three potential reservoirs for TGEV between seasonal epidemics proposed are (1) herds with enzootic TGE; (2) hosts other than swine and (3) carrier pigs. Dogs, cats and foxes could be the possible carriers, facilitating its spread on farms, as the shedding of the virus occurs for variable periods (Haelterman 1962; McClurkin et al. 1970) with the excreted virus (by dogs) remaining infectious for pigs (Haelterman 1962; Reynolds et al. 1980).

The increased concentration of wintering starlings (*Sturnus vulgaris*) in feedlots was suggested to contribute to the mechanical spread of TGEV between farms during the cold season. In a study, TGEV was seen in the droppings of starlings for as long as 32 h after feeding TGEV (Pilchard 1965). Likewise, housefly (*Musca domestica*) is also proposed as a mechanical vector for TGEV (Pilchard 1965). TGEV antigen is seen in flies on a swine herd and TGEV shedding up to 3 days from experimentally feeding flies (Gough and Jorgenson 1983). Notably, surveys done in Central Europe confirmed the presence of TGEV antibodies in nearly 30% of the feral pigs (Sedlak et al. 2008). Although TGEV shedding is detectable for up to 104 dpi (Underdahl et al. 1975), it is undefined yet whether infectious virus particles are shed or transmitted at that time. Adding of sentinel pigs to a herd at 3, 4 and 5 months after a past TGE outbreak resulted in no new disease in the introduced pigs (Derbyshire et al. 1969).

4.6.2 PRCV

Swine population density, farm distances and seasons influence PRCV epidemiology (Pensaert 1989; Have 1990). Pigs get PRCV infection at any age through contact or airborne transmission. The risk of spreading PRCV increases in zones of high swine density, where the virus can journey several miles.

4.6.3 PEDV

As for other enteric viral infections, in PEDV also direct or indirect faecal-oral transmission is the main route of virus transmission. Acute outbreaks in non-immune farms often occur within 4–5 days after newly purchased pig arrival. The virus enters farms mostly via infected pigs, but also by contaminated feed, trucks, boots or other fomites. Farm workers may also act as a vehicle for virus transmission to naïve pigs (Dee et al. 2014, 2016; Schumacher et al. 2016). Evidence of PEDV aerosol transmission is reported in some (Alonso et al. 2014) but not other studies (Niederwerder et al. 2016). In four-week-old pigs infected with emerging non-S INDEL PEDV strain, infectious virus excretion lasted for 14–16 days (Crawford et al. 2015). Nevertheless, a few pigs shed PEDV RNA, at 42 days post-initial oral exposure, but non-infectious virus particles were seen in faeces.

Similar to TGEV, after initial outbreaks on the breeding farms, PEDV can become endemic if sufficient litters of pigs are produced and weaned, allowing maintaining the virus. Of note, a report from South Korea showed 9.75% PEDV infection rate in wild boars (Lee et al. 2016b), although their role in maintenance and transmission of PEDV is unclear.

4.6.4 PDCoV

The main method of PDCoV transmission is the faecal–oral route. Faeces, vomit and other contaminated fomites are the major sources of the virus. Experimentally induced PDCoV diarrhoea lasted for ~5–10 days, with faecal virus RNA shedding lasting for up to 19 days (Hu et al. 2016; Ma et al. 2015). Pigs generally continue shedding PDCoV RNA in the faeces after recovery from disease; therefore, another possible reservoir for PDCoV may be subclinically infected or convalescent carriers.

4.7 Prevention and Control

4.7.1 TGEV

Treatment of clinically affected newborn piglets is usually ineffective in field situations; however, electrolyte/glucose solution supplementation of piglets that are 1 week or older may reduce their mortality (Bohl 1981). Extra heat, deep bedding and antibiotic solutions (to treat secondary infections) generally can improve piglet health.

Enhanced biosecurity measures should be maintained to decrease a chance of introduction of infected animals, and contaminated vehicles from TGEV-affected farms to susceptible herds. TGEV infection can be spread not only with infected live animals but also with unprocessed tissues of slaughtered TGEV-infected animals (Forman 1991).

Many methods for immunising sows to induce lactogenic immunity and consequent protection of neonatal piglets have been attempted (Chattha et al. 2015; Saif and Jackwood 1990; Bohl and Saif 1975). Several viral vaccines (virulent, attenuated, inactivated and recombinant subunit) with different routes of administration (oral, intra-nasal, subcutaneous, intramuscular and intra-mammary) (Saif and Sestak 2006; Moxley and Olson 1989) have been evaluated in the past. To note, intramuscular, parenteral or intra-mammary administration of pregnant sows with live attenuated, inactivated or subunit vaccines did not offer complete protection but were found to be effective to reduce piglet mortality rates (Brim et al. 1994). Unlike natural intestinal infection with the virulent virus, attenuated viruses do not stimulate the gut-MG-sIgA axis sufficiently for the induction of immunity similar to that observed following this. There are two commercial vaccines based on a live-modified TGEV strain for combined oral-intramuscular administration produced by Merck Animal Health: PROSYSTEM® TGE/Rota and PROSYSTEM® TREC. These vaccines can effectively stimulate a response in previously exposed pigs, but do not protect the naïve population.

Herd immunity can be enhanced by exposing all the sows to virulent TGEV (using intestinal contents or gut tissues of affected pigs) to boost lactogenic (milk) immunity (Bohl and Saif 1975; Bohl et al. 1972). This practice is called feedback and results in the rapid development of immunity in pregnant sows (particularly in those due to farrow 2 weeks or more after the start of the outbreak) and reduces losses in newborn piglets. However, it may also result in dissemination of other pathogens (potentially present in TGEV-containing faeces/intestinal contents) to adjacent herds. On small-scale farms, herd immunity is accomplished, and TGEV infection is self-limiting. In contrast, in larger farms ($\leq$200 sows) with a continuous farrowing system and continual influx of susceptible animals, TGEV infection frequently becomes endemic after the primary outbreak (Saif and Sestak 2006). Elimination of endemic TGE in a herd can be attempted using the feedback method. After this is done, no weaning of piglets should occur during the following 3–4 weeks so that there remains no susceptible host in the herd while TGEV is circulating on the farm.

4.7.2 PEDV

Due to the lack of PEDV-specific antivirals, the treatment is focused on alleviating the diarrhoeal disease. PEDV-infected pigs must get enough water to reduce dehydration, which exacerbates the severity of the disease. Temporary withholding of feed may benefit fattening pigs during the acute stage of the disease.

As in the case with TGEV, appropriate biosecurity measures should be applied to avoid the introduction of PED onto farms. Present epidemiological knowledge indicates that virus is spread between farms mainly through animal and human traffic, and contaminated feed (https://www.aphis.usda.gov/animal_health/animal_dis_spec/swine/downloads/secd_final_report.pdf). Careful disposal of the dead stock is recommended.

In contrast to the present situation in Asia, in Europe, PEDV infection (with mostly mild S INDEL strains in circulation) is considered to be of marginal economic importance and therefore does not warrant the development of a vaccine (Lee 2015). However, severe classical PEDV outbreaks in Asia have necessitated the development of PEDV vaccines to prevent and control the infection. In China, the CV777-based inactivated and attenuated PEDV vaccines were approved in 1995 and 1998, respectively (Wang et al. 2016b). Soon after, attenuated vaccines based on classical PEDV strains KPEDV-9 and DR13 were commercialised in 1999 and 2004, respectively, in Korea (Kweon et al. 1999; Song et al. 2007). Since 1997 a commercial attenuated PEDV vaccine based on cell culture-adapted classical PEDV P-5V strain (Nisseiken Co. Ltd., Japan) is administered to sows in Japan (Sato et al. 2011). These vaccines based on classical PEDV strains appeared to satisfactorily control PED in Asia until the highly virulent non-S INDEL PEDV strains have emerged (Lee 2015). As demonstrated in the field, the classical PEDV vaccines have failed to protect pigs from severe diarrhoeal disease associated with the emerging highly virulent non-S INDEL PEDV strains (Lee 2015).

The deliberate exposure of pregnant sows (feedback method) to PEDV promotes the rapid development of lactogenic immunity and thus shortens the course and the severity of the disease on the farm (Chattha et al. 2015). However, as mentioned in the TGEV section, this method may contribute to the spread of other infectious agents throughout the farm. We have recently demonstrated that high dose of virulent PEDV administered to sows can substantially increase their piglet survival rate as compared to low-dose and mock-infected sows (Fig. 4.3) (Langel et al. 2016). This novel finding suggests that the current feedback-based control strategies can be improved by ensuring the uniform administration of high-dose PEDV to pregnant sows.

Since the outbreaks of 2013, the USA has conditionally licensed two PEDV vaccines targeting emerging non-S INDEL PEDV strains: alphavirus-based vaccine (Harris vaccines™, now Merck Animal Health) and an inactivated vaccine (Zoetis) (2014). The first vaccine was developed in June 2014 using a replication-deficient Venezuelan equine encephalitis (VEE) virus packaging system expressing the S protein of an emerging non-S INDEL PEDV strain (Crawford et al. 2016).

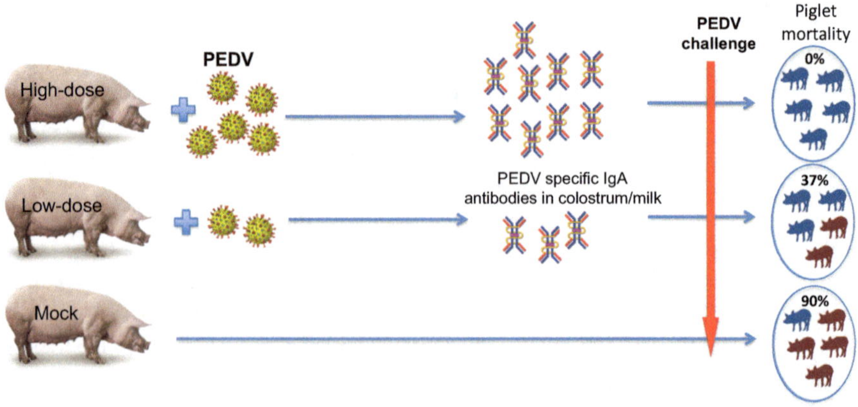

Fig. 4.3 Mucosal immune responses and lactogenic immunity may be influenced by PEDV dose given to pregnant swine. Gilts received high and high, low PEDV dose or mock at 3–4 weeks prepartum. All piglets were PEDV-challenged at 3–5 days post-partum (Langel et al. 2016)

The second vaccine developed in September 2015 was an inactivated whole-virus (non-S INDEL PEDV) vaccine plus an adjuvant (Crawford et al. 2016). In October 2016, an inactivated vaccine based on non-S INDEL PEDV strain AJ1102 was licensed in China (Wang et al. 2016b). In South Korea, an inactivated vaccine candidate based on non-S INDEL strain KNU-141112 was demonstrated to be protective in sows and their suckling piglets (Baek et al. 2016). However, the efficacy of these vaccines/vaccine candidates in the field is not assessed. To date, reverse genetics platforms have been generated for both classical and emerging non-S INDEL PEDV strains using different approaches (Beall et al. 2016; Jengarn et al. 2015; Li et al. 2013) and can be used for the future rational design of safe and effective PEDV vaccines.

4.7.3 PDCoV

The disease-preventive measures adopted for TGEV and PEDV control and prevention can be useful for PDCoV infection too. In the absence of any suitable vaccines or antivirals to control PDCoV disease, reliable regime includes symptomatic action giving bicarbonate liquids and ad lib water to alleviate acidosis and dehydration in suckling pigs. Antibiotics administration may be beneficial in the case complicated by concurrent/secondary bacterial infection. In the event of heavy mortality, feedback techniques must be opted to stimulate lactogenic immunity and reduce mortality. Additionally, during PDCoV epidemics, a strict biosecurity plan must be implemented to lessen PDCoV spread via infected fomites. The systems all in/all out and thorough disinfection (using phenolic disinfectants, bleach, peroxides, aldehydes or iodophors) can break the disease cycle.

Acknowledgements All the authors of the manuscript thank and acknowledge their respective universities and institutes.

Conflict of interest: There is no conflict of interest.

References

Akimkin V, Beer M, Blome S et al (2016) New chimeric porcine coronavirus in swine feces, Germany, 2012. Emerg Infect Dis 22:1314–1315

Alonso C, Goede DP, Morrison RB et al (2014) Evidence of infectivity of airborne porcine epidemic diarrhea virus and detection of airborne viral RNA at long distances from infected herds. Vet Res 45:73

Annamalai T, Saif LJ, Lu Z et al (2015) Age-dependent variation in innate immune responses to porcine epidemic diarrhea virus infection in suckling versus weaned pigs. Vet Immunol Immunopathol 168:193–202

Anon (2014) USDA to require reports of PED. J Am Vet Med Assoc 244:1234

Atanasova K, Van Gucht S, Barbé F, Lefebvre DJ, Chiers K, Van Reeth K (2008) Lung cell tropism and inflammatory cytokine-profile of porcine respiratory coronavirus infection. Open Vet Sci J 2:117–126

Baek PS, Choi HW, Lee S et al (2016) Efficacy of an inactivated genotype 2b porcine epidemic diarrhea virus vaccine in neonatal piglets. Vet Immunol Immunopathol 174:45–49

Ballesteros ML, Sanchez CM, Enjuanes L (1997) Two amino acid changes at the N-terminus of transmissible gastroenteritis coronavirus spike protein result in the loss of enteric tropism. Virology 227:378–388

Barman N, Barman B, Sarma K et al (2003) Indian J Anim Sci 73:576–578

Beall A, Yount B, Lin CM et al (2016) Characterization of a pathogenic full-length cDNA clone and transmission model for porcine epidemic diarrhea virus strain PC22A. MBio 7:e01451-15

Belsham GJ, Rasmussen TB, Normann P et al (2016) Characterization of a novel chimeric swine enteric coronavirus from diseased pigs in Central Eastern Europe in 2016. Transbound Emerg Dis 63:595–601

Bernard S, Bottreau E, Aynaud JM et al (1989) Natural infection with the porcine respiratory coronavirus induces protective lactogenic immunity against transmissible gastroenteritis. Vet Microbiol 21:1–8

Bjustrom-Kraft J, Woodard K, Gimenez-Lirola L et al (2016) Porcine epidemic diarrhea virus (PEDV) detection and antibody response in commercial growing pigs. BMC Vet Res 12:99

Bohl E (1979) Diagnosis of diarrhea in pigs due to transmissible gastroenteritis virus or rotavirus. In: Bricout F, Scherrer R (eds) Viral enteritis in humans and animals. INSERM, Paris, pp 341–343

Bohl EH (1981) Transmissible gastroenteritis. In: Leman AD, Glock RD, Mengeling WL, Penny RHC, Scholl E, Straw B (eds) Diseases of swine, 5th edn. University Press, Ames, IA, pp 195–208

Bohl EH, Kumagai T (1965) The use of cell cultures for the study of swine transmissible gastroenteritis virus. In: United States Livestock Sanitary Association meeting

Bohl EH, Saif LJ (1975) Passive immunity in transmissible gastroenteritis of swine: immunoglobulin characteristics of antibodies in milk after inoculating virus by different routes. Infect Immun 11:23–32

Bohl EH, Gupta RK, Olquin MV et al (1972) Antibody responses in serum, colostrum, and milk of swine after infection or vaccination with transmissible gastroenteritis virus. Infect Immun 6:289–301

Bohl EH, Kohler EM, Saif LJ et al (1978) Rotavirus as a cause of diarrhea in pigs. J Am Vet Med Assoc 172:458–463

Boniotti MB, Papetti A, Lavazza A et al (2016) Porcine epidemic diarrhea virus and discovery of a recombinant swine enteric coronavirus, Italy. Emerg Infect Dis 22:83–87
Brim TA, VanCott JL, Lunney JK et al (1994) Lymphocyte proliferation responses of pigs inoculated with transmissible gastroenteritis virus or porcine respiratory coronavirus. Am J Vet Res 55:494–501
Brim TA, VanCott JL, Lunney JK et al (1995) Cellular immune responses of pigs after primary inoculation with porcine respiratory coronavirus or transmissible gastroenteritis virus and challenge with transmissible gastroenteritis virus. Vet Immunol Immunopathol 48:35–54
Brown I, Cartwright S (1986) New porcine coronavirus? Vet Rec 119:282–283
Brown TT Jr (1981) Laboratory evaluation of selected disinfectants as virucidal agents against porcine parvovirus, pseudorabies virus, and transmissible gastroenteritis virus. Am J Vet Res 42:1033–1036
Butler DG, Gall DG, Kelly MH et al (1974) Transmissible gastroenteritis. Mechanisms responsible for diarrhea in an acute viral enteritis in piglets. J Clin Invest 53:1335–1342
Callebaut P, Debouck P, Pensaert M (1982) Enzyme-linked immunosorbent assay for the detection of the coronavirus-like agent and its antibodies in pigs with porcine epidemic diarrhea. Vet Microbiol 7:295–306
Callebaut P, Pensaert MB, Hooyberghs J (1989) A competitive inhibition ELISA for the differentiation of serum antibodies from pigs infected with transmissible gastroenteritis virus (TGEV) or with the TGEV-related porcine respiratory coronavirus. Vet Microbiol 20:9–19
Cartwright SF, Harris HM, Blandford TB et al (1965) A cytopathic virus causing a transmissible gastroenteritis in swine. I. Isolation and properties. J Comp Pathol 75:387–396
Carvajal A, Lanza I, Diego R et al (1995) Evaluation of a blocking ELISA using monoclonal antibodies for the detection of porcine epidemic diarrhea virus and its antibodies. J Vet Diagn Investig 7:60–64
Cepica A, Derbyshire JB (1984) Antibody-dependent and spontaneous cell-mediated cytotoxicity against transmissible gastroenteritis virus infected cells by lymphocytes from sows, fetuses and neonatal piglets. Can J Comp Med 48:258–261
Chae C, Kim O, Choi C et al (2000) Prevalence of porcine epidemic diarrhoea virus and transmissible gastroenteritis virus infection in Korean pigs. Vet Rec 147:606–608
Chattha KS, Roth JA, Saif LJ (2015) Strategies for design and application of enteric viral vaccines. Annu Rev Anim Biosci 3:375–395
Chen W, Yan M, Yang L et al (2005) SARS-associated coronavirus transmitted from human to pig. Emerg Infect Dis 11:446–448
Chen Q, Li G, Stasko J et al (2014) Isolation and characterization of porcine epidemic diarrhea viruses associated with the 2013 disease outbreak among swine in the United States. J Clin Microbiol 52:234–243
Chen Q, Gauger P, Stafne M et al (2015) Pathogenicity and pathogenesis of a United States porcine deltacoronavirus cell culture isolate in 5-day-old neonatal piglets. Virology 482:51–59
Chen Q, Gauger PC, Stafne MR et al (2016a) Pathogenesis comparison between the United States porcine epidemic diarrhea virus prototype and S-INDEL-variant strains in conventional neonatal piglets. J Gen Virol 97:1107–1121
Chen X, Tu C, Qin T et al (2016b) Retinoic acid facilitates inactivated transmissible gastroenteritis virus induction of CD8(+) T-cell migration to the porcine gut. Sci Rep 6:24152
Cheun-Arom T, Temeeyasen G, Srijangwad A et al (2015) Complete genome sequences of two genetically distinct variants of porcine epidemic diarrhea virus in the eastern region of Thailand. Genome Announc 3:e00634-15
Costantini V, Lewis P, Alsop J et al (2004) Respiratory and fecal shedding of porcine respiratory coronavirus (PRCV) in sentinel weaned pigs and sequence of the partial S-gene of the PRCV isolates. Arch Virol 149:957–974
Coussement W, Ducatelle R, Debouck P et al (1982) Pathology of experimental CV777 coronavirus enteritis in piglets. I. Histological and histochemical study. Vet Pathol 19:46–56

Cox E, Hooyberghs J, Pensaert MB (1990) Sites of replication of a porcine respiratory coronavirus related to transmissible gastroenteritis virus. Res Vet Sci 48:165–169

Cox E, Pensaert MB, Callebaut P (1993) Intestinal protection against challenge with transmissible gastroenteritis virus of pigs immune after infection with the porcine respiratory coronavirus. Vaccine 11:267–272

Crawford K, Lager K, Miller L et al (2015) Evaluation of porcine epidemic diarrhea virus transmission and the immune response in growing pigs. Vet Res 46:49

Crawford K, Lager KM, Kulshreshtha V et al (2016) Status of vaccines for porcine epidemic diarrhea virus in the United States and Canada. Virus Res 226:108–116

De Diego M, Laviada MD, Enjuanes L et al (1992) Epitope specificity of protective lactogenic immunity against swine transmissible gastroenteritis virus. J Virol 66:6502–6508

de Groot R, Ziebuhr J, Poon L et al (2008) Revision of the family *Coronaviridae*. Taxonomic proposal of the Coronavirus Study Group to the ICTV Executive Committee. http://talk.ictvonline.org/media/p/1230.aspx

Debouck P, Pensaert M, Coussement W (1981) The pathogenesis of an enteric infection in pigs, experimentally induced by the coronavirus-like agent, CV 777. Vet Microbiol 6:157–165

Dee S, Clement T, Schelkopf A et al (2014) An evaluation of contaminated complete feed as a vehicle for porcine epidemic diarrhea virus infection of naive pigs following consumption via natural feeding behavior: proof of concept. BMC Vet Res 10:176

Dee S, Neill C, Singrey A et al (2016) Modeling the transboundary risk of feed ingredients contaminated with porcine epidemic diarrhea virus. BMC Vet Res 12:51

Derbyshire JB, Jessett DM, Newman G (1969) An experimental epidemiological study of porcine transmissible gastroenteritis. J Comp Pathol 79:445–452

Diep NV, Norimine J, Sueyoshi M et al (2017) Novel porcine epidemic diarrhea virus (PEDV) variants with large deletions in the spike (S) gene coexist with PEDV strains possessing an intact S gene in domestic pigs in Japan: a new disease situation. PLoS One 12:e0170126

Ding Z, Fang L, Jing H et al (2014) Porcine epidemic diarrhea virus nucleocapsid protein antagonizes beta interferon production by sequestering the interaction between IRF3 and TBK1. J Virol 88:8936–8945

Dong BQ, Liu W, Fan XH et al (2007) Detection of a novel and highly divergent coronavirus from Asian leopard cats and Chinese ferret badgers in southern China. J Virol 81:6920–6926

Dong N, Fang L, Zeng S et al (2015) Porcine deltacoronavirus in mainland China. Emerg Infect Dis 21:2254–2255

Doyle LP, Hutchings LM (1946) A transmissible gastroenteritis in pigs. J Am Vet Med Assoc 108:257–259

Enjuanes L, Van der Zeijst B (1995) Molecular basis of transmissible gastroenteritis virus epidemiology. In: Siddell SG (ed) The Coronaviridae. Plenum Press, New York, pp 337–376

Forman AJ (1991) Infection of pigs with transmissible gastroenteritis virus from contaminated carcases. Aust Vet J 68:25–27

Frederick GT, Bohl EH, Cross RF (1976) Pathogenicity of an attenuated strain of transmissible gastroenteritis virus for newborn pigs. Am J Vet Res 37:165–169

Garwes DJ, Stewart F, Cartwright SF et al (1988) Differentiation of porcine coronavirus from transmissible gastroenteritis virus. Vet Rec 122:86–87

Gerber PF, Opriessnig T (2015) Detection of immunoglobulin (Ig) A antibodies against porcine epidemic diarrhea virus (PEDV) in fecal and serum samples. MethodsX 2:368–373

Gerber PF, Gong Q, Huang YW et al (2014) Detection of antibodies against porcine epidemic diarrhea virus in serum and colostrum by indirect ELISA. Vet J 202:33–36

Gong L, Li J, Zhou Q et al (2017) A new bat-HKU2-like coronavirus in swine, China, 2017. Emerg Infect Dis 23. https://doi.org/10.3201/eid2309.170915

Gonzalez JM, Gomez-Puertas P, Cavanagh D et al (2003) A comparative sequence analysis to revise the current taxonomy of the family Coronaviridae. Arch Virol 148:2207–2235

Gough PM, Jorgenson RD (1983) Identification of porcine transmissible gastroenteritis virus in house flies (Musca domestica Linneaus). Am J Vet Res 44:2078–2082

Haas B, Ahl R, Bohm R et al (1995) Inactivation of viruses in liquid manure. Rev Sci Tech 14:435–445
Haelterman EO (1962) Epidemiological studies of transmissible gastroenteritis of swine. In: United States Livestock Sanitary Association meeting
Halbur PG, Paul PS, Vaughn EM et al (1993) Experimental reproduction of pneumonia in gnotobiotic pigs with porcine respiratory coronavirus isolate AR310. J Vet Diagn Investig 5:184–188
Have P (1990) Infection with a new porcine respiratory coronavirus in Denmark: serologic differentiation from transmissible gastroenteritis virus using monoclonal antibodies. Adv Exp Med Biol 276:435–439
Hill H, Biwer J, Woods R, Wesley R (1990) Porcine respiratory coronavirus isolated from two U.S. swine herds. In: American Association of Swine Practitioners meeting
Hofmann M, Wyler R (1988) Propagation of the virus of porcine epidemic diarrhea in cell culture. J Clin Microbiol 26:2235–2239
Hofmann M, Wyler R (1990) Enzyme-linked immunosorbent assay for the detection of porcine epidemic diarrhea coronavirus antibodies in swine sera. Vet Microbiol 21:263–273
Hooper BE, Haelterman EO (1966a) Concepts of pathogenesis and passive immunity in transmissible gastroenteritis of swine. J Am Vet Med Assoc 149:1580–1586
Hooper BE, Haelterman EO (1966b) Growth of transmissible gastroenteritis virus in young pigs. Am J Vet Res 27:286–291
Hou Y, Lin CM, Yokoyama M et al (2017) Deletion of a 197-amino-acid region in the N-terminal domain of spike protein attenuates porcine epidemic diarrhea virus in piglets. J Virol 91:e00227-17
Hu H, Jung K, Vlasova AN et al (2015) Isolation and characterization of porcine deltacoronavirus from pigs with diarrhea in the United States. J Clin Microbiol 53:1537–1548
Hu H, Jung K, Vlasova AN et al (2016) Experimental infection of gnotobiotic pigs with the cell-culture-adapted porcine deltacoronavirus strain OH-FD22. Arch Virol 161:3421–3434
Hwang EK, Kim JH, Jean YH et al (1994) Current occurrence of porcine epidemic diarrhea in Korea. RDA J Agric Sci 36:587–596
Ishikawa K, Sekiguchi H, Ogino T et al (1997) Direct and rapid detection of porcine epidemic diarrhea virus by RT-PCR. J Virol Methods 69:191–195
Janetanakit T, Lumyai M, Bunpapong N et al (2016) Porcine deltacoronavirus, Thailand, 2015. Emerg Infect Dis 22:757–759
Jengarn J, Wongthida P, Wanasen N et al (2015) Genetic manipulation of porcine epidemic diarrhea virus (PEDV) recovered from a full length infectious cDNA clone. J Gen Virol 96:2206–2218
Jung K, Alekseev KP, Zhang X et al (2007) Altered pathogenesis of porcine respiratory coronavirus in pigs due to immunosuppressive effects of dexamethasone: implications for corticosteroid use in treatment of severe acute respiratory syndrome coronavirus. J Virol 81:13681–13693
Jung K, Wang Q, Scheuer KA et al (2014) Pathology of US porcine epidemic diarrhea virus strain PC21A in gnotobiotic pigs. Emerg Infect Dis 20:662–665
Jung K, Annamalai T, Lu Z et al (2015a) Comparative pathogenesis of US porcine epidemic diarrhea virus (PEDV) strain PC21A in conventional 9-day-old nursing piglets vs. 26-day-old weaned pigs. Vet Microbiol 178:31–40
Jung K, Hu H, Eyerly B et al (2015b) Pathogenicity of 2 porcine deltacoronavirus strains in gnotobiotic pigs. Emerg Infect Dis 21:650–654
Jung K, Hu H, Saif LJ (2016a) Porcine deltacoronavirus induces apoptosis in swine testicular and LLC porcine kidney cell lines in vitro but not in infected intestinal enterocytes in vivo. Vet Microbiol 182:57–63
Jung K, Hu H, Saif LJ (2016b) Porcine deltacoronavirus infection: etiology, cell culture for virus isolation and propagation, molecular epidemiology and pathogenesis. Virus Res 226:50–59
Kemeny LJ, Wiltsey VL, Riley JL (1975) Upper respiratory infection of lactating sows with transmissible gastroenteritis virus following contact exposure to infected piglets. Cornell Vet 65:352–362

Kim O, Chae C (2003) Experimental infection of piglets with a Korean strain of porcine epidemic diarrhoea virus. J Comp Pathol 129:55–60
Kim L, Chang KO, Sestak K et al (2000) Development of a reverse transcription-nested polymerase chain reaction assay for differential diagnosis of transmissible gastroenteritis virus and porcine respiratory coronavirus from feces and nasal swabs of infected pigs. J Vet Diagn Investig 12:385–388
Kim SY, Song DS, Park BK (2001) Differential detection of transmissible gastroenteritis virus and porcine epidemic diarrhea virus by duplex RT-PCR. J Vet Diagn Investig 13:516–520
Kim SH, Kim IJ, Pyo HM et al (2007) Multiplex real-time RT-PCR for the simultaneous detection and quantification of transmissible gastroenteritis virus and porcine epidemic diarrhea virus. J Virol Methods 146:172–177
Kim SH, Lee JM, Jung J et al (2015) Genetic characterization of porcine epidemic diarrhea virus in Korea from 1998 to 2013. Arch Virol 160:1055–1064
Kim YK, Cho YY, An BH et al (2016) Molecular characterization of the spike and ORF3 genes of porcine epidemic diarrhea virus in the Philippines. Arch Virol 161:1323–1328
Knuchel M, Ackermann M, Muller HK et al (1992) An ELISA for detection of antibodies against porcine epidemic diarrhoea virus (PEDV) based on the specific solubility of the viral surface glycoprotein. Vet Microbiol 32:117–134
Kodama Y, Ogata M, Simizu Y (1980) Characterization of immunoglobulin A antibody in serum of swine inoculated with transmissible gastroenteritis virus. Am J Vet Res 41:740–745
Ksiazek TG, Erdman D, Goldsmith CS et al (2003) A novel coronavirus associated with severe acute respiratory syndrome. N Engl J Med 348:1953–1966
Kuwahara H, Nunoya T, Samejima T et al (1988) Passage in piglets of a coronavirus associated with porcine epidemic diarrhea. J Jpn Vet Med Assoc 41:169–173
Kweon C, Kwon BJ, Jung TS et al (1993) Isolation of porcine epidemic diarrhea virus (PEDV) infection in Korea. Korean J Vet Res 33:249–254
Kweon CH, Kwon BJ, Kang YB et al (1994) Cell adaptation of KPEDV-9 and serological survey on porcine epidemic diarrhea virus (PEDV) infection in Korea. Korean J Vet Res 34:321–326
Kweon CH, Kwon BJ, Lee JG et al (1999) Derivation of attenuated porcine epidemic diarrhea virus (PEDV) as vaccine candidate. Vaccine 17:2546–2553
Langel SN, Paim FC, Lager KM et al (2016) Lactogenic immunity and vaccines for porcine epidemic diarrhea virus (PEDV): historical and current concepts. Virus Res 226:93–107
Lanza I, Shoup DI, Saif LJ (1995) Lactogenic immunity and milk antibody isotypes to transmissible gastroenteritis virus in sows exposed to porcine respiratory coronavirus during pregnancy. Am J Vet Res 56:739–748
Laude H, Gelfi J, Aynaud JM (1981) In vitro properties of low- and high-passaged strains of transmissible gastroenteritis coronavirus of swine. Am J Vet Res 42:447–449
Laude H, Van Reeth K, Pensaert M (1993) Porcine respiratory coronavirus: molecular features and virus-host interactions. Vet Res 24:125–150
Lee C (2015) Porcine epidemic diarrhea virus: an emerging and re-emerging epizootic swine virus. Virol J 12:193
Lee S, Park GS, Shin JH et al (2014) Full-genome sequence analysis of a variant strain of porcine epidemic diarrhea virus in South Korea. Genome Announc 2:e00753-14
Lee JH, Chung HC, Nguyen VG et al (2016a) Detection and phylogenetic analysis of porcine deltacoronavirus in Korean swine farms, 2015. Transbound Emerg Dis 63:248–252
Lee DU, Kwon T, Je SH et al (2016b) Wild boars harboring porcine epidemic diarrhea virus (PEDV) may play an important role as a PEDV reservoir. Vet Microbiol 192:90–94
Li C, Li Z, Zou Y et al (2013) Manipulation of the porcine epidemic diarrhea virus genome using targeted RNA recombination. PLoS One 8:e69997
Li G, Chen Q, Harmon KM et al (2014) Full-length genome sequence of porcine deltacoronavirus strain USA/IA/2014/8734. Genome Announc 2:e00278-14
Li Z, He W, Lan Y et al (2016) The evidence of porcine hemagglutinating encephalomyelitis virus induced nonsuppurative encephalitis as the cause of death in piglets. PeerJ 4:e2443

Lin CN, Chung WB, Chang SW et al (2014) US-like strain of porcine epidemic diarrhea virus outbreaks in Taiwan, 2013-2014. J Vet Med Sci 76:1297–1299
Lin CM, Gao X, Oka T et al (2015a) Antigenic relationships among porcine epidemic diarrhea virus and transmissible gastroenteritis virus strains. J Virol 89:3332–3342
Lin CM, Annamalai T, Liu X et al (2015b) Experimental infection of a US spike-insertion deletion porcine epidemic diarrhea virus in conventional nursing piglets and cross-protection to the original US PEDV infection. Vet Res 46:134
Lin CM, Saif LJ, Marthaler D et al (2016) Evolution, antigenicity and pathogenicity of global porcine epidemic diarrhea virus strains. Virus Res 226:20–39
Lohse L, Krog JS, Strandbygaard B et al (2017) Experimental infection of young pigs with an early European strain of porcine epidemic diarrhoea virus and a recent US strain. Transbound Emerg Dis 64:1380–1386
Lorbach JN, Wang L, Nolting JM et al (2017) Porcine hemagglutinating encephalomyelitis virus and respiratory disease in exhibition swine, Michigan, USA, 2015. Emerg Infect Dis 23:1168–1171
Ma G, Feng Y, Gao F et al (2005) Biochemical and biophysical characterization of the transmissible gastroenteritis coronavirus fusion core. Biochem Biophys Res Commun 337:1301–1307
Ma Y, Zhang Y, Liang X et al (2015) Origin, evolution, and virulence of porcine deltacoronaviruses in the United States. MBio 6:e00064
Ma Y, Zhang Y, Liang X et al (2016) Two-way antigenic cross-reactivity between porcine epidemic diarrhea virus and porcine deltacoronavirus. Vet Microbiol 186:90–96
Madson DM, Magstadt DR, Arruda PH et al (2014) Pathogenesis of porcine epidemic diarrhea virus isolate (US/Iowa/18984/2013) in 3-week-old weaned pigs. Vet Microbiol 174:60–68
Martelli P, Lavazza A, Nigrelli AD et al (2008) Epidemic of diarrhoea caused by porcine epidemic diarrhoea virus in Italy. Vet Rec 162:307–310
Marthaler D, Jiang Y, Collins J et al (2014a) Complete genome sequence of strain SDCV/USA/Illinois121/2014, a porcine deltacoronavirus from the United States. Genome Announc 2:e00218-14
Marthaler D, Raymond L, Jiang Y et al (2014b) Rapid detection, complete genome sequencing, and phylogenetic analysis of porcine deltacoronavirus. Emerg Infect Dis 20:1347–1350
Masuda T, Murakami S, Takahashi O et al (2015) New porcine epidemic diarrhoea virus variant with a large deletion in the spike gene identified in domestic pigs. Arch Virol 160:2565–2568
Masuda T, Tsuchiaka S, Ashiba T et al (2016) Development of one-step real-time reverse transcriptase-PCR-based assays for the rapid and simultaneous detection of four viruses causing porcine diarrhea. Jpn J Vet Res 64:5–14
McClurkin AW, Norman JO (1966) Studies on transmissible gastroenteritis of swine. II. Selected characteristics of a cytopathogenic virus common to five isolates from transmissible gastroenteritis. Can J Comp Med Vet Sci 30:190–198
McClurkin AW, Stark SL, Norman JO (1970) Transmissible gastroenteritis (TGE) of swine: the possible role of dogs in the epizootiology of TGE. Can J Comp Med 34:347–349
Moon HW (1978) Mechanisms in the pathogenesis of diarrhea: a review. J Am Vet Med Assoc 172:443–448
Moon HW, Kemeny LJ, Lambert G et al (1975) Age-dependent resistance to transmissible gastroenteritis of swine. III. Effects of epithelial cell kinetics on coronavirus production and on atrophy of intestinal villi. Vet Pathol 12:434–445
Moxley RA, Olson LD (1989) Clinical evaluation of transmissible gastroenteritis virus vaccines and vaccination procedures for inducing lactogenic immunity in sows. Am J Vet Res 50:111–118
Niederwerder MC, Nietfeld JC, Bai J et al (2016) Tissue localization, shedding, virus carriage, antibody response, and aerosol transmission of porcine epidemic diarrhea virus following inoculation of 4-week-old feeder pigs. J Vet Diagn Investig 28:671–678

Ogawa H, Taira O, Hirai T et al (2009) Multiplex PCR and multiplex RT-PCR for inclusive detection of major swine DNA and RNA viruses in pigs with multiple infections. J Virol Methods 160:210–214

Oh JS, Song DS, Yang JS et al (2005) Comparison of an enzyme-linked immunosorbent assay with serum neutralization test for serodiagnosis of porcine epidemic diarrhea virus infection. J Vet Sci 6:349–352

Oka T, Saif LJ, Marthaler D et al (2014) Cell culture isolation and sequence analysis of genetically diverse US porcine epidemic diarrhea virus strains including a novel strain with a large deletion in the spike gene. Vet Microbiol 173:258–269

Okda F, Liu X, Singrey A et al (2015) Development of an indirect ELISA, blocking ELISA, fluorescent microsphere immunoassay and fluorescent focus neutralization assay for serologic evaluation of exposure to North American strains of porcine epidemic diarrhea virus. BMC Vet Res 11:180

Okda F, Lawson S, Liu X et al (2016) Development of monoclonal antibodies and serological assays including indirect ELISA and fluorescent microsphere immunoassays for diagnosis of porcine deltacoronavirus. BMC Vet Res 12:95

O'Toole D, Brown I, Bridges A et al (1989) Pathogenicity of experimental infection with 'pneumotropic' porcine coronavirus. Res Vet Sci 47:23–29

Park JE, Shin HJ (2014) Porcine epidemic diarrhea virus infects and replicates in porcine alveolar macrophages. Virus Res 191:143–152

Paudel S, Park JE, Jang H et al (2014) Comparison of serum neutralization and enzyme-linked immunosorbent assay on sera from porcine epidemic diarrhea virus vaccinated pigs. Vet Q 34:218–223

PED vaccine gains conditional approval. J Am Vet Med Assoc 2014;245:267

Pensaert M (1989) Transmissible gastroenteritis virus (respiratory variant). In: Pensaert M (ed) Virus infections of porcines, vol 2. Elsevier Science Publishers, Amsterdam, pp 154–165

Pensaert M, Haelterman EO, Burnstein T (1970) Transmissible gastroenteritis of swine: virus-intestinal cell interactions. I. Immunofluorescence, histopathology and virus production in the small intestine through the course of infection. Arch Gesamte Virusforsch 31:321–334

Pensaert M, Callebaut P, Vergote J (1986) Isolation of a porcine respiratory, non-enteric coronavirus related to transmissible gastroenteritis. Vet Q 8:257–261

Pensaert M, Cox E, van Deun K et al (1993) A sero-epizootiological study of porcine respiratory coronavirus in Belgian swine. Vet Q 15:16–20

Pilchard EI (1965) Experimental transmission of transmissible gastroenteritis virus by starlings. Am J Vet Res 26:1177–1179

Pospischil A, Hess RG, Bachmann PA (1981) Light microscopy and ultrahistology of intestinal changes in pigs infected with epizootic diarrhoea virus (EVD): comparison with transmissible gastroenteritis (TGE) virus and porcine rotavirus infections. Zentralbl Veterinarmed B 28:564–577

Pritchard GC (1987) Transmissible gastroenteritis in endemically infected breeding herds of pigs in East Anglia, 1981-85. Vet Rec 120:226–230

Puranaveja S, Poolperm P, Lertwatcharasarakul P et al (2009) Chinese-like strain of porcine epidemic diarrhea virus, Thailand. Emerg Infect Dis 15:1112–1115

Ren X, Li P (2011) Development of reverse transcription loop-mediated isothermal amplification for rapid detection of porcine epidemic diarrhea virus. Virus Genes 42:229–235

Reynolds DJ, Garwes DJ, Lucey S (1980) Differentiation of canine coronavirus and porcine transmissible gastroenteritis virus by neutralization with canine, porcine, and feline sera. Vet Microbiol 5:283–290

Saeng-Chuto K, Lorsirigool A, Temeeyasen G et al (2017) Different lineage of porcine deltacoronavirus in Thailand, Vietnam and Lao PDR in 2015. Transbound Emerg Dis 64:3–10

Saif LJ (1999) Enteric viral infections of pigs and strategies for induction of mucosal immunity. Adv Vet Med 41:429–446

Saif LJ, Bohl EH (1979) Role of SIgA in passive immunity of swine to enteric viral infections. In: Ogra P, Dayton D (eds) Immunology of breast milk. Raven Press, New York, NY, pp 237–248

Saif L, Jackwood DJ (1990) Enteric virus vaccines: theoretical considerations, current status, and future approaches. In: Saif L, Theil KW (eds) Viral diarrheas of man and animals. CRC Press, Inc., Boca Raton, FL, pp 313–329

Saif LJ, Sestak K (2006) Transmissible gastroenteritis virus and porcine respiratory coronavirus. In: Straw BE, Zimmerman JJ, D'Allaire D, Taylor DJ (eds) Diseases of swine, 9th edn. Blackwell Publishing Company, Ames, IA, pp 489–516

Saif LJ, Bohl EH, Gupta RK (1972) Isolation of porcine immunoglobulins and determination of the immunoglobulin classes of transmissible gastroenteritis viral antibodies. Infect Immun 6:600–609

Saif LJ, Bohl EH, Kohler EM et al (1977) Immune electron microscopy of transmissible gastroenteritis virus and rotavirus (reovirus-like agent) of swine. Am J Vet Res 38:13–20

Saif LJ, van Cott JL, Brim TA (1994) Immunity to transmissible gastroenteritis virus and porcine respiratory coronavirus infections in swine. Vet Immunol Immunopathol 43:89–97

Sanchez CM, Izeta A, Sanchez-Morgado JM et al (1999) Targeted recombination demonstrates that the spike gene of transmissible gastroenteritis coronavirus is a determinant of its enteric tropism and virulence. J Virol 73:7607–7618

Sato T, Takeyama N, Katsumata A et al (2011) Mutations in the spike gene of porcine epidemic diarrhea virus associated with growth adaptation in vitro and attenuation of virulence in vivo. Virus Genes 43:72–78

Schumacher LL, Woodworth JC, Jones CK et al (2016) Evaluation of the minimum infectious dose of porcine epidemic diarrhea virus in virus-inoculated feed. Am J Vet Res 77:1108–1113

Sedlak K, Bartova E, Machova J (2008) Antibodies to selected viral disease agents in wild boars from the Czech Republic. J Wildl Dis 44:777–780

Sestak K, Lanza I, Park SK et al (1996) Contribution of passive immunity to porcine respiratory coronavirus to protection against transmissible gastroenteritis virus challenge exposure in suckling pigs. Am J Vet Res 57:664–671

Sestak K, Zhou Z, Shoup DI et al (1999a) Evaluation of the baculovirus-expressed S glycoprotein of transmissible gastroenteritis virus (TGEV) as antigen in a competition ELISA to differentiate porcine respiratory coronavirus from TGEV antibodies in pigs. J Vet Diagn Investig 11:205–214

Sestak K, Meister RK, Hayes JR et al (1999b) Active immunity and T-cell populations in pigs intraperitoneally inoculated with baculovirus-expressed transmissible gastroenteritis virus structural proteins. Vet Immunol Immunopathol 70:203–221

Shimizu M, Shimizu Y (1979) Lymphocyte proliferative response to viral antigen in pigs infected with transmissible gastroenteritis virus. Infect Immun 23:239–243

Shoup DI, Swayne DE, Jackwood DJ et al (1996) Immunohistochemistry of transmissible gastroenteritis virus antigens in fixed paraffin-embedded tissues. J Vet Diagn Investig 8:161–167

Simkins RA, Weilnau PA, Van Cott J et al (1993) Competition ELISA, using monoclonal antibodies to the transmissible gastroenteritis virus (TGEV) S protein, for serologic differentiation of pigs infected with TGEV or porcine respiratory coronavirus. Am J Vet Res 54:254–259

Sinha A, Gauger P, Zhang J et al (2015) PCR-based retrospective evaluation of diagnostic samples for emergence of porcine deltacoronavirus in US swine. Vet Microbiol 179:296–298

Song DS, Oh JS, Kang BK et al (2007) Oral efficacy of Vero cell attenuated porcine epidemic diarrhea virus DR13 strain. Res Vet Sci 82:134–140

Song D, Zhou X, Peng Q et al (2015) Newly emerged porcine deltacoronavirus associated with diarrhoea in swine in China: identification, prevalence and full-length genome sequence analysis. Transbound Emerg Dis 62:575–580

Stadler J, Zoels S, Fux R et al (2015) Emergence of porcine epidemic diarrhea virus in southern Germany. BMC Vet Res 11:142

Stepanek J, Mensik J, Franz J, Hornich M (1979) Epizootiology, diagnosis and prevention of viral diarrhea in piglets under intensive husbandry conditions. In: 21st World Veterinary congress, Moscow

Stevenson GW, Hoang H, Schwartz KJ et al (2013) Emergence of porcine epidemic diarrhea virus in the United States: clinical signs, lesions, and viral genomic sequences. J Vet Diagn Investig 25:649–654

Stone SS, Kemeny LJ, Woods RD et al (1977) Efficacy of isolated colostral IgA, IgG, and IgM(A) to protect neonatal pigs against the coronavirus of transmissible gastroenteritis. Am J Vet Res 38:1285–1288

Su M, Li C, Guo D et al (2016) A recombinant nucleocapsid protein-based indirect enzyme-linked immunosorbent assay to detect antibodies against porcine deltacoronavirus. J Vet Med Sci 78:601–606

Sueyoshi M, Tsuda T, Yamazaki K et al (1995) An immunohistochemical investigation of porcine epidemic diarrhoea. J Comp Pathol 113:59–67

Sun RQ, Cai RJ, Chen YQ et al (2012) Outbreak of porcine epidemic diarrhea in suckling piglets, China. Emerg Infect Dis 18:161–163

Sun D, Wang X, Wei S et al (2016) Epidemiology and vaccine of porcine epidemic diarrhea virus in China: a mini-review. J Vet Med Sci 78:355–363

Suzuki T, Murakami S, Takahashi O et al (2015) Molecular characterization of pig epidemic diarrhoea viruses isolated in Japan from 2013 to 2014. Infect Genet Evol 36:363–368

Suzuki T, Shibahara T, Yamaguchi R et al (2016) Pig epidemic diarrhoea virus S gene variant with a large deletion non-lethal to colostrum-deprived newborn piglets. J Gen Virol 97:1823–1828

Takahashi K, Okada K, Ohshima K (1983) An outbreak of swine diarrhea of a new-type associated with coronavirus-like particles in Japan. Nihon Juigaku Zasshi 45:829–832

Thachil A, Gerber PF, Xiao CT et al (2015) Development and application of an ELISA for the detection of porcine deltacoronavirus IgG antibodies. PLoS One 10:e0124363

Thomas JT, Chen Q, Gauger PC et al (2015) Effect of porcine epidemic diarrhea virus infectious doses on infection outcomes in naive conventional neonatal and weaned pigs. PLoS One 10:e0139266

Underdahl NR, Mebus CA, Torres-Medina A (1975) Recovery of transmissible gastroenteritis virus from chronically infected experimental pigs. Am J Vet Res 36:1473–1476

van Nieuwstadt AP, Zetstra T (1991) Use of two enzyme-linked immunosorbent assays to monitor antibody responses in swine with experimentally induced infection with porcine epidemic diarrhea virus. Am J Vet Res 52:1044–1050

van Nieuwstadt AP, Cornelissen JB, Zetstra T (1988) Comparison of two methods for detection of transmissible gastroenteritis virus in feces of pigs with experimentally induced infection. Am J Vet Res 49:1836–1843

van Nieuwstadt AP, Zetstra T, Boonstra J (1989) Infection with porcine respiratory coronavirus does not fully protect pigs against intestinal transmissible gastroenteritis virus. Vet Rec 125:58–60

VanCott JL, Brim TA, Simkins RA et al (1993) Isotype-specific antibody-secreting cells to transmissible gastroenteritis virus and porcine respiratory coronavirus in gut- and bronchus-associated lymphoid tissues of suckling pigs. J Immunol 150:3990–4000

VanCott JL, Brim TA, Lunney JK et al (1994) Contribution of antibody-secreting cells induced in mucosal lymphoid tissues of pigs inoculated with respiratory or enteric strains of coronavirus to immunity against enteric coronavirus challenge. J Immunol 152:3980–3990

Vlasova AN, Marthaler D, Wang Q et al (2014) Distinct characteristics and complex evolution of PEDV strains, North America, May 2013–February 2014. Emerg Infect Dis 20:1620–1628

Vui DT, Tung N, Inui K et al (2014) Complete genome sequence of porcine epidemic diarrhea virus in Vietnam. Genome Announc 2

Wang L, Byrum B, Zhang Y (2014a) New variant of porcine epidemic diarrhea virus, United States, 2014. Emerg Infect Dis 20:917–919

Wang L, Byrum B, Zhang Y (2014b) Detection and genetic characterization of deltacoronavirus in pigs, Ohio, USA, 2014. Emerg Infect Dis 20:1227–1230
Wang L, Byrum B, Zhang Y (2014c) Porcine coronavirus HKU15 detected in 9 US states, 2014. Emerg Infect Dis 20:1594–1595
Wang D, Fang L, Shi Y et al (2015) Porcine epidemic diarrhea virus 3C-like protease regulates its interferon antagonism by cleaving NEMO. J Virol 90:2090–2101
Wang L, Hayes J, Sarver C et al (2016a) Porcine deltacoronavirus: histological lesions and genetic characterization. Arch Virol 161:171–175
Wang D, Fang L, Xiao S (2016b) Porcine epidemic diarrhea in China. Virus Res 226:7–13
Wesley RD, Woods RD (1996) Induction of protective immunity against transmissible gastroenteritis virus after exposure of neonatal pigs to porcine respiratory coronavirus. Am J Vet Res 57:157–162
Wesley RD, Woods RD, Hill HT et al (1990) Evidence for a porcine respiratory coronavirus, antigenically similar to transmissible gastroenteritis virus, in the United States. J Vet Diagn Investig 2:312–317
Wesley RD, Woods RD, McKean JD et al (1997) Prevalence of coronavirus antibodies in Iowa swine. Can J Vet Res 61:305–308
Witte KH (1971) Isolation of the virus of transmissible gastroenteritis (TGE) from naturally infected piglets in cell culture. Zentralbl Veterinarmed B 18:770–778
Woo PC, Lau SK, Lam CS et al (2012) Discovery of seven novel mammalian and avian coronaviruses in the genus deltacoronavirus supports bat coronaviruses as the gene source of alphacoronavirus and betacoronavirus and avian coronaviruses as the gene source of gammacoronavirus and deltacoronavirus. J Virol 86:3995–4008
Xuan H, Xing D, Wang D et al (1984) Culture of porcine epidemic diarrhea virus by using pig intestinal monolayer cell. J People's Liber Army Vet Univ 4:202–208
Yaeger M, Funk N, Hoffman L (2002) A survey of agents associated with neonatal diarrhea in Iowa swine including Clostridium difficile and porcine reproductive and respiratory syndrome virus. J Vet Diagn Investig 14:281–287
Yu X, Shi L, Lv X et al (2015) Development of a real-time reverse transcription loop-mediated isothermal amplification method for the rapid detection of porcine epidemic diarrhea virus. Virol J 12:76
Zhang J (2016) Porcine deltacoronavirus: overview of infection dynamics, diagnostic methods, prevalence and genetic evolution. Virus Res 226:71–84
Zhang Q, Shi K, Yoo D (2016a) Suppression of type I interferon production by porcine epidemic diarrhea virus and degradation of CREB-binding protein by nsp1. Virology 489:252–268
Zhang J, Tsai YL, Lee PY et al (2016b) Evaluation of two singleplex reverse transcription-insulated isothermal PCR tests and a duplex real-time RT-PCR test for the detection of porcine epidemic diarrhea virus and porcine deltacoronavirus. J Virol Methods 234:34–42
Zhao S, Gao Q, Qin T et al (2014) Effects of virulent and attenuated transmissible gastroenteritis virus on the ability of porcine dendritic cells to sample and present antigen. Vet Microbiol 171:74–86

Chapter 5
Torque Teno Virus

Souvik Ghosh, Alyssa Kleymann, Yashpal Singh Malik, and Nobumichi Kobayashi

Abstract Torque teno viruses (TTVs), family *Anelloviridae*, are small, non-enveloped viruses with a circular, negative-sense, single-stranded DNA genome. Although TTVs have been reported in a wide variety of animals, porcine TTVs (or Torque teno sus virus, TTSuVs) have been more studied compared to those from other non-human host species, especially because of their association with economically important diseases of pigs, notably porcine circovirus infections. TTSuVs are highly prevalent and widely distributed in both healthy and diseased pigs and have been found to contaminate cell lines, human drugs, veterinary products and retail pork products. Although the current belief is that TTSuVs by itself cannot induce severe disease in pigs, they have been proposed to serve as complicating agents/co-factors in triggering or exacerbating certain porcine diseases, especially porcine circovirus-2 systemic disease (PCV2-SD). However, there are contradicting reports on the role of TTSuV in PCV2-SD and other porcine viral infections. This book chapter is a comprehensive and updated review of TTSuVs. Various aspects of TTSuVs, such as virus diversity and classification, epidemiology, pathogenesis, diagnostics and immunisation, have been discussed.

Keywords *Anelloviridae* · Disease association · Pigs · Porcine circovirus-associated disease · Porcine circovirus-2 systemic disease · Torque teno virus · Torque teno sus virus

---

S. Ghosh (✉)
Department of Biomedical Sciences, One Health Center for Zoonoses and Tropical Veterinary Medicine, Ross University School of Veterinary Medicine, Basseterre, St. Kitts and Nevis, West Indies
e-mail: sghosh@rossu.edu

A. Kleymann
Department of Biomedical Sciences, Ross University School of Veterinary Medicine, Basseterre, St. Kitts and Nevis, West Indies

Y. S. Malik
Division of Biological Standardization, ICAR-Indian Veterinary Research Institute, Izatnagar, Uttar Pradesh, India

N. Kobayashi
Department of Hygiene, Sapporo Medical University, Sapporo, Hokkaido, Japan

Y. S. Malik et al. (eds.), *Emerging and Transboundary Animal Viruses*, Livestock Diseases and Management, https://doi.org/10.1007/978-981-15-0402-0_5

Abbreviations

ICTV	International Committee on Taxonomy of Viruses
ORF	Open reading frame
PCV2	Porcine circovirus-2
PCV2-SD	PCV2 systemic disease
PCVAD	Porcine circovirus-associated disease
PRRSV	Porcine reproductive and respiratory syndrome virus
ssDNA	Single-stranded DNA
TTSuV	Torque teno sus virus
TTSuV1a	*Torque teno sus virus 1a*
TTSuV1b	*Torque teno sus virus 1b*
TTSuVk2a	*Torque teno sus virus k2a*
TTSuVk2b	*Torque teno sus virus k2b*
TTV	Toque teno virus
UTR	Untranslated region

5.1 Preamble

Torque teno virus (TTV) was first reported in the serum of an old human patient suffering from post-blood transfusion non-A to G hepatitis (Nishizawa et al. 1997). Since the earlier human isolates were found to display structural and molecular similarities with circoviruses, TTVs were initially assigned to the family *Circoviridae* (Miyata et al. 1999). Eventually, based on differences in genomic features and nucleotide sequence identities between TTVs and circoviruses, the International Committee on Taxonomy of Viruses (ICTV) classified TTVs into a floating new genus '*Anellovirus*' in 2005, and thereafter a new family '*Anelloviridae*' in 2009 (Biagini et al. 2005, 2012). Although the first reported virus was provisionally named TTV after the initials of the sampled patient (T.T.), the acronym TTV (Torque teno virus), derived from the Latin words 'torque' (necklace) and 'tenuis' (narrow), actually refers to the genomic organisation of the virus (Biagini et al. 2012; Nishizawa et al. 1997).

TTVs have been detected in a wide variety of other host species (badgers, bats, camels, cats, cattle, dogs, horses, non-human primates, opossums, pigs, pigeons, pine martens, poultry, rodents, sea lions, sea turtles, tupaias and wild boars) (de Souza et al. 2018; Li et al. 2015; Manzin et al. 2015; Zhang et al. 2017) (Table 5.1). Although TTVs have been related to various disease conditions and co-infections, especially in humans and pigs, they are ubiquitous in healthy hosts, raising doubts on the pathogenic potential of these viruses (Kekarainen and Segalés 2012; Manzin et al. 2015; Meng 2012).

Table 5.1 Classification of the family *Anelloviridae*

Genus[a]	Virus species Number[a,b]; *name/s*[a,b] (number of unclassified viruses[c])	Host/s[a,b,c]
Alphatorquevirus	30; *Torque teno virus 1–29, Opossum torque teno virus* (5)	African green monkey, chimpanzee, humans, opossum
Betatorquevirus	12; *Torque teno mini virus 1–12* (4)	African green monkey, chimpanzee, humans
Deltatorquevirus	2; *Torque teno tupaia virus, Torque teno calomys tener virus*	Common tree shrew (tupaia), rodent
Epsilontorquevirus	1; *Torque teno tamarin virus*	Tamarin
Etatorquevirus	2; *Torque teno felis virus, Torque teno felis virus 2*	Cats
Gammatorquevirus	15; *Torque teno midi virus 1–15*	African green monkey, chimpanzee, humans
Gyrovirus	1; *Chicken anaemia virus* (8)	Ashy storm petrel, chicken, humans
Iotatorquevirus	2; *Torque teno sus virus 1a, Torque teno sus virus 1b*	Pigs
Kappatorquevirus	2; *Torque teno sus virus k2a, Torque teno sus virus k2b*	Pigs
Lambdatorquevirus	6; *Torque teno seal virus 1–3, 8 and 9, Torque teno zalophus virus 1*	Seals, sea lions
Mutorquevirus	1; *Torque teno equus virus 1*	Horses
Nutorquevirus	2; *Torque teno seal virus 4 and 5*	Seals
Thetatorquevirus	1; *Torque teno canis virus*	Dogs
Zetatorquevirus	1; *Torque teno douroucouli virus*	Night monkeys (douroucouli)
Proposed genus[b]		
Omegatorquevirus	6; *Rodent Torque teno virus 3–8*	Rodent
Sigmatorquevirus	3; *Torque teno carollia perspicillata virus, Torque teno desmodus rotundus virus, Torque teno didelphis albiventris virus*	Bats, opossum

[a]Based on ICTV, https://talk.ictvonline.org/ictv-reports/ictv_9th_report/ssdna-viruses-2011/w/ssdna_viruses/139/anelloviridae (accessed 16 April 2019)
[b]Based on de Souza et al. (2018)
[c]Based on https://www.ncbi.nlm.nih.gov/genomes/GenomesGroup.cgi?taxid=687329 (accessed 16 April 2019)

Since TTVs have been associated with economically important diseases of pigs (Ghosh et al. 2018; Kekarainen and Segalés 2012; Meng 2012), this book chapter primarily focuses on porcine Torque teno viruses (currently known as Torque teno sus viruses, TTSuVs) among the various animal TTVs.

5.2 Virus Taxonomy

In 2009, the ICTV assigned the TTVs to a new family of single-stranded DNA (ssDNA) viruses, the '*Anelloviridae*' ('anello', Latin for ring) (Biagini et al. 2012). The family *Anelloviridae* has a wide host range and includes 14 genera (*Alphatorquevirus, Betatorquevirus, Deltatorquevirus, Epsilontorquevirus, Etatorquevirus, Gammatorquevirus, Gyrovirus, Iotatorquevirus, Kappatorquevirus, Lambdatorquevirus, Mutorquevirus, Nutorquevirus, Thetatorquevirus* and *Zetatorquevirus*) and two newly proposed genera (*Omegatorquevirus* and *Sigmatorquevirus*) (Table 5.1) (de Souza et al. 2018; ICTV 2018).

Torque teno sus viruses are grouped into two genera, genus *Iotatorquevirus* (species *Torque teno sus virus 1a*, TTSuV1a and *Torque teno sus virus 1b*, TTSuV1b) and genus *Kappatorquevirus* (species *Torque teno sus virus k2a*, TTSuVk2a and *Torque teno sus virus k2b*, TTSuVk2b) (Cornelissen-Keijsers et al. 2012; ICTV 2018) (Table 5.1).

5.3 Structure, Genome Organisation and Replication

Torque teno virions are usually ~30–32 nm in diameter and non-enveloped, and possess an icosahedral capsid (Biagini et al. 2012; Kekarainen and Segalés 2012). The viral genome consists of a circular, negative-sense, ssDNA molecule, ranging from ~2000 to ~3900 bp in size depending on the host species (approximately 2800 bp in TTSuVs) (Biagini et al. 2012; Cornelissen-Keijsers et al. 2012; Cortey et al. 2011; Okamoto et al. 2002). All TTVs, including TTSuVs, preserve a short stretch of highly conserved sequence in the untranslated region (UTR) of the viral genome (Biagini et al. 2012; Kekarainen and Segalés 2012).

At least three open reading frames (ORFs) have been deduced from the nucleotide sequences of TTSuV strains (Cornelissen-Keijsers et al. 2012; Cortey et al. 2011; Kekarainen and Segalés 2012). Open reading frame 1 is believed to encode the putative viral capsid protein, while ORF2 may encode a protein with characteristics of tyrosine phosphatase and/or has also been associated with suppression of NF-κb pathways (Biagini et al. 2012; Kekarainen and Segalés 2012; Zheng et al. 2007). Open reading frame 3 is believed to be generated by splicing and encodes a protein with unknown function (Cortey et al. 2011; Kekarainen and Segalés 2012). Alternative splicing patterns and generation of different protein isoforms have been reported for TTSuVs (Martínez-Guinó et al. 2011).

The lack of a proper cell culture propagation system has made it difficult to study the major aspects of the TTV replication cycle (Meng 2012). However, the detection of TTVs in various tissues, secretions and excretions, and transfection studies suggest virus replication in different cell types (Biagini et al. 2012; Manzin et al. 2015; Meng 2012; Nieto et al. 2013). Based on the observation that the deduced amino acid sequence of TTV ORF1 contains conserved motifs similar to those seen in Rep proteins of other animal and plant viruses with circular ssDNA genomes, a rolling circle mechanism of replication has been proposed for TTVs (Biagini et al. 2012).

5.4 Torque Teno Sus Virus

Although TTVs have been reported in a wide variety of animals, porcine TTVs (or TTSuVs) have been more studied compared to those from other non-human host species, especially because of their association with economically important diseases of pigs, notably porcine circovirus infections (Kekarainen and Segalés 2012; Manzin et al. 2015).

5.4.1 Virus Diversity

The nucleotide substitution rates in TTSuVs are higher than those observed in typical DNA viruses and comparable to those of RNA viruses (Cortey et al. 2011). Given the apparent lack of selection pressures (vaccination and removal of diseased animals), natural selection and drift are believed to be the main driving forces behind the evolution of TTSuVs (Cortey et al. 2012). Based on nucleotide sequence identities and phylogenetic analysis, the ICTV has classified TTSuVs into at least two genera and four species (Cornelissen-Keijsers et al. 2012; ICTV 2018). Virus species TTSuV1a and TTSuV1b belong to genus *Iotatorquevirus*, while species TTSuVk2a and TTSuVk2b have been placed under genus *Kappatorquevirus*. Phylogenetically, an additional TTSuV1 species (TTSuV1c) has been proposed in a recent study (Ramos et al. 2018).

Nucleotide sequence identities of around 50% have been observed between TTSuV1 and TTSuV2 (Kekarainen and Segalés 2012). Higher intraspecies variation has been observed among TTSuV1 strains (nucleotide sequence divergence of >30%) compared to TTSuV2 strains (<15%) (Cortey et al. 2011). Global trade of animals has been implicated in altering and eroding the genetic composition of TTSuV populations that may have been circulating in pig herds since their domestication (Cortey et al. 2012).

5.4.2 Epidemiology

TTSuV1 and TTSuVk2a are highly prevalent and widely distributed in both healthy and diseased pigs, with prevalence rates ranging from 16.8% to 100% in different countries across the world (Manzin et al. 2015). On the other hand, limited information is available on TTSuVk2b so far. A surveillance study on pig sera from 17 countries revealed TTSuVk2b prevalence rates of 0% to 100%, while two subsequent studies (one from Italy and the other from the USA) reported overall TTSuVk2b prevalence rates of 11.5% and 24.7%, respectively (Blois et al. 2014; Cornelissen-Keijsers et al. 2012; Rogers et al. 2017). Co-infection with strains belonging to different TTSuV species have been frequently reported in pigs worldwide (Cornelissen-Keijsers et al. 2012; Ghosh et al. 2018; Kekarainen and Segalés 2012; Ramos et al. 2018).

Although the major route of transmission of anelloviruses appears to be faeco-oral, TTSuV is believed to be transmitted by both horizontal and vertical routes, as viral DNA has been detected in porcine foetal blood and tissues, colostrum and semen (Aramouni et al. 2010; Kekarainen et al. 2007; Martínez-Guinó et al. 2009, 2010; Sibila et al. 2009b; Tshering et al. 2012). TTSuV1 DNA has been widely detected in sera of humans, domestic animals and wildlife, and seroconversion to TTSuV1 has been reported in cattle, horses and ovines (Singh and Ramamoorthy 2018; Ssemadaali et al. 2016). Moreover, TTSuVs have been shown to contaminate certain cell lines, human drugs, retail pork products and veterinary vaccines (Leblanc et al. 2014; Meng 2012; Monini et al. 2016). These observations point towards an increased risk of widespread transmission of TTSuVs in pigs and heterologous host species including zoonosis.

There are contradicting reports on the frequency of TTSuV infection and viral load in relation to different porcine age groups. While a few studies have proposed no association between frequency of detection, or viral loads and the age of the animal, other research groups have provided evidence for increasing prevalence with age, and higher prevalence of at least one genus of TTSuV in nursery and young pigs (Blois et al. 2014; de Castro et al. 2015; de Menezes Cruz et al. 2016; Ramos et al. 2018; Teixeira et al. 2015; Xiao et al. 2012). Persistent infection and viremia with TTSuV are common in pigs (Aramouni et al. 2010; Kekarainen et al. 2007; Kekarainen and Segalés 2012; Martínez-Guinó et al. 2009, 2010; Sibila et al. 2009a, b).

5.4.3 Pathogenesis and Clinical Disease

The pathogenicity of TTSuVs remains to be clearly elucidated (Meng 2012; Ramos et al. 2018). Since the virus is widespread and persistent in healthy pigs, the current belief is that TTSuVs by itself cannot induce severe disease (Kekarainen and Segalés 2012; Manzin et al. 2015). Furthermore, a few researchers have even raised the possibility of TTVs being a part of the host commensal microbiota (Griffiths 1999; Manzin et al. 2015). TTSuVs gained attention because of their association with economically important porcine diseases, especially those caused by circoviruses (Kekarainen and Segalés 2012).

Porcine circovirus-2 (PCV2) is the essential etiological agent of PCV-2 systemic disease (PCV2-SD, formerly known as a post-weaning multisystemic wasting syndrome) as well as other disease conditions in pigs, collectively known as the porcine circovirus-associated disease (PCVAD) (Segalés et al. 2013). PCV2-SD is considered as one of the most economically important diseases for the pig industry (Meng 2012; Segalés et al. 2013).

The prevalence rates and viral loads of TTSuVk2a were shown to be significantly higher, and anti-TTSuV2 antibody titres lower in PCV2-SD-affected pigs compared to healthy pigs (Aramouni et al. 2011; Huang et al. 2011; Nieto et al. 2013; Kekarainen et al. 2006). On the other hand, the TTSuV1 viral loads were found to be similar in all the study groups (Kekarainen et al. 2006). However, experimental inoculation of TTSuV1 and PCV2 induced PCV2-SD in gnotobiotic pigs (Ellis et al. 2008). In another study, higher TTSuVk2b loads were observed in the sera of PCV2-SD pigs than in healthy animals (Cornelissen-Keijsers et al. 2012). Both TTSuV1 and TTSuV2 species have been suggested to attribute to the development of PCV-2-associated lymphoid lesions by alternating the host immune system (Lee et al. 2015).

In addition to PCV2, the association between TTSuV and other porcine pathogens (classical swine fever, hepatitis E, *M. hyopneumoniae*, porcine reproductive and respiratory syndrome virus (PRRSV) and porcine parvovirus) has been speculated (Opriessnig and Halbur 2012; Pérez et al. 2011; Savic et al. 2010; Zhang et al. 2014b).

Based on these observations, it has been proposed that TTSuVs may act as complicating agents/co-factors in triggering or exacerbating certain porcine diseases (Kekarainen and Segalés 2012; Meng 2012; Zhang et al. 2012). On the other hand, there are contradicting studies that did not find any correlation between TTSuVs and PCV2-SD-affected pigs, raising doubts on the role of TTSuVs in PCVAD (Lee et al. 2010; Ramos et al. 2018; Rogers et al. 2017; Teixeira et al. 2015; Vargas-Ruiz et al. 2017).

5.4.4 Diagnosis

TTSuVs have been widely detected in both healthy and diseased animals, and therefore it may not be possible to perform preliminary diagnosis based on clinical symptoms, if any. PCR and qPCR assays that can detect and differentiate TTSuV species have been successfully employed in most screening studies, with the latter method offering the additional advantage of estimating viral DNA loads in various samples including retail pork products (Ghosh et al. 2018; Cornelissen-Keijsers et al. 2012; Huang et al. 2010; Leblanc et al. 2014; Monini et al. 2016; Ramos et al. 2018; Rogers et al. 2017; Segalés et al. 2009; Teixeira et al. 2015; Vargas-Ruiz et al. 2018). The PCR/qPCR screening assays are primarily based on the UTR, as this region appears to be conserved across TTSuV genomes, with, however, enough diversity to distinguish viral species (Huang et al. 2010; Segalés et al. 2009). Since TTSuVs are frequently reported in co-infections with other pathogens, a multiplex qPCR assay that can simultaneously detect TTSuV species and other porcine viruses has been developed (Pérez et al. 2012).

Other assays that have been used to detect and/or differentiate TTSuV species include metagenomics, ELISAs, Western blots, a fluorescent microbead-based 4-plex immunoassay (for simultaneous detection of IgG antibodies against TTSuV1,

TTSuV2, PRRSV-1 and PRRSV-2) and in situ hybridisation using multi-strained pooled probes (Huang et al. 2011; Giménez-Lirola et al. 2014; Lee et al. 2014; Nieto et al. 2015; Zhang et al. 2014a).

5.4.5 Immunisation

Currently, there is no licensed vaccine available against TTSuV. In one study, pigs immunised with a combination of DNA (at 3 and 5 weeks of age) and protein (at 7 weeks of age) vaccines against TTSuVk2a developed specific antibodies and showed a significant reduction of TTSuVk2a viremia during natural infection (Jiménez-Melsió et al. 2015). However, DNA loads of the other TTSuV species were not affected by the vaccination strategy.

Acknowledgements Resources for writing this book chapter were funded by the One Health Center for Zoonoses and Tropical Veterinary Medicine, Ross University School of Veterinary Medicine, St. Kitts.

Conflict of interest: None.

References

Aramouni M, Segalés J, Cortey M, Kekarainen T (2010) Age-related tissue distribution of swine Torque teno sus virus 1 and 2. Vet Microbiol 146(3–4):350–353. https://doi.org/10.1016/j.vetmic.2010.05.036

Aramouni M, Segalés J, Sibila M, Martin-Valls GE, Nieto D, Kekarainen T (2011) Torque teno sus virus 1 and 2 viral loads in postweaning multisystemic wasting syndrome (PMWS) and porcine dermatitis and nephropathy syndrome (PDNS) affected pigs. Vet Microbiol 153(3–4):377–381. https://doi.org/10.1016/j.vetmic.2011.05.046

Biagini P, Todd D, Bendinelli M, Hino S, Mankertz A, Mishiro S, Niel C, Okamoto H, Raidal S, Ritchie BW, Teo GC (2005) Anellovirus. In: Fauquet CM, Mayo MA, Maniloff J, Desselberger U, Ball LA (eds) Virus taxonomy: eighth report of the international committee on taxonomy of viruses. Elsevier/Academic Press, London, pp 335–341

Biagini P, Bendinelli M, Hino S, Kakkola L, Mankertz A, Niel C, Okamoto H, Raidal S, Teo CG, Todd D (2012) Family Anelloviridae. In: King AMQ, Adams MJ, Carstens EB, Lefkowitz EJ (eds) Virus taxonomy: classification and nomenclature of viruses: ninth report of the international committee on taxonomy of viruses. Academic Press, London, pp 331–341

Blois S, Mallus F, Liciardi M, Pilo C, Camboni T, Macera L, Maggi F, Manzin A (2014) High prevalence of co-infection with multiple Torque teno sus virus species in Italian pig herds. PLoS One 9(11):e113720. https://doi.org/10.1371/journal.pone.0113720

Cornelissen-Keijsers V, Jiménez-Melsió A, Sonnemans D, Cortey M, Segalés J, van den Born E, Kekarainen T (2012) Discovery of a novel Torque teno sus virus species: genetic characterization, epidemiological assessment and disease association. J Gen Virol 93(Pt 12):2682–2691. https://doi.org/10.1099/vir.0.045518-0

Cortey M, Macera L, Segalés J, Kekarainen T (2011) Genetic variability and phylogeny of Torque teno sus virus 1 (TTSuV1) and 2 (TTSuV2) based on complete genomes. Vet Microbiol 148(2–4):125–131. https://doi.org/10.1016/j.vetmic.2010.08.013

Cortey M, Pileri E, Segalés J, Kekarainen T (2012) Globalisation and global trade influence molecular viral population genetics of Torque teno sus viruses 1 and 2 in pigs. Vet Microbiol 156(1–2):81–87. https://doi.org/10.1016/j.vetmic.2011.10.026

de Castro AMMG, Baldin CM, Favero CM, Gerber PF, Cipullo RI, Richtzenhain LJ (2015) Torque teno sus virus 1 and 2 viral loads in faeces of porcine circovirus 2-positive pigs. Acta Vet Brno 84:91–95

de Menezes Cruz AC, Silveira RL, Baez CF, Varella RB, de Castro TX (2016) Clinical aspects and weight gain reduction in swine infected with porcine circovirus type 2 and Torque teno sus virus in Brazil. Vet Microbiol 195:154–157. https://doi.org/10.1016/j.vetmic.2016.09.012

de Souza WM, Fumagalli MJ, de Araujo J, Sabino-Santos G Jr, Maia FGM, Romeiro MF, Modha S, Nardi MS, Queiroz LH, Durigon EL, Nunes MRT, Murcia PR, Figueiredo LTM (2018) Discovery of novel anelloviruses in small mammals expands the host range and diversity of the Anelloviridae. Virology 514:9–17. https://doi.org/10.1016/j.virol.2017.11.001

Ellis JA, Allan G, Krakowka S (2008) Effect of coinfection with genogroup 1 porcine Torque teno virus on porcine circovirus type 2-associated postweaning multisystemic wasting syndrome in gnotobiotic pigs. Am J Vet Res 69(12):1608–1614. https://doi.org/10.2460/ajvr.69.12.1608

Ghosh S, Soto E, Illanes O, Navarro R, Aung MS, Malik YS, Kobayashi N, Fuentealba C (2018) High detection rates of Torque teno sus virus in co-infection with important viral pathogens in porcine kidneys on St. Kitts Island, Lesser Antilles. Transbound Emerg Dis 65(5):1175–1181. https://doi.org/10.1111/tbed.12960

Giménez-Lirola LG, Gerber PF, Rowland RR, Halbur PG, Huang YW, Meng XJ, Opriessnig T (2014) Development and validation of a 4-plex antibody assay for simultaneous detection of IgG antibodies against Torque teno sus virus 1 (TTSuV1), TTSuV2, and porcine reproductive and respiratory syndrome virus types 1 and 2. Res Vet Sci 96(3):543–550. https://doi.org/10.1016/j.rvsc.2014.02.014

Griffiths P (1999) Time to consider the concept of a commensal virus? Rev Med Virol 9(2):73–74

Huang YW, Dryman BA, Harrall KK, Vaughn EM, Roof MB, Meng XJ (2010) Development of SYBR green-based real-time PCR and duplex nested PCR assays for quantitation and differential detection of species- or type-specific porcine Torque teno viruses. J Virol Methods 170(1–2):140–146. https://doi.org/10.1016/j.jviromet.2010.09.018

Huang YW, Harrall KK, Dryman BA, Beach NM, Kenney SP, Opriessnig T, Vaughn EM, Roof MB, Meng XJ (2011) Expression of the putative ORF1 capsid protein of Torque teno sus virus 2 (TTSuV2) and development of Western blot and ELISA serodiagnostic assays: correlation between TTSuV2 viral load and IgG antibody level in pigs. Virus Res 158(1–2):79–88. https://doi.org/10.1016/j.virusres.2011.03.013

ICTV (2018) Family Anelloviridae, Virus Taxonomy: 2018b Release. https://talk.ictvonline.org/ictv-reports/ictv_9th_report/ssdna-viruses-2011/w/ssdna_viruses/139/anelloviridae. Accessed 12 Apr 2019

Jiménez-Melsió A, Rodriguez F, Darji A, Segalés J, Cornelissen-Keijsers V, van den Born E, Kekarainen T (2015) Vaccination of pigs reduces Torque teno sus virus viremia during natural infection. Vaccine 33(30):3497–3503. https://doi.org/10.1016/j.vaccine.2015.05.064

Kekarainen T, Segalés J (2012) Torque teno sus virus in pigs: an emerging pathogen? Transbound Emerg Dis 59(Suppl 1):103–108. https://doi.org/10.1111/j.1865-1682.2011.01289.x.

Kekarainen T, Sibila M, Segalés J (2006) Prevalence of swine Torque teno virus in post-weaning multisystemic wasting syndrome (PMWS)-affected and non-PMWS-affected pigs in Spain. J Gen Virol 87(Pt 4):833–837

Kekarainen T, López-Soria S, Segalés J (2007) Detection of swine Torque teno virus genogroups 1 and 2 in boar sera and semen. Theriogenology 68(7):966–971

Leblanc D, Houde A, Gagné MJ, Plante D, Bellon-Gagnon P, Jones TH, Muehlhauser V, Wilhelm B, Avery B, Janecko N, Brassard J (2014) Presence, viral load and characterization of Torque teno sus viruses in liver and pork chop samples at retail. Int J Food Microbiol 178:60–64. https://doi.org/10.1016/j.ijfoodmicro.2014.03.005

Lee SS, Sunyoung S, Jung H, Shin J, Lyoo YS (2010) Quantitative detection of porcine Torque teno virus in porcine circovirus-2-negative and porcine circovirus-associated disease-affected pigs. J Vet Diagn Investig 22(2):261–264

Lee Y, Lin CM, Jeng CR, Pang VF (2014) Detection of Torque teno sus virus 1 and 2 in porcine tissues by in situ hybridization using multi-strained pooled probes. Vet Microbiol 172(3–4):390–399. https://doi.org/10.1016/j.vetmic.2014.06.002

Lee Y, Lin CM, Jeng CR, Chang HW, Chang CC, Pang VF (2015) The pathogenic role of Torque teno sus virus 1 and 2 and their correlations with various viral pathogens and host immunocytes in wasting pigs. Vet Microbiol 180(3–4):186–195. https://doi.org/10.1016/j.vetmic.2015.08.027

Li L, Giannitti F, Low J, Keyes C, Ullmann LS, Deng X, Aleman M, Pesavento PA, Pusterla N, Delwart E (2015) Exploring the virome of diseased horses. J Gen Virol 96(9):2721–2733. https://doi.org/10.1099/vir.0.000199

Manzin A, Mallus F, Macera L, Maggi F, Blois S (2015) Global impact of Torque teno virus infection in wild and domesticated animals. J Infect Dev Ctries 9(6):562–570. https://doi.org/10.3855/jidc.6912

Martínez-Guinó L, Kekarainen T, Segalés J (2009) Evidence of Torque teno virus (TTV) vertical transmission in swine. Theriogenology 71(9):1390–1395. https://doi.org/10.1016/j.theriogenology.2009.01.010

Martínez-Guinó L, Kekarainen T, Maldonado J, Aramouni M, Llorens A, Segalés J (2010) Torque teno sus virus (TTV) detection in aborted and slaughterhouse collected foetuses. Theriogenology 74(2):277–281. https://doi.org/10.1016/j.theriogenology.2010.02.011

Martínez-Guinó L, Ballester M, Segalés J, Kekarainen T (2011) Expression profile and subcellular localization of Torque teno sus virus proteins. J Gen Virol 92(Pt 10):2446–2457. https://doi.org/10.1099/vir.0.033134-0

Meng XJ (2012) Emerging and re-emerging swine viruses. Transbound Emerg Dis 59(Suppl 1):85–102. https://doi.org/10.1111/j.1865-1682.2011.01291.x

Miyata H, Tsunoda H, Kazi A, Yamada A, Khan MA, Murakami J, Kamahora T, Shiraki K, Hino S (1999) Identification of a novel GC-rich 113-nucleotide region to complete the circular, single-stranded DNA genome of TT virus, the first human circovirus. J Virol 73(5):3582–3586

Monini M, Vignolo E, Ianiro G, Ostanello F, Ruggeri FM, Di Bartolo I (2016) Detection of Torque teno sus virus in pork bile and liver sausages. Food Environ Virol 8(4):283–288

Nieto D, Kekarainen T, Aramouni M, Segalés J (2013) Torque teno sus virus 1 and 2 distribution in tissues of porcine circovirus type 2-systemic disease affected and age-matched healthy pigs. Vet Microbiol 163(3–4):364–367. https://doi.org/10.1016/j.vetmic.2013.01.005

Nieto D, Martínez-Guinó L, Jiménez-Melsió A, Segalés J, Kekarainen T (2015) Development of an indirect ELISA assay for the detection of IgG antibodies against the ORF1 of Torque teno sus viruses 1 and 2 in conventional pigs. Vet Microbiol 180(1–2):22–27. https://doi.org/10.1016/j.vetmic.2015.08.023

Nishizawa T, Okamoto H, Konishi K, Yoshizawa H, Miyakawa Y, Mayumi M (1997) A novel DNA virus (TTV) associated with elevated transaminase levels in posttransfusion hepatitis of unknown etiology. Biochem Biophys Res Commun 241(1):92–97

Okamoto H, Takahashi M, Nishizawa T, Tawara A, Fukai K, Muramatsu U, Naito Y, Yoshikawa A (2002) Genomic characterization of TT viruses (TTVs) in pigs, cats and dogs and their relatedness with species-specific TTVs in primates and tupaias. J Gen Virol 83(Pt 6):1291–1297

Opriessnig T, Halbur PG (2012) Concurrent infections are important for expression of porcine circovirus associated disease. Virus Res 164(1–2):20–32. https://doi.org/10.1016/j.virusres.2011.09.014

Pérez LJ, de Arce HD, Frias MT, Perera CL, Ganges L, Núñez JI (2011) Molecular detection of Torque teno sus virus in lymphoid tissues in concomitant infections with other porcine viral pathogens. Res Vet Sci 91(3):e154–e157. https://doi.org/10.1016/j.rvsc.2011.02.012

Pérez LJ, Perera CL, Frías MT, Núñez JI, Ganges L, de Arce HD (2012) A multiple SYBR Green I-based real-time PCR system for the simultaneous detection of porcine circovirus type 2, porcine parvovirus, pseudorabies virus and Torque teno sus virus 1 and 2 in pigs. J Virol Methods 179(1):233–241. https://doi.org/10.1016/j.jviromet.2011.11.009

Ramos N, Mirazo S, Botto G, Teixeira TF, Cibulski SP, Castro G, Cabrera K, Roehe PM, Arbiza J (2018) High frequency and extensive genetic heterogeneity of TTSuV1 and TTSuVk2a in PCV2-infected and non-infected domestic pigs and wild boars from Uruguay. Vet Microbiol 224:78–87. https://doi.org/10.1016/j.vetmic.2018.08.029

Rogers AJ, Huang YW, Heffron CL, Opriessnig T, Patterson AR, Meng XJ (2017) Prevalence of the novel Torque teno sus virus species k2b from pigs in the United States and lack of association with post-weaning multisystemic wasting syndrome or mulberry heart disease. Transbound Emerg Dis 64(6):1877–1883. https://doi.org/10.1111/tbed.12586

Savic B, Milicevic V, Bojkovski J, Kureljusic B, Ivetic V, Pavlovic I (2010) Detection rates of the swine Torque teno viruses (TTVs), porcine circovirus type 2 (PCV2) and hepatitis E virus (HEV) in the livers of pigs with hepatitis. Vet Res Commun 34(7):641–648. https://doi.org/10.1007/s11259-010-9432-z

Segalés J, Martínez-Guinó L, Cortey M, Navarro N, Huerta E, Sibila M, Pujols J, Kekarainen T (2009) Retrospective study on swine Torque teno virus genogroups 1 and 2 infection from 1985 to 2005 in Spain. Vet Microbiol 134(3–4):199–207. https://doi.org/10.1016/j.vetmic.2008.08.002

Segalés J, Kekarainen T, Cortey M (2013) The natural history of porcine circovirus type 2: from an inoffensive virus to a devastating swine disease? Vet Microbiol 165(1–2):13–20. https://doi.org/10.1016/j.vetmic.2012.12.033

Sibila M, Martínez-Guinó L, Huerta E, Llorens A, Mora M, Grau-Roma L, Kekarainen T, Segalés J (2009a) Swine Torque teno virus (TTV) infection and excretion dynamics in conventional pig farms. Vet Microbiol 139(3–4):213–218. https://doi.org/10.1016/j.vetmic.2009.05.017

Sibila M, Martínez-Guinó L, Huerta E, Mora M, Grau-Roma L, Kekarainen T, Segalés J (2009b) Torque teno virus (TTV) infection in sows and suckling piglets. Vet Microbiol 137(3–4):354–358. https://doi.org/10.1016/j.vetmic.2009.01.008

Singh G, Ramamoorthy S (2018) Potential for the cross-species transmission of swine Torque teno viruses. Vet Microbiol 215:66–70. https://doi.org/10.1016/j.vetmic.2017.12.017

Ssemadaali MA, Effertz K, Singh P, Kolyvushko O, Ramamoorthy S (2016) Identification of heterologous Torque teno viruses in humans and swine. Sci Rep 6:26655. https://doi.org/10.1038/srep26655

Teixeira TF, Cibulski SP, dos Santos HF, Wendlant A, de Sales Lima FE, Schmidt C, Franco AC, Roehe PM (2015) Torque teno sus virus 1 (TTSuV1) and 2 (TTSuV2) viral loads in serum of postweaning multisystemic wasting syndrome (PMWS)-affected and healthy pigs in Brazil. Res Vet Sci 101:38–41. https://doi.org/10.1016/j.rvsc.2015.05.016

Tshering C, Takagi M, Deguchi E (2012) Detection of Torque teno sus virus types 1 and 2 by nested polymerase chain reaction in sera of sows at parturition and of their newborn piglets immediately after birth without suckling colostrum and at 24 hr after suckling colostrum. J Vet Med Sci 74(3):315–319

Vargas-Ruiz A, Ramírez-Álvarez H, Sánchez-Betancourt JI, Quintero-Ramírez V, Rangel-Rodríguez IC, Vázquez-Perez JA, García-Camacho LA (2017) Retrospective study of the relationship of Torque teno sus virus 1a and Torque teno sus virus 1b with porcine circovirus associated disease. Can J Vet Res 81(3):178–185

Vargas-Ruiz A, García-Camacho LA, Ramírez-Alvarez H, Rangel-Rodriguez IC, Alonso-Morales RA, Sánchez-Betancourt JI (2018) Molecular characterization of the ORF2 of Torque teno sus virus 1a and Torque teno sus virus 1b detected in cases of postweaning multisystemic wasting syndrome in Mexico. Transbound Emerg Dis 65(6):1806–1815. https://doi.org/10.1111/tbed.12956

Xiao CT, Giménez-Lirola L, Huang YW, Meng XJ, Halbur PG, Opriessnig T (2012) The prevalence of Torque teno sus virus (TTSuV) is common and increases with the age of growing pigs in the United States. J Virol Methods 183(1):40–44. https://doi.org/10.1016/j.jviromet.2012.03.026

Zhang Z, Wang Y, Fan H, Lu C (2012) Natural infection with Torque teno sus virus 1 (TTSuV1) suppresses the immune response to porcine reproductive and respiratory syndrome virus (PRRSV) vaccination. Arch Virol 157(5):927–933. https://doi.org/10.1007/s00705-012-1249-3

Zhang B, Tang C, Yue H, Ren Y, Song Z (2014a) Viral metagenomics analysis demonstrates the diversity of viral flora in piglet diarrhoeic faeces in China. J Gen Virol 95(Pt 7):1603–1611. https://doi.org/10.1099/vir.0.063743-0.

Zhang Z, Wang Y, Dai W, Dai D (2014b) Detection and distribution of Torque teno sus virus 1 in porcine reproductive and respiratory syndrome virus positive/negative pigs. Vet Microbiol 172(3–4):367–370. https://doi.org/10.1016/j.vetmic.2014.05.008

Zhang Z, Dai W, Dai D (2017) Molecular characterization of pigeon Torque teno virus (PTTV) in Jiangsu province. Comput Biol Chem 69:10–18. https://doi.org/10.1016/j.compbiolchem.2017.04.012

Zheng H, Ye L, Fang X, Li B, Wang Y, Xiang X, Kong L, Wang W, Zeng Y, Ye L, Wu Z, She Y, Zhou X (2007) Torque teno virus (SANBAN isolate) ORF2 protein suppresses NF-kappaB pathways via interaction with IkappaB kinases. J Virol 81(21):11917–11924

Chapter 6
Teschovirus

Yashpal Singh Malik, Sudipta Bhat, Anastasia N. Vlasova, Fun-In Wang, Nadia Touil, Souvik Ghosh, Kuldeep Dhama, Mahendra Pal Yadav, and Raj Kumar Singh

Abstract Teschoviruses are emerging pathogens, belonging to the family *Picornaviridae*, and infects porcine population only. Among all, porcine teschoviruses (PTVs) are of high prominence leading to clinical illness and consequent economic loss to the livestock sector. These are associated with extremely lethal non-suppurative polioencephalomyelitis (Teschen disease) and are distributed world over. Its milder form, Talfan disease, inflicts low morbidity and mortality and general clinical disease. The first epizootics of Teschen disease occurred in 1929 in the Czech Republic. Mature virions are small (23–30 nm) and stable in environmental conditions (pH range 2–9, heat, lipid solvents). Genetic variations in the major surface protein, VP1, lead to the evolution of several new types. As of now, 13

Y. S. Malik (✉) · S. Bhat · K. Dhama · R. K. Singh
ICAR-Indian Veterinary Research Institute, Izatnagar, Uttar Pradesh, India

A. N. Vlasova
Department of Veterinary Preventive Medicine, Food Animal Health Research Program, CFAES, Ohio Agricultural Research and Development Center, College of Veterinary Medicine, The Ohio State University, Wooster, OH, USA

F.-I. Wang
School of Veterinary Medicine, National Taiwan University, Taipei, Taiwan

N. Touil
Laboratoire de Biosécurité et de Recherche, Hôpital Militaire d'Instruction Med V, Rabat, Morocco

S. Ghosh
Department of Biomedical Sciences, One Health Center for Zoonoses and Tropical Veterinary Medicine, Ross University School of Veterinary Medicine, Basseterre, St. Kitts and Nevis, West Indies

M. P. Yadav
ICAR-Indian Veterinary Research Institute, Izatnagar, Uttar Pradesh, India

Sardar Vallabhbhai Patel University of Agriculture & Technology, Meerut, Uttar Pradesh, India

Y. S. Malik et al. (eds.), *Emerging and Transboundary Animal Viruses*, Livestock Diseases and Management, https://doi.org/10.1007/978-981-15-0402-0_6

genotypes in PTV are approved, namely PTV-1–13. Among all, PTV-1 is highly virulent and causes severe mortality and morbidity in the porcine population, domestic as well as wild. Ingestion is the main spreading route of infection, though intranasal infection and virus secretion in the urine are also noted. Along with encephalitis, PTVs are also responsible for reproductive disorders, diarrhoea, pneumonia, pericarditis and myocarditis. PTVs are also found as co-infection with several bacterial, viral and parasitic pathogens. Several conventional and modern diagnostics tools are available for their detection. Alternative of serological typing, VP1 and VP2 gene-based molecular typing is now preferred to know the epidemiological pattern. Although initially vaccines were used for its eradication in Europe, due to the sporadic reports of mild PTV infection in several countries, the approach was discontinued. Moreover, due to the presence of multiple serotypes, developing a multivalent PTV vaccine to protect against all strains is a major challenge.

Keywords Teschoviruses · Encephalomyelitis · Mortality · Economical loss · Mutation · Genotypes · Diagnosis · Epidemiology · Vaccine

6.1 Preamble

Teschovirus is also known by several other names including Teschen disease, Talfan disease, poliomyelitis suum, benign enzootic paresis, Klobauk disease and contagious porcine paralysis. Teschovirus-induced encephalitis in pigs was first narrated as Teschen disease, a virulent extremely lethal polioencephalomyelitis with high mortality, in the township of Teschen in the Czech Republic over 90 years ago, in 1929 (Trefny 1930). Porcine Teschovirus (PTV) is detected ubiquitously in the pig population worldwide (Knowles 2015). While the majority of infections are subclinical, there are several clinical conditions which affect different body systems, including nervous (polioencephalomyelitis), reproductive, enteric, respiratory and integumentary. From 1929 to 1950s, severe PTV outbreaks have been reported in European and neighbouring countries (Knowles 2015). Afterwards, the disease became very rare and disappeared from Western Europe. Lately, two fresh PTV outbreaks occurred in 2009 in Haiti and subsequently in 2011 in Canada (Deng et al. 2012; Salles et al. 2011). To date, 13 different types of PTV are noted, and additional types of PTV-14–22 proposed, circulating throughout the world (Knowles 2015). Different PTV types are associated with a variety of clinical symptoms in their natural hosts (pig and wild boar). PTV-1 is considered as one of the most virulent viruses responsible for fatal outbreaks in pigs of all age groups. Nowadays, less pathogenic strains are commonly found in most swine populations causing asymptomatic infections in young animals.

6.2 Epidemiology of the Disease

6.2.1 The Causative Agent

6.2.1.1 Classification

Until 1999, porcine teschoviruses (PTVs) assigned to the genus *Enterovirus*, known as porcine enteroviruses (PEVs) (family: *Picornaviridae*, order: *Picornavirales*). PEVs are classified into three genetic groups: (1) PEV types 1–7 and 11–13, (2) PEV type 8 and (3) PEV types 9 and 10. Before reclassification, the original 11 PEV serotypes were divided into three subgroups (I, II and III) based on physicochemical properties, serological assays, types of cytopathic effect (CPE) produced in the porcine kidney (PK) cells and diverse cell culture host ranges (Knowles et al. 1979). In 1999, more recently complete genome analysis of available PEV group I (former PEV1–7 and PEV11–13) sequences revealed that they were different from other Enteroviruses. Based on the genetic differences they were classified as a new genus, *Teschovirus* containing a single species PTV (the name was derived from Teschen disease) of multiple serotypes (Kaku et al. 2001; Zell et al. 2001). The species name is also renamed to *Teschovirus A* to remove the ambiguity reference to natural host, but the virus name remains PTV. Lately, nine additional genetic types in PTVs (PTV-14–PTV-22) from China are proposed (Yang et al. 2018). Most severe clinical disease (teschovirus encephalomyelitis) relates to PTV-1, whereas other types are associated with a milder form of the disease (Talfan disease).

6.2.1.2 Virus Characteristics and Genome Organisation

PTV virions are small (25–30 nm) and non-enveloped. Icosahedral capsid is composed of 60 protomers, each of which consists of three surface proteins (VP1, 2 and 3) and one inner protein (VP4). Mutations in the surface protein sequences lead to diversity in protomers, which is responsible for the idiosyncrasies such as antigenicity, receptor recognition, buoyant density and pH stability.

The viral genome contains a 7.1 kb positive-sense single-stranded RNA that codes for a single long open reading frame (ORF) of a polyprotein, which is further processed to form 12 individual viral proteins. Genome linked with a Vpg protein at 5′ end followed by 5′ noncoding region (NCR), a leader protein (L), four structural proteins (VP1–4) and seven non-structural proteins (2A–C, 3A–D), 3′ NCR and poly A tail (Fig. 6.1). Among the four structural proteins, VP1 is the most important for molecular epidemiology and genotyping. Neutralising epitope(s) primarily reside in VP1, although VP2 is also involved.

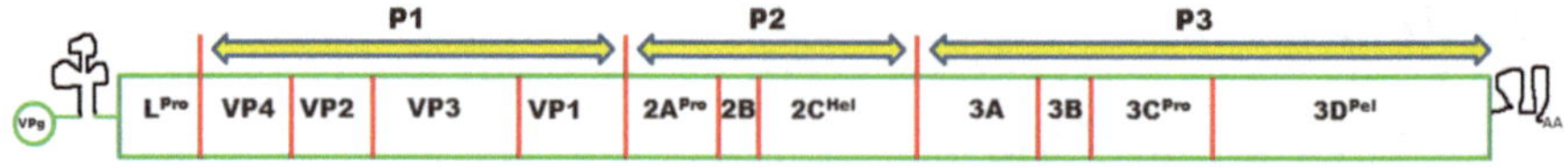

Fig. 6.1 A schematic representation of Teschovirus genome structure. The genes, viz. VPg (genome-linked viral protein), NCR (noncoding RNA), L (leader protein), VP (viral protein), P1 (structural protein), P2–P3 (non-structural protein), pro (protease), hel (helicase) and pol (polymerase) arrangements, are shown

6.2.1.3 Virion Resistance

Mature PTV virions are very stable and can resist several environmental conditions including a pH range of 2–9 and further in liquid manure it remains viable for a long time. Heat, lipid solvents and several disinfectants are not able to inactivate PTV virions (Derbyshire and Arkell 1971). In the presence of halide ions, sodium chlorite, heat and 70% ethanol inactivate PTV effectively.

6.2.1.4 Serotypes/Strain Variability

PTVs have 13 known genotypes based on diversity of VP1 gene sequence or through cross-neutralisation tests. PTV serotypes 1–10 were previously known as porcine enterovirus group I. Former PEV-1–7 have been renamed PTV-1–7, and PEV-11–13 were designated as PTV-8–10 (Kaku et al. 2001). In 2011, PTV-12 was identified for the first time in Spanish pig population, which is presumed to be the result of VP1 gene mutation (Cano-Gomez et al. 2011). Likewise, PTV-13 was for the first time detected in faeces of wild boar in Hungary (Boros et al. 2012). Very recently, nine more genotypes (PTV-14–22) of PTV have been proposed from a study in China in 2018. However, these strains remain unverified by serological methods (Yang et al. 2018). Negative selection and homologous recombination are two significant causes of genetic diversity in structural and non-structural genes served as the major driving mechanisms of PTV evolution (Lin et al. 2012). Two-step evolution is observed in PTVs (Zell et al. 2001). The first step led to the generation of three groups that have undergone further diversification and emergence of 13 distinct serotypes (Fig. 6.2). Another feature regarding evolutionary changes of PTVs is the gradual changes in their virulence, generating less pathogenic variants.

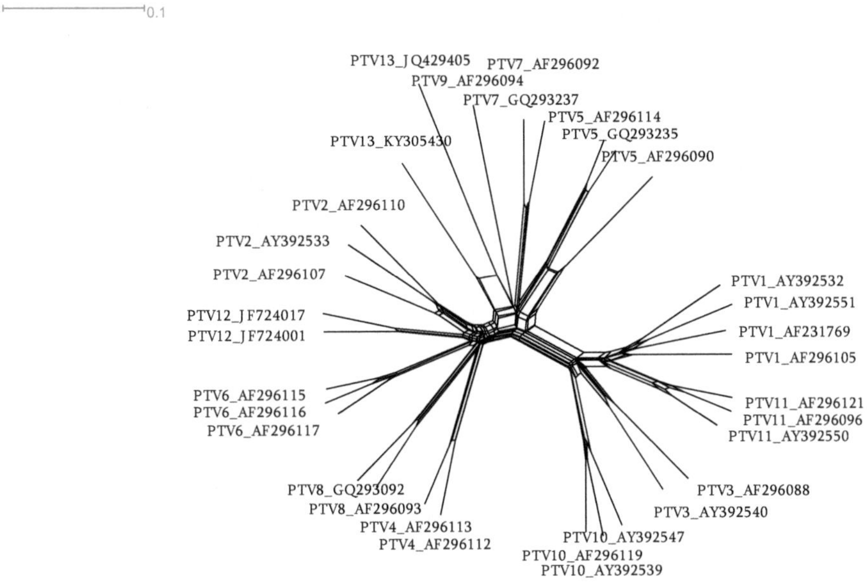

Fig. 6.2 Unrooted maximum-likelihood phylogenetic tree based on VP1 amino acid sequences of the representative members of the 13 PTV types. The tree was developed using Splits Tree4 programme (Huson and Bryant 2006)

6.2.2 Host Species

The only known host for the PTVs is the porcine but wild boars were also found to be infected with PTV. Still, limited literature is available related to wild suid susceptibility (Cano-Gómez et al. 2013). Growing (suckling or weaned) pigs are more prone to infection with PTV.

6.2.3 Geographical Distribution

The mild form of PTV-associated encephalomyelitis is reported throughout the world, whereas the fatal teschovirus encephalomyelitis is presently a rare disease, with most recent outbreaks reported in Madagascar and Central and Eastern Europe. Outbreaks of teschovirus encephalomyelitis have been recorded in the World Organisation for Animal Health (OIE) during 1996, 1999 and 2005 by Belarus; in 2002–2004 by Moldavia; in 2002 by Romania; in 2004 by Russia; in 1996–2005 by Ukraine; in 1997 and 2000–2002 by Latvia; in 1996–2000, 2002 and 2004–2005 by Madagascar; in 2001 by Uganda; in 2002 by Japan; in 2000 and 2004 by Taiwan; and in 2009 and 2011 by Haiti and Canada, respectively (Fig. 6.3).

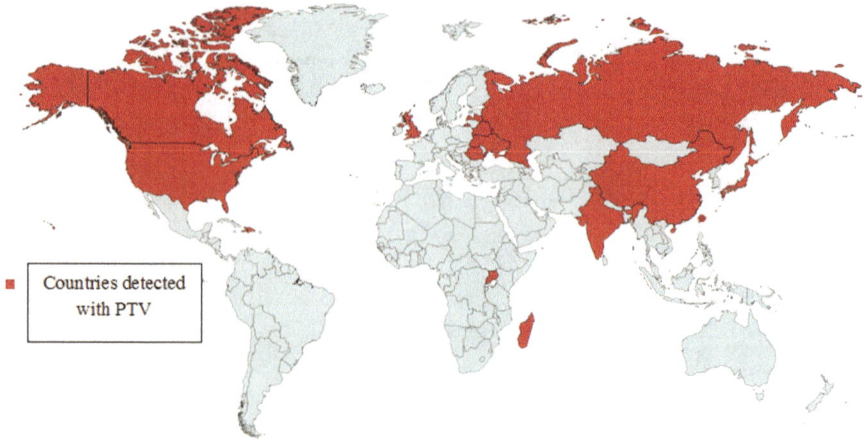

Fig. 6.3 A worldwide map of porcine Teschovirus detection is depicted (red-coloured) based on the published reports on the disease in different parts of the world

6.2.4 Mortality and Morbidity

All age groups of commercial porcine populations are ubiquitously infected, and sometimes enzootic or endemic. The milder form, Talfan disease, leads to low morbidity and mortality and in general clinical disease is limited to younger and post-weaning animals. Higher mortality is presumably associated with a virgin epidemic, where the porcine populations have no antibody and the invaded strains have at least moderate virulence, such as the recent outbreak in Haiti in 2009, wherein 40% mortality and 60% morbidity were reported (Deng et al. 2012). In 2011, Canada reported 100% mortality (Salles et al. 2011).

6.2.5 Transmission

Ingestion is the commonest route of PTV infection, and intestinal tract and associated lymph nodes are the main multiplication sites. The virus shedding occurs in the faeces and oral secretions of convalescent animals for up to 7 weeks. Due to the stable nature, PTVs resist in the environment for more than 5 months at 15 °C and readily spread on fomites (Horak et al. 2016). Intranasal infection is also highly suggested (Chiu et al. 2013, 2014) as PTVs detected in the cranial cerebrum, including the olfactory bulb. Urinary shedding of infectious PTV virion also proved as a potent transmission in the endemic field situation and slurry (faeces mixed with urine) makes the survival and transmission easier (Tsai et al. 2016). Till date, no zoonotic transmissions have been reported.

6.3 Pathogenesis

After ingestion through the oral route, PTVs primarily replicate in the tonsils and intestinal tract. Tonsils have an important role in virus entry, survival and shedding of infections. Viruses are known to replicate to relatively higher titres in large intestine and ileum than the other portions of the intestine (Long 1985; Chiu et al. 2013).

All the virulent strains are known to spread through blood and access to the central nervous system (CNS). One to two days post-infection (dpi) is characterised by increased body temperature and later by diarrhoea. Both flaccid and spastic paralysis are reported by 10–11 dpi (Long 1985). Respiratory paralysis (asphyxiation) is the main cause of death of affected animals (Knowles 2015). Experimental infection in pigs through the intranasal route can develop CNS signs. In reproductive disorders, the virus is thought to reach the placenta via the blood. Experimental infection in pregnant gilts through nasal and oral route has resulted in foetal infection (Chiu et al. 2014).

6.4 Clinical Signs

Subclinical diseases are most common with PTV infections. Clinical signs varied with the virulence of different serotypes and are mainly associated with highly virulent PTV-1. Co-infections by several serotypes are frequent. The incubation period for teschovirus encephalomyelitis is 14 days (Knowles 2015). In experimentally infected piglets, the highly virulent 'Zabreh' strain of PTV-1 produces clinical signs in 5–7 days.

In teschovirus encephalomyelitis before paralysis/paresis, several clinical signs are observed which include fever, anorexia, listlessness and locomotor ataxia. As early as 2–3 dpi, caudal ataxia is seen to be leading which progresses to paresis or paralysis. Commonly, 3–4 days after the onset of clinical symptoms, death occurs (Yamada et al. 2014). Polioencephalomyelitis rarely progresses to complete paralysis. SMEDI syndrome (stillbirth [S], mummified foetus [M], embryonic death [ED], infertility [I]) is associated with reproductive disorders (Dunne et al. 1965). Similar reproductive syndromes are also associated with parvovirus infection, which occurs more commonly than PTV. The infection causes embryonic death and mummification in early to mid-gestation (40–70 days) but at later stages it may result in stillbirth (Lin et al. 2012). Experimental and field infection also reported the link with abortion (Kirkbride and McAdaragh 1978). Though isolation of virus is done from the male reproductive tract, experimental intrauterine inoculation of sperm containing virus did not infect embryos or prevent conception. However, virus isolation was reported both from healthy and diarrhoeic pigs, but experimentally PTVs can induce

Table 6.1 Natural or experimental clinical syndromes associated with porcine Teschovirus infection (Alexandersen et al. 2012)

Syndrome	Associated PTV serotypes
Polioencephalomyelitis	PTV-1, PTV-2, PTV-3, PTV-5, PTV-12, PTV-13
Reproductive disorders	PTV-1, PTV-3, PTV-6
Diarrhoea	PTV-1, PTV-2, PTV-3, PTV-5
Pneumonia	PTV-1, PTV-2, PTV-3
Pericarditis and myocarditis	PTV-2, PTV-3

diarrhoea in a host (free from other pathogens). There are not many reports available on respiratory infections caused by PTVs. Recently a congenital microphthalmic syndrome in a pig with typical non-suppurative myelitis in the lumbar spinal cord from the Czech Republic has been reported (Andrysikova et al. 2018). Table 6.1 shows the association of some serotypes with specific disease conditions but many serotypes are not related to any clinical signs.

6.5 Diagnosis

6.5.1 History

The affected pigs showing fever, followed by ataxia and paralysis/paresis, are suggestive of PTV infection. If there are gilts or sows with stillborn or mummified foetuses it is suggestive of PTV-induced reproductive disorder.

6.5.2 Samples to Collect

In all the diagnostic tests, the brain and spinal cord are the most preferred samples, except in serology (Alexandersen et al. 2012). Histopathologic lesions are found mainly in the cerebrum, cerebellum, diencephalon, medulla oblongata, cervical and lumbar spinal cords. Samples from pigs that died very recently or were sacrificed for necropsy are ideal for virus isolation. Paired serum samples are preferred for assessing seroconversions by either virus neutralisation or ELISA (Hübschle et al. 1983).

6.5.3 *Post-mortem Lesions*

No gross lesions are associated with PTV-induced polioencephalomyelitis. Throughout the CNS non-suppurative polioencephalomyelitis with perivascular lymphocytic cuffing is usually found upon histological examination of affected tissues. Neuronal degeneration (swelling, chromatolysis and necrosis) and axonal degeneration are often present in the late stages of disease (Yamada et al. 2007). Grossly, pericarditis is serofibrinous with a cloudy pericardial effusion that quickly forms a coagulum upon standing. Occasionally focal myocardial necrosis with cellular infiltrate is also present.

6.5.4 *Virus Isolation*

Porcine origin primary and secondary kidney cells are very susceptible to PTV and mostly used for isolation. The virus also replicates well in some established cell lines, like IBRS-2 (porcine kidney cells). The cultured PTVs have been identified with virus neutralisation (VN) or immunofluorescence antibody (IFA) assays (Knowles 2015). VN and IFA assays are time consuming and, respectively, take 72 and 12 h.

6.5.5 *Nucleic Acid Detection*

The gold standard test for the PTV diagnosis is the detection of the nucleic acids. 'Palmquist RT-PCR' enables the detection and differentiation of PTV from porcine sapeloviruses (PSVs) with a single primer pair and multiplex 'Zell RT-PCR' with three primer pairs can differentiate among PTV, PSV and porcine enteroviruses (PEVs) (Palmquist et al. 2002; Zell et al. 2000). These RT-PCR tests are very specific, sensitive and rapid, and do not cross-react with pseudorabies virus, porcine reproductive and respiratory syndrome virus (PRRSV), porcine parvovirus, porcine coronavirus, porcine reovirus or picorna-like virus.

A rapid detection system using reverse transcription loop-mediated isothermal amplification (RT-LAMP) was developed in China (Wang et al. 2011). The nucleic acid-based multiplex PCR assay for the detection of common porcine viral infections was developed for easy detection of viral diseases. A probe-based real-time RT-PCR reports high sensitivity and specificity, allowing a detection limit of ten copies (Zhang et al. 2013).

Table 6.2 RT-PCR assays used for detection and genotyping of PTVs

Assays	Primer sequences	Targeted gene	References
RT-PCR	5′-AGTTTTGGATTATCTTGTGCCC-3′	5′NTR	Zell et al. (2000)
	5′-CCAGCCGCGACCCTGTCAGGCAGCAC-3′		
nRT-PCR	5′-TGAAAGACCTGCTCTGGCGCGAG-3′		
	5′-GCTGGTGGGCCCCAGAGAAATCTC-3′		
Duplex PCR for PTV and PSV	5′-GTGGCGACAGGGTACAGAAGAG-3	5′UTR	Palmquist et al. (2002)
	5′-GGCCAGCCGCGACCCTGTCAG-3′		
RT-PCR (PTV-1 specific)	ATGCCTTTGAGACCTGTTAATGA	VP3-VP1 of PTV-1	Zell et al. (2000)
	CAACATTAGTCATCTTTGTAATTGT		
RT-PCR (genotyping)	GCATCHAAYGARAAYCC	VP1	La Rosa et al. (2006)
	CCAAAYCCAAARTCYTG		
RT-PCR (genotyping)	CACCARYTGCTTAARTGYKGTTGG	VP2	Kaku et al. (2007)
	CACAGGGTTGCTGAAGARTTTGT		
RT-LAMP	CACATCAATGACACGGGTTTTCCC	3DPol	Wang et al. (2011)
	TCGCCTTCTTTCACAAGAATCCCG		
Multiplex RT-PCR (PTV, PRRSV, CSFV)	GTGGCGACAGGGTACAGAAGAG	5′UTR	Liu et al. (2011)
	GGCCAGCCGCGACCCTGTCAG		
Real-time RT-PCR	5′-CTCCTGACTGGGCAATGGG-3′	5′UTR	Zhang et al. (2013)
	5′-TGTCAGGCAGCACAAGTCCA-3′		

As there are several types of PTV exit all over the world, the genotyping of PTVs has a great impact on understanding the epidemiology of PTVs. The neutralising epitopes identified in both VP1 and VP2 are used for typing and molecular epidemiology of PTVs. Some of the details of primers and targeted gene are mentioned in Table 6.2.

6.5.6 Antibody Detection

PTV is a ubiquitous virus; therefore, a single positive serological test does not indicate the presence of the infection. However, clinical signs associated with a four-fold rise in antibody titre in paired serum samples are considered positive for PTV. An ELISA is also available for detection and typing of PTV infections.

6.5.7 *Differential Diagnosis*

The PTV disease needs to be differentiated from the following disease:

Viral diseases	Pseudorabies (Aujeszky's disease) Classical swine fever (hog cholera, acute form) Japanese encephalitis Haemagglutinating encephalomyelitis Rabies Porcine reproductive and respiratory syndrome (PRRS) virus (highly virulent strains)
Bacterial diseases	Meningoencephalitis induced by *Streptococcus suis* Oedema disease induced by *Escherichia coli* enterotoxemia
Poisonings	Salt (water deprivation), lead, insecticides

6.6 Prevention and Control

As with most viral infections, control measures for PTV depend upon prevention rather than treatment. Hosts infected with a mild form of PTV may survive if their appetite returns after the transient paresis phase. Teschovirus encephalomyelitis needs reporting to the authorities (Knowles 2015; http://www.oie.int/animal-health-in-the-world/information-on-aquatic-and-terrestrial-animal-diseases/).

6.6.1 *Immunity*

IgG and IgM antibody-mediated humoral immune response is the primary anti-PTV immune response. IgA antibodies generated within the intestinal tract have shown to be protective when the virus enters orally. Persistent virus infection may develop in the intestinal tract due to failure of antibody production. Maternal antibodies are effective to prevent viremia and transplacental spread of PTV. PTVs spread slowly through intrauterine causing foetal deaths at different developmental stages (Wang and Pensaert 1989). Foetal anti-PTV antibodies are mainly IgM type followed by IgG, which starts developing by 70 days and matures by 90 days (Wang et al. 1973). These antibodies can protect the foetus from infection during this stage. Maternal antibody acquired from colostrum can protect after weaning (Wang and Pensaert 1989).

6.6.2 *Cross-Protection*

Antibody-mediated immunity has a more important role in protection against PTV-induced disease (Alexandersen et al. 2012). Due to high PTV serotype/genotype diversity, cross-protection is not likely to occur.

6.6.3 *Preventive Measures*

Successful control methods for PTV-induced encephalomyelitis include movement controls, quarantine, slaughter and ring vaccination.

(a) Vaccines: Commercial attenuated and inactivated vaccines were available in central Europe and Madagascar during the higher incidence of clinical disease of PTVs but discontinued later as the disease became rare. As the disease is caused by several PTV serotypes, developing a vaccine from multiple serotypes is quite challenging. Vaccination may be economical to control the virulent form of Teschen disease and to protect valuable breeder stocks, sows or pigs.
(b) Restrictions in imports from PTV-affected countries could help in limiting the spread of the virulent PTV-1 strain. Quarantine and slaughter would likely be effective control measures.
(c) Introduce new breeding stock into the unit more than 1 month before breeding to expose them to enzootic or endemic PTV strains and allow the development of immunity.
(d) Closed-herd system reduces the risk of introducing extraneous viruses, but it is not possible to eliminate this risk since the relatively resistant PTVs transmit by a variety of fomites.

Acknowledgements All the authors of the manuscript thank and acknowledge their respective universities and institutes. Dr. YS Malik acknowledges ICAR-Education Division, Ministry of Agriculture and Farmers Welfare, GoI, New Delhi for National Fellowship.

Conflict of interest: There is no conflict of interest.

References

Alexandersen S, Knowles NJ, Dekker A, Belsham G, Zhang Z, Koenen F (2012) Picornaviruses. In: Zimmerman J, Karriker LA, Ramirez A, Schwartz KJ, Stevenson GW (eds) Diseases of swine, 10th edn. Blackwell Publishing Press, Ames, IA, pp 587–620

Andrysikova R, Sydler T, Kümmerlen D, Pendl W, Graage R, Moutelikova R, Prodelalova J, Voelter K (2018) Congenital microphthalmic syndrome in a swine. Case Rep Vet Med 2018:1–7, 2051350, 6 p

Boros A, Nemes C, Pankovics P, Kapusinszky B, Delwart E, Reuter G (2012) Porcine teschovirus in wild boars in Hungary. Arch Virol 157(8):1573–1578

Cano-Gomez C, Palero F, Buitrago MD, García-Casado MA, Fernández-Pinero J, Fernández-Pacheco P, Agüero M, Gómez-Tejedor C, Jiménez-Clavero MÁ (2011) Analyzing the genetic diversity of teschoviruses in Spanish pig populations using complete VP1 sequences. Infect Genet Evol 11(8):2144–2150

Cano-Gómez C, García-Casado MA, Soriguer R, Palero F, Jiménez-Clavero MA (2013) Teschoviruses and sapeloviruses in faecal samples from wild boar in Spain. Vet Microbiol 165(1–2):115–122

Chiu S, Hu S, Chang C, Chang CY, Huang CC, Pang VF, Wang FI (2013) The role of porcine teschovirus in causing diseases in endemically infected pigs. Vet Microbiol 161(1/2):88–95

Chiu S, Yang CL, Chen YM, Hu SC, Chiu KC, Lin YC, Chang CY, Wang FI (2014) Multiple models of porcine teschovirus pathogenesis in endemically infected pigs. Vet Microbiol 168(1):69–77

Deng MY, Millien M, Jacques-Simon R, Flanagan JK, Bracht AJ, Carrillo C, Barrette RW, Fabian A, Mohamed F, Moran K, Rowland J, Swenson SL, Jenkins-Moore M, Koster L, Thomsen BV, Mayr G, Pyburn D, Morales P, Shaw J, Burrage T, White W, McIntosh MT, Metwally S (2012) Diagnosis of porcine teschovirus encephalomyelitis in the Republic of Haiti. J Vet Diagn Investig 24(4):671–678

Derbyshire JB, Arkell S (1971) Activity of some chemical disinfectants against Talfan virus and porcine adenovirus type-2. Br Vet J 127(3):137–142

Dunne HW, Gobble JL, Hokanson JF, Kradel DC, Bubash GR (1965) Porcine reproductive failure associated with a newly identified "SMEDI" group of picorna viruses. Am J Vet Res 26(115):1285–1297

Horak S, Killoran K, Leedom Larson KR (2016) Porcine teschovirus. Swine Health Information Center and Center for Food Security and Public Health, Ames, IA

Hübschle OJ, Rajanarison I, Koko M, Rakotondramary E, Rasiofomanana P (1983) ELISA for testing swine sera for antibodies against Teschen virus. Dtsch Tierarztl Wochenschr 90(3):86–88

Huson DH, Bryant D (2006) Application of phylogenetic networks in evolutionary studies. Mol Biol Evol 23(2):254–267

Kaku Y, Sarai A, Murakami Y (2001) Genetic reclassification of porcine enteroviruses. J Gen Virol 82(2):417–424

Kaku Y, Murakami Y, Sarai A, Wang Y, Ohashi S, Sakamoto K (2007) Antigenic properties of porcine teschovirus 1 (PTV-1) Talfan strain and molecular strategy for serotyping of PTVs. Arch Virol 152(5):929–940

Kirkbride CA, McAdaragh JP (1978) Infectious agents associated with fetal and early neonatal death and abortion in swine. J Am Vet Med Assoc 172(4):480–483

Knowles NJ (2015) Teschovirus encephalomyelitis (previously enterovirus encephalomyelitis or Teschen/Talfan disease). In: Manual of diagnostic tests and vaccines for terrestrial animals. OIE, Paris

Knowles NJ, Buckley LS, Pereira HG (1979) Classification of porcine enteroviruses by antigenic analysis and cytopathic effects in tissue-culture—description of 3 new serotypes. Arch Virol 62(3):201–208

La Rosa G, Muscillo M, Di Grazia A, Fontana S, Iaconelli M, Tollis M (2006) Validation of RT-PCR assays for molecular characterization of porcine teschoviruses and enteroviruses. J Vet Med B Infect Dis Vet Public Health 53(6):257–265

Lin W, Cui S, Zell R (2012) Phylogeny and evolution of porcine teschovirus 8 isolated from pigs in China with reproductive failure. Arch Virol 157(7):1387–1391

Liu S, Zhao Y, Hu Q, Lv C, Zhang C, Zhao R, Hu F, Lin W, Cui S (2011) A multiplex RT-PCR for rapid and simultaneous detection of porcine teschovirus, classical swine fever virus, and porcine reproductive and respiratory syndrome virus in clinical specimens. J Virol Methods 172(1–2):88–92

Long JF (1985) Pathogenesis of porcine polioencephalomyelitis. In: Comparative pathobiology of viral diseases, vol I. CRC Press, Boca Raton, FL, pp 179–197

Palmquist JM, Munir S, Taku A, Kapur V, Goyal SM (2002) Detection of porcine teschovirus and enterovirus type II by reverse transcription-polymerase chain reaction. J Vet Diagn Investig 14(6):476–480

Salles MWS, Scholes SFE, Dauber M, Strebelow G, Wojnarowicz C, Hassard L, Acton AC, Bollinger TK (2011) Porcine teschovirus polioencephalomyelitis in western Canada. J Vet Diagn Investig 23(2):367–373

Trefny L (1930) Hromadna onemocnemi vepru na Tesinsku. Zverolek Obz 23:235–236

Tsai ATH, Kuo CC, Kuo YC, Yang JL, Chang CY, Wang FI (2016) The urinary shedding of porcine teschovirus in endemic field situation. Vet Microbiol 182:150–155

Wang JT, Pensaert MB (1989) Porcine enterovirus (reproductive disorders). In: Virus infections of porcines. Elsevier, Amsterdam, pp 235–239

Wang JT, Dunne HW, Griel LC, Hokanson JF, Murphy DM (1973) Mortality, antibody development, and viral persistence in porcine fetuses incubated in utero with SMEDI (entero-) virus. Am J Vet Res 34(6):785–791

Wang B, Wang Y, Tian ZJ, An TQ, Peng JM, Tong GZ (2011) Development of a reverse transcription loop-mediated isothermal amplification assay for detection of porcine teschovirus. J Vet Diagn Investig 23(3):516–518

Yamada M, Kaku Y, Nakamura K, Yoshii M, Yamamoto Y, Miyazaki A, Tsunemitsu H, Narita M (2007) Immunohistochemical detection of porcine teschovirus antigen in the formalin-fixed paraffin-embedded specimens from pigs experimentally infected with porcine teschovirus. J Vet Med Physiol Pathol Clin Med 54(10):571–574

Yamada M, Miyazaki A, Yamamoto Y, Nakamura K, Ito M, Tsunemitsu H, Narita M (2014) Experimental teschovirus encephalomyelitis in gnotobiotic pigs. J Comp Pathol 150(2–3):276–286

Yang T, Li R, Yao Q, Zhou X, Liao H, Ge M, Yu X (2018) Prevalence of porcine teschovirus genotypes in Hunan, China: identification of novel viral species and genotypes. J Gen Virol 99(9):1261–1267

Zell R, Krumbholz A, Henke A, Birch-Hirschfeld E, Stelzner A, Doherty M, Hoey E, Dauber M, Prager D, Wurm R (2000) Detection of porcine enteroviruses by nRT-PCR: differentiation of CPE groups I-III with specific primer sets. J Virol Methods 88(2):205–218

Zell R, Dauber M, Krumbholz A, Henke A, Birch-Hirschfeld E, Stelzner A, Prager D, Wurm R (2001) Porcine teschoviruses comprise at least eleven distinct serotypes: molecular and evolutionary aspects. J Virol 75(4):1620–1631

Zhang C, Wang Z, Hu F, Liu Y, Qiu Z, Zhou S, Cui S, Wang M (2013) The survey of porcine teschoviruses in field samples in China with a universal rapid probe real-time RT-PCR assay. Trop Anim Health Prod 45(4):1057–1061

Chapter 7
Animal Flaviviruses

Shailendra K. Saxena, Swatantra Kumar, and Amrita Haikerwal

Abstract The burden of *Flavivirus* has made the whole world susceptible towards its infection. Japanese encephalitis virus (JEV) is the causative agent of Japanese encephalitis, a paediatric disease causing mortality and morbidity in children but is not limited to them. Animals are playing a crucial role in *Flavivirus* life cycle such as pigs are amplifying host for JEV and migratory birds are reservoirs for WNV. The resurgence of JEV majorly occurs at the time of ambient temperature and precipitation facilitating favourable breeding habitats of mosquitoes generally in paddy fields. WNV also causes neurological diseases like encephalitis in humans and mostly in horses. Despite the differences in pathogenesis and tropism, flaviviruses exhibit similarity in the overall genome organisation. CD209, C-type lectin receptor, also known as DC-SIGN, is crucial for virus attachment both for JEV and WNV. Host immune response during *Flavivirus* infection involves both innate and adaptive immune responses for clearance of the virus. Several vaccine candidates are available for human use; however, vaccine for livestock is underdeveloped. To improve the life expectancy of animals live attenuated virus or inactivated virus is used for the vaccination. This chapter majorly focuses on replication cycle, incidences, and prevalence in livestock, transmission cycle, risk factors, immunobiology, diagnostics, prevention, and control strategies.

Keywords Flavivirus · Japanese encephalitis virus · West Nile virus · Transmission cycle · Immunobiology and risk factors

---

S. K. Saxena (✉)
Center for Advanced Research (CFAR), King George's Medical University (KGMU), Lucknow, India

CSIR-Centre for Cellular and Molecular Biology, Hyderabad, India
e-mail: shailen@kgmcindia.edu

S. Kumar
Department of Center for Advanced Research (CFAR), King George's Medical University (KGMU), Lucknow, India

A. Haikerwal
Center for Advanced Research (CFAR), King George's Medical University (KGMU), Lucknow, India

Y. S. Malik et al. (eds.), *Emerging and Transboundary Animal Viruses*, Livestock Diseases and Management, https://doi.org/10.1007/978-981-15-0402-0_7

Abbreviations

CDC	Centres for Disease Control and Prevention
DCs	Dendritic cells
DENV	Dengue virus
JEV	Japanese encephalitis virus
MIF	Macrophage migration inhibitory factor
NTPase	Nucleotide 5′ triphosphatase
PRNT	Plaque reduction neutralisation assay
TNF-α	Tumour necrosis factor alpha
WHO	World Health Organization
WNV	West Nile virus
YFV	Yellow fever virus

7.1 Preamble

Vector-borne viruses from *Flaviviridae* family are the leading cause of disease and death among animals and humans. *Flavivirus*, *Pestivirus*, and *Hepacivirus* are the three categorised genus under *Flaviviridae* (Simmonds et al. 2017). The term "Flavivirus" is originated from the Latin word *flavus* means yellow defining jaundice caused by yellow fever virus. More than 70 viruses have been classified as *Flavivirus*, where 16 are tick borne, 40 are mosquito borne, and rest do not possess known vectors (Mukhopadhyay et al. 2005). Most prominent mosquito-borne flaviviruses are Japanese encephalitis virus (JEV), Zika virus (ZIKV), dengue virus (DENV), West Nile virus (WNV), and yellow fever virus (YFV). Flaviviruses can be divided into distinct serological groups such as DENV serogroup, JEV serogroup, and less serological cohesive YFV groups. Based on molecular phylogenetics, Flaviviruses can be further subdivided into clades, clusters, and species (Kuno et al. 1998). Flaviviruses are also categorised as neurotropic viruses infecting the central nervous system with prominent neurological impairment as encephalitis in case of JEV (Solomon 2004) and DENV (Garg et al. 2017) while Guillain–Barré syndrome and microcephaly during ZIKV infection in humans (Broutet et al. 2016). Animals are playing a crucial role in Flavivirus life cycle such as pigs are amplifying hosts for JEV (Mansfield et al. 2017) and migratory birds are virus reservoirs for WNV (Pérez-Ramírez et al. 2014).

Despite the differences in pathogenesis and tropism, flaviviruses exhibit similarity in overall genome organisation though they have differences in translation mechanism (Neufeldt et al. 2018). The size of the Flavivirus virions is about 500 Å in diameter that comprises an ~11 kb single positive-sense RNA strand genome sheltered within capsid protein and enclosed by 180 copies of envelope glycoproteins.

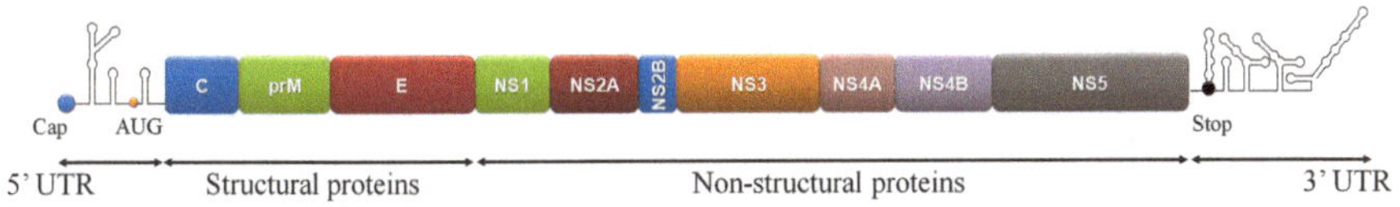

Fig. 7.1 Flavivirus genome. Flavivirus genome is ~10.5 kb long encoding three structural proteins: capsid (C), premembrane (prM), and envelope (E) and seven non-structural proteins: NS1, NS2A, NS2B, NS3, NS4A, NS4B, and NS5. The genome comprises untranslated region (UTR) at 3′ and 5′ ends. The flavivirus RNA undergoes cap-dependent translation via rough endoplasmic reticulum (RER)

The flaviviral genome has a single open reading frame encoding a long polyprotein of 3400 amino acids with type 1 cap on the 5′ end. The N-terminus of the polypeptide is devoted to structural protein synthesis—capsid (C), pre-membrane (PrM), and envelope (E) which constitute the virus particle—whereas the remainder of the genome is dedicated for seven non-structural proteins—NS1, NS2A, NS2B, NS3, NS4A, NS4B, and NS5 (Fig. 7.1). NS1 protein is a secretory in flaviviruses (Muller and Young 2013), and NS2A is crucial in the early replication process and generation of infectious particles. The NS2B acts as a cofactor and helps in the NS3 recruitment to the ER membranes (Li et al. 2016). NS3 possesses more than one activity with protease, nucleotide 5′ triphosphatase (NTPase), and helicase activity (Li et al. 2014). NS4A has a membrane curvature-inducing activity and is an integral membrane protein (Miller and Krijnse-Locker 2008) whereas NS4B has no enzymatic activity though it's crucial for virus replication by interacting with NS3 (Apte-Sengupta et al. 2014). The N-terminal of NS5 has guanine-N7-methyltransferase, guanylyltransferase, and nucleoside-2-*O*-methyltransferase activity responsible for methylation of the viral genome and 5 RNA capping. The C-terminal domain of NS5 has RNA-dependent RNA polymerase (RdRp) and causes viral RNA synthesis (Zou et al. 2011).

7.2 Flavivirus Replication

Upon ingestion of blood meal by infected mosquito bites, Flavivirus enters into primary host cells such as skin dendritic cells during JEV/WNV infection (Lannes et al. 2017). The attachment factors or receptors are required for binding of virions on the target cells following interaction with secondary receptors for internalisation. CD209, C-type lectin receptor (DC-SIGN), is crucial for virus attachment both for JEV and WNV (Wang et al. 2016). Flavivirus enters into the host cells via clathrin-mediated endocytosis, exposing the virus to acidic endosomal compartments that results in viral and endosomal membrane fusion due to conformational rearrangement of envelope glycoproteins (Nour et al. 2013). The fusion of the membranes induces viral RNA release into the cytoplasm, which undergoes cap-dependent translation via ribosomes at rough endoplasmic reticular membrane leading to polypeptide synthesis. The polyprotein is processed into three structural and seven

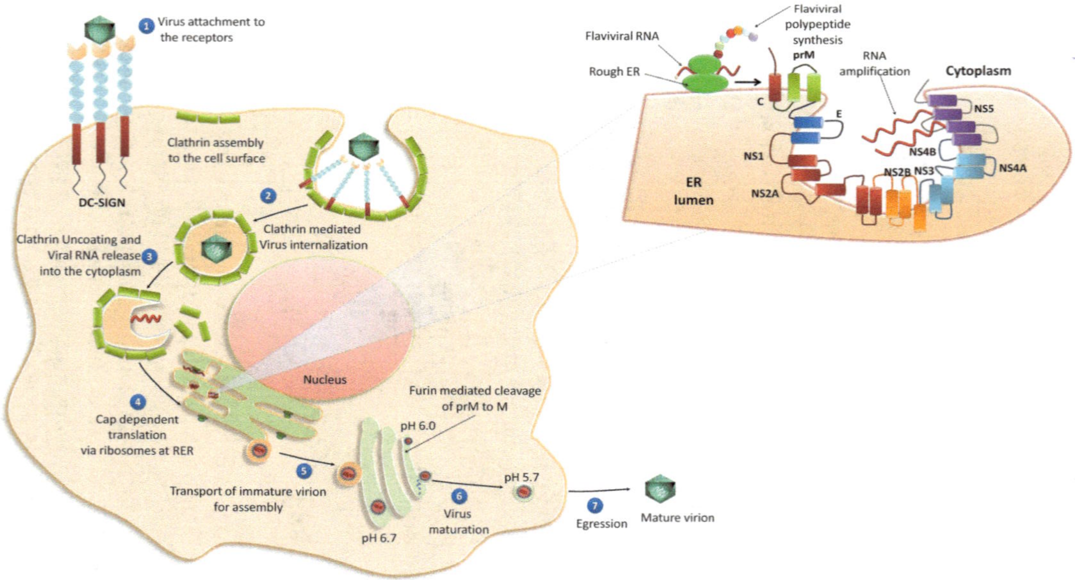

Fig. 7.2 Flavivirus replication. (1) DC-SIGN is crucial for virus attachment both for JEV and WNV; (2) clathrin assembly to the cell surface; (3) Flavivirus enters into the host cells via clathrin-mediated endocytosis; (4) uncoating of clathrin in conjunction with viral and endosomal membrane fusion leading to viral RNA release into the cytoplasm; (5) cap-dependent translation via ribosomes at rough endoplasmic reticular membrane leading to polypeptide synthesis; (6) on the surface of the endoplasmic reticulum (ER), virus assembly occurs; (7) immature virions transported to Golgi complex for virus maturation and virus maturation result in cleavage of prM to M via furin, a Golgi apparatus-resident protease; (8) mature infectious virions released from the infected cells via egression. The inset in the figure represents the cap-dependent translation of flaviviral RNA via ribosomes at rough endoplasmic reticular membrane leading to polypeptide synthesis. The polyprotein is cleaved into three structural and seven non-structural proteins by viral proteases and host signal peptidases. The cleaved proteins are associated with the intracellular membrane of ER where the non-structural proteins coordinate the ER membrane invaginations, and the sites of RNA amplification via negative-strand RNA (−) synthesis

non-structural proteins by viral proteases and host signal peptidases. The cleaved proteins are associated with the intracellular membrane of ER where the non-structural proteins coordinate the invaginations of the ER membrane and the sites of RNA amplification via negative-strand RNA (−) (RNA) synthesis (Paul and Bartenschlager 2015). The generated (+) RNA strand by viral replicase complex will be encapsulated in viral particles via virus encapsidation process following budding into the ER lumen, which occurs in the regions opposite to the replication sites (Shi and Suzuki 2018). Maturation of virus results in cleavage of prM to M via furin, a Golgi apparatus-resident protease (Fig. 7.2). Infectious mature viruses exit the cells via a conventional process known as egress (Murray et al. 2008).

7.3 Japanese Encephalitis Virus

7.3.1 Origin of Japanese Encephalitis Virus

JEV is a positive-sense single-stranded RNA virus from genus *Flavivirus* and family *Flaviviridae*. A phylogenetic study reveals the origin of JEV in the area of Malay Archipelago from the ancestral virus (Solomon et al. 2003). Evolution of JEV has probably transpired 1000 years ago into various genotypes (I–V) and subsequently widespread across Asia. The first case of Japanese encephalitis (JE) with the clinical presentation was reported in 1871 in Japan, and afterwards, around half a century later, more than 6000 clinical cases were identified (WHO 2019). Consequently, several outbreaks have been reported in the years 1927, 1934, and 1935. The aetiological agent was isolated from the human brain in 1935 and 10 years later was proven as JEV when the isolate was transfected into monkey (Erlanger et al. 2009). The first case of JE in several countries was reported such as Korea in 1933; China in 1940; the Philippines in 1950; Malaysia in 1952; and India in 1955. In 1938, the life cycle of JEV has been demonstrated with the involvement of wading birds, *Culex tritaeniorhynchus*, as a vector and pigs as amplifying host (van den Hurk et al. 2009).

7.3.2 Genotype of JEV

JEV can be categorised into five genotypes as GI–GV (Schuh et al. 2014). The most predominant genotype is GIII which is widely distributed among Asian countries as Japan, People's Republic of China, South Korea, Taiwan, the Philippines, India, Nepal, and Sri Lanka. Nonetheless, in the past decades, the GI has replaced the GIII in several countries, including China, South Korea, and Thailand (Fulmali et al. 2011). Furthermore, GI has been categorised into GI-a prevalent in mostly tropical regions, and GI-b reported in temperate regions. The GV has re-emerged after around 60 years of undetected virus circulation (Schuh et al. 2013a). GIV has been confined to Indonesia only, and the reason remains undefined (Schuh et al. 2013b) (Fig. 7.3).

Fig. 7.3 Geographical distribution of JEV genotypes: GIII is widely distributed among Asian countries as Japan, People's Republic of China, South Korea, Taiwan, the Philippines, India, Nepal, and Sri Lanka. In the past decades, the GI has replaced the GIII in several countries including China, South Korea, and Thailand and GIV has been confined to Indonesia only

7.3.3 Incidence and Prevalence of JEV in Livestock

JEV is a zoonotic viral infectious disease and is the foremost cause of viral encephalitis in Southeast Asia, East Asia, and Australia with approximately 68,000 clinical cases and 24,000 deaths yearly (Campbell et al. 2011). According to the World Health Organization (WHO), 24 countries in the Western Pacific Regions and Southeast Asia have endemic JEV transmission, and about 3 billion people are at risk for JEV infection (WHO 2019). The epidemic and endemic areas of JE are described to have the prevalence of the extensive host range. A systemic review of the global epidemiology of human JEV infection suggests that the actual incidence of JE is approximately ten times elevated than the WHO reports (Wang and Liang 2015). Surveillance systems monitoring the viremia profiles of domestic animals are imperative for understanding the host range, vector transmission, and ecological distribution of JEV. Seroprevalence of JE in livestock animals in Malaysia shows highest seropositivity in dogs and pigs (Kumar et al. 2018). An epidemiological survey of JE incidence in China shows the highest seropositivity in pigs (Zhang et al. 2017).

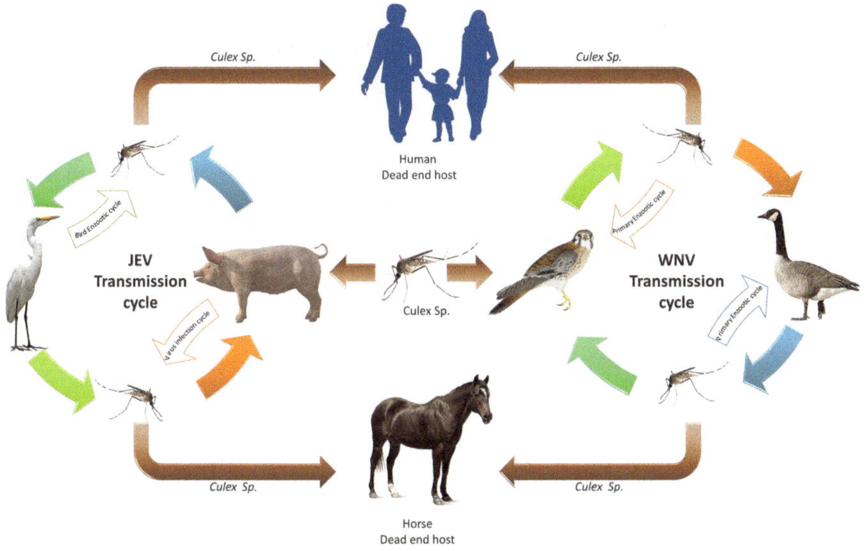

Fig. 7.4 Flavivirus transmission. Transmission cycle initiates with an infected mosquito biting (*Culex* sp.) to the amplifying hosts. Pigs and egrets act as the amplifying host and reservoir for JEV whereas wandering birds are the amplifying host and reservoir for WNV. Human and horse are the dead-end hosts in both the transmission cycle

Periodic cases of JE in equine have been documented in diverse countries where the mortality rate ranges from 5% to 30%. The epidemiological data of Korea shows the variable incidence rate among the domestic animals where wild birds represent 86.7%, cattle shows 51.7%, horse shows 49.7%, and goat shows 12.1% incident rate (Wang and Liang 2015). Similar surveillance programme of JE shows 10% seropositivity among equines in different regions of India (Gulati et al. 2011).

7.3.4 Transmission of JEV

Frequent cases of human JEV infection in the endemic areas can be associated with intense pig husbandry and agricultural rice fields, which describe the pig-associated rural cycle of JEV infection (Su et al. 2014). Pigs play a crucial role in the JEV life cycle where it acts as amplifying hosts with a high level of viremia (Cappelle et al. 2016). The clinical presentations of JEV infection among the host are diverse as mild symptoms represented by pigs, and severe disease reported in human and horse delineating the dead-end host (Fig. 7.4) (Impoinvil et al. 2013).

Moreover, another vertebrate such as dogs, ducks, reptiles, and chickens shows higher seroprevalence, suggesting the diverse host range of JEV (Oliveira et al. 2017). JEV is primarily transmitted through infected mosquito bites where *Culex tritaeniorhynchus*, *Culex vishnui*, and *Culex pseudovishnui* are the most significant mosquito species. Vertical transmission of JEV entails other species of mosquitoes including *Culex annulus*, *Culex pipiens*, *Aedes albopictus*, and *Aedes togoi* (Rosen et al. 1989). The association between mosquito and water birds defines the ecological perspectives of JEV and is known as a bird-associated life cycle. The prevalence of JE in the low-pig-density areas over other livestock suggests the importance of diverse host range for virus transmission. The role of domesticated birds in the virus transmission is significant where ducks and chickens represent sufficient viremia to infect mosquitoes (Lord et al. 2015). Furthermore, under experimental conditions, the infected *Culex tritaeniorhynchus* choose to feed cattle over pigs, and it has also been demonstrated in the natural infection of JEV in India (Philip Samuel et al. 2008). Vector-free transmission of JEV has been demonstrated where oronasal secretions from pigs make others susceptible to virus infection (Ricklin et al. 2016a).

7.3.5 Risk Factors Associated with JEV Infection

The geographic expansion of JE indicates that around 50% of the world population is living in JE endemic areas. The emergence of JE can be explained by several biological, ecological, and socioeconomic factors. Biological factors include vector population dynamics, sero-cross-reactivity with other flaviviruses, and vector-free transmission (Kilpatrick and Randolph 2012). Ecological factors include diverse agricultural practices, increased mean summer temperatures, increased migration of water birds, and increased pig farming (Tian et al. 2015). The role of seasonal variability is imperative in Taiwan, where the occurrence of JE cases intensifies between May and August during 1991–2005 (Lin et al. 2017).

Similarly, seasonal rainfall is responsible for an increase in the mosquito population and can be correlated with the modulation of various meteorological parameters (Ramesh et al. 2015). Socioeconomic factors include international travel, inadequate public health systems, and peri-urban growth (Luo et al. 1995). Apart from the mentioned risk factors, the absence of effective mass vaccination programme in the endemic areas may result in the emergence of JE cases.

7.3.6 Immunobiology of JEV

Host immune response during JEV infection involves both innate and adaptive immune responses for clearance of the virus. Humoral immune response has been well established, whereas cell-mediated immune response is the current area of research worldwide. The role of T cell for the effective maintenance of antibody-mediated response suggests the association of humoral and cell-mediated immunity to prevent immunopathogenesis and to understand the persistence of JEV in CNS (Larena et al. 2011). Ironically, the host-mediated immune response against the virus is critically responsible for clinical pathology, which is termed as encephalitis. In several studies neutralising antibodies have been demonstrated to be protective against JEV infection. However, antibody-mediated immune response is protective in case of anti-NS1 IgG1 antibodies (Li et al. 2012).

During the acute period of JEV infection, neutralising antibodies are responsible for the abolishment of viral replication and viral spread whereas these antibodies are known to reduce the cytopathic effect in the chronic infection resulting in low tissue damage. Greater extent of antigenic resemblance among flaviviruses causes the antibody-dependent enhancement of infection (Pierson et al. 2008).

The involvement of lymphocyte during JEV infection in humans has been identified; however, the precise role is not well established and is the current area of research worldwide. Adoptive intracerebral inoculation of JEV-primed T cells along with the lethal viral dose in mice suggests the involvement of T cells in protection (Murali-Krishna et al. 1996). Clinical studies are suggesting the principal role of NS3 viral proteins for eliciting the $CD8^+$ and $CD4^+$ T-cell responses. The rationale behind $CD8^+$ and $CD4^+$ T-cell responses in JEV-infected individuals reveals the source of antigens from the amino acid sequences ranging from 193 to 324 (Kumar et al. 2004). A recent study using a synthetic peptide library suggesting the human T memory cell response during JEV infection is differentially targeted. The $CD8^+$ T-cell response is predominant in healthy donors exposed with JEV, whereas the $CD4^+$ T-cell response shows dominance in recovered JE patients (Turtle et al. 2016).

Furthermore, the function of IFN-γ in the host defence mechanism during viral infection and T cells is involved in the IFN-γ production for clearance of virus from CNS (Larena et al. 2013). Also, the microenvironment of cytokines during early JEV infection plays a crucial role in the host immune response. Dendritic cells (Dcs) are the principal source of proinflammatory cytokines and upon JEV infection Dcs secrete higher levels of IL-8, IL-6, and TNF-α and more efficiently expand the pre-existing Treg cells (Gupta et al. 2014). To establish a successful infection, JEV modulates both the adaptive and innate immune responses. The viral genome encodes 3400-amino acid-long polyprotein which undergoes processing via virus proteases and host signalase and gives rise to three structural and seven non-structural proteins. The viral proteins are involved in modulating the host defence mechanisms, and the incidents are characterised as the viral immunopathogenesis (Kumar et al. 2016). The immunopathogenesis of JEV in animals is not well understood; however, studies on pigs are suggesting that the viral tropism is not limited to the CNS only; rather it involves secondary lymphoid organs as well (Ricklin et al. 2016b).

7.3.7 Diagnostics for JEV Infection

JEV infection in most of the animals is asymptomatic that creates challenges for identification and prevalence of JEV in livestock population. Various aspects of diagnosis are taken into consideration which combines the clinical, serological, and pathological evaluation of infection (Table 7.1). The perfect diagnosis in equines can be achieved upon isolation of virus from the CNS tissue of infected horse (WHA 2016). In vitro isolation of virus involves primary cell culture and secondary cell lines for the propagation of isolated virus samples. Plaque assay can be utilised for determination of viral titre and infectivity rate and pathogenicity of the isolated strain. The handling of the infectious samples should be processed under biosafety

Table 7.1 JEV and WNV diagnosis

S. no.	Test	Detects
JEV diagnosis		
1	ELISA	JEV antigen
2	PRNT assay	Titre of neutralising antibody
3	Indirect immunofluorescence test (IIFT)	Specific reactivity of antibodies with viral antigen
4	Haemagglutination inhibition (HI) test	Detect and quantify the concentration of virus by titration
5	Complement fixation test (CFT)	Antibody titre against viral antigen
6	Reverse transcription-polymerase chain reaction (RT-PCR)	Viral RNA
7	Reverse transcription-loop-mediated isothermal amplification (RT-LAMP) assay	Virions
WNV diagnosis		
1	VecTOR test	Viral antigen
2	Antigen capture ELISA	Viral antigen
3	Rapid analyte measurement platform (RAMP)	Viral antigen
4	Virus isolation in suckling mouse	Infectious virus
5	Virus isolation in Vero cell culture	Infectious virus
6	Conventional RT-PCR	Viral RNA
7	Nucleic acid sequence-based amplification (NASBA)	Viral RNA
8	TaqMan (real-time RT-PCR)	Viral RNA
9	Reverse transcription-loop-mediated isothermal amplification (RT-LAMP)	Viral RNA

level 3 laboratories. The serology-based diagnosis is unreliable and may result in false-positive data due to sero-cross-reactivity among flaviviruses or previous exposure of animals to JEV infection (Hobson-Peters 2012). Therefore, the serological based diagnosis has to be validated in the context of viral titre. The plaque reduction neutralisation assay (PRNT) is considered to be an effective serological based diagnosis for JEV infection in animals (OIE 2012).

Furthermore, ELISAs can be utilised where IgM-specific antibodies can be detected for an acute case of JEV infection (Litzba et al. 2010). In animals, the anti-NS1 antibody can be detected to differentiate the natural infection and vaccinated animals (Konishi et al. 2006). One of the most common diagnostic platforms is the nucleic acid-based detection of JEV. The three available nucleic acid-based assays are reverse transcription polymerase chain reaction, real-time PCR, and reverse transcription loop-mediated isothermal amplification (RT-LAMP). All the three assays are highly specific in case of detection of JEV in blood sample (Dhanze et al. 2015). All the genotypes show at least 12% of sequence variation, and some of the nucleotide-based assays have been developed to differentiate among the genotypes, specifically GI and GIII by using RT-PCR (Chen et al. 2014).

Table 7.2 JEV and WNV vaccines

S. no.	Vaccine name	Status	Human/ animal	Remark
JEV vaccine				
1	Anyang300 vaccine	Available	Animals	Live JEV vaccine using Anyang300 strain
2	JEV G1 vaccine	Clinical trial, phase I	Animals	Inactivated JEV G1 (KV1899 strain) containing recombinant porcine GM-CSF
3	JENCEVAC	Phase III trial completed	Human	Mouse brain purified inactivated JE vaccine
4	IXIARO	Based on SA-14-14-2, approved for use in some countries	Human	Vero cell-inactivated vaccine
5	ChimeriVax	Based on phase I–III clinical trial, approved in some countries	Human	Live, attenuated recombinant virus constructed from yellow fever (YF) 17D
6	JENVAC	Phase III trial completed	Human	Vero cell-derived vaccine developed using an Indian strain of JEV
WNV vaccine				
1	West Nile innovator	Commercially available	Animals	Whole inactivated WNV
2	Vetera vaccine	Commercially available	Animals	Whole inactivated WNV
3	West Nile-innovator DNA	Discontinued	Animals	Plasmid DNA PrM/E
4	Recombitek	Commercially available	Animals	Canarypox expressing PrM/E
5	Prevenile	Recalled	Animals	YF17D backbone expressing WNV PrM/E
6	ChimeriVax-WN02	Clinical trials, phase II	Humans	Chimeric YF17D backbone expressing WNV PrM/E
7	Chimeric WN/ DEN4-3′delta30	Clinical trials, phase I	Humans	Chimeric DV4 backbone expressing WNV PrM/E
8	Clinical trial VRC303	Clinical trials, phase I	Humans	Plasmid DNA expressing PrM/E
9	WN-80E	Clinical trials, phase I	Humans	Soluble E lacking the transmembrane domain

7.3.8 Prevention and Control Strategies for JE

Several vaccine candidates (Table 7.2) are available for human use; however, vaccine for livestock is underdeveloped. Live attenuated virus or inactivated virus is administered to improve the life expectancy of animals. Anyang300 (live attenuated vaccine) has been administered in the swine in South Korea from the last 30 years, which shows the reduction of JE in the swine (Nah et al. 2015). Another approach utilises the DNA vaccine encoding viral structural proteins or DNA vaccine has been used in combination with the inactivated virus (Konishi et al. 2000). Several imperative advantages are associated with this approach, such as preventing foetal abnormalities, generation of immunological memory, and production of neutralising antibodies.

Furthermore, vaccination in horses shows higher efficacy in preventing the occurrence of disease. An alternative and effective approach is to control the vector population in the endemic areas for preventing the transmission of the virus (Benelli et al. 2016). Insecticides such as pyrethrins in the form of fogging can be applied in the animal housing areas, including the use of mosquito-proof screen for preventing access of vectors to the animals (Karunaratne and Hemingway 2000).

7.4 West Nile Virus

7.4.1 Origin of West Nile Virus

West Nile virus is a member of genus *Flavivirus* in family *Flaviviridae* and shares similar immunological properties with Japanese encephalitis virus and St. Louis encephalitis virus. The virus was first characterised as a distinct pathogen in 1937 isolated from the blood of a febrile woman in the West Nile region of Uganda (Smithburn et al. 1940). Transmission of WNV occurs between ornithophilic mosquito vectors and migratory birds as primary and reservoir hosts, where humans and several other mammals act as incidental or dead-end hosts. With around 20% of symptomatic cases among humans and horses, 1% of cases in humans represent with neurological manifestation along with 1–10% fatality rate (Danis et al. 2011) while in horses 90% of cases have neurological diseases and approximately 30–40% fatality rates (Ward et al. 2006). Thus, the distribution of WNV extended throughout Africa, the Middle East countries, Eurasia, and then North and South America, affecting the health of the public, domestic animals, and wildlife (Karabatsos 1978). During 1999–2001 in the United States, 149 clinical cases associated with WNV in humans were reported with encephalitis and meningitis, there were 814 cases of equine encephalitis, and 11,932 birds died with major fatality occurring in crows (Komar et al. 2003). ArboNET, an arboviral surveillance system governed by Centers for Disease Control and Prevention (CDC, USA), monitors the associated avian deaths, which also include WNV. Apart from a mosquito bite, the transmission route of WNV reported among birds might be through contact or oral transmission.

7.4.2 Incidence and Prevalence of WNV

WNV was not considered as a potential threat to public health at the time of its discovery; however, during the 1950s the Middle East epidemic of fever and encephalitis associated with WNV determined its epidemiology and ecology.

The clinical manifestations of WNV were identified in 1951's Israel outbreak where mostly young children were infected with major symptoms like headache, abdominal pain, fever, and vomiting (Hurlbut et al. 1956). Numerous outbreaks of WNV in Egypt during the 1950s led to the detection of high seroprevalence rate among the population which instigated an extended study including sero-surveys in animals, WNV vector identification, and experimental infection of arthropods, birds, humans, and equines. These findings greatly helped in understanding the epidemiological and clinical aspects of WNV. Sero-surveys among animals suggested a wide range of hosts such as birds and other mammals including equines.

Interestingly, a finding demonstrated that the virus could be specifically isolated from mosquitoes and not from other arthropods, establishing mosquitoes as the primary vectors. In 1957, for the first time severe neurological manifestations were reported in an elderly group during an outbreak in Israel (Hayes and Gubler 2006). In 1974, South Africa outbreak patients with meningitis or encephalitis were reported. However, symptomatic patients with neurological manifestations were sporadic, and the majority of patients developed the mild febrile disease. WNV outbreaks occurred intermittently in Russia, India, Spain, and South Africa; however, large outbreaks were insignificant in number during the late 1970s and 1980s. In 1999, the WNV strain was detected in North America, specifically Queens and New York City where more than 150 species of dead birds were reported WNV positive. The virus was isolated from the carcasses of dead birds, and the number was confirmed using WNV-specific RNA sequences. The New York City 1999 strain was virulent in American crows (*Corvus brachyrhynchos*). In 2003, WNV strains were found to be attenuated in birds present in Texas and Mexico, showing phenotypic variation in WNV strains in Western countries (Davis et al. 2004). In the year 2006, cases of WNV have been reported in Assam, India, during surveillance of acute encephalitis syndrome with 11.65% of positive cases. Furthermore, WNV has been reported in clinical samples in West Bengal, India (Khatun and Chatterjee 2017).

7.4.3 Transmission of WNV

WNV transmission involves mosquito-bird-mosquito cycle where mosquitoes majorly belong to *Culex* sp. (Turell et al. 2001) (Fig. 7.3). However, in the United States alone, WNV has been isolated from 29 different species of mosquito belonging to ten genera. *Culex pipiens* was abundantly found as an epizootic vector in birds in Northern region of the United States. Similarly, *C. quinquefasciatus*, *C. nigripalpus,* and *C. tarsalis* were the epidemic vectors of WNV in other parts of North America (Sardelis et al. 2001). In Africa, *C. univittatus* was significantly the most important vector of WNV transmission to human beings. WNV has been isolated from both soft and hard ticks in some parts of Eastern hemisphere, yet ticks are not considered as an important epizootic vector. Birds are the amplifying hosts and are considered as natural reservoirs for WNV; 111 bird's species were infected in

North America alone with WNV (Centers for Disease Control and Prevention (CDC) 2002). These infected avian species initially developed a high viral load to increase WNV transmission to feeding mosquitoes; however, these infected birds generally survived their viremia and generated immunity against it. However, a few species of the birds were reported to become ill and died. Introduction of WNV strains in Mediterranean and European countries was due to infected migratory birds. Various mammalian species are receptive of getting naturally or experimentally infected with WNV. A wide range of mammalian species such as human beings, few species of bats, horses, squirrels, chipmunks, cats, skunks, and rabbit had been naturally infected with WNV.

7.4.4 Immunobiology of WNV

With the mosquito bite, the WNV replicates for a limited time in Langerhans cells (LCs) of the epidermis (Garcia-Tapia et al. 2007), further infecting dendritic cells (DCs) in lymphoid tissues where they secrete cytokines such as INF-α and INF-β (Liu et al. 2006). During the transient viremia or acute phase of infection, WNV could also be isolated from blood. These dendritic cells also secrete tumour necrosis factor alpha (TNF-α) (Diamond et al. 2003), and macrophage migration inhibitory factor (MIF) (Arjona et al. 2007) that induces alterations in the integrity of BBB leading to leakage of plasma and cellular proteins in the brain, thus resulting in neuroinvasion of WNV. This induces production of cytokines/chemokines and migration of immune cells in the brain, leading to severe inflammation and death of neurons, further causing the disruption of BBB and inhibiting the differentiation and proliferation of neural progenitors. WNV can persist in neurons for a long time to evoke both innate and adaptive immune response.

7.4.4.1 Innate Immune Response

With the WNV infection, LCs are activated and DCs overexpress MHC II molecules, CD80, E-cadherin, and CD54 (Byrne et al. 2001). These cells within 24–48 h eventually migrate to draining lymph nodes, initiating activation of innate immune response comprising secretion of various cytokines and chemokines, complement system activation, proliferation of leukocytes in lymph nodes, processing, and presentation of antigen to T cells. Pathogen recognition receptors (PRRs) such as toll-like receptors (TLR-3, TLR-7), a retinoic-acid-inducible gene I (RIG-I), and melanoma-differentiation-associated gene 5 (MDA5) play an important role in recognising and responding by binding with an RNA virus. Thus, their binding leads to stimulation of transcription factors like IFN regulatory factors (IRF) 3 and 7, and expression of IFN-stimulated genes (ISGs) that induce antiviral pathways to restrict WNV infection. IFN stimulates Janus kinase/signal transducer and activator of transcription (JAK/STAT) pathway that activates various components with antiviral activity against WNV (Platanias 2005). ISGs potentiate the production of IFN-β inhibiting viral translation. However, several non-structural viral proteins such as NS2A, NS2B, NS3, NS4A, NS4B, and NS5 hinder IFN signalling at multiple stages within the infected cell. The expression of IFN activates T cells to inhibit WNV infection. γδ T cells are connecting link between innate and adaptive immune responses that produce Th1- or Th2-type cytokines but these cells recognise a limited range of antigens as compared to αβ T cells (Poccia et al. 2005). T cells (γδ) generate IFN-γ, which plays a pivotal role in restricting virus propagation by several methods such as its antiviral activity activates T_h cells and overexpression of MHC-I, and activates phagocytic cells.

7.4.4.2 Adaptive Immunity

WNV-infected DCs activate B cell that leads to the production of immunoglobulins such as IgM and IgG. These immunoglobulins are specific to WNV trigger viral clearance from the bloodstream. WNV virions majorly comprise two antibody epitopes such as pre-membrane or membrane (prM or M) and envelope (E) proteins. The neutralising antibodies against WNV are targeted towards envelope and subsets of antibodies are against prM or M protein (Nybakken et al. 2005). They inhibit the entry of the virus in cells by preventing viral attachment and fusion with cells. NS1, a non-structural glycoprotein, binds with the host cell surface receptors. A study reported that antibodies against NS1 when given prophylactically or in a dose-dependent manner protect mice from WNV encephalitis. Cytotoxic T cells play a significant role in immunity against the clearance of viral load from the peripheral system during encephalitic flavivirus infections. The cytotoxic T lymphocytes (CTLs) when exposed to WNV-infected cells proliferate, release cytokines, and kill infected cells (Camenga et al. 1974). Studies have shown that cellular and humoral immune response works in combination to reduce viral blood in blood and prevent CNS infection.

7.4.4.3 Immune Responses Within CNS

The mechanism of WNV entry into the CNS is unclear; it might involve retrograde axonal transport (Samuel et al. 2007) from neurons to the spinal cord or crossing of the blood–brain barrier (BBB) (Samuel and Diamond 2006). WNV might breach the BBB by infecting macrophages or endothelial cells, or disruption of BBB by upregulation of cytokines (TNF-α, MIF) (Wang et al. 2004). During the viral attack in CNS, WNV causes neuronal death or injury, specifically in targeted regions like spinal cord neurons, cortical neurons, midbrain neurons, and cerebellar neurons. Within the cerebellum, WNV mediates a regional upregulation of CXCL10, which attracts the T cells with CXCR3 receptors, i.e. $CD8^+$ T cells to reduce the viral load (Zhang et al. 2008). Various studies in mice have reported that chemokines CCL3–5 bind to CCR5 receptor and are highly induced within the brain; in an experiment targeted deletion of CCR5 led to decreased trafficking of leukocytes, increased viral load, and neuronal deaths. However, there is a limited understanding of T-cell infiltration in perivascular space during WNV encephalitis (Bouffard et al. 2004).

7.4.5 Diagnostics for WNV

WNV is considered as a threat to public health. Thus accurate surveillance of a number of human and animal cases is imperative. Various serological diagnostic methods have been developed for detection of WNV in human cases, as well as in mosquitoes and non-human vertebrate hosts. Infections with WNV and other arboviruses that cause encephalitis are mostly asymptomatic or non-specific, as several pathogens could be responsible for such clinical symptoms or manifestations. Laboratory-based diagnostic tests should be considered for the differentially diagnosed pathogens. Numerous diagnostic tests are available for the detection of WNV from blood, serum, CSF, and tissue of human, other mammals, birds, and reptiles (Centers for Disease Control and Prevention 2013). The confirmatory test for WNV identification is the IgM assay, which presently could be detected by four commercially available FDA-approved kits. Numerous tests are available to detect WNV, WNV proteins, or WNV RNA from human samples, and few of them could be used in other vertebrates and mosquitoes for its detection (Table 7.1).

7.4.6 Risk Factors Associated with WNV Infection

The risk factors contributing towards amplification of WNV are dependent on climate conditions, including the optimum level of rainfall, precipitation and ambient temperature, and geographic distribution of mosquito vectors. Several studies have confirmed the correlation among ambient temperature, precipitation, and rising number of *Culex pipiens* population (Paz and Albersheim 2008). Heavy rainfall and floods destroy the breeding habitats of mosquitoes from stagnant water whereas in drought condition WNV transmission increases as birds (amplifying host) and mosquitoes share the same water sources during its scarcity. On the other hand, risk factors associated with clinical manifestations and diagnosis for WNV infections are critical as the infection is mostly asymptomatic. Only 1% of cases of WNV in humans are characterised by neurological sequelae such as meningitis, acute flaccid paralysis, and encephalitis (Gray and Webb 2014). The diagnosis of WNV infection could be challenging as its clinical manifestation resembles with various other viruses such as Japanese encephalitis virus, Kunjin, and St. Louis encephalitis. Specific tests based on IgM and IgG antibodies verify the identity of the virus. Several risk factors are host specific such as in conditions of comorbidity with hepatitis C infection, history of cardiovascular diseases, immunosuppression, and chronic renal diseases are mostly vulnerable to WNV infections. Immigration of infected individuals may also be one of the reasons for the spreading of WNV infection in new regions.

7.4.7 Prevention and Control for WNV

Prevention and control of zoonotic arboviral diseases could be accomplished by establishing a comprehensive, integrated vector management (IVM) programme. A robust surveillance system to monitor the abundance of infected adult mosquitoes and larval mosquitoes is manifested to control the rising infection rate due to vectors. The source or habitat of mosquitoes should be modified to reduce their number. Application of mosquito repellents, impregnated mosquito nets, larvivorous fishes, oils, space sprayers, and insecticides, in addition to prevention of water logging, solid waste management, and disinfection of pits, leads to a reduction in point source of mosquito production (Haikerwal et al. 2017). The most promising method to prevent WNV infection is vaccination in both animals and humans. However, there is no commercially available licensed vaccine (Brandler and Tangy 2013). West Nile vaccines for animals are commercially available for horses and other mammals. Human vaccines could be subcategorised as vaccines under clinical trials (Table 7.2) and preclinical trials. Various vaccine candidates are under preclinical trials and are tested in animal models such as subunit vaccine, DNA vaccine, inactivated virus vaccines, recombinant viral vector vaccines, chimeric vaccines, and live attenuated vaccines.

7.5 Conclusions

The burden of Flavivirus has made the whole world susceptible to its infection. Thus a detailed discussion is carried out regarding JEV and WNV; most prominently they are prevalent in Southeast Asian countries and Western countries, respectively. Various important characteristics, such as the background of the virus, incidence and prevalence, transmission cycle, immunobiology, diagnostics, risk factors, and prevention and control, were examined to understand the complexity of these viruses. JEV is the causative agent of Japanese encephalitis, a paediatric disease causing mortality and morbidity in children but not limited to them. The surveillance systems have reported cases of JEV infection in animals such as pigs, equine, wild birds, cattle, and goat. The resurgence of JEV majorly occurs at the time of ambient temperature and precipitation facilitating favourable breeding habitats of mosquitoes generally in paddy fields. The transmission cycle of JEV involves pigs as the amplification host, water and migratory birds as the reservoirs, and vector mosquitoes. Similar to JEV, WNV also causes neurological diseases like encephalitis in humans and mostly in horses. Transmission of WNV occurs between ornithophilic mosquito vectors and migratory birds as primary and reservoir hosts, where humans and several other mammals act as incidental or dead-end hosts. Thus, the surveillance system has detected a large number of avian hosts of WNV. These infected migratory birds had spread the virus from Uganda to Middle Eastern countries, then Europe, and North American countries. Both viruses could coexist due to structural similarities. Thus, the emergence of these viruses with a wide range of vector population and establishment of virus in new geographic places could be prevented by considering all the factors and an internationally collaborative approach.

7.6 Future Perspectives

Globalisation, deforestation, climate change, global trade, and travel, expanding the human population, and microbial evolution are the leading causes of emergence and resurgence of these flaviviruses. The natural as well as anthropogenic cause of vector establishment needs to be uprooted. Currently, breeding places of mosquitoes are not limited to temperate, tropical, or subtropical regions; globalisation leading to advanced transportation has brought the whole world at risk of Flavivirus infection. The prevention and control programme of zoonotic arboviral diseases could be accomplished by establishing a comprehensive integrated vector management programme. A robust surveillance system to monitor the abundance of infected adult mosquitoes and larval mosquitoes could be manifested to control the rising infection rate. The source or habitat of mosquitoes should be modified to reduce their number. Application of mosquito repellents, impregnated mosquito nets, larvivorous fishes, oils, space sprayers, and insecticides, in addition to prevention of water logging, solid waste management, and disinfection of pits, leads to a reduction in point source of mosquito production. The most promising method to prevent WNV and JEV infection is vaccination in both animals and humans. West Nile vaccines and JEV vaccines for animals are commercially available for horses and other mammals. Nevertheless, to mitigate these diseases, we must compressively understand the immunobiology and pathogenesis of flaviviruses in various hosts, including the natural host/reservoir animals.

Acknowledgements All the authors of the manuscript thank and acknowledge their respective universities and institutes.

Conflict of interest: There is no conflict of interest.

References

Apte-Sengupta S, Sirohi D, Kuhn RJ (2014) Coupling of replication and assembly in flaviviruses. Curr Opin Virol 9:134–142

Arjona A, Foellmer HG, Town T, Leng L, McDonald C, Wang T, Wong SJ, Montgomery RR, Fikrig E, Bucala R (2007) Abrogation of macrophage migration inhibitory factor decreases West Nile virus lethality by limiting viral neuroinvasion. J Clin Invest 117(10):3059–3066

Benelli G, Jeffries CL, Walker T (2016) Biological control of mosquito vectors: past, present, and future. Insects 7(4):E52

Bouffard JP, Riudavets MA, Holman R, Rushing EJ (2004) Neuropathology of the brain and spinal cord in human West Nile virus infection. Clin Neuropathol 23(2):59–61

Brandler S, Tangy F (2013) Vaccines in development against West Nile virus. Viruses 5(10):2384–2409
Broutet N, Krauer F, Riesen M, Khalakdina A, Almiron M, Aldighieri S, Espinal M, Low N, Dye C (2016) Zika virus as a cause of neurologic disorders. N Engl J Med 374(16):1506–1509
Byrne SN, Halliday GM, Johnston LJ, King NJ (2001) Interleukin-1β but not tumor necrosis factor is involved in West Nile virus-induced Langerhans cell migration from the skin in C57BL/6 mice. J Investig Dermatol 117(3):702–709
Camenga DL, Nathanson N, Cole GA (1974) Cyclophosphamide-potentiated West Nile viral encephalitis: relative influence of cellular and humoral factors. J Infect Dis 130(6):634–641
Campbell GL, Hills SL, Fischer M, Jacobson JA, Hoke CH, Hombach JM, Marfin AA, Solomon T, Tsai TF, Tsu VD, Ginsburg AS (2011) Estimated global incidence of Japanese encephalitis: a systematic review. Bull World Health Organ 89(10):766–774, 774A–774E
Cappelle J, Duong V, Pring L, Kong L, Yakovleff M, Prasetyo DB, Peng B, Choeung R, Duboz R, Ong S, Sorn S, Dussart P, Tarantola A, Buchy P, Chevalier V (2016) Intensive circulation of Japanese encephalitis virus in peri-urban sentinel pigs near Phnom Penh, Cambodia. PLoS Negl Trop Dis 10(12):e0005149
Centers for Disease Control and Prevention (2013) West Nile virus in the United States: guidelines for surveillance, prevention, and control. https://www.cdc.gov/westnile/resources/pdfs/wnvguidelines.pdf. Accessed 28-2-2018
Centers for Disease Control and Prevention (CDC) (2002) West Nile virus activity—United States, 2001. MMWR Morb Mortal Wkly Rep 51(23):497–501
Chen YY, Lin JW, Fan YC, Chiou SS (2014) Detection and differentiation of genotype I and III Japanese encephalitis virus in mosquitoes by multiplex reverse transcriptase-polymerase chain reaction. Transbound Emerg Dis 61(1):37–43
Danis K, Papa A, Theocharopoulos G, Dougas G, Athanasiou M, Detsis M, Baka A, Lytras T, Mellou K, Bonovas S, Panagiotopoulos T (2011) Outbreak of West Nile virus infection in Greece, 2010. Emerg Infect Dis 17(10):1868–1872
Davis CT, Beasley DW, Guzman H, Siirin M, Parsons RE, Tesh RB, Barrett AD (2004) Emergence of attenuated West Nile virus variants in Texas, 2003. Virology 330(1):342–350
Dhanze H, Bhilegaonkar KN, Ravi Kumar GV, Thomas P, Chethan Kumar HB, Suman Kumar M, Rawat S, Kerketta P, Rawool DB, Kumar A (2015) Comparative evaluation of nucleic acid-based assays for detection of Japanese encephalitis virus in swine blood samples. Arch Virol 160(5):1259–1266
Diamond MS, Sitati EM, Friend LD, Higgs S, Shrestha B, Engle M (2003) A critical role for induced IgM in the protection against West Nile virus infection. J Exp Med 198(12):1853–1862
Erlanger TE, Weiss S, Keiser J, Utzinger J, Wiedenmayer K (2009) Past, present, and future of Japanese encephalitis. Emerg Infect Dis 15(1):1–7
Fulmali PV, Sapkal GN, Athawale S, Gore MM, Mishra AC, Bondre VP (2011) Introduction of Japanese encephalitis virus genotype I, India. Emerg Infect Dis 17(2):319–321
Garcia-Tapia D, Hassett DE, Mitchell WJ Jr, Johnson GC, Kleiboeker SB (2007) West Nile virus encephalitis: sequential histopathological and immunological events in a murine model of infection. J Neurovirol 13(2):130–138
Garg RK, Malhotra HS, Jain A, Kumar N (2017) Dengue encephalopathy: very unusual neuroimaging findings. J Neurovirol 23(5):779–782
Gray TJ, Webb CE (2014) A review of the epidemiological and clinical aspects of West Nile virus. Int J Gen Med 7:193–203
Gulati BR, Singha H, Singh BK, Virmani N, Khurana SK, Singh RK (2011) Serosurveillance for Japanese encephalitis virus infection among equines in India. J Vet Sci 12(4):341–345
Gupta N, Hegde P, Lecerf M, Nain M, Kaur M, Kalia M, Vrati S, Bayry J, Lacroix-Desmazes S, Kaveri SV (2014) Japanese encephalitis virus expands regulatory T cells by increasing the expression of PD-L1 on dendritic cells. Eur J Immunol 44(5):1363–1374
Haikerwal A, Bhatt ML, Saxena SK (2017) Reducing the global burden of dengue: steps toward preventive methods. Arch Prev Med 2(1):28–33

Hayes EB, Gubler DJ (2006) West Nile virus: epidemiology and clinical features of an emerging epidemic in the United States. Annu Rev Med 57:181–194

Hobson-Peters J (2012) Approaches for the development of rapid serological assays for surveillance and diagnosis of infections caused by zoonotic flaviviruses of the Japanese encephalitis virus serocomplex. J Biomed Biotechnol 2012:379738

Hurlbut HS, Rizk F, Taylor RM, Work TH (1956) A study of the ecology of West Nile virus in Egypt. Am J Trop Med Hyg 5(4):579–620

Impoinvil DE, Baylis M, Solomon T (2013) Japanese encephalitis: on the One Health agenda. Curr Top Microbiol Immunol 365:205–247

Karabatsos N (1978) Supplement to international catalogue of arboviruses including certain other viruses of vertebrates. Am J Trop Med Hyg 27(2 Pt 2 Suppl):372–440

Karunaratne SH, Hemingway J (2000) Insecticide resistance spectra and resistance mechanisms in populations of Japanese encephalitis vector mosquitoes, *Culex tritaeniorhynchus* and Cx. gelidus, in Sri Lanka. Med Vet Entomol 14(4):430–436

Khatun T, Chatterjee S (2017) Emergence of West Nile virus in West Bengal, India: a new report. Trans R Soc Trop Med Hyg 111(4):178–184

Kilpatrick AM, Randolph SE (2012) Drivers, dynamics, and control of emerging vector-borne zoonotic diseases. Lancet 380(9857):1946–1955

Komar N, Langevin S, Hinten S, Nemeth N, Edwards E, Hettler D, Davis B, Bowen R, Bunning M (2003) Experimental infection of North American birds with the New York 1999 strain of West Nile virus. Emerg Infect Dis 9(3):311–322

Konishi E, Yamaoka M, Kurane I, Mason PW (2000) Japanese encephalitis DNA vaccine candidates expressing premembrane and envelope genes induce virus-specific memory B cells and long-lasting antibodies in swine. Virology 268(1):49–55

Konishi E, Shoda M, Kondo T (2006) Analysis of yearly changes in levels of antibodies to Japanese encephalitis virus nonstructural 1 protein in racehorses in central Japan shows high levels of natural virus activity still exist. Vaccine 24(4):516–524

Kumar P, Sulochana P, Nirmala G, Haridattatreya M, Satchidanandam V (2004) Conserved amino acids 193-324 of non-structural protein 3 are a dominant source of peptide determinants for CD4+ and CD8+ T cells in a healthy Japanese encephalitis virus-endemic cohort. J Gen Virol 85(Pt 5):1131–1143

Kumar S, Chitti SV, Kant R, Saxena SK (2016) Insights into the immunopathogenesis during Japanese encephalitis virus infection. J Immune Serum Biol 3(1):1–4

Kumar K, Arshad SS, Selvarajah GT, Abu J, Toung OP, Abba Y, Bande F, Yasmin AR, Sharma R, Ong BL, Rasid AA, Hashim N, Peli A, Heshini EP, Shah AKMK (2018) Prevalence and risk factors of Japanese encephalitis virus (JEV) in livestock and companion animal in high-risk areas in Malaysia. Trop Anim Health Prod 50(4):741–752

Kuno G, Chang GJ, Tsuchiya KR, Karabatsos N, Cropp CB (1998) Phylogeny of the genus Flavivirus. J Virol 72(1):73–83

Lannes N, Summerfield A, Filgueira L (2017) Regulation of inflammation in Japanese encephalitis. J Neuroinflammation 14(1):158

Larena M, Regner M, Lee E, Lobigs M (2011) Pivotal role of antibody and subsidiary contribution of CD8+ T cells to recovery from infection in a murine model of Japanese encephalitis. J Virol 85(11):5446–5455

Larena M, Regner M, Lobigs M (2013) Cytolytic effector pathways and IFN-γ help protect against Japanese encephalitis. Eur J Immunol 43(7):1789–1798

Li Y, Counor D, Lu P, Duong V, Yu Y, Deubel V (2012) Protective immunity to Japanese encephalitis virus associated with anti-NS1 antibodies in a mouse model. Virol J 9:135

Li K, Phoo WW, Luo D (2014) Functional interplay among the flavivirus NS3 protease, helicase, and cofactors. Virol Sin 29(2):74–85

Li XD, Deng CL, Ye HQ, Zhang HL, Zhang QY, Chen DD, Zhang PT, Shi PY, Yuan ZM, Zhang B (2016) Transmembrane domains of NS2B contribute to both viral RNA replication and particle formation in Japanese encephalitis virus. J Virol 90(12):5735–5749

Lin CL, Chang HL, Lin CY, Chen KT (2017) Seasonal patterns of Japanese encephalitis and associated meteorological factors in Taiwan. Int J Environ Res Public Health 14(11):E1317

Litzba N, Klade CS, Lederer S, Niedrig M (2010) Evaluation of serological diagnostic test systems assessing the immune response to Japanese encephalitis vaccination. PLoS Negl Trop Dis 4(11):e883

Liu WJ, Wang XJ, Clark DC, Lobigs M, Hall RA, Khromykh AA (2006) A single amino acid substitution in the West Nile virus nonstructural protein NS2A disables its ability to inhibit alpha/beta interferon induction and attenuates virus virulence in mice. J Virol 80(5):2396–2404

Lord JS, Gurley ES, Pulliam JR (2015) Rethinking Japanese encephalitis virus transmission: a framework for implicating host and vector species. PLoS Negl Trop Dis 9(12):e0004074

Luo D, Ying H, Yao R, Song J, Wang Z (1995) Socio-economic status and micro-environmental factors in relation to the risk of Japanese encephalitis: a case-control study. Southeast Asian J Trop Med Public Health 26(2):276–279

Mansfield KL, Hernández-Triana LM, Banyard AC, Fooks AR, Johnson N (2017) Japanese encephalitis virus infection, diagnosis and control in domestic animals. Vet Microbiol 201:85–92

Miller S, Krijnse-Locker J (2008) Modification of intracellular membrane structures for virus replication. Nat Rev Microbiol 6(5):363–374

Mukhopadhyay S, Kuhn RJ, Rossmann MG (2005) A structural perspective of the flavivirus life cycle. Nat Rev Microbiol 3(1):13–22

Muller DA, Young PR (2013) The flavivirus NS1 protein: molecular and structural biology, immunology, role in pathogenesis and application as a diagnostic biomarker. Antiviral Res 98(2):192–208

Murali-Krishna K, Ravi V, Manjunath R (1996) Protection of adult but not newborn mice against lethal intracerebral challenge with Japanese encephalitis virus by adoptively transferred virus-specific cytotoxic T lymphocytes: requirement for L3T4+ T cells. J Gen Virol 77(Pt 4):705–714

Murray CL, Jones CT, Rice CM (2008) Architects of assembly: roles of Flaviviridae non-structural proteins in virion morphogenesis. Nat Rev Microbiol 6(9):699–708

Nah JJ, Yang DK, Kim HH, Song JY (2015) The present and future of veterinary vaccines for Japanese encephalitis in Korea. Clin Exp Vaccine Res 4(2):130–136

Neufeldt CJ, Cortese M, Acosta EG, Bartenschlager R (2018) Rewiring cellular networks by members of the Flaviviridae family. Nat Rev Microbiol 16(3):125–142

Nour AM, Li Y, Wolenski J, Modis Y (2013) Viral membrane fusion and nucleocapsid delivery into the cytoplasm are distinct events in some flaviviruses. PLoS Pathog 9(9):e1003585

Nybakken GE, Oliphant T, Johnson S, Burke S, Diamond MS, Fremont DH (2005) Structural basis of West Nile virus neutralization by a therapeutic antibody. Nature 437(7059):764–769

OIE (2012) Manual of diagnostic tests and vaccines for terrestrial animals, 7th edn. World Health Organization for Animal Health (OIE), Paris

Oliveira ARS, Cohnstaedt LW, Strathe E, Hernández LE, McVey DS, Piaggio J, Cernicchiaro N (2017) Meta-analyses of the proportion of Japanese encephalitis virus infection in vectors and vertebrate hosts. Parasit Vectors 10(1):418

Paul D, Bartenschlager R (2015) Flaviviridae replication organelles: Oh, what a tangled web we weave. Annu Rev Virol 2(1):289–310

Paz S, Albersheim I (2008) Influence of warming tendency on Culex pipiens population abundance and on the probability of West Nile fever outbreaks (Israeli Case Study: 2001–2005). Ecohealth 5(1):40–48

Pérez-Ramírez E, Llorente F, Jiménez-Clavero MÁ (2014) Experimental infections of wild birds with West Nile virus. Viruses 6(2):752–781

Philip Samuel P, Arunachalam N, Hiriyan J, Tyagi BK (2008) Host feeding pattern of Japanese encephalitis virus vector mosquitoes (Diptera: Culicidae) from Kuttanadu, Kerala, India. J Med Entomol 45(5):927–932

Pierson TC, Fremont DH, Kuhn RJ, Diamond MS (2008) Structural insights into the mechanisms of antibody-mediated neutralization of flavivirus infection: implications for vaccine development. Cell Host Microbe 4(3):229–238

Platanias LC (2005) Mechanisms of type-I- and type-II-interferon-mediated signalling. Nat Rev Immunol 5(5):375–386
Poccia F, Agrati C, Martini F, Capobianchi MR, Wallace M, Malkovsky M (2005) Antiviral reactivities of gammadelta T cells. Microbes Infect 7(3):518–528
Ramesh D, Muniaraj M, Samuel PP, Thenmozhi V, Venkatesh A, Nagaraj J, Tyagi BK (2015) Seasonal abundance & role of predominant Japanese encephalitis vectors Culex tritaeniorhynchus & Cx. gelidus Theobald in Cuddalore district, Tamil Nadu. Indian J Med Res 142(Suppl):S23–S29
Ricklin ME, García-Nicolás O, Brechbühl D, Python S, Zumkehr B, Nougairede A, Charrel RN, Posthaus H, Oevermann A, Summerfield A (2016a) Vector-free transmission and persistence of Japanese encephalitis virus in pigs. Nat Commun 7:10832
Ricklin ME, Garcìa-Nicolàs O, Brechbühl D, Python S, Zumkehr B, Posthaus H, Oevermann A, Summerfield A (2016b) Japanese encephalitis virus tropism in experimentally infected pigs. Vet Res 47:34
Rosen L, Lien JC, Shroyer DA, Baker RH, Lu LC (1989) Experimental vertical transmission of Japanese encephalitis virus by Culex tritaeniorhynchus and other mosquitoes. Am J Trop Med Hyg 40(5):548–556
Samuel MA, Diamond MS (2006) Pathogenesis of West Nile Virus infection: a balance between virulence, innate and adaptive immunity, and viral evasion. J Virol 80(19):9349–9360
Samuel MA, Wang H, Siddharthan V, Morrey JD, Diamond MS (2007) Axonal transport mediates West Nile virus entry into the central nervous system and induces acute flaccid paralysis. Proc Natl Acad Sci U S A 104(43):17140–17145
Sardelis MR, Turell MJ, Dohm DJ, O'Guinn ML (2001) Vector competence of selected North American Culex and Coquillettidia mosquitoes for West Nile virus. Emerg Infect Dis 7(6):1018–1022
Schuh AJ, Ward MJ, Brown AJ, Barrett AD (2013a) Phylogeography of Japanese encephalitis virus: genotype is associated with climate. PLoS Negl Trop Dis 7(8):e2411
Schuh AJ, Guzman H, Tesh RB, Barrett AD (2013b) Genetic diversity of Japanese encephalitis virus isolates obtained from the Indonesian archipelago between 1974 and 1987. Vector Borne Zoonotic Dis 13(7):479–488
Schuh AJ, Ward MJ, Leigh Brown AJ, Barrett AD (2014) Dynamics of the emergence and establishment of a newly dominant genotype of Japanese encephalitis virus throughout Asia. J Virol 88(8):4522–4532
Shi G, Suzuki T (2018) Molecular basis of encapsidation of hepatitis C virus genome. Front Microbiol 9:396
Simmonds P, Becher P, Bukh J, Gould EA, Meyers G, Monath T, Muerhoff S, Pletnev A, Rico-Hesse R, Smith DB, Stapleton JT, ICTV Report Consortium (2017) ICTV virus taxonomy profile: Flaviviridae. J Gen Virol 98(1):2–3
Smithburn KC, Hughes TP, Burke AW, Paul JH (1940) A neurotropic virus isolated from the blood of a native of Uganda1. Am J Trop Med Hyg 1(4):471–492
Solomon T (2004) Flavivirus encephalitis. N Engl J Med 351(4):370–378
Solomon T, Ni H, Beasley DW, Ekkelenkamp M, Cardosa MJ, Barrett AD (2003) Origin and evolution of Japanese encephalitis virus in Southeast Asia. J Virol 77(5):3091–3098
Su CL, Yang CF, Teng HJ, Lu LC, Lin C, Tsai KH, Chen YY, Chen LY, Chang SF, Shu PY (2014) Molecular epidemiology of Japanese encephalitis virus in mosquitoes in Taiwan during 2005–2012. PLoS Negl Trop Dis 8(10):e3122
Tian HY, Bi P, Cazelles B, Zhou S, Huang SQ, Yang J, Pei Y, Wu XX, Fu SH, Tong SL, Wang HY, Xu B (2015) How environmental conditions impact mosquito ecology and Japanese encephalitis: an eco-epidemiological approach. Environ Int 79:17–24
Turell MJ, Sardelis MR, Dohm DJ, O'Guinn ML (2001) Potential North American vectors of West Nile virus. Ann N Y Acad Sci 951:317–324
Turtle L, Bali T, Buxton G, Chib S, Chan S, Soni M, Hussain M, Isenman H, Fadnis P, Venkataswamy MM, Satishkumar V, Lewthwaite P, Kurioka A, Krishna S, Shankar MV, Ahmed R, Begum A,

Ravi V, Desai A, Yoksan S, Fernandez S, Willberg CB, Kloverpris HN, Conlon C, Klenerman P, Satchidanandam V, Solomon T (2016) Human T cell responses to Japanese encephalitis virus in health and disease. J Exp Med 213(7):1331–1352

van den Hurk AF, Ritchie SA, Mackenzie JS (2009) Ecology and geographical expansion of Japanese encephalitis virus. Annu Rev Entomol 54:17–35

Wang H, Liang G (2015) Epidemiology of Japanese encephalitis: past, present, and future prospects. Ther Clin Risk Manag 11:435–448

Wang T, Town T, Alexopoulou L, Anderson JF, Fikrig E, Flavell RA (2004) Toll-like receptor 3 mediates West Nile virus entry into the brain causing lethal encephalitis. Nat Med 10(12):1366–1373

Wang P, Hu K, Luo S, Zhang M, Deng X, Li C, Jin W, Hu B, He S, Li M, Du T, Xiao G, Zhang B, Liu Y, Hu Q (2016) DC-SIGN as an attachment factor mediates Japanese encephalitis virus infection of human dendritic cells via interaction with a single high-mannose residue of viral E glycoprotein. Virology 488:108–119

Ward MP, Schuermann JA, Highfield LD, Murray KO (2006) Characteristics of an outbreak of West Nile virus encephalomyelitis in a previously uninfected population of horses. Vet Microbiol 118(3–4):255–259

WHA (2016) Fact sheet: EXOTIC-Japanese encephalitis. https://wildlifehealthaustralia.com.au/Portals/0/Documents/FactSheets/Exotic/EXOTIC%20-%20Japanese%20Encephalitis%20Mar%202016%20(2.0).pdf

WHO (2019) Japanese encephalitis. Fact sheet. http://www.who.int/mediacentre/factsheets/fs386/en/. Accessed 6-3-2018

Zhang B, Chan YK, Lu B, Diamond MS, Klein RS (2008) CXCR3 mediates region-specific antiviral T cell trafficking within the central nervous system during West Nile virus encephalitis. J Immunol 180(4):2641–2649

Zhang H, Rehman MU, Li K, Luo H, Lan Y, Nabi F, Zhang L, Iqbal MK, Zhu S, Javed MT, Chamba Y, Li JK (2017) Epidemiologic survey of Japanese encephalitis virus infection, Tibet, China, 2015. Emerg Infect Dis 23(6):1023–1024

Zou G, Chen YL, Dong H, Lim CC, Yap LJ, Yau YH, Shochat SG, Lescar J, Shi PY (2011) Functional analysis of two cavities in flavivirus NS5 polymerase. J Biol Chem 286(16):14362–14372

Chapter 8
Orbiviruses

Sushila Maan, Manjunatha N. Belaganahalli, Narender S. Maan, Houssam Attoui, and Peter P. C. Mertens

Abstract The family *Reoviridae* contains 15 genera with the genus *Orbivirus* being the largest of the currently classified ones. It contains *Bluetongue virus* (the *Orbivirus* type species), 27 virus species with *bluetongue virus* as type species and (at least) 8 currently unclassified 'orbiviruses'. With a few exceptions, the orbiviruses infect, replicate in and are transmitted by ticks, mosquitoes, *Culicoides* or other blood-sucking insect vectors like phlebotomine flies. Collectively, they infect a varied vertebrate host species, including humans, domesticated and wild ruminants, felines, canines, equines, marsupials, sloths, bats and birds. Bluetongue virus (BTV), epizootic haemorrhagic disease virus (EHDV) and African horse sickness virus (AHSV) are the most economically important orbiviruses, based on their global impact on ruminant-based livestock industries (BTV and EHDV), and on horses primarily in Africa (AHSV). Complete genome sequences are currently accessible for 27 formally recognised species of *Orbivirus*, a complete reference database and comprehensive phylogenetic analysis for the entire genus *Orbivirus*, to further appreciate the taxonomic relationships, epidemiology, replication mechanisms and evolutionary process of these viruses. The phylogenetic trees show branching patterns of orbiviruses as per their vector grouping, supporting the assumption that they have evolved with their vectors through 'co-speciation'. These analyses helped identification of novel *Orbivirus* species and grouped the previously unclassified viruses into existing species. The bioinformatic analyses suggest that

S. Maan (✉) · N. S. Maan
Lala Lajpat Rai University of Veterinary and Animal Sciences, Hisar, Haryana, India

M. N. Belaganahalli
One Health Institute, School of Veterinary Medicine, University of California, Davis, Davis, CA, USA

H. Attoui
Institut National de la Recherche Agronomique (INRA), National Institute for Agricultural Research, UMR1161 Virologie, INRA-ANSES-ENVA, Paris, France

P. P. C. Mertens
School of Veterinary Medicine and Science, University of Nottingham, Sutton Bonington Campus, Loughborough, UK

The Pirbright Institute, Woking, Surrey, UK

Y. S. Malik et al. (eds.), *Emerging and Transboundary Animal Viruses*, Livestock Diseases and Management, https://doi.org/10.1007/978-981-15-0402-0_8

the evolution of orbiviruses has occurred at a slower rate than other RNA arboviruses and non-vector-borne reoviruses.

Keywords Orbivirus · Bluetongue virus · BTV · Double-stranded (ds) RNA · Sequencing · FLAC · Anchor-primer · Reassortment · Second- and third-generation sequencing · Sequence database

8.1 Preamble

The orbiviruses are a big cluster of dsRNA viruses and classified within the genus *Orbivirus*, family *Reoviridae*. Twenty-seven discrete *Orbivirus* species have been documented by the International Committee for the Taxonomy of Viruses (ICTV), containing a total of over 161 distinct serotypes. Based primarily on studies of the orbivirus type species, *bluetongue virus*, the orbiviruses are approximately 60–80 nm in diameter, spherical in appearance, containing a 10-segmented genome (Mertens et al. 2005b). The orbiviruses are 'real' arboviruses (they replicate in both their arthropod vectors and vertebrate hosts) and are transmitted by haematophagous arthropod vectors, comprising *Culicoides*, ticks, sandflies and mosquitoes (although vectors for some orbiviruses have not yet been identified). Some orbiviruses can also be transmitted vertically by their vertebrate hosts, or horizontally via an oral track (Batten et al. 2014; Menzies et al. 2008; Prasad et al. 2007).

Collectively the orbiviruses have a varied host choice including humans, domesticated and wild ruminants, felines, canines, equines, marsupials, sloths, bats and birds. Although some orbiviruses can infect humans, they include several species that are primarily responsible for many economically important diseases of domesticated and wild animals, comprising bluetongue virus (BTV), African horse sickness virus (AHSV), epizootic haemorrhagic disease virus (EHDV) and equine encephalosis virus (EEV). Some orbiviruses that infect wildlife species may also influence the sustainability of wild animal populations. The ability of the orbiviruses to replicate in diverse groups of vectors and hosts, along with large numbers of viruses (serotypes), may help to explain their ubiquity in different ecosystems. They may therefore also pose a greater risk of 'emergence' than other genera of viruses.

The introduction of orbiviruses into disease-free areas and naive populations of susceptible host species can cause high levels of morbidity and mortality. This was demonstrated by an outbreak of BTV-8 in northern Europe in 2006 (Elbers et al. 2008a; Wilson et al. 2007), by tammar wallaby sudden death syndrome caused by Eubenangee viruses, and by epidemic blindness in kangaroos caused by Wallal and Warrego viruses (Hooper et al. 1999; Reddacliff et al. 1999). In addition, novel orbiviruses like *Middle Point orbivirus* (MPOV) (Cowled et al. 2007) and *Stretch Lagoon orbiviruses* (SLOV) have been confirmed serologically in domestic animals and equines in Australia (Cowled et al. 2009), suggesting that these viruses are likely to 'emerge' as great threats to domesticated animal and human health. However, it is also possible that along with the orbiviruses that have already been recognised, further new and unclassified orbiviruses could emerge to cause significant problems.

8.2 Orbivirus History

The Book of Exodus describes the fifth plague of Egypt as a very severe 'plague of livestock'. This is possibly the first record of a real epizootic, which could have been caused by orbiviruses (Marr and Malloy 1996).

A French biologist Francois de Vaillanthad first described bluetongue (BT) as a disease in cattle and sheep in the late eighteenth century and named it as 'tong-sikte' (Gutsche 1979). However, the first detailed scientific descriptions of clinical signs in sheep were given by Hutcheon (1902). Spreull (1902) initially named the disease as 'malarial catarrhal fever' but later he suggested changing the name of the disease to 'bluetongue' (BT) due to the cyanotic appearance of the mouth and tongue in some of the more severely affected sheep (Spreull 1905). The first confirmed BT outbreak other than Africa occurred in Cyprus in 1943, with very high mortality rates in the naive sheep population, though there are reports of unconfirmed outbreaks in Cyprus since 1924 (Gambles 1949; Polydorou 1985).

Another important orbiviral disease is African horse sickness (AHS) that has caused an epidemic in Yemen during 1327 (Moule 1896). The virus is expected to have come from Africa (Theal 1899). AHSV has probably been present in South Africa from very old times, although the disease AHS was first recognised in 1719, killing over 1700 animals (Henning 1956). AHS can have very high case fatality rates (>95%) and is thought to have prevented the use of horses by early Dutch settlers in southern Africa, during the seventeenth and eighteenth centuries. Until recently, AHSV was confined mainly to sub-Saharan Africa, in zebra, its primary wildlife host. However, periodic incursions have also been seen in North Africa, and the Iberian and Arabian peninsulas. During the period 1959–1961, the virus has been detected in countries of the Middle East and Asia, causing heavy mortality in horses (Mellor and Hamblin 2004; Rafyi 1961).

Epizootic haemorrhagic disease (EHD) had been present in wild ruminants throughout the south-eastern USA since 1890, and was known among woods men and hunters as 'blacktongue' but was first officially identified in 1955 from a fatal epizootic in white-tailed deer (*Odocoileus virginianus*) (Shope et al. 1955).

8.3 Classification/Taxonomy of Orbivirus

After 1950, many bluetongue-like viruses were isolated, either from vertebrate hosts or from insect vectors, and were often named after the location of their first isolation (Karabatsos 1985). In 1971, Murphy et al. (1971) and Borden et al. (1971b) concluded that the taxonomic parameters (morphological, physicochemical and serological) of these bluetongue-related viruses were distinct from those of other arthropod-borne viruses and any other recognised virus group, suggesting that collectively they represent a new and distinct group. The surface structure of the orbivirus core particle is visible by negatively stained electron microscopy (EM), as a series of 'doughnut'-shaped capsomeres interlocked with each other (Borden et al. 1971b; Murphy et al. 1971). Therefore, the name '*Orbivirus*' (in Latin 'orbis' mean 'ring' or 'circle') was proposed for the new genus of 'bluetongue-like' arthropod-borne viruses, with bluetongue virus as the 'type' virus (Borden et al. 1971b). Later, in 1976 the International Committee on Taxonomy of Viruses (ICTV) officially recognised the genus *Orbivirus* (Fenner 1976) and *Bluetongue virus*, the economically most important member of the genus, remains as the 'type' species.

The genus *Orbivirus* represents one of 15 genera in the family *Reoviridae.* The family *Reoviridae* contains icosahedral viruses, with genomes comprised of 9–12 linear segments of dsRNA, and a total genome size of 19–32 kb (Table 8.1) (Attoui et al. 2005b; Mertens et al. 2005a). The genera within the family *Reoviridae* are grouped into two subfamilies, namely the *Spinareovirinae* (Latin, ***Spina*** meaning 'thorn' or 'spike') and *Sedoreovirinae* (Latin **Sedo** meaning 'smooth'), based on the presence or absence of 'turrets' or 'spikes' on the surface of the viral subcore 'T2' layer. The genus *Orbivirus* which belongs to subfamily *Sedoreovirinae* is the largest genus within the family *Reoviridae* and currently contains 22 recognised virus species, as well as 13 unclassified 'orbiviruses' (Cowled et al. 2007, 2009; Karabatsos 1985; Mertens et al. 2005b). Viruses classified within each *Orbivirus* species share common antigens that are detectable by complement fixation (CF), agar gel immunodiffusion (AGID), fluorescent antibody (FA) tests or enzyme-linked immunosorbent assays (ELISA) (Gorman et al. 1983; Mertens et al. 2005b). Serological cross-reactions, due to shared common antigens, formed an initial basis for the division of different orbivirus isolates into distinct 'serogroups' (Borden et al. 1971a; Della-Porta 1985; Moore and Lee 1972), which equate to the different virus species currently recognised by the International Committee for the Taxonomy of Viruses (ICTV) (Mertens et al. 2005b). However, in some cases, e.g. between the BTV and EHDV serogroups, a low-level or 'one-way' cross-reaction has been observed in some serological assays, and in cross-hybridisation studies (Della-Porta et al. 1985; Huismans et al. 1979; Moore 1974), making serological identification of individual viruses for diagnosis and classification difficult (Borden et al. 1971b; Della-Porta et al. 1985; Moore 1974; Moore and Lee 1972).

As virus 'species' is the most fundamental taxonomic category in all biological classifications, the ICTV has set a number of different 'polythetic parameters' for identification of different genera and species in the *Reoviridae* family (Attoui et al. 2011; Mertens et al. 2005b; van Regenmortel and Mahy 2004). The primary determinant of virus species in the family *Reoviridae* is an ability to exchange genome segments between different strains during co-infection (a process known as reassortment), thereby generating viable progeny viruses. However, in the absence of reassortment data of specific reovirus strains, other parameters either singly or in combination can be used to identify members of the same virus species. The members of different species have therefore been identified by other parameters, including serological cross-reactions between conserved viral proteins; similarities in conserved RNAs and

Table 8.1 The genera of the family *Reoviridae*

Genus	Structure	No. of genome segments (genome size)	No. of species (+ unassigned species)	Hosts [vectors]
Spinareovirinae				
Aquareovirus	Turreted	11 (~23.7 kb)	6 (+5)	Molluscs, finfish, Crustacea
Coltivirus	Turreted	12 (~29 kb)	2 (+1)	Mammals (including humans) [ticks]
Cypovirus	Turreted	10 (~24.8–33.3 kb)[a]	21 (+2)	Insects (Lepidoptera, Diptera and Hymenoptera), a single isolate was reported from a freshwater daphnid
Dinovernavirus	Turreted	9 (~23.35 kb)	1	Mosquitoes
Fijivirus	Turreted	10 (~28.7 kb)	5	Plants (Graminae, Liliaceae) (delphacid planthoppers)
Idnoreovirus	Turreted	10 +1[b] (~25.1 kb)	5 (+1)	Insects (Hymenoptera)
Mycoreovirus	Turreted	11 or 12 (~24.4 kb)	3	Fungi
Oryzavirus	Turreted	10 (~26.1 kb)	2	Plants (Gramineae) [delphacid planthoppers]
Orthoreovirus	Turreted	10 (~23.5 kb)	5	Birds, reptiles, mammals
Sedoreovirinae				
Cardoreovirus	Non-turreted	12 (~20.15 kb)	1 (+2)	Crabs
Orbivirus	Non-turreted	10 (~19.2 kb)	22 (+14)	Mammals (including humans), birds [*Culicoides*, mosquitoes, phlebotomines, ticks]
Mimoreovirus	Non-turreted	11 (~25.56 kb)	1	Phytoplankton
Phytoreovirus	Non-turreted	12 (~25.1 kb)	3 (+1)	Plants (Cicadellid leafhoppers)
Rotavirus	Non-turreted	11 (~18.5 kb)	5 (+2)	Birds, mammals
Seadornavirus	Non-turreted	12 (~21 kb)	3	Mammals (including humans) [mosquitoes]
Unassigned reoviruses		10, 11 or 12	5	

[a]In some cases, the size of the genome for different *Cypovirus* species was estimated only by electrophoretic analyses

[b]Individual *Idnoreovirus* particles may contain a 11th genome segment depending on the sex and ploidy of the host wasp from which they are derived. Reproduced with some modification from Attoui et al. (2009a)

proteins, which can be detected by sequence analyses; or RNA cross-hybridisation. Within each *Orbivirus* species and to a lesser extent within each genus conserved termini (hexanucleotides) exist on the all genome segments. Similarities in the migration patterns of the viral genome segments during electrophoresis (in agarose gels) reflect their relative sizes which are conserved within each *Orbivirus* species. Similarities in host range, clinical signs and identity of insect vector species can also help to identify viruses within the different virus species. Each of the *Orbivirus* species also contains several distinct virus serotypes, which can be identified by the specificity of their reactions in serum or virus neutralisation tests (SNT or VNT), thereby also confirming their virus species (Gould and Eaton 1990; Mertens et al. 2005b). Sequence comparisons, based on cross-hybridisation of conserved genome segments, or RT-PCR-based assays (e.g. Seg-1, encoding the viral polymerase, or Seg-3, encoding the sub-core shell T2 protein), have been used to identify related orbiviruses. Based on these criteria, until very recently, 22 *Orbivirus* species have been recognised together with 15 unassigned viruses that may denote additional new species (Attoui et al. 2011; Cowled et al. 2009; Martins et al. 2007; Mertens et al. 2005b; Vieira Cde et al. 2009).

However, with the recent development of more rapid and reliable sequencing technologies, and creation of relevant databases, phylogenetic comparisons of nucleotide sequences of 'conserved' genes with those of previously characterised viruses have become a primary method for identifying novel viruses, including novel orbivirus isolates (Anthony et al. 2007; Aradaib 2009; Aradaib et al. 2009; Attoui et al. 2005a; Belaganahalli 2012; Cowled et al. 2007; Maan et al. 2011c, 2012; Shaw et al. 2007) (Table 8.2). These studies have provided information concerning the classification of *Orbivirus* species and have identified virus topotypes (geographic variants), which could not be generated by earlier serological assays. Phylogenetic comparisons of the more variable genome segments, encoding the outer capsid proteins, have also helped to identify and provide information concerning the distinct serotypes, within each of the *Orbivirus* species (Table 8.2) (Anthony et al. 2009c; Maan et al. 2007).

8.4 Properties of the Orbiviruses

8.4.1 Physicochemical and Physical Properties of the Orbiviruses

The molecular weight (Mr) of orbivirus particle is ~10.8×10^7 and that of 'core' is ~6.7×10^7. The buoyant densities in CsCl are 1.36 and 1.40 g/cm^3 for virions and cores. The S_{20W} is 550S (virions) and 470S (cores) (Mertens et al. 2005b). Orbiviruses are comparatively resistant to treatment with detergents and solvents, though their sensitivity varies for specific detergents and with virus species. Sodium dodecyl sulphate usually disrupts the virus particle and destroys infectivity (Gorman 1978; Mertens et al. 2005b). *Orbivirus* infectivity is stable at pH 8–9, but virions exhibit a marked decrease in infectivity outside the pH range of 6.5–10.2, due to the loss of

Table 8.2 List of *Orbivirus* species

Sl. no.	Species	Abbreviation	No. of serotypes	Year[a]	Vertebrate hosts	Vectors
1	*African Horse sickness virus*	AHSV	9	1899	Equines and dog	*Culicoides* spp.
2	*Bluetongue virus*	BTV	26	1905	Domestic and wild ruminants	*Culicoides* spp.
3	*Changuinola virus*	CGLV	12	1960	Sloth, armadello, anteater, rodents and Human (Mosquito catcher)	Phlebotomines flies
4	*Chenuda virus*	CNUV	7	1954		
5	*Chobar Gorge virus*	CGV	2	1970	Bats (isolation) and domestic animals (serology)	Ticks
6	*Corriparta virus*	CORV	6	1960	Shorebird. Human and domestic fowls (serology)	Culex mosquitoes
7	*Epizootic haemorrhagic disease virus*	EHDV	7	1955	White tailed deer, cattle	*Culicoides* spp.
8	*Equine encephalosis virus*	EEV	7	1967	Equines	*Culicoides* spp.
9	*Eubenangee virus*	EUBV	4	1963	Kangaroos, wallabies and cattle	Mosquiotes and *Culicoides* spp.
10	*Great island virus*	GIV	34[b]	1971	Seabirds, domestic animals, horses and humans	Ticks
11	*Ieri virus*	IERIV	3	1955	Birds	Mosquitoes
12	*Lebombo virus*	LEBV	1	1956	Man, rodent	Mosquitoes
13	*Orungo virus*	ORUV	4	1959	Man (isolation), sheep, cattle, goat and camels (serology)	Mosquitoes
14	*Palyam virus*	PALV	13	1956	Domestic and wild ruminants	*Culicoides*, Mosquitoes and ticks
15	*Peruvian horse sickness virus*	PHSV	1	1997	Horse	Mosquitoes
16	*St Croix river virus*	SCRV	1	1994	Ticks?	Ticks
17	*Umatilla virus*	UMAV	4	1969	Birds	Mosquitoes
18	*Warrego virus*	WARV	3	1969	Marsupials (isolation) and cattle (serology)	*Culicoides* spp. mosquitoes
19	*Wallal virus*	WALV	3	1970	Marsupials	*Culicoides* spp.
20	*Wad Medani virus*	WMV	2	1952	Domestic animals and rodents	Ticks
21	*Wongorr virus*	WGRV	8	1970	Cattle and marsupials	*Culicoides* spp. Mosquitoes
22	*Yunnan orbivirus*	YUOV	2	1997	Cattle, sheep and donkeys	Mosquitoes
23	Kemorovo virus[c]	KEMV	2	1962	Human, birds, horses, cattle, seabirds and small mammals	Ticks
24	Pata virus[c]	PATAV	1	1968	Unknown hosts	Mosquitoes or *Culicoides*?
25	Andasibe virus[c]	ANDV	1	1979	Unknown hosts	Mosquitoes
26	Japanaut virus[c]	JAPV	1	1965	Bats and some unknown hosts	Mosquitoes
27	Matucare virus[c]	MATV	1	1963	Unknown hosts	Ticks
	Tentative species					
1	Codajas virus	COV		1984	Unknown hosts	Mosquitoes
2	Ife virus	IFEV		1971	Bats, rodents, birds, ruminants	Mosquitoes

(continued)

Table 8.2 (continued)

3	Itupiranga virus	ITUV	1976	Unknown hosts	Mosquitoes
4	Kammavanpettai virus	KMPV	1963	Birds	Unknown
5	Lake Clarendon virus	LCV	1981	Birds, cattle (serology)	Ticks
6	Breu Branco virus		1988	Unknown	Mosquitoes
7	Minacu virus		1996	Unknown hosts	Mosquitoes
8	Golok virus		1981	Unknown hosts	Mosquitoes

[a]Year of first isolation of prototype serotype in the species. Novel species identified in this study are highlighted in purple colour. Kemerovo species, highlighted in green colour is classified as distinct species based on Belhouchet et al. (2010) and Dilcher et al. (2012)
[b]KEMV and TRBV are removed from *Great Island virus* species
[c]Species identified recently (Belaganahalli 2012)

outer coat proteins, particularly at the lower pH range (Mertens et al. 2005b). At even lower pH values (less than 5.0), virions are progressively disrupted, and at pH 3.0 virus infectivity is abolished (Mertens et al. 2005b). Orbiviruses can be distinguished from orthoreoviruses and rotaviruses by their acid lability, slight solvent sensitivity and serological properties (Borden et al. 1971b; Gorman 1978; Roy 2005). Orbiviruses stored at 4 °C in 0.1 M Tris–HCl pH 8.0 can maintain high levels of infectivity for 1 year. However, at 60 °C virus infectivity is rapidly inactivated. Freezing disrupts virus particle, thereby reducing its infectivity by about 90% (Mertens et al. 2005b). Freeze-drying with an appropriate combination of sugars, including trehalose, will also stabilise the virus, preventing loss of infectivity at high temperatures, even up to 100 °C. The distinctive physicochemical properties of these viruses include (1) the segmented dsRNA genome, (2) relative resistance to deoxycholate and lipid solvents and (3) lability at acidic pH.

8.4.2 Biological Properties of the Orbiviruses

Based primarily on studies of bluetongue virus, the replication of orbiviruses in most mammalian cell lines is usually lytic with cytopathic effect (CPE) (Karabatsos 1985; Mertens et al. 1987). Virus particles are released from infected mammalian cells either by extrusion by cell membrane damage or by budding, which results in the formation of membrane-enveloped virus particles (MEVP) (Celma and Roy 2009; Hyatt et al. 1989, 1993; Mertens et al. 2005b). The virus can also cause cell death and lysis, releasing the intracellular virus from mammalian cells. However, in a range of insect cell cultures, BTV replicates with no widespread cell lysis or cytopathic effect (CPE) (Mertens et al. 1996, 2005b). The mechanism by which they leave insect cells is less certain but appears to involve the small viral membrane protein NS3 (encoded by Seg-10) (Hyatt et al. 1989, 1993).

Infectious subviral particles (ISVPs), in which VP2 is cleaved, can be generated by enzymatic treatment of BTV with chymotrypsin or trypsin (Mertens et al. 1987, 1996). These ISVPs have significantly higher specific infectivity for insect cell lines: KC cells (from *Culicoides sonorensis*) and C6/36 cells (derived from *Aedes albopictus*) than for mammalian cells (BHK-21) (Mertens et al. 1987, 1996).

As opposed to intact purified virus these BTV ISVPs don't have haemagglutinating activity as well as cannot aggregate but have significantly higher infectivity for adult vector insects and insect vector cell lines (KC cells) (Mertens et al. 2005b). The orbivirus core particle also has much higher infectivity for insect cells (KC cells) than for mammalian cells (Mertens et al. 1996).

8.5 Orbivirus Genome and Their Encoded Proteins

The orbivirus genome comprises of ten linear dsRNA segments ranging in size from ~3.9 to ~0.8 kb (segments 1–10), and is packaged in exactly equimolar ratios (Huismans et al. 1979; Mertens et al. 2005b; Verwoerd et al. 1970). In contrast, the separation of the dsRNA genome segments during polyacrylamide gel electrophoresis (PAGE) is only partly dependent on molecular weight but is also influenced by the primary nucleotide sequence and RNA structure, while separation during AGE is solely dependent on molecular weight (Maan 2004). This system can be used to produce an AGE 'electropherotype' which is characteristic for an individual *Orbivirus* species, reflecting a high level of conservation in the sizes/molecular weights of the individual genome segments (Belaganahalli et al. 2015; Gonzalez and Knudson 1988). The BTV genome segments can be assigned to the proteins they encode based upon their consistent migration in 1% agarose gels and translation of denatured dsRNA in vitro (Mertens et al. 1984).

The hexanucleotide termini are often fully conserved within isolates of each *Orbivirus* species but show less conservation between different *Orbivirus* species (Mertens and Diprose 2004; Mertens and Sangar 1985; Mertens et al. 2005b; Rao et al. 1983). The first and last two nucleotides of each orbivirus genome segment are inverted complements, which may be involved in controlling the efficiency of viral mRNA translation (Roy 1989).

Recent studies have identified short segment-specific conserved regions thought to be involved in cross-hybridisation between different genome segments during genome assembly and virus replication (Boyce et al. 2016).

Like other *Reoviridae* family members, most of the orbivirus genome segments are considered to be monocistronic. However, Seg-10 of BTV encodes two related proteins (NS3 and NS3a) starting at different in-frame initiation codons (French et al. 1989; Mertens et al. 1984, 2005b; Wade-Evans 1990; Wade-Evans et al. 1992; Wu et al. 1992). Small non-structural proteins are also translated from alternate ORFs on Seg-9 and Seg-10, generating NS4 and NS5 proteins, respectively (Belhouchet et al. 2010; Ratinier et al. 2016; Stewart et al. 2015).

Twelve distinct virus-specific proteins have been identified in orbivirus-infected cells (Belhouchet et al. 2011; Gorman et al. 1981, 1983; Ratinier et al. 2011). There are seven BTV structural proteins (VP1–VP7) (organised as a three-layered icosahedral capsid structure) and five distinct non-structural proteins (NS1–NS5) (Belhouchet et al. 2011; Mertens et al. 2005b; Stewart et al. 2015; Stewart and Roy 2010; Verwoerd et al. 1972). VP1–VP7 structural proteins are encoded by BTV genome segments-1, -2, -3, -4, -6, -7 and -9, respectively, while the non-structural proteins (NS1–NS5) are encoded by Seg-5, -8, -10, -9 and -10, respectively (Belhouchet et al. 2011; Huismans 1979; Huismans et al. 1979; Mecham and Dean 1988; Mertens et al. 1984, 2005b; Owens et al. 2004; Ratinier et al. 2011; Verwoerd et al. 1972).

The orbivirus structural proteins play many different and important functions during virus replication, including receptor binding and cell entry; transcription, capping and methylation of viral mRNA; capsid formation; genome encapsidation; and −ve-strand RNA synthesis (Owens et al. 2004). Two of the seven structural proteins (VP2 and VP5) make up the outer capsid. The specificity of interactions of these outer capsid proteins (especially VP2) with neutralising antibodies determines the virus serotype (Huismans et al. 1987; Inumaru and Roy 1987). Consequently, VP2 and VP5 are both highly variable showing amino acid (aa) sequence variations, particularly in VP2 that correlates with virus serotype (Maan et al. 2007).

VP3 and VP7 form the inner (sub-core) and outer layers of the BTV core particle. The self-assembly of 120 copies of VP3 to form the innermost sub-core capsid shell determines the overall size and conformation of the outer layers of the virus particle (Grimes et al. 1998; Mertens and Diprose 2004). VP3 is also an RNA-binding protein, interacting with the viral RNA genome and with remaining minor structural proteins (VP1, VP4 and VP6). Seven hundred eighty copies of VP7 are arranged as the core surface layer, outside the VP3 sub-core layer (Grimes et al. 1995). VP7 is also an immunodominant and serogroup-specific antigen, representing a prime target for serogroup/species- or type-specific serological assays (Gumm and Newman 1982; Huismans and Erasmus 1981). VP7 can also bind dsRNA and can mediate cell entry (in the absence of VP2 and VP5), resulting in a high specific infectivity of BTV core particles for insect cells (Diprose et al. 2002; Huismans and Erasmus 1981; Mertens et al. 1996).

The orbivirus transcriptase enzyme complex consists of VP1, VP4 and VP6 (packaged as up to 12 copies within the central space of the virus core), together with the genomic dsRNAs (Stewart and Roy 2010). VP1 is the viral RNA-dependant RNA polymerase (RdRp), VP4 is a capping enzyme and transmethylase (CaP) and VP6 is a helicase (Hel). Each of these minor proteins is highly conserved (de Waal and Huismans 2005; Mertens and Diprose 2004; Ramadevi et al. 1998; Ramadevi and Roy 1998; Stauber et al. 1997).

All of the BTV non-structural proteins (NS) (present in infected cells) are thought to play important roles in virus replication, intracellular sorting and transport, genome packaging, capsid assembly, virus release and control of cellular responses to infection (Owens et al. 2004; Roy 1996; Kusari and Roy 1986; Mertens et al. 1984; Sangar and Mertens 1983). NS1 is expressed at high levels in infected cells and forms tubules that become attached to the intermediate filaments of the cytoskeleton and are characteristic of orbivirus infections, which is shown by Eaton et al. (1988). Therefore, NS1 is also identified as the 'tubule' protein (TuP). Tubules may have a direct role in virus transport, regulation of viral gene expression, cellular pathogenesis and morphogenesis (Boyce et al. 2012; Huismans 1979; Huismans and Els 1979; Owens et al. 2004; Urakawa and Roy 1988).

NS2 is a single-stranded (ss) RNA-binding protein and is the only virus-specific phosphoprotein, forming a major component of the viral inclusion body (VIB), the site of virus assembly (Theron et al. 1996a, b; Thomas et al. 1990; Uitenweerde et al. 1995). NS2 is an ATPase and is thought to have roles in RNA packaging and translation (Butan and Tucker 2010; Taraporewala et al. 2001). NS2, which is phosphorylated by protein kinase 2, found in the cell cytoplasm, aggregates to form VIB, while unphosphorylated NS2 remains dispersed. It has been suggested that NS2 may play a role in shuttling viral mRNAs between the VIB (where they are synthesised) and the host cell cytoplasm for translation, depending on its phosphorylation state.

NS3 and NS3a are both translated from different in-frame initiation sites on a single ORF of Seg-10. They have been connected with determination of virulence, vector competence and virus egress particularly from infected insect cells, where they are abundantly synthesised (Celma and Roy 2009, 2011; Hyatt et al. 1993; Martin et al. 1998; O'Hara et al. 1998). NS3 is only poorly translated if at all in mammalian cells and was originally identified only after in vitro translation of Seg-10 (Mertens et al. 1984). NS3 and NS3a are both membrane proteins. They have two hydrophobic transmembrane domains, and in AHSV it is the second most variable protein after VP2 (Bansal et al. 1998; Jensen and Wilson 1995; van Niekerk et al. 2001).

Recently, two novel non-structural proteins (NS4 and NS5) have been identified. NS4, which is encoded from an alternate, overlapping ORF of Seg-9, has nucleolar localisation signals and appears to play a role in combatting the innate immune response, although its importance is not yet fully elucidated (Belhouchet et al. 2011; Ratinier et al. 2011). It was also suggested that BTV NS4 plays a role in counteracting host antiviral responses during BTV-8 infection (Belhouchet et al. 2011; Ratinier et al. 2011).

8.6 Structure of the Orbiviruses

BTV has been studied extensively at the molecular, genetic and structural levels, and exemplifies the most characterised orbivirus (Maclachlan et al. 2009; Mellor et al. 2009; Osburn 1994; Roy 2008; Stuart and Grimes 2006). In contrast to the characteristic and indistinct appearance of the outer capsid when viewed by negative staining and electron microscopy (EM), cryo-electron microscopy (cryo-EM) and image analysis of BTV reveal a well-ordered outer shell that differs markedly from other known reoviruses (Els and Verwoerd 1969; Hewat et al. 1992b; Nason et al. 2004).

The BTV outer capsid has icosahedral symmetry and exhibits a unique organisation made up of two distinct motifs: 60 'triskelions' formed by trimers of VP2 and 120 globular structures formed by trimers of VP5 (Hassan et al. 2001; Hassan and Roy 1999; Nason et al. 2004). The VP5 trimers sit neatly centred above each of the six-membered rings of VP7 trimers in the outer layer of the BTV core, while the 'sail-shaped' spikes of the VP2 triskelion motifs are located above 180 of the VP7 trimers and cover all of them. VP2 and VP5 together form a continuous layer around the inner shell, except for 12 holes on the five-fold axes (Hewat et al. 1992a, b). The interactions between VP5 and VP7 are extensive and may be relatively strong, while the interactions between VP2 and VP7 are less extensive and may be weaker, suggesting that assembly of VP5 precedes that of VP2. VP2 is also removed more easily from the surface of the particle. However, the classical theory of 'quasi-equivalence' does not strictly allow a triangulation number of 2 ($T = 2$) for the structure and assembly of icosahedral viruses (Caspar and Klug 1962). It has therefore also been proposed (as an alternative) that the inner layer of the reoviruses is arranged with $T = 1$ symmetry, as 60 dimers of the sub-core protein (λ1 (Hel) of orthoreovirus) (Reinisch et al. 2000). However, the sub-core shell protein, VP3 of BTV, also appears to be arranged as 12 'dish-shaped' decamers, one centred at each of the 12 five-fold axes of symmetry of the icosahedral particle, which interact and assemble to form the entire sub-core shell via their zigzag outer edges (Grimes et al. 1998).

A novel principle called 'geometric quasi-equivalence' has therefore been projected to elucidate the formation of virus particles assembled in this manner (Grimes et al. 1998). This feature of the BTV sub-core seems to be shared by many other dsRNA viruses, particularly the other reoviruses, and appears to indicate a shared solution to specific problems related with packaging and replication of dsRNA genomes and possibly a common ancestry.

In contrast, the $T = 13$ symmetry of the VP7 core surface layer denotes a 'typical' example of quasi-equivalence (Grimes et al. 1998; Mertens 2001; Mertens and Attoui 2009). The majority of the BTV core surface layer is composed of ring-shaped capsomers composed of six trimers of the core lattice protein VP7. Rings composed of five trimers of VP7 allow the core surface layer to curve around the sub-core, to 'complete' and 'close' the icosahedral core surface shell. In cells infected with the related AHSV, VP7 can assemble to form large, flat hexagonal arrays that are composed entirely of six-membered rings of VP7 (Burroughs et al. 1994).

The inner surface of the sub-core shell/T2 proteins interacts with the viral genome segments and with the transcriptase complexes (TCs) that are situated at the five-fold axes of the icosahedral structure. The TCs are composed of the three minor structural proteins, VP1 (Pol) (Urakawa et al. 1989), VP4 (CaP) (Ramadevi et al. 1998; Ramadevi and Roy 1998) and VP6 (Hel) (Stauber et al. 1997). The viral RNA-dependent RNA polymerase (RdRp) and viral capping enzyme (VP4) are associated with the interior base of channels that extend through the centre of the pentamers, i.e. the fivefold axes of the core (Patton and Spencer 2000; Ramadevi et al. 1998).

The TCs appear to fit into spaces in the packaged RNA layers (Gouet et al. 1999). Crystallographic and cryo-EM analysis of multiple reoviruses has shown that the dsRNA genome is arranged as partially ordered concentric layers in a liquid crystal structure in the core (Gouet et al. 1999; Prasad et al. 1996; Reinisch et al. 2000). This would retain the RNA molecules in a highly organised and closely packed form but still allow a high level of flexibility and movement within the particle. It has also been suggested that each segment may exist as a tightly wound spiral around one of the 12 RdRp-capping complexes at the five-fold axes of the core (Gouet et al. 1999).

The icosahedral nature of the core restricts the number of RdRp-capping complexes to a maximum of 12, possibly restricting the maximum number of segments to 12. Indeed, since most reoviruses contain less than 12 segments, some of the 12 potential sites in the core may be unoccupied by TCs and dsRNA segments (Patton and Spencer 2000).

The structural organisation of the transcriptionally active reovirus core particle suggests that only a single mRNA species would be synthesised by each RNA segment/TC complex and then be exported via the 'channel' situated at the adjacent five-fold axes of the core structure. The independent functioning of each of the polymerase units of the core is reflected by the production of the different mRNAs in approximately (but not exactly) equal amounts by weight. In contrast, the genome segments are packaged in exactly equimolar levels during RNA replication (Patton and Spencer 2000). The non-equimolar synthesis of the different mRNAs reflects a near-constant rate of RNA chain elongation for each segment but more rapid completion and re-initiation of the smaller transcripts, leading to their synthesis in larger relative molar amounts (Mertens and Diprose 2004; Patton and Spencer 2000).

Despite their similarities at high magnification, certain ultrastructural features differ between the different *Orbivirus* species. For example, viral tubules (composed of NS1) from AHSV are 18 nm wide, whereas tubules from EHDV are 54 nm wide and tubules from BTV are 68 nm wide (Huismans and Els 1979). Gould and Hyatt (1994) reported that differences can also be seen between different strains within a single species and that the arrangement of tubules can also vary between different orbiviruses. They demonstrated that tubules of BTV-1 (Australia) are arranged in bundles, whereas those of BTV-10 (USA) are aligned in parallel arrays. However, the diameter of virions is highly dependent on the method of sample preparation and size as determined by EM is considered as an unreliable diagnostic feature (Gould and Hyatt 1994). Since the tubule protein NS1 is highly conserved within each geographic region and does not vary with serotype, these variations in appearance or organisation may reflect the larger sequence variations that exist between different topotypes of BTV or EHDV, rather than between different serotypes within the same geographic region/topotype (Anthony et al. 2009b; Maan et al. 2010b).

8.7 Orbivirus Infection and Replication

Studies on BTV have shown that the orbiviruses can replicate in several insect and mammalian cell types (Darpel 2007). In insects, BTV (and other orbiviruses) that are ingested as part of a viraemic blood meal can infect the midgut epithelium (DeMaula et al. 2002; Mellor 2000). After an initial round of replication, the virus is released through the basal laminar into the insect's haemocoel, where it is disseminated and can infect secondary organs involving the salivary glands. Further, replication of virus in the salivary glands is thought to release virus into the salivary ducts, where it can be injected into the host along with saliva proteins during the next blood meal

(Anthony 2007; Darpel et al. 2011; Mellor 2000; Mellor and Boorman 1995; Mellor et al. 2000). The replication of BTV in mammalian cell cultures usually results in a shut-off of protein synthesis, cell cycle arrest and virus release by budding and by lysis, usually leading to widespread CPE and cell death within 48–72 h postinfection (Beaton et al. 2002; Owens et al. 2004; Shaw et al. 2013). However, gamma delta T cells and dendritic cells can become persistently infected with BTV, do not exhibit shut-off of host-cell protein synthesis and may play a significant role in the movement, dissemination and persistence of BTV within the vertebrate host (Hemati et al. 2009; Mertens and Attoui 2009; Takamatsu et al. 2003).

Insect cells get persistently infected with orbiviruses, with mildly noticeable cytopathology and slower release of virus compared to mammalian cells. These is also an absence for shut-off of host cell protein synthesis, although, after an extended period, viral protein synthesis and virus particle assembly may slow and even cease, possibly due to immune responses such as ‘silencing’.

The virus release mechanisms in insect cells involving cell membrane penetration or fusion of vacuoles containing multiple virus particles with the cell membrane appear to be mediated by NS3 (Fu et al. 1999; Hyatt et al. 1993; Mellor 2000).

The replication of BTV is summarised in Fig. 8.1. Outer capsid protein VP2 of BTV-10 mediates virion attachment to the cell surface and subsequent internalisa-

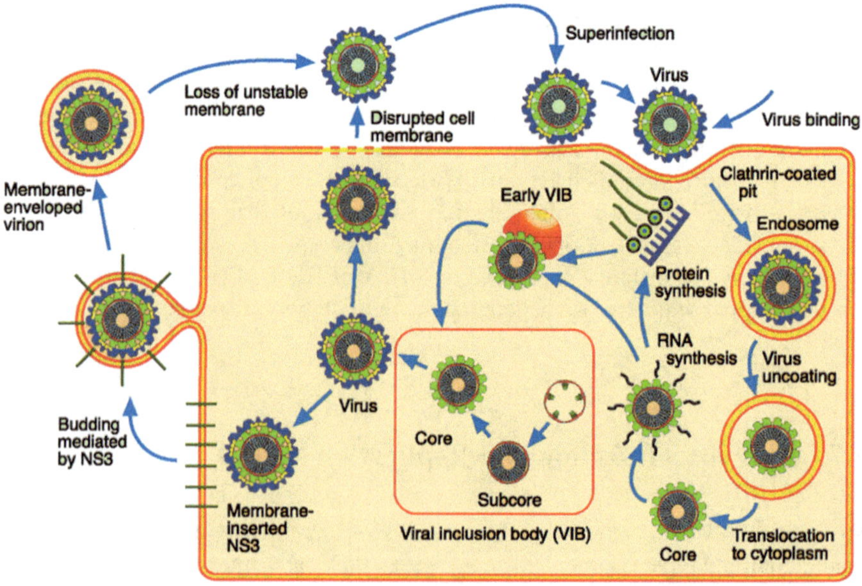

Fig. 8.1 BTV and other orbivirus replication cycle. In summary, the virus binds to the surface via VP2 and is internalised. Transcription and translation occur to produce viral mRNAs and proteins. Progeny particles are assembled, the dsRNA segments of the genome are generated through negative-strand RNA synthesis and then outer capsid proteins are added before the progeny virions are released from the cell, either via a disrupted membrane or by budding (reproduced with permission from Mertens 2001)

tion via clathrin-dependent endocytic pathway (Forzan et al. 2007; Hassan and Roy 1999; Mertens 2001). However, recent studies with BTV-1 indicate that entry does not consistently involve the clathrin-dependent pathway and can also be mediated by macropinocytosis (Gold et al. 2010). The low-pH environment of the endosome dissociates VP2 and induces conformation changes and extension of the VP5 fusion protein, causing penetration of the endosomal membrane, delivering the transcriptionally active core into the cell cytoplasm (Forzan et al. 2004). BTV core particles can also bind to cells (particularly insect cells) via the VP7 trimers on the core surface (Mertens et al. 1996; Schwartz-Cornil et al. 2008). Within the core, VP1 (RdRp) repeatedly and simultaneously transcribes all ten genome segments into mRNA copies (Boyce et al. 2004; Diprose et al. 2001). The newly synthesised mRNAs are capped by the guanylyl-transferase and transmethylase activities of VP4 (Sutton et al. 2007) and then extruded through pores in the VP3 (T2) layer at the five-fold axes (Mertens and Diprose 2004).

The cytoplasm of *Orbivirus*-infected cells contains large granular matrices known as 'viral inclusion bodies' (VIBs) containing large amounts of viral RNA, non-structural viral protein NS2 (ViP) and protein components of the virus core, as well as nascent progeny virus particles (Brookes et al. 1993). Late in infection, the NS2 proteins are located mainly around the periphery of the VIB, suggesting an exchange between the VIB and the host cell cytoplasm, possibly suggesting a role in the transport of viral mRNAs. The ten-messenger or positive-sense RNAs synthesised by the parental virus core act as templates for the synthesis of the viral proteins and can also interact with the nascent virus particles within VIBs, acting as templates for negative RNA strand synthesis during RNA packaging (Diprose et al. 2001).

In the persistently infected insect vector cells, VIBs gradually increase in size, but may eventually become devoid of virus particles, suggesting that cellular defences may prevent further virus replication (Mertens 2001; Mertens and Attoui 2009). Viral mRNAs are synthesised by parental or progeny core particles within the VIB and act as templates for viral protein synthesis. They also become associated as groups of ten mRNAs (one for each segment) with the nascent virus particles, via a poorly understood assembly mechanism that appears to involve specific short sequences that are complementary between different pairs of segments (Boyce et al. 2016). The mRNAs are encapsidated, involving interactions with the helicase VP6, the ssRNA-binding NS2 protein, VP1 and VP4 (Kar et al. 2007). The nascent particles are subsequently thought to mature via a process that involves the addition of core surface protein VP7 (T13) and synthesis of negative RNA strands (by VP1) to produce the dsRNA segments (Boyce et al. 2004).

It has been suggested that each dsRNA segment individually associates with a different transcription complex (VP1, VP4 and VP6) located at the inner side of VP3 along a fivefold axis, making a 'flower shape' in cryo-electron microscopy (Nason et al. 2004; Schwartz-Cornil et al. 2008).

In the event of superinfection by two different BTVs (serotypes or strains), exchange/reassortment of genome segments can occur, which may be due to the fusion of different VIBs formed by different incoming parental viral particles (Schwartz-Cornil et al. 2008). Some genome segments appear to be exchanged more often than others (Gould and Hyatt 1994), which may reflect functional constraints that require certain segments to be 'paired'. Although assembly of the VP3 sub-core shell provides a scaffold, dictating the size and organisation of the rest of the virus particle, it is relatively fragile and unstable. However, the addition of VP7 trimers generates the more stable core particle, which effectively masks the viral dsRNA from the host-cell cytoplasm, preventing activation of dsRNA-dependent defence mechanisms. The newly formed cores then transcribe additional mRNAs (Schwartz-Cornil et al. 2008).

The outer capsid components (VP2 and VP5) are added as the progeny particles leave the VIBs to enter the host cell cytoplasm, forming the fully mature progeny virions. Mature particles are transported within the cytoplasm, perhaps on microtubules, involving VP2/vimentin interactions (Bhattacharya et al. 2007). Release of virions from the infected cell can occur via cell membrane destabilisation mediated by the NS3 viroporin activity (Han and Harty 2004). This mechanism may be more important in insect cells where NS3/3a synthesises at much higher levels than in mammalian cell (Guirakhoo et al. 1995). In some mammalian cells (e.g. BHK21 cells) the BTV particles can exit via budding or as a result of cell death and lysis (Mertens 2001; Schwartz-Cornil et al. 2008).

8.8 Epidemiology and Disease of Orbiviruses

Orbiviruses replicate and cause variable diseases both in their mammalian and arthropod hosts, although they don't have significant detrimental effect on the vectors, which get persistently infected (Fu et al. 1999; Anthony 2007; Attoui et al. 2009a).

BTV, EHDV and AHSV are presently the most significant orbiviruses causing diseases in domestic and wild ruminants and equines, respectively, which are OIE listed (Maclachlan and Guthrie 2010; OIE 2010b). However, other orbiviruses are also potentially important, either regionally or globally (involving equine encephalosis virus (EEV), Palyam virus (PALV) and Peruvian horse sickness virus (PHSV)). At present, BTV has worldwide distribution involving ever-growing new serotypes (Hofmann et al. 2008; Maan et al. 2011b). It is anticipated that there might be many other unexplored arboviruses and other orbiviruses in certain endemic regions such as South America.

8.8.1 Bluetongue Virus

BTV is infectious to a wide range of ruminants, but the overt and severe disease is primarily seen in improved naive populations of sheep breeds (e.g. Polled Dorset) and in some species of deer (e.g. brocket deer and white-tailed deer) (Mellor 1994).

BTV has been reported to infect certain carnivorous species, by injection or via an oral route (Alexander et al. 1994; Brown et al. 1996; Oura and El Harrak 2011). Although BTV is transmitted primarily by *Culicoides* biting midges multiple serotypes have also been isolated from a single sample of bull semen, suggesting the testes as an immunoprivileged site of infection and persistence. Heifers can be infected through insemination with BTV-infected sperm (Bowen et al. 1983; Luedke et al. 1977). BTV transmission through oral and transplacental (vertical transmission) route has been reported in European cattle infected with BTV-8 (Backx et al. 2009; Santman-Berends et al. 2010b).

BTV infection in cattle, goats and many wild ruminant species is classically asymptomatic or subclinical (Barratt-Boyes and MacLachlan 1995; Parsonson 1990; Verwoerd and Erasmus 2004) and cattle are often regarded as a silent or reservoir hosts, maintaining and spreading the virus, possibly through vertical transmission and overwintering (Darpel et al. 2009; Luedke et al. 1977; Nevill 1971; Santman-Berends et al. 2010a, b). However, in 2007, severe (fatal) clinical disease was also observed with low frequency in cattle infected with BTV-8 (Elbers et al. 2008a, 2009), including high fever (reaching 42 °C), nasal discharge, despair, heavy salivation, lameness, haemorrhaging of the coronary bands (coronitis), increased respiration, oral lesions, swollen muzzle and necrosis of the tongue ('bluetongue') (Gard et al. 1987; MacLachlan 1994; Maclachlan et al. 2009; Moulton 1961; Verwoerd and Erasmus 2004). Early prenatal infection in cattle can lead to failed pregnancies, or embryonic death resulting in abortions or stillbirths (Osburn 1994; Tabachnick et al. 1996). However, later infections in utero can lead to teratogenic effects (including dummy calves) and viraemia at birth (Darpel et al. 2009).

The introduction of orbiviruses into disease-free naive population of susceptible animals can cause high levels of morbidity and mortality (Elbers et al. 2008b; Wilson et al. 2007), as demonstrated by the outbreak caused by BTV-8 in northern Europe during 2006 (Maan et al. 2008). The economic significance of BT has led to mandatory restraints on the transport of ruminants, their germplasm and other animal products from BT-endemic countries or regions to BT-free areas (Darpel et al. 2007; OIE 2007, 2010a; Osburn 1994). These trade restrictions due to outbreaks of BT and AHS cause heavy financial burden.

Worldwide and periodic dispersal of the orbiviruses concurs with the distribution of their biological vectors, which are affected by climate conditions (Purse et al. 2005; Wilson et al. 2009a). Recent outbreaks of BTV in Europe have been linked to climate changes in the region (Purse et al. 2005, 2008) suggesting an increased risk from other arboviruses (particularly other orbiviruses) that are transmitted by the same *Culicoides* vectors (e.g. AHSV and EHDV) (Attoui et al. 2009a; Gale et al. 2010). Simultaneous infection of bluetongue and peste-des-petits ruminants (PPRV) has been reported in India (Maan et al. 2018). The emergence of orbiviruses into new regions depends on the capability of the virus involved to infect and be transmitted by resident vector populations. Recently, some orbiviruses have extended their global distribution and it is becoming more difficult to foresee which vectors are capable of carrying specific orbiviruses worldwide.

The vectors of BTV (*Culicoides* species) are most active between 18 and 29 °C and a comparatively small increase in temperature in this effective range can raise their activity and numbers, effectively increasing their efficiency as vectors. The distribution of bluetongue is essentially limited between 50° North and 30° South in America and between 40° North and 35° South in the remaining part of the world. The outbreaks of BTV in southern and central Europe after 2006 are among the strongest paradigms of disease circulation being disturbed by climate change. Increases in average temperatures in the region have led to changes in the dispersal of *C. imicola* (one of the main BTV vector species) in the region. Also novel vector species populations in central and northern Europe, beyond the distribution limits of *C. imicola*, have been colonised as vectors for BTV, since 2006, allowing BT outbreaks to spread for the first time in these regions.

8.8.2 Epizootic Haemorrhagic Disease Virus

EHDV is closely related to BTV, and they are both transmitted by similar *Culicoides* spp. EHDV is prevalent worldwide (Maclachlan and Guthrie 2010). Cattle are asymptomatic reservoirs for EHDV (Gibbs and Lawman 1977; Uren 1986). However, an outbreak of Ibaraki (caused by a strain of EHDV-2) killed >39,000 cattle during 1959 in Japan (Anthony 2007). Since then EHDV has been isolated from North America (USA), South America (Guyana and Suriname), Israel, the Caribbean countries (Jamaica, Antigua, Barbados, Grenada, Trinidad and Tobago), Australia, Japan (Ibaraki virus) and Africa (Bahrain, Sudan, Oman, Algeria and Morocco) (Attoui et al. 2009a).

Latest outbreaks in cattle (during 2004–2009) in Israel, North Africa, USA and Reunion Island have increased the economic importance of EHDV (Anthony et al. 2009c). Clinical EHD has also been observed in wild ungulates (white-tailed deer, mule deer, bighorn sheep and pronghorn antelope) (Hoff et al. 1973; Hoff and Trainer 1974; Karstad et al. 1961; Noon et al. 2002a, b; Shope et al. 1955). Further outbreaks of EHDV with significant morbidity and mortality in cattle have been reported since the outbreak of Ibaraki virus in Japan during 1959 (Abdy et al. 1999; McLaughlin et al. 2003; Ohashi et al. 2002; Omori et al. 1969a, b).

The clinical signs of EHD include high fever, anorexia, respiratory distress, facial oedema extending up to neck and tongue, conjunctivitis, excessive salivation and nasal discharge, and haemorrhages in skin, heart, gastrointestinal tract and mucous membranes. Occasionally the tongue may become cyanotic, and the disease was known locally as 'blacktongue' in the USA. Classical findings in dead animals include bloody diarrhoea and/or haematuria and dehydration (Maclachlan and Guthrie 2010; Savini et al. 2011).

8.8.3 African Horse Sickness Virus

AHS is an ancient disease that was initially recognised as early as 1657, but the big initial outbreak occurred in 1719 in Southern Africa followed by several others during 1854–1855 killing over 70,000 horses (Mellor and Hamblin 2004). AHSV-9 is most prevalent and has been causing all epidemics outside Africa (Attoui et al. 2009a; Mellor and Boorman 1995). The only exception is the Spanish-Portuguese outbreak that occurred in 1987–1990, which was due to AHSV-4, resulting from the importation of infected zebra from Namibia (Mellor and Boorman 1995; Sanchez-Vizcaino 2004). As BTV and AHSV have similar vectors responsible for their transmission, the recent outbreaks of BT across southern, central and northern Europe suggest that this entire region should now be regarded as 'at-risk' to any future African horse sickness outbreaks, although the vector competence of vector populations in different geographic regions may vary for different strains of either BTV or AHSV.

AHS is usually peracute to acute and in naive animals can cause >95% mortality. Horses are more susceptible than mules, while African donkeys and zebras can act as reservoirs for the virus (Maclachlan and Guthrie 2010). AHS is characterised by high temperature, inappetence, oedema of the lungs, subcutaneous and intermuscular tissues, transudation into body cavities and haemorrhages of the serosal surfaces (Maclachlan and Guthrie 2010).

Dogs don't seem to play any major role in the epidemiology of AHS but can get infected with a highly fatal form of AHS mainly due to ingestion of infected carcass material from horses (Bevan 1911; McIntosh 1955; Van Rensberg et al. 1981). There is also serological evidence for widespread infections of other African carnivore species (Alexander et al. 1995). Apart from zebra, no other wildlife species or domesticated animal has been shown to play a substantial role in the epidemiology of AHS (Barnard 1997; Davies and Otieno 1977). Unlike some strains of BTV, AHSV infection is not known to cause reproductive problems in pregnant mares (Maclachlan and Guthrie 2010).

8.8.4 Equine Encephalosis Virus

Equine encephalosis (EE) is generally a mild infection of horses (Guthrie et al. 2009). Equine encephalosis virus (EEV) was first identified in 1967, in southern African horse, and is enzootic in Africa. After 1967, seasonal outbreaks of EE took an epidemic proportions in southern Africa (Theodoridis et al. 1979). Early abortion during the first few months of gestation due to EEV infection can potentially lead to a misdiagnosis as infertility in a mare (Attoui et al. 2009a). The disease was referred to as 'equine ephemeral fever' by Sir Arnold Theiler in the early 1900s (Theiler 1910). EEV has recently occurred for the first time in Israel, causing pyrexia, unrest, anorexia, increased breathing, congestion of mucosae and oedema of the face, neck and legs, but with no detectable increase in mortality (Maclachlan and Guthrie 2010; Mildenberg et al. 2009). Recently, EEV was first reported in India (Yadav et al. 2018).

8.8.5 Peruvian Horse Sickness Virus

Peruvian horse sickness virus (PHSV) is a mosquito-transmitted orbivirus isolated in 1997 from a horse that died of encephalitis in South America (Attoui et al. 2009b). The virus was also isolated from cattle, sheep and dogs. A closely related virus was consequently isolated from diseased horses in the Northern Territory of Australia during 1999 (Elsey virus, ELSV). The epidemiology of this virus is poorly understood, and it is unclear how nearly identical viruses are present in such far-off regions of the world.

8.8.6 Palyam Virus

Palyam viruses have been associated with infections of domesticated animals (cattle and goats) and a variety of haematophagous arthropod vectors distributed in the tropical and subtropical regions of the world. The pathogenicity of this serogroup had not been well studied, although restrictions were imposed by the USDA to prevent work on Nyabira virus (a serotype of Palyam serogroup), which is regarded as a potential pathogen of livestock (Swanepoel and Blackburn 1976; Whistler and Swanepoel 1988). However, several members of the Palyam serogroup have been isolated recurrently from aborted bovine foetuses. An epizootic of congenital abnormalities in calves caused by Chuzan virus was also observed in Japan (1985–1986; Goto et al. 1988a, b; Miura et al. 1988a, b). Chuzan virus was later isolated from native Korean goats, indicating that it has a more extended host and geographical range, and may pose a future threat to domesticated livestock (Yang et al. 2008). Although Chuzan virus was isolated on several more occasions from 1987 to 1996, there have been no further major outbreaks of the disease, although small-scale sporadic disease outbreaks have occurred in the Kyushu district in the late 1990s (Yamakawa et al. 2000).

8.8.7 Eubenangee Virus

The species *Eubenangee virus* (EUBV) includes arthropod-borne orbiviruses that have been associated with sudden deaths and high mortality rates in tammar wallabies and kangaroos in Australia (Kirkland 2005; Rose et al. 2000, 2012). Tilligerry virus (TILV) was isolated along with Ross River virus, which is the causative agent of the epidemic polyarthritis at Nelson Bay in Australia (Marshall et al. 1980). The disease producing a potential of Tilligerry virus has not been fully explored, but such viruses should be considered during the investigation of diseases of unknown aetiology (Marshall et al. 1980).

8.8.8 Wallal Virus and Warrego Virus

The *Wallal virus* (WALV) and *Warrego virus* (WARV) species include viruses that have caused epidemic blindness in kangaroos in South Eastern Australia (Hooper et al. 1999; Reddacliff et al. 1999). Both viruses were isolated in 1994 from populations of western grey kangaroos (*Macropus fuliginosus*), red kangaroos (*Macropus rufus*), eastern grey kangaroos (*Macropus giganteus*) and euros or wallaroo (*Macropus robustus*) in Australia. These animals suffered from blindness due to non-suppurative choroiditis, retinal degeneration, inflammation and secondary degeneration of the optic nerve (Hooper et al. 1999; Reddacliff et al. 1999).

8.8.9 Yunnan Orbivirus

In 1997, Yunnan orbivirus (YUOV) was first isolated from mosquitoes in China (Attoui et al. 2005a; Okamoto et al. 2010). YUOV is closely related to Rioja virus (RIOV), which was isolated along with PHSV in Peru during 1997 (Attoui et al. 2009b). These two viruses have been grouped into the same species, *Yunnan orbivirus* (Attoui et al. 2009b).

In 1998, Middle Point orbivirus (MPOV) was isolated in Australia from healthy cattle and was characterised in 2007 (Cowled et al. 2007). MPOV and YUOV belong to the same virus species as indicated by high amino acid identity (96%) in their VP2 (T2) protein. Therefore YUOV and PHSV both have a widely separated circulation (South America, China and Australia) with the unknown point of origin.

8.8.10 Changuinola Virus

Changuinola virus (CGLV) is one of the 12 identified serotypes in the *Changuinola* virus species and was isolated from the blood of a human, presenting mild febrile illness (Gorman et al. 1983; Karabatsos 1985). The virus was also isolated from phlebotomine sandflies in Panama, and antibodies were detected in rodents. CGLV replicates in insect cell lines with no CPE and is pathogenic for newborn mice or hamsters following intra-cerebral inoculation (Karabatsos 1985). Although virus-specific antibodies were also detected in rodents and other mammals, the prevalence of human infection has not been studied.

8.8.11 Corriparta Virus

Corriparta virus (CORV) was isolated from *Culex* mosquitoes in Australia in 1960 (Doherty et al. 1963). Later serologically related viruses were isolated from Australia, Africa and South America (Gonzalez and Knudson 1987). Neutralising antibodies to CORV have been detected in domestic and wild birds, cattle, marsupials, horses and human (Boughton et al. 1990; Doherty et al. 1963, 1970; Whitehead et al. 1968).

8.8.12 Ieri Virus

Ieri virus (IERIV) was first isolated in 1955 from a pool of female *Psorophora albipes* mosquitoes gathered from evergreen season forest in Trinidad. The virus was successfully isolated in mice brain and BHK cells (Spence et al. 1967). This species contains three serotypes Ieri virus (IERIV), Arkonam virus (ARKV) and Gomoka virus (GOMV). ARKV was first isolated in 1957 from a pool of mosquitoes collected in India, and the evidence of neutralising antibodies against this virus in human was also reported (Dandawate and Shope 1975). Later, the virus was also isolated from mosquitoes collected from Sri Lanka (Peiris et al. 1994). GOMV was first isolated from mosquitoes collected in 1970 at equatorial forest in the Central African Republic, and the virus was also isolated from birds (Karabatsos 1985). This indicated that IERIV viruses have the potential to infect birds and humans.

8.8.13 Great Island Virus

Great Island species members are tick-borne and have been isolated from birds (Mertens et al. 2000). More than 20 strains of Kemerovo virus were isolated from blood and cerebrospinal fluid (CSF) of human patients with meningoencephalitis in the Kemerovo region of Russia and *Ixodes persulcatus* ticks in 1962 (Monath and Guirakhoo 1996). *Some Great Island virus* (GIV) species, including Kemerovo virus (KEMV), Lipovnik virus (LPV) and Tribec viruses (TRBV), have been responsible for fever or neurological infection of humans, in the former USSR (Kemerovo) and Central Europe (Lipovnik and Tribec) (Belhouchet et al. 2010; Dilcher et al. 2012; Libikova et al. 1970, 1978). Lipovnik and Tribec viruses were associated with 'central European encephalitis' (CEE), with >50% of CEE patients having antibodies to Lipovnik virus (LPV) (Attoui et al. 2009a; Libikova et al. 1978; Monath and Guirakhoo 1996). LPV is also supposed to be associated in some chronic neurological diseases and multiple sclerosis. Greater than 20 strains of Kemerovo virus were isolated in 1962 from the Kemerovo area of Russia from *Ixodes persulcatus* ticks and from individuals with meningoencephalitis. The virus, which was also isolated from birds, can infect Vero or BHK-21 cells (Attoui et al. 2009a).

8.8.14 Chenuda Virus

Antibodies against a Kemerovo-related virus were detected in Oklahoma and Texas, in patients with 'Oklahoma tick fever'. Sixgun City virus is one of the seven tick-borne serotypes of the Chenuda virus species (CNUV-1 to CNUV-7) that have been isolated from birds (Mertens et al. 2000). Several Oklahoma tick fever patients also had antibodies to Sixgun City virus, together with antibodies to Kemerovo virus, although no virus was isolated. However, virus isolation was not performed, and the precise cause of the fever is therefore not definitively known (Fields et al. 1985; Monath and Guirakhoo 1996).

8.8.15 Lebombo Virus

Lebombo virus-1 (LEBV-1), the only known serotype of the *Lebombo virus* species, was isolated in Ibadan and Nigeria, in 1968, from a child with fever (Attoui et al. 2009a; Fields et al. 1985; Monath and Guirakhoo 1996). It is pathogenic for suckling mice and has been isolated from both rodents and mosquitoes (Mansonia and Aedes species) in Africa (Fields et al. 1985). The virus replicates in C6/36 cells (noCPE) and in Vero and LLC-MK2 cells.

8.8.16 Orungo Virus

The species *Orungo virus* (ORUV) has four distinct serotypes (ORUV-1 to ORUV-4), isolated from Anopheles, Aedes and Culex mosquitoes (Fields et al. 1985). ORUV was first isolated in Uganda during 1959 from the blood of a human patient with initial symptoms of fever and diarrhoea (Mohd Jaafar et al. 2014). ORUV is widely distributed in Africa in humans, ruminants, monkeys and mosquitoes (Attoui et al. 2009a; Fields et al. 1985; Monath and Guirakhoo 1996). This virus causes lethal encephalitis in suckling mice and hamsters and replicates in adult *Aedes aegypti* mosquitoes after intrathoracic inoculation (Karabatsos 1985). High rates of co-infection of this virus with yellow fever have been reported, showing their similar topographical distribution and transmission by Aedes mosquitoes (Attoui et al. 2009a).

8.8.17 Other Orbiviruses

Stretch Lagoon orbivirus (SLOV) was originally isolated from a pool of mosquitoes collected in 2002 in Australia. Antibodies to SLOV have been detected in horses, donkeys and goats (Cowled et al. 2009). This virus was also isolated from the mosquitoes collected near urban areas of eastern Australia (Jansen et al. 2009). Minacu virus (from South America) was first isolated in Brazil in 1996. Pathogenesis studies in mice showed extensive neuro- and viscerotropisms (Martins et al. 2007). In 1988, the Breu Branco virus (also from South America) was isolated from a pool of mosquitoes, and genetic characterisation suggests that it may also represent a different *Orbivirus* species (Vieira Cde et al. 2009). Published data indicated that a large number of other uncharacterised orbiviruses exist in South America, particularly in the Amazonian rainforests, an area that contains recognised centres of biodiversity. Initial studies also suggested that several of these viruses are likely to represent further new species (Tables 8.1 and 2). Some of the bat-originated orbiviruses (Ife, Japanaut, Fomede and Chobar Gorge viruses) may have the potential to 'emerge' as major threats to human and livestock population (Calisher et al. 2006).

Many emerging viruses are zoonotic (Wang and Crameri 2014). They can be transmitted from wild or domestic reservoir animals to humans, either by insect vectors or by exposure to the droppings or tissues of infected animals. Several orbiviruses have been isolated from humans (Brown et al. 1991; Karabatsos 1985; Libikova et al. 1970, 1978).

8.9 Relationships Among Orbiviruses

8.9.1 Serological Relationships of the Orbivirus *Species*

The orbiviruses are a diverse group that has been differentiated into 27 distinct species that also represent distinct serological groups, with no common 'generic' antigens. Some of the orbivirus proteins are highly conserved, sharing common antigens/epitopes within each group that can be detected in complement fixation (CF) tests, fluorescent antibody (FA) tests and ELISA (Mertens et al. 2005b; Verwoerd et al. 1979). The core surface 'T13' protein (VP7 of BTV) is recognised as 'immunodominant' and is the major serogroup-specific antigen of some *Orbivirus* species (Gumm and Newman 1982), and has therefore provided a primary target for the development of serogroup-specific serological diagnostic assays (Gumm and Newman 1982; Huismans and Erasmus 1981). However, low-level or one-way cross-reactions have been observed among some of the more closely related but distinct *Orbivirus* species in ELISA, CF and agar gel precipitation tests (e.g. between BTV, EHDV and EUBV) (Borden et al. 1971b; Gorman 1979; Gorman and Taylor 1978; Moore 1974; Moore and Lee 1972). This cross-reactivity has been used as a measure of the evolutionary distance between the different species (Gorman 1986). However, partly due to difficulties in obtaining reference antisera to all of the recognised *Orbivirus* species, some virus isolates have not been assigned to recognised serogroups/species by serological methods, and therefore remain as 'unclassified' viruses within the genus (Gorman et al. 1983; Karabatsos 1985).

In many *Orbivirus* species/serogroups, multiple distinct 'serotypes' have been demarcated, based on the specificity of reactions in serological neutralisation tests that target the outer surface proteins of the virus (VP2 and VP5 of BTV). These important diagnostic tools include 'virus neutralisation tests' (VNT) (based on the ability of a specific virus isolate to be neutralised by reference antisera to known serotypes) and 'serum neutralisation tests' (SNT) (based on the ability of antiserum from a previously infected or immunised animal to neutralise reference strains of known serotypes). Using these methods, distinct serotypes have been identified for BTV, AHSV, CGLV, KEMV, PALV and CORV (Dandawate 1974; Dandawate et al. 1969; Davies and Blackburn 1971; Gonzalez and Knudson 1987; Gorman 1979; Hamblin et al. 1991; House et al. 1990; Karabatsos 1985; Libikova and Buckley 1971; Myers et al. 1971; Travassos da Rosa et al. 1984).

The cell attachment and outer capsid protein, VP2 of BTV, contains major neutralisation determinants. However, the smaller outer capsid protein VP5 of BTV, which is involved in membrane penetration during cell entry, also contains some neutralising epitopes and may influence serotype, possibly through conformational interactions with VP2 (Cowley and Gorman 1989, 1990; DeMaula et al. 2000; Mertens et al. 1989). VP2 of BTV is approximately twice the size of the cell attachment protein of some tick-borne orbiviruses, suggesting that it is the product of a concatemerisation event during the evolution of BTV and other closely related (*Culicoides* transmitted) orbiviruses (Anthony et al. 2011; Dilcher et al. 2012). In this context neutralising epitopes have been identified within the amino-terminal half of VP2, several of which appear to be shared between AHSV, BTV and EHDV (Bentley et al. 2000; Martinez-Torrecuadrada and Casal 1995; Martinez-Torrecuadrada et al. 2001).

8.9.2 The Genetic Relatedness of Different Orbiviruses

Serological assays cannot be used to quantitatively determine the genetic closeness of different orbiviruses in individual genome segments. CF tests which have been used in serogroup detection and identification are limited to antigenic site comparison on a restricted number of virus proteins. SNT determines the relatedness of relevant epitopes, corresponding to only part of the genes encoding the cell attachment proteins. However, comparisons of orbivirus genome segments by RNA-RNA hybridisation, oligonucleotide mapping and particularly sequencing and phylogenetic analysis can provide accurate and quantitative evaluations of genetic relatedness in every genome segment. This can be used to reveal the extent and degree of genetic diversity between specific orbivirus strains and species.

The consistent pattern of genome segment migration during AGE for members of a single *Orbivirus* species is also reflected by a greater level of sequence conservation between members of the same species, than with the members of other distinct *Orbivirus* species (Gonzalez and Knudson 1988; Mecham and Johnson 2005). However, significant differences in RNA migration patterns have been observed within the EHDV species (Anthony et al. 2009a). These include a smaller Seg-9 in the eastern strains as compared to the western topotype strains (generating a migration pattern for Seg-7, -8 and -9 that is more similar to BTV in the eastern strains).

Similarities in electropherotype can exist between members of distinct but related species/serogroups (Gorman et al. 1981; Mertens et al. 2005b). However, there is as yet no single study that compares the PAGE and/or AGE electropherotype for representative isolates of all *Orbivirus* species.

The variability in the genome segment sizes of the *Culicoides*-borne orbiviruses (CBOs), mosquito-borne orbiviruses (MBOs) and tick-borne orbiviruses (TBOs) has been documented earlier (Attoui et al. 2005a, 2009b; Belaganahalli et al. 2011; Belhouchet et al. 2010; Moss et al. 1992). In contrast to the CBOs and TBOs, which have distinct but broadly conserved electropherotypes after 1% AGE (with either 3-3-4 or 2-4-4 patterns, respectively), the MBOs showed much more variable migration patterns in their small, medium and large segments, suggesting greater variation in the sizes of cognate genome segment and proteins. The only exception is ORUV, which has an RNA migration pattern similar to that of the CBOs, suggesting that it could also be transmitted by *Culicoides*. The TBOs such as CGV, CNUV and WMV as a group exhibited a 2-4-4 (2-4-3-1) pattern although each serogroup/species has a distinct pattern. Although the genome segments of some MBOs also exhibit a 2-4-4 pattern, their grouping and genome size are different from that of the TBOs. The tick-borne orbiviruses can, therefore, be differentiated from insect-borne orbiviruses (IBOs) (mosquito and *Culicoides*) based on their distinct 1% AGE electropherotype.

Sequence differences and size heterogeneities between the different *Orbivirus* species have been acquired by a combination of insertions, deletions, reassortment, recombination and concatemerisation events, over a very long period of time (Anthony et al. 2009a, c; 2011; Cao et al. 2008; He et al. 2010; Hundley et al. 1987; Murao et al. 1996; Troupin et al. 2011). The extent of this heterogeneity is illustrated by differences between the TBOs and CBOs, where OC1 of the TBOs is approximately half the size of OC1 of CBOs. Therefore, it is possible that OC1 of CBO is the result of gene duplication event of OC1 of a TBO, suggesting the TBOs as ancestors of the CBOs (Belhouchet et al. 2010; Dilcher et al. 2012). This also suggests that the observed change in the size of OC1 may be associated in some way with the change in vector specificity. Different levels of sequence variation suggest that selection acts with differing intensities on the different orbivirus genes. The orbivirus OC1 exhibits the highest variability both within and between species and is therefore likely to be under the highest level of selective pressure, as suggested for EHDV (Anthony et al. 2009c).

8.9.3 Phylogenetic Relationships Between the Different Orbivirus *Species*

8.9.3.1 Phylogenetic Relationship Based on Sub-core T2 Protein

In all of the tick-borne orbiviruses [GIV, SCRV, PHSV, YUOV and CORV], the sub-core shell 'T2' protein is recognised as VP2 rather than as VP3. The VP3 (T2) and VP2 (T2) groups form two discrete phylogenetic lineages, the insect-borne group and tick-borne group, respectively (Attoui et al. 2009b; Moss et al. 1992; Parkes and Gould 1996). The genome coding assignments of BTV and GIV are equivalent for genome segments 1, 7, 8, 9 and 10. The GIV genome segments 2, 3, 4, 5 and 6 are homologous to segments 3, 4, 6, 2 and 5 of BTV, respectively (Attoui et al. 2009a; Belhouchet et al. 2010; Moss et al. 1987, 1992).

The aa sequence of the orbivirus T2 protein (seg-3/VP3 of BTV) determines the conformation and structure of the virus particle and shows a high level of sequence identity (>83%) within each *Orbivirus* species (Attoui et al. 2001, 2009a; Belaganahalli et al. 2011; Belhouchet et al. 2010). The genome segment encoding the T2 protein being most conserved regions can be used to identify various *Orbivirus* species (serogroups) (Attoui et al. 2001, 2009a); it has therefore taken some priority in sequencing efforts (Attoui et al. 2001, 2005a, 2009b; Cowled et al. 2007, 2009; Maan et al. 2008; Pritchard et al. 1995). Consequently, the T2 gene sequence is the only sequence available for several orbiviruses in GenBank (and partial sequences are available for WALV, WGRV, EUBV and CORV) (Belaganahalli et al. 2011; Belhouchet et al. 2010). Therefore, T2 is a logical starting point in sequencing novel orbiviruses to determine their virus species (Attoui et al. 2001; Belhouchet et al. 2010; Pritchard et al. 1995).

8.9.3.2 Phylogenetic Relationship Based on VP1 (Pol or RdRp) Protein

The genome segment encoding the viral polymerase (Seg-1/VP1 of BTV) is also very highly conserved. It is therefore also a useful region to target for sequencing and phylogenetic comparison at a family/genus/species level since all viruses within the family *Reoviridae* have a homologous gene. Phylogenetic analysis of the amino acid sequences of the viral RdRp groups turreted and non-turreted viruses in separate clusters. The orbiviruses are grouped as separate cluster showing >30% aa identity in their polymerase, 13–15% aa identity to other non-turreted genera and <8% identity to the turreted viruses (Attoui et al. 2009a).

The RdRp amino acid sequence has been determined for members of several *Orbivirus* species, including BTV, EHDV, AHSV, PALV (Chuzan strain), YUOV, PHSV, EEV and SCRV. A phylogenetic tree built with the RdRp sequences suggests that SCRV is closest to an ancestral virus from which all other orbiviruses originated (Attoui et al. 2009a).

8.9.3.3 Phylogenetic Relationship Based on Outer Capsid Proteins

The genes encoding the orbivirus outer coat proteins are highly variable and may show partial or high levels of serotype-specific variability (as shown for VP2 and VP5 of BTV, and VP3 (OC1) of some other viruses). Studies which included complete nucleotide sequence analysis of BTV Seg-2 from different serotypes have confirmed that the variations detected correlate with virus serotype (Maan et al. 2007), providing data for the development of type-specific molecular assays

(Mertens et al. 2007). Phylogenetic analysis of Seg-2 sequences from strains within individual BTV serotypes has also identified differences that are associated with the topographical origins of these viruses. These differences divide BTV strains into 'eastern' and 'western' groups or 'Seg-2 topotypes', with evidence of further division into geographical subgroups (Maan et al. 2007; Mertens et al. 2007). These studies and novel assay systems have helped to detect 11 distinct serotypes of BTV (1, 2, 4, 6, 8, 9, 11, 14, 16, 25 and 27) that have recently invaded or been detected in Europe, along with their routes of entry (http://www.reoviridae.org/dsRNA_virus_proteins/ReoID/BTV-mol-epidem.htm).

Up-to-date molecular phylogenetics provides a better and quantitative evaluation of the taxonomic relationships of the orbiviruses both within and between virus species (Anthony 2007; Anthony et al. 2009c; Attoui et al. 2001, 2005a; Belaganahalli et al. 2011; Belhouchet et al. 2010; Maan et al. 2007).

The phylogenetic trees (Figs. 8.2 and 8.3) show typical branching arrangements that reflect vector grouping. Previous phylogenetic analysis of the orbivirus VP1(Pol) sequence and arthropod sequences (using mitochondrial genes) indicated that the tick-borne viruses provided a familial root for the insect-borne viruses (Attoui et al. 2009a; Belhouchet et al. 2010) supporting the assumption that orbiviruses evolve along with their vectors through the process known as 'co-speciation'. Analyses concerning codon bias and genome composition will strengthen this hypothesis.

The amino acid sequence identities for all proteins were calculated for all orbiviruses, using p-distance and pairwise deletion parameter (Table 8.3). These values may have implications for the assignment of these viruses to vector groups (Belaganahalli 2012).

8.9.4 *Evolutionary Relationship of the Orbiviruses*

It has been suggested that different orbiviruses are related, having been derived originally from a single common ancestor. However, they have acquired multiple point mutations, insertions, deletions and duplications (concatemerisations), over a very long period. This has divided the orbiviruses into distinct species, topotypes, serotypes, lineages and other groupings that can be identified by sequencing and phylogenetic analyses (Attoui et al. 2009a). The evolutionary feat of the RNA viruses may be due to their highly mutable genomes and reassortment, giving them the ability to incorporate different positive mutations in different genome segments, within a single virus genome. RNA viruses can also generate genetic variability by generation of defective interfering (DI) RNAs, homologous and non-homologous recombination, insertion, duplications, deletion and concatemerisation events (Anthony et al. 2011; Bonneau et al. 2001; Eaton and Gould 1987; Mohd Jaafar et al. 2008; Qiu and Scholthof 2001), helping them to adapt to different environments and econiches (Domingo and Holland 1997; Elena and Sanjuan 2005; Fields and Joklik 1969).

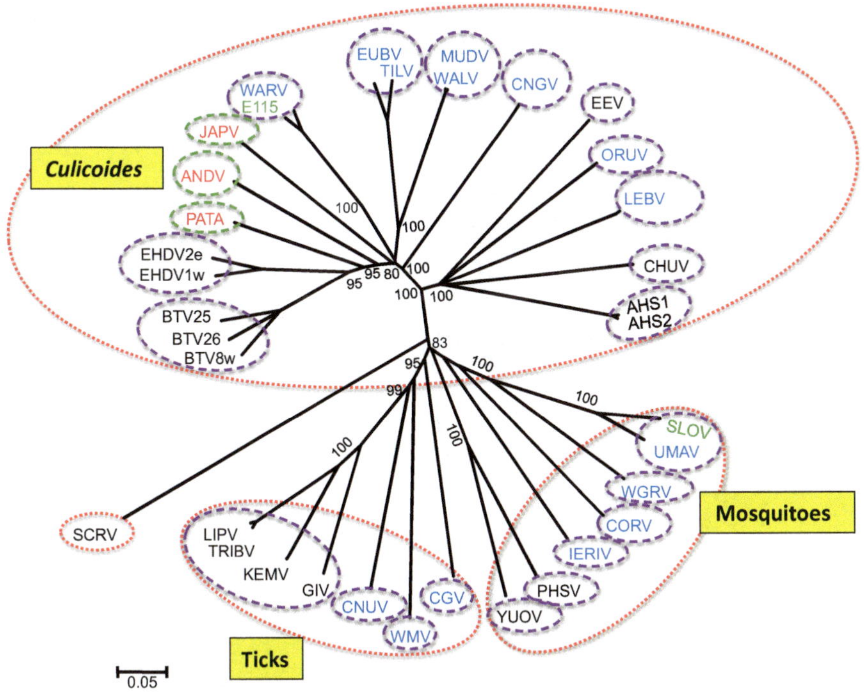

Fig. 8.2 Neighbour-joining phylogenetic tree showing grouping of orbivirus VP1 amino acid sequences. The tree was generated using distance matrices and the p-distance determination algorithm in MEGA 5 (1000 bootstrap replicates) (Tamura et al. 2011). The numbers at nodes indicate bootstrap confidence values after 1000 replications. Virus names are shown in red (represents novel species), green (grouped into existing species) and blue font and are sequenced recently (Belaganahalli 2012). '(e)' and '(w)' indicate eastern and western strains, respectively

There is little information on the molecular evolutionary rate of the dsRNA viruses, including the orbiviruses. Bayesian evolutionary estimates of BTV suggest that Seg-2, -3, -6, and -10 have evolved at the mean rate of $6.9–0.52 \times 10^{-4}$ substitutions/site/year (subs/site/year) which is lower than +ve-sense RNA viruses ($24.2–3.3 \times 10^{-4}$ subs/site/year) and non-vector-borne dsRNA viruses of the family *Reoviridae* ($19.3–8.4 \times 10^{-4}$ substitutions/site/year), though they are comparable to other vector-borne –ve-sense ssRNA viruses ($12.3–0.7 \times 10^{-4}$ subs/site/year) (Carpi et al. 2010; Hanada et al. 2004).

However, studies on tick-borne viruses of the genus *Coltivirus*, family *Reoviridae*, suggested that they have evolved at the rate of 10^{-8} to 10^{-9} subs/site/year which is similar to rates seen for dsDNA genomes (Attoui et al. 2002a, b; Holland et al. 1982). This indicates that the tick-borne viruses evolve more slowly than other arthropod-borne viruses, possibly due to the ability of ticks to lay dormant for long periods, potentially reducing the virus replication frequency, further helping to reduce the rate of change (Attoui et al. 2002b).

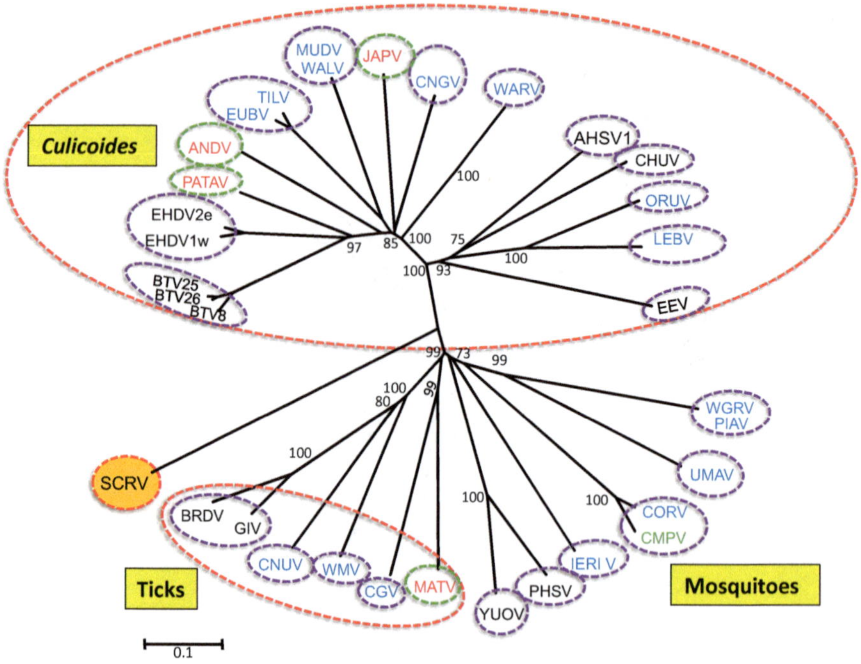

Fig. 8.3 Neighbour-joining phylogenetic tree showing grouping of orbivirus T2 amino acid sequences. The tree was generated using distance matrices and the p-distance determination algorithm in MEGA 5 (1000 bootstrap replicates) (Tamura et al. 2011). The numbers at nodes indicate bootstrap confidence values after 1000 replications. Virus names are shown in red (represents novel species), green (grouped into existing species) and blue font and are sequenced in this study (Belaganahalli 2012). '(e)' and '(w)' indicate eastern and western strains, respectively

Table 8.3 Suggested cut-off amino acid identity values for classification of *Orbivirus* species

Protein	Minimum % aa identity within species	Maximum % aa identity between species
VP1	79[a] (between KEMV and TRBV)	73.5 (bt BTV and EHDV)
OCP1	22.1 (Maan et al. 2010b)	29.28 (bt PATAV and BTV)
T2	87.8 (Maan et al. 2010b)	82.8[a] (bt GIV and KEMV)
CaP (VP4)	80.1 (Maan et al. 2010b)	66.46 (bt BTV and EHDV)
NS1 (TuP)	76.6 (Maan et al. 2010b)	58.8[a] (bt TRBV and GIV)
OCP2 (VP5)	54.2 (Maan et al. 2010b)	63.43 (bt BTV and EHDV)
T13	79.9[b]	77.6[a] (bt TRBV and GIV)
NS2 (ViP)	67.7 (Maan et al. 2010b)	57.5[a] (bt TRBV and GIV)
Hel (VP6)	53.1 (Maan et al. 2010b)	48.7[a] (bt TRBV and GIV)
NS3	63.7 (van Niekerk et al. 2001)	59.29 (bt PATAV and BTV)

bt = between

[a]Values were taken from Belhouchet et al. (2010) and Dilcher et al. (2012)

[b]Value calculated based on available BTV sequences

It has been suggested that the stabilising effect of dsRNA (over ssRNA), a requirement of host alternation, intracellular encapsidation of the dsRNA genome within a transcriptionally active core (to avoid exposure of their genetic material to the cytoplasm of the infected cell) and reduced viral replication frequency due to persistent infection of hosts/vectors during its enzootic cycle, collectively imposes strong purifying selection pressure on the viral genome segments that could affect the overall rate of evolution (Carpi et al. 2010).

The need for intramolecular base-pairing of the viral mRNAs, to maintain specific structures required for translation, packaging, replication and other functions, may also serve to reduce the rate at which even neutral mutations can be accumulated. The segmented nature of the virus genome may in itself provide some protection against the potentially damaging effects of a high mutation rate and its potential to reduce overall virus 'fitness'. By breaking the genome into distinct pieces, there is a reduction in the target size for lethal mutations. Reassortment and high rates of selection (with bottlenecks, such as provided during transmission to and from arthropod vectors) would rapidly remove non-viable progeny viruses, and could specifically remove segments containing deleterious mutations from the virus population.

Despite these constraints, high levels of genetic variations have been observed in some *Orbivirus* species (Bonneau et al. 2000, 2001; de Mattos et al. 1996; Gould and Pritchard 1991). Some changes reflect adaptation in response to selective pressures, particularly in proteins such as VP2 and VP5 of BTV (which are targeted by the host's immune system), and VP7 and possibly NS3 which may have a role in infection of insects and insect cells (Bonneau et al. 2001; Carpi et al. 2010).

The evolutionary dynamics of the arboviruses are influenced by the number, volume and delay between the blood meals during the arthropod life span, as well as by host range, vector range and potential for vertical transmission (Cook and Holmes 2006; Zanotto et al. 1996). The mosquito-borne arboviruses evolve approximately 2.5 times faster than tick-borne arboviruses (Gould et al. 2003; Zanotto et al. 1995). This may reflect a significantly slower virus turnover in ticks due to their longer life cycle (2–5 years), the potential for long periods of inactivity and a limited number of blood meal (three blood meals per life cycle). In contrast, the mosquitoes have a shorter life span and take several blood meals (Gould et al. 2003). Also, the titre of mosquito-borne viruses increases very rapidly in both vector and hosts during their sylvatic replication cycle. The entire replication cycle and transmission of the mosquito-borne arboviruses are far more dynamic (i.e. a greater number of replication cycles) than in ticks, which contributes to their increased rate of evolution.

The ecology of *Culicoides* and phlebotomine flies is similar to that of the mosquitoes (they are all insects), and it is therefore considered likely that the evolution rate of the viruses that they transmit would also be similar. Bayesian analysis of the orbiviruses also shows that insect-borne orbiviruses evolve two times faster than tick-borne orbiviruses.

8.10 Orbivirus Diagnosis

8.10.1 Virological and Immunological (Serological) Approaches

Diagnosis of *Orbivirus* conventionally involved virus isolation followed by their serological characterisation. However, these assays can be slow in generating results, which can be sometimes unreliable (Akita et al. 1992; Tabachnick et al. 1996; Wade-Evans et al. 1990; Prasad et al. 2000).

Virus isolation has previously involved intra-cerebral inoculation into newborn mice, or intravenous injection of embryonated chicken eggs, followed by passage in cell culture (Karabatsos 1985). However, recent studies have shown a high level of success in the direct isolation of *Culicoides*-transmitted orbiviruses in KC cells (derived from the North American vector species *Culicoides sonorensis*). These cells are also highly tolerant of blood or blood cells, possibly reflecting their derivation from a haematophagous insect species, facilitating the recovery of even low-titre viruses directly from viraemic blood samples. Direct isolation in cell culture also reduces the use of animals in experimental studies supporting the objectives of the 'three Rs' (replacement, refinement and reduction of the use of animals in research).

Virus species-specific (serogroup-specific) diagnosis of orbiviruses initially involved virus isolation, CF or FA tests for virus antigen detection, and CF or agar gel immunodiffusion (AGID) tests for antibody detection (Boulanger et al. 1967; Della-Porta et al. 1983; Jochim and Chow 1969; Pearson and Jochim 1979). ELISA was subsequently developed for both virus-specific antigen (Crafford et al. 2003; Hawkes et al. 2000; Hosseini et al. 1998; Thevasagayam et al. 1996) and antibody detection (Afshar 1994; Afshar et al. 1987, 1997; Crafford et al. 2011; Hamblin et al. 1990; Thevasagayam et al. 1995). Neutralisation assays (VNT or SNT) can be used for identification of virus serotype for diagnostic confirmation (Blackburn and Swanepoel 1988; House et al. 1990; Howell et al. 1970; Parker et al. 1975).

Serological methods, such as ELISA, can be used to identify orbivirus-specific antibodies in endemic or outbreak areas, or to monitor infection and seroconversion in sentinel herds (Anderson 1984; Batten et al. 2009; Crafford et al. 2011; Laviada et al. 1992, 1995; OIE 2008). However, the widespread use of vaccination (as seen during the outbreaks caused by BTV-1 and -8 in northern Europe) precludes surveillance using serological assays to detect virus-specific antibodies, as both live vaccines and the current inactivated vaccines produce antibodies to all of the structural and the non-structural viral proteins. Although attempts have been made to design a 'DIVA' assay (to distinguish infected from vaccinated animals) based on non-structural proteins (Barros et al. 2009; Laviada et al. 1995), the results obtained are unreliable and may reflect differences in the serological properties of these proteins between the specific vaccine and field strains tested, rather than generalised differentiation between vaccination and infection.

Earlier diagnostic tools to detect and identify the different orbiviruses were based on serology, needed resources (reference strains and antisera for all viruses), time (often taking weeks to complete) and considerable effort before the final confirmation. In the event of an outbreak, these requirements and resulting delays made it difficult to take immediate preventive measures, such as selection and deployment of appropriate vaccines. However, genetic characterisation of representative strains of the different *Orbivirus* species (including closely related isolates within individual species) provides a basis to understand their genetic relationships and to design differential diagnostic tests.

Complete genomes of many isolates of BTV, EHDV and AHSV have been characterised with high priority because of their economic importance. Sequence databases for these viruses have helped the development of quick and dependable molecular diagnostic assays for serogroup and serotype determination (Aguero et al. 2008; Anthony et al. 2007; Bremer and Viljoen 1998; Fernandez-Pinero et al. 2009; Maan et al. 2010a, 2011a, 2012; Mertens et al. 2007; Orru et al. 2006; Shaw et al. 2007). Correlations of the OC1 sequence with serotype (Anthony et al. 2009c; Maan et al. 2007); RdRp sequence with genus, species and topotype (Attoui et al. 2005a; Belaganahalli et al. 2011; Belhouchet et al. 2010); and T2 sequences with species and topotype (Attoui et al. 2001, 2005a; Nomikou et al. 2009; Pritchard et al. 1995) have been established for the different orbiviruses. Sequence data has therefore been recognised as a vital tool for the taxonomic classification and reclassification of the orbiviruses at genus, species, serotype and topotype levels (Attoui et al. 2005a; Maan et al. 2007, 2010b; Mertens et al. 2005b).

8.10.2 Molecular Approaches to Virus Identification and Diagnosis

Recent advances in molecular and sequencing techniques have helped in the development of very precise, sensitive, quick and reproducible diagnostic tools for the identification of orbiviruses. Nucleic acid hybridisation techniques were originally used for detection of orbiviruses (Brown et al. 1988a, b; de Mattos et al. 1989; Gonzalez and Knudson 1987, 1988; Gould 1988; Huismans and Cloete 1987; Koekemoer and Dijk 2004; Mohammed et al. 1996; Roy et al. 1985; Squire et al. 1985a, b; Venter et al. 1991; Zientara et al. 1998). Both radioactive and non-radioactive methods were used to label nucleic acid probes for use in these hybridisation assays (Akita et al. 1993).

Detection methods for orbivirus RNA, based on RT-PCR, were established in 1990 (Dangler et al. 1990). Various additional amendments of these methods were later developed (Katz et al. 1993; MacLachlan et al. 1994). More fast and more subtle diagnostic assays for both orbivirus species and individual serotypes were developed at the beginning of the twenty-first century based on RT-PCR (Anthony et al. 2007; Aradaib et al. 2003, 2005, 2009; Billinis et al. 2001; Maan et al. 2010a; Zientara et al. 2002; Palacios et al. 2011; Prasad et al. 1999).

Many 'conventional' RT-PCR assays require agarose gel electrophoresis to detect and identify the amplified cDNA products. This makes the assay labour intensive and increases the possibility of post-PCR contamination of samples unless strict procedures are adhered to. In contrast, real-time RT-PCR assays use a sealed-tube setup and detection methods, significantly increasing throughput and reducing contamination risks. Therefore, many serogroup- and type-specific real-time RT-PCR assays have now been invented based on different segments for different orbiviruses (Batten et al. 2010; Hoffmann et al. 2009; Jimenez-Clavero et al. 2006; Monaco et al. 2011; Orru et al. 2006; Quan et al. 2010; Rodriguez-Sanchez et al. 2008; Shaw et al. 2007; Toussaint et al. 2007; Wilson et al. 2009b, c). Phylogenetic studies confirmed that sequence differences in Seg-2 are associated with BTV serotype, which was used to develop type-specific RT-PCR assays for BTV (Mertens et al. 2007). These analyses and recently developed sequencing technologies have helped in the identification of additional serotype incursions into newer geographical areas (Hofmann et al. 2008; Maan et al. 2011b). In addition these databases can also be used to trace the movement, origins and parental strains of viruses (Anthony et al. 2009a, 2010; Batten et al. 2008; Maan et al. 2008, 2010b; Nomikou et al. 2009; Oura 2011; Ozkul et al. 2009).

8.11 Immunology of Orbiviruses

Comparatively few studies on the immune response to the type species of *Orbivirus* the *bluetongue virus* have been available; hence there is limited knowledge about viral and mammalian host interactions which is needed for the development of effective vaccines.

BTV-infected ruminants develop numerous antiviral immune responses. These include serogroup-specific antibodies directed against several of the more conserved virus proteins, e.g. VP7, an immunodominant protein.

Neutralising antibodies raised against single serotype of BTV (serotype-specific antibodies) usually protect against homologous BTV strain challenge. However, sequential infection with BTV serotypes generates a broader spectrum of cross-reactive neutralising antibody responses. But concurrent inoculation with three different BTV serotypes did not raise heterotypic antibodies and resulted in the replication of two of the three administered serotypes. Therefore, vaccination by live attenuated polyvalent vaccines may not develop an immune response against some of the inoculated serotypes.

Studies on cellular immune responses against BTV are very limited. Bluetongue virus can proficiently induce IFN in vitro and in vivo (MacLachlan and Thompson 1985). However, the influence of the IFN response on virus clearance and adaptive immune responses is unknown. Animals infected with BTV or EHDV develop lymphopenia (Ellis et al. 1990; Quist et al. 1997).

Cytotoxic T cells (CTLs) are vital for defence against intracellular pathogens.

Anti-BTV CTLs (having serotype cross-reactivity) have been demonstrated in BTV-infected mice (Jeggo and Wardley 1982, 1985) and sheep. The cross-reactivity of BTV-infected sheep T cells did not correlate with the cross-reactions shown by neutralising antibodies, which may be due to the recognition of short, linear peptides by T cells via the MHC pathway while neutralising antibodies primarily recognise conformational epitopes (Takamatsu and Jeggo 1989).

Major ovine recognition antigens for CTLs are BTV NS1, and VP2 to a lesser extent. VP5, VP7 and NS3 proteins are the minor antigens (Andrew et al. 1995; Janardhana et al. 1999; Jones et al. 1997).

8.12 Emergence and Re-emergence of Orbiviruses

Arboviruses including orbiviruses and other pathogens are emerging and re-emerging. But in today's world, the major difference is that emergence and spreading of arboviruses are faster and geographically extensive, mainly due to massive global trade and transportation, adaptation of arthropod vectors to urbanisation, failure to contain mosquito population density increases and land perturbation (Gould et al. 2017).

8.13 Surveillance, Prevention and Control of Orbiviruses

The surveillance strategies should be fit for the purpose and should incorporate appropriate diagnostic approaches. The disease control and prevention of orbiviruses, particularly at the international level, should be as per the severity of disease and based on risk/cost-benefit assessment (Papadopoulos et al. 2009).

8.14 Conclusions

The genus *Orbivirus* includes viruses that are transmitted by *Culicoides*, phlebotomines, mosquitoes and ticks. This is similar to the flaviviruses, which represent a paradigm for the evolution of other arboviruses concerning their mode of transmission (Cook and Holmes 2006) providing understandings into the source and transmission of emerging and re-emerging viruses. Similarly, the orbiviruses present a practical model to study the evolution of vector-borne segmented genome dsRNA viruses and their mode of transmission.

The sequence analyses of the orbiviruses have allowed us to draw the following conclusions: (1) The genus *Orbivirus* is monophyletic (Figs. 8.2 and 8.3) and SCRV roots all other orbiviruses. (2) Culicoides-, mosquito- and tick-borne orbiviruses represent separate phylogenetic lineages. Phlebotomine-borne viruses group with the *Culicoides*-borne viruses. (3) Phylogenies (trees) relating to the orbiviruses, constructed based on different genes, were almost entirely congruently implying that orbiviruses within each species maintain a certain degree of homology in all of their genome segments and proteins, which does not allow reassortment between species (possible evidence of functional constraints). (4) VP1 gene contains sufficient phylogenetic signals to identify topotype and classification of orbivirus both at species and genus levels. (5) Maintenance of conservation at both species and genus levels suggests that VP1 of the orbiviruses is a suitable candidate for molecular evolutionary studies of the orbiviruses. (6) Orbiviruses have evolved along with their vectors, which transmit them, a phenomenon called as co-speciation.

Acknowledgements All the authors of the manuscript thank and acknowledge their respective universities and institutes.

Conflict of interest: There is no conflict of interest.

References

Abdy MJ, Howerth EE, Stallknecht DE (1999) Experimental infection of calves with epizootic hemorrhagic disease virus. Am J Vet Res 60(5):621–626

Afshar A (1994) Bluetongue: laboratory diagnosis. Comp Immunol Microbiol Infect Dis 17(3–4):221–242

Afshar A, Thomas FC, Wright PF, Shapiro JL, Shettigara PT, Anderson J (1987) Comparison of competitive and indirect enzyme-linked immunosorbent assays for detection of bluetongue virus antibodies in serum and whole blood. J Clin Microbiol 25(9):1705–1710

Afshar A, Anderson J, Nielsen KH, Pearson JE, Trotter HC, Woolhouse TR, Thevasagayam JA, Gall DE, Lapointe JM, Gustafson GA (1997) Evaluation of a competitive ELISA for detection of antibodies to epizootic hemorrhagic disease virus of deer. J Vet Diagn Invest 9(3):309–311

Aguero M, Gomez-Tejedor C, Angeles Cubillo M, Rubio C, Romero E, Jimenez-Clavero A (2008) Real-time fluorogenic reverse transcription polymerase chain reaction assay for detection of African horse sickness virus. J Vet Diagn Invest 20(3):325–328

Akita GY, Chinsangaram J, Osburn BI, Ianconescu M, Kaufman R (1992) Detection of bluetongue virus serogroup by polymerase chain reaction. J Vet Diagn Invest 4(4):400–405

Akita GY, Glenn J, Castro AE, Osburn BI (1993) Detection of bluetongue virus in clinical samples by polymerase chain reaction. J Vet Diagn Invest 5(2):154–158

Alexander KA, MacLachlan NJ, Kat PW, House C, O'Brien SJ, Lerche NW, Sawyer M, Frank LG, Holekamp K, Smale L et al (1994) Evidence of natural bluetongue virus infection among African carnivores. Am J Trop Med Hyg 51(5):568–576

Alexander KA, Kat PW, House J, House C, O'Brien SJ, Laurenson MK, McNutt JW, Osburn BI (1995) African horse sickness and African carnivores. Vet Microbiol 47(1–2):133–140

Anderson J (1984) Use of monoclonal antibody in a blocking ELISA to detect group specific antibodies to bluetongue virus. J Immunol Methods 74(1):139–149

Andrew M, Whiteley P, Janardhana V, Lobato Z, Gould A, Coupar B (1995) Antigen specificity of the ovine cytotoxic T lymphocyte response to bluetongue virus. Vet Immunol Immunopathol 47(3–4):311–322

Anthony SJ (2007) Genetic studies of epizootic haemorrhagic disease virus (EHDV). DPhil thesis submitted in partial fulfilment for the degree of Doctor of Philosophy, University of Oxford

Anthony S, Jones H, Darpel KE, Elliott H, Maan S, Samuel A, Mellor PS, Mertens PP (2007) A duplex RT-PCR assay for detection of genome segment 7 (VP7 gene) from 24 BTV serotypes. J Virol Methods 141(2):188–197

Anthony SJ, Maan N, Maan S, Sutton G, Attoui H, Mertens PP (2009a) Genetic and phylogenetic analysis of the core proteins VP1, VP3, VP4, VP6 and VP7 of epizootic haemorrhagic disease virus (EHDV). Virus Res 145(2):187–199

Anthony SJ, Maan N, Maan S, Sutton G, Attoui H, Mertens PP (2009b) Genetic and phylogenetic analysis of the non-structural proteins NS1, NS2 and NS3 of epizootic haemorrhagic disease virus (EHDV). Virus Res 145(2):211–219

Anthony SJ, Maan S, Maan N, Kgosana L, Bachanek-Bankowska K, Batten C, Darpel KE, Sutton G, Attoui H, Mertens PP (2009c) Genetic and phylogenetic analysis of the outer-coat proteins VP2 and VP5 of epizootic haemorrhagic disease virus (EHDV): comparison of genetic and serological data to characterise the EHDV serogroup. Virus Res 145(2):200–210

Anthony SJ, Darpel KE, Maan S, Sutton G, Attoui H, Mertens PP (2010) The evolution of two homologues of the core protein VP6 of epizootic haemorrhagic disease virus (EHDV), which correspond to the geographical origin of the virus. Virus Genes 40(1):67–75

Anthony SJ, Darpel KE, Belaganahalli MN, Maan N, Nomikou K, Sutton G, Attoui H, Maan S, Mertens PP (2011) RNA segment 9 exists as a duplex concatemer in an Australian strain of epizootic haemorrhagic disease virus (EHDV): genetic analysis and evidence for the presence of concatemers as a normal feature of orbivirus replication. Virology 420(2):164–171

Aradaib IE (2009) PCR detection of African horse sickness virus serogroup based on genome segment three sequence analysis. J Virol Methods 159(1):1–5

Aradaib IE, Smith WL, Osburn BI, Cullor JS (2003) A multiplex PCR for simultaneous detection and differentiation of North American serotypes of bluetongue and epizootic hemorrhagic disease viruses. Comp Immunol Microbiol Infect Dis 26(2):77–87

Aradaib IE, Mohamed ME, Abdalla TM, Sarr J, Abdalla MA, Yousof MA, Hassan YA, Karrar AR (2005) Serogrouping of United States and some African serotypes of bluetongue virus using RT-PCR. Vet Microbiol 111(3–4):145–150

Aradaib IE, Mohamed ME, Abdalla MA (2009) A single-tube RT-PCR for rapid detection and differentiation of some African isolates of palyam serogroup orbiviruses. J Virol Methods 161(1):70–74

Attoui H, Stirling JM, Munderloh UG, Billoir F, Brookes SM, Burroughs JN, de Micco P, Mertens PP, de Lamballerie X (2001) Complete sequence characterization of the genome of the St Croix River virus, a new orbivirus isolated from cells of Ixodes scapularis. J Gen Virol 82(Pt 4):795–804

Attoui H, Fang Q, Mohd Jaafar F, Cantaloube JF, Biagini P, de Micco P, de Lamballerie X (2002a) Common evolutionary origin of aquareoviruses and orthoreoviruses revealed by genome characterization of Golden shiner reovirus, Grass carp reovirus, Striped bass reovirus and golden ide reovirus (genus Aquareovirus, family Reoviridae). J Gen Virol 83(Pt 8):1941–1951

Attoui H, Mohd Jaafar F, Biagini P, Cantaloube JF, de Micco P, Murphy FA, de Lamballerie X (2002b) Genus Coltivirus (family Reoviridae): genomic and morphologic characterization of Old World and New World viruses. Arch Virol 147(3):533–561

Attoui H, Mohd Jaafar F, Belhouchet M, Aldrovandi N, Tao S, Chen B, Liang G, Tesh RB, de Micco P, de Lamballerie X (2005a) Yunnan orbivirus, a new orbivirus species isolated from Culex tritaeniorhynchus mosquitoes in China. J Gen Virol 86(Pt 12):3409–3417

Attoui H, Mohd Jaafar F, Belhouchet M, Biagini P, Cantaloube JF, de Micco P, de Lamballerie X (2005b) Expansion of family Reoviridae to include nine-segmented dsRNA viruses: isolation and characterization of a new virus designated Aedes pseudoscutellaris reovirus assigned to a proposed genus (Dinovernavirus). Virology 343(2):212–223

Attoui H, Maan SS, Anthony SJ, Mertens PPC (2009a) Bluetongue virus, other orbiviruses and other reoviruses: their relationships and taxonomy. In: Mellor PS, Baylis M, Mertens PPC (eds) Bluetongue monograph, 1st edn. Elsevier/Academic Press, London, pp 23–552

Attoui H, Mendez-Lopez MR, Rao S, Hurtado-Alendes A, Lizaraso-Caparo F, Jaafar FM, Samuel AR, Belhouchet M, Pritchard LI, Melville L, Weir RP, Hyatt AD, Davis SS, Lunt R, Calisher CH, Tesh RB, Fujita R, Mertens PP (2009b) Peruvian horse sickness virus and Yunnan orbivirus, isolated from vertebrates and mosquitoes in Peru and Australia. Virology 394(2):298–310

Attoui H, Mertens PPC, Becnel J, Belaganahalli M, Bergoin M, Brussaard CP, Chappell JD, Ciarlet M, del Vas M, Dermody TS, Dormitzer PR, Duncan R, Fcang Q, Graham R, Guglielmi KM, Harding RM, Hillman B, Maan S, Makkay A, Marzachì C, Matthijnssens J et al (2011) Reoviridae. Elsevier Academic Press, London

Backx A, Heutink R, van Rooij E, van Rijn P (2009) Transplacental and oral transmission of wild-type bluetongue virus serotype 8 in cattle after experimental infection. Vet Microbiol 138(3–4):235–243

Bansal OB, Stokes A, Bansal A, Bishop D, Roy P (1998) Membrane organization of bluetongue virus nonstructural glycoprotein NS3. J Virol 72(4):3362–3369

Barnard BJ (1997) Antibodies against some viruses of domestic animals in southern African wild animals. Onderstepoort J Vet Res 64(2):95–110

Barratt-Boyes SM, MacLachlan NJ (1995) Pathogenesis of bluetongue virus infection of cattle. J Am Vet Med Assoc 206(9):1322–1329

Barros SC, Cruz B, Luis TM, Ramos F, Fagulha T, Duarte M, Henriques M, Fevereiro M (2009) A DIVA system based on the detection of antibodies to non-structural protein 3 (NS3) of bluetongue virus. Vet Microbiol 137(3–4):252–259

Batten CA, Maan S, Shaw AE, Maan NS, Mertens PP (2008) A European field strain of bluetongue virus derived from two parental vaccine strains by genome segment reassortment. Virus Res 137(1):56–63

Batten CA, Sanders AJ, Bachanek-Bankowska K, Bin-Tarif A, Oura CA (2009) Bluetongue virus: European Community proficiency test (2007) to evaluate ELISA and RT-PCR detection methods with special reference to pooling of samples. Vet Microbiol 135(3–4):380–383

Batten CA, van Rijn PA, Oura CA (2010) Detection of the European 'field' strain of bluetongue virus serotype 6 by real-time RT-PCR. Vet Microbiol 141(1–2):186–188

Batten C, Darpel K, Henstock M, Fay P, Veronesi E, Gubbins S, Graves S, Frost L, Oura C (2014) Evidence for transmission of bluetongue virus serotype 26 through direct contact. PLoS One 9(5):e96049

Beaton AR, Rodriguez J, Reddy YK, Roy P (2002) The membrane trafficking protein calpactin forms a complex with bluetongue virus protein NS3 and mediates virus release. Proc Natl Acad Sci U S A 99(20):13154–13159

Belaganahalli MN (2012) Full genome sequencing and phylogenetic analyses of different Orbivirus species. PhD thesis, University of London.

Belaganahalli MN, Maan S, Maan NS, Tesh R, Attoui H, Mertens PP (2011) Umatilla virus genome sequencing and phylogenetic analysis: identification of stretch lagoon orbivirus as a new member of the Umatilla virus species. PLoS One 6(8):e23605

Belaganahalli MN, Maan S, Maan NS, Brownlie J, Tesh R, Attoui H, Mertens PP (2015) Genetic characterization of the tick-borne orbiviruses. Viruses 7(5):2185–2209

Belhouchet M, Mohd Jaafar F, Tesh R, Grimes J, Maan S, Mertens PP, Attoui H (2010) Complete sequence of Great Island virus and comparison with the T2 and outer-capsid proteins of Kemerovo, Lipovnik and Tribec viruses (genus Orbivirus, family Reoviridae). J Gen Virol 91(Pt 12):2985–2993

Belhouchet M, Mohd Jaafar F, Firth AE, Grimes JM, Mertens PP, Attoui H (2011) Detection of a fourth orbivirus non-structural protein. PLoS One 6(10):e25697

Bentley L, Fehrsen J, Jordaan F, Huismans H, du Plessis DH (2000) Identification of antigenic regions on VP2 of African horse sickness virus serotype 3 by using phage-displayed epitope libraries. J Gen Virol 81(Pt 4):993–1000

Bevan LEW (1911) The transmission of African horse sickness to the dog by feeding. Vet J 67:402–408

Bhattacharya B, Noad RJ, Roy P (2007) Interaction between Bluetongue virus outer capsid protein VP2 and vimentin is necessary for virus egress. Virol J 4:7
Billinis C, Koumbati M, Spyrou V, Nomikou K, Mangana O, Panagiotidis CA, Papadopoulos O (2001) Bluetongue virus diagnosis of clinical cases by a duplex reverse transcription-PCR: a comparison with conventional methods. J Virol Methods 98(1):77–89
Blackburn NK, Swanepoel R (1988) Observations on antibody levels associated with active and passive immunity to African horse sickness. Trop Anim Health Prod 20(4):203–210
Bonneau KR, Zhang NZ, Wilson WC, Zhu JB, Zhang FQ, Li ZH, Zhang KL, Xiao L, Xiang WB, MacLchlan NJ (2000) Phylogenetic analysis of the S7 gene does not segregate Chinese strains of bluetongue virus into a single topotype. Arch Virol 145(6):1163–1171
Bonneau KR, Mullens BA, MacLachlan NJ (2001) Occurrence of genetic drift and founder effect during quasispecies evolution of the VP2 and NS3/NS3A genes of bluetongue virus upon passage between sheep, cattle, and Culicoides sonorensis. J Virol 75(17):8298–8305
Borden EC, Murphy FA, Nathanson N, Monath TP (1971a) Effect of antilymphocyte serum on Tacaribe virus infection in infant mice. Infect Immun 3(3):466–471
Borden EC, Shope RE, Murphy FA (1971b) Physicochemical and morphological relationships of some arthropod-borne viruses to bluetongue virus—a new taxonomic group. Physicochemical and serological studies. J Gen Virol 13(2):261–271
Boughton CR, Hawkes RA, Naim HM (1990) Arbovirus infection in humans in NSW: seroprevalence and pathogenicity of certain Australian bunyaviruses. Aust N Z J Med 20(1):51–55
Boulanger P, Ruckerbauer GM, Bannister GL, Gray DP, Girard A (1967) Studies on bluetongue. 3. Comparison of two complement-fixation methods. Can J Comp Med Vet Sci 31(7):166–170
Bowen RA, Howard TH, Entwistle KW, Pickett BW (1983) Seminal shedding of bluetongue virus in experimentally infected mature bulls. Am J Vet Res 44(12):2268–2270
Boyce M, Wehrfritz J, Noad R, Roy P (2004) Purified recombinant bluetongue virus VP1 exhibits RNA replicase activity. J Virol 78(8):3994–4002
Boyce M, Celma CC, Roy P (2012) Bluetongue virus non-structural protein 1 is a positive regulator of viral protein synthesis. Virol J 9:178
Boyce M, McCrae MA, Boyce P, Kim JT (2016) Inter-segment complementarity in orbiviruses: a driver for co-ordinated genome packaging in the Reoviridae? J Gen Virol 97(5):1145–1157
Bremer CW, Viljoen GJ (1998) Detection of African horse sickness virus and discrimination between two equine orbivirus serogroups by reverse transcription polymerase chain reaction. Onderstepoort J Vet Res 65(1):1–8
Brookes SM, Hyatt AD, Eaton BT (1993) Characterization of virus inclusion bodies in bluetongue virus-infected cells. J Gen Virol 74(Pt 3):525–530
Brown SE, Gonzalez HA, Bodkin DK, Tesh RB, Knudson DL (1988a) Intra- and inter-serogroup genetic relatedness of orbiviruses. II. Blot hybridization and reassortment in vitro of epizootic haemorrhagic disease serogroup, bluetongue type 10 and Pata viruses. J Gen Virol 69(Pt 1):135–147
Brown SE, Morrison HG, Buckley SM, Shope RE, Knudson DL (1988b) Genetic relatedness of the Kemerovo serogroup viruses: I. RNA-RNA blot hybridization and gene reassortment in vitro of the Kemerovo serocomplex. Acta Virol 32(5):369–378
Brown SE, Morrison HG, Karabatsos N, Knudson DL (1991) Genetic relatedness of two new Orbivirus serogroups: Orungo and Lebombo. J Gen Virol 72(Pt 5):1065–1072
Brown CC, Rhyan JC, Grubman MJ, Wilbur LA (1996) Distribution of bluetongue virus in tissues of experimentally infected pregnant dogs as determined by in situ hybridization. Vet Pathol 33(3):337–340
Burroughs JN, O'Hara RS, Smale CJ, Hamblin C, Walton A, Armstrong R, Mertens PP (1994) Purification and properties of virus particles, infectious subviral particles, cores and VP7 crystals of African horse sickness virus serotype 9. J Gen Virol 75(Pt 8):1849–1857
Butan C, Tucker P (2010) Insights into the role of the non-structural protein 2 (NS2) in Bluetongue virus morphogenesis. Virus Res 151(2):109–117

Calisher CH, Childs JE, Field HE, Holmes KV, Schountz T (2006) Bats: important reservoir hosts of emerging viruses. Clin Microbiol Rev 19(3):531–545

Cao D, Barro M, Hoshino Y (2008) Porcine rotavirus bearing an aberrant gene stemming from an intergenic recombination of the NSP2 and NSP5 genes is defective and interfering. J Virol 82(12):6073–6077

Carpi G, Holmes EC, Kitchen A (2010) The evolutionary dynamics of bluetongue virus. J Mol Evol 70(6):583–592

Caspar DL, Klug A (1962) Physical principles in the construction of regular viruses. Cold Spring Harb Symp Quant Biol 27:1–24

Celma CC, Roy P (2009) A viral nonstructural protein regulates bluetongue virus trafficking and release. J Virol 83(13):6806–6816

Celma CC, Roy P (2011) Interaction of calpactin light chain (S100A10/p11) and a viral NS protein is essential for intracellular trafficking of nonenveloped bluetongue virus. J Virol 85(10):4783–4791

Cook S, Holmes EC (2006) A multigene analysis of the phylogenetic relationships among the flaviviruses (Family: Flaviviridae) and the evolution of vector transmission. Arch Virol 151(2):309–325

Cowled C, Melville L, Weir R, Walsh S, Hyatt A, Van Driel R, Davis S, Gubala A, Boyle D (2007) Genetic and epidemiological characterization of Middle Point orbivirus, a novel virus isolated from sentinel cattle in northern Australia. J Gen Virol 88(Pt 12):3413–3422

Cowled C, Palacios G, Melville L, Weir R, Walsh S, Davis S, Gubala A, Lipkin WI, Briese T, Boyle D (2009) Genetic and epidemiological characterization of Stretch Lagoon orbivirus, a novel orbivirus isolated from Culex and Aedes mosquitoes in northern Australia. J Gen Virol 90(Pt 6):1433–1439

Cowley JA, Gorman BM (1989) Cross-neutralization of genetic reassortants of bluetongue virus serotypes 20 and 21. Vet Microbiol 19(1):37–51

Cowley JA, Gorman BM (1990) Effects of proteolytic enzymes on the infectivity, haemagglutinating activity and protein composition of bluetongue virus type 20. Vet Microbiol 22(2–3):137–152

Crafford JE, Guthrie AJ, van Vuuren M, Mertens PP, Burroughs JN, Howell PG, Hamblin C (2003) A group-specific, indirect sandwich ELISA for the detection of equine encephalosis virus antigen. J Virol Methods 112(1–2):129–135

Crafford JE, Guthrie AJ, Van Vuuren M, Mertens PP, Burroughs JN, Howell PG, Batten CA, Hamblin C (2011) A competitive ELISA for the detection of group-specific antibody to equine encephalosis virus. J Virol Methods 174(1–2):60–64

Dandawate CN (1974) Antigenic relationship among Palyam, Kasba and Vellore viruses—a new serogroup of arboviruses. Indian J Med Res 62(3):326–331

Dandawate CN, Shope RE (1975) Studies on physicochemical and biological properties of two ungrouped arboviruses: Minnal and Arkonam. Indian J Med Res 63(8):1180–1187

Dandawate CN, Rajagopalan PK, Pavri KM, Work TH (1969) Virus isolations from mosquitoes collected in North Arcot district, Madras state, and Chittoor district, Andhra Pradesh between November 1955 and October 1957. Indian J Med Res 57(8):1420–1426

Dangler CA, de Mattos CA, de Mattos CC, Osburn BI (1990) Identifying bluetongue virus ribonucleic acid sequences by the polymerase chain reaction. J Virol Methods 28(3):281–292

Darpel KE (2007) The bluetongue virus 'ruminant host–insect vector' transmission cycle; the role of Culicoides saliva proteins in infection. PhD thesis submitted in partial fulfilment for the degree of Doctor of Philosophy, University of London

Darpel KE, Batten CA, Veronesi E, Shaw AE, Anthony S, Bachanek-Bankowska K, Kgosana L, bin-Tarif A, Carpenter S, Muller-Doblies UU, Takamatsu HH, Mellor PS, Mertens PP, Oura CA (2007) Clinical signs and pathology shown by British sheep and cattle infected with bluetongue virus serotype 8 derived from the 2006 outbreak in northern Europe. Vet Rec 161(8):253–261

Darpel KE, Batten CA, Veronesi E, Williamson S, Anderson P, Dennison M, Clifford S, Smith C, Philips L, Bidewell C, Bachanek-Bankowska K, Sanders A, Bin-Tarif A, Wilson AJ, Gubbins

S, Mertens PP, Oura CA, Mellor PS (2009) Transplacental transmission of bluetongue virus 8 in cattle, UK. Emerg Infect Dis 15(12):2025–2028

Darpel KE, Langner KF, Nimtz M, Anthony SJ, Brownlie J, Takamatsu HH, Mellor PS, Mertens PP (2011) Saliva proteins of vector Culicoides modify structure and infectivity of bluetongue virus particles. PLoS One 6(3):e17545

Davies FG, Blackburn NK (1971) The typing of bluetongue virus. Res Vet Sci 12(2):181–183

Davies FG, Otieno S (1977) Elephants and zebras as possible reservoir hosts for African horse sickness virus. Vet Rec 100(14):291–292

de Mattos CC, de Mattos CA, Osburn BI, Dangler CA, Chuang RY, Doi RH (1989) Recombinant DNA probe for serotype-specific identification of bluetongue virus 17. Am J Vet Res 50(4):536–541

de Mattos CC, de Mattos CA, MacLachlan NJ, Giavedoni LD, Yilma T, Osburn BI (1996) Phylogenetic comparison of the S3 gene of United States prototype strains of bluetongue virus with that of field isolates from California. J Virol 70(8):5735–5739

de Waal PJ, Huismans H (2005) Characterization of the nucleic acid binding activity of inner core protein VP6 of African horse sickness virus. Arch Virol 150(10):2037–2050

Della-Porta AJ (1985) Classification of orbiviruses: a need for supergroups of genera. Prog Clin Biol Res 178:267–274

Della-Porta AJ, Sellers RF, Herniman KA, Littlejohns IR, Cybinski DH, St George TD, McPhee DA, Snowdon WA, Campbell J, Cargill C, Corbould A, Chung YS, Smith VW (1983) Serological studies of Australian and Papua New Guinean cattle and Australian sheep for the presence of antibodies against bluetongue group viruses. Vet Microbiol 8(2):147–162

Della-Porta AJ, Parsonson IM, McPhee DA (1985) Problems in the interpretation of diagnostic tests due to cross-reactions between orbiviruses and broad serological responses in animals. Prog Clin Biol Res 178:445–453

DeMaula CD, Bonneau KR, MacLachlan NJ (2000) Changes in the outer capsid proteins of bluetongue virus serotype ten that abrogate neutralization by monoclonal antibodies. Virus Res 67(1):59–66

DeMaula CD, Leutenegger CM, Jutila MA, MacLachlan NJ (2002) Bluetongue virus-induced activation of primary bovine lung microvascular endothelial cells. Vet Immunol Immunopathol 86(3–4):147–157

Dilcher M, Hasib L, Lechner M, Wieseke N, Middendorf M, Marz M, Koch A, Spiegel M, Dobler G, Hufert FT, Weidmann M (2012) Genetic characterization of Tribec virus and Kemerovo virus, two tick-transmitted human-pathogenic Orbiviruses. Virology 423(1):68–76

Diprose JM, Burroughs JN, Sutton GC, Goldsmith A, Gouet P, Malby R, Overton I, Zientara S, Mertens PP, Stuart DI, Grimes JM (2001) Translocation portals for the substrates and products of a viral transcription complex: the bluetongue virus core. EMBO J 20(24):7229–7239

Diprose JM, Grimes JM, Sutton GC, Burroughs JN, Meyer A, Maan S, Mertens PP, Stuart DI (2002) The core of bluetongue virus binds double-stranded RNA. J Virol 76(18):9533–9536

Doherty RL, Carley JG, Mackerras MJ, Marks EN (1963) Studies of arthropod-borne virus infections in Queensland. III. Isolation and characterization of virus strains from wild-caught mosquitoes in North Queensland. Aust J Exp Biol Med Sci 41:17–39

Doherty RL, Whitehead RH, Wetters EJ, Gorman BM, Carley JG (1970) A survey of antibody to 10 arboviruses (Koongol group, Mapputta group and ungrouped) isolated in Queensland. Trans R Soc Trop Med Hyg 64(5):748–753

Domingo E, Holland JJ (1997) RNA virus mutations and fitness for survival. Annu Rev Microbiol 51:151–178

Eaton BT, Gould AR (1987) Isolation and characterization of orbivirus genotypic variants. Virus Res 6(4):363–382

Eaton BT, Hyatt AD, White JR (1988) Localization of the nonstructural protein NS1 in bluetongue virus-infected cells and its presence in virus particles. Virology 163(2):527–537

Elbers A, Backx A, van der Spek A, Ekker M, Leijs P, Steijn K, van Langen H, van Rijn P (2008a) [Epidemiology of bluetongue virus serotype 8 outbreaks in the Netherlands in 2006]. Tijdschr Diergeneeskd 133(6):222–229

Elbers AR, Popma J, Oosterwolde S, van Rijn PA, Vellema P, van Rooij EM (2008b) A cross-sectional study to determine the seroprevalence of bluetongue virus serotype 8 in sheep and goats in 2006 and 2007 in the Netherlands. BMC Vet Res 4:33

Elbers AR, van der Spek AN, van Rijn PA (2009) Epidemiologic characteristics of bluetongue virus serotype 8 laboratory-confirmed outbreaks in The Netherlands in 2007 and a comparison with the situation in 2006. Prev Vet Med 92(1–2):1–8

Elena SF, Sanjuan R (2005) Adaptive value of high mutation rates of RNA viruses: separating causes from consequences. J Virol 79(18):11555–11558

Ellis JA, Luedke AJ, Davis WC, Wechsler SJ, Mecham JO, Pratt DL, Elliott JD (1990) T lymphocyte subset alterations following bluetongue virus infection in sheep and cattle. Vet Immunol Immunopathol 24(1):49–67

Els HJ, Verwoerd DW (1969) Morphology of bluetongue virus. Virology 38(2):213–219

Fenner F (1976) The classification and nomenclature of viruses. Summary of results of meetings of the International Committee on Taxonomy of Viruses in Madrid, September 1975. J Gen Virol 31(3):463–470

Fernandez-Pinero J, Fernandez-Pacheco P, Rodriguez B, Sotelo E, Robles A, Arias M, Sanchez-Vizcaino JM (2009) Rapid and sensitive detection of African horse sickness virus by real-time PCR. Res Vet Sci 86(2):353–358

Fields BN, Joklik WK (1969) Isolation and preliminary genetic and biochemical characterization of temperature-sensitive mutants of reovirus. Virology 37(3):335–342

Fields BN, Knipe DM, Howley PM (1985) Fields virology, 3rd edn. Lippincott-Raven, Philadelphia, PA

Forzan M, Wirblich C, Roy P (2004) A capsid protein of nonenveloped Bluetongue virus exhibits membrane fusion activity. Proc Natl Acad Sci U S A 101(7):2100–2105

Forzan M, Marsh M, Roy P (2007) Bluetongue virus entry into cells. J Virol 81(9):4819–4827

French TJ, Inumaru S, Roy P (1989) Expression of two related nonstructural proteins of bluetongue virus (BTV) type 10 in insect cells by a recombinant baculovirus: production of polyclonal ascitic fluid and characterization of the gene product in BTV-infected BHK cells. J Virol 63(8):3270–3278

Fu H, Leake CJ, Mertens PP, Mellor PS (1999) The barriers to bluetongue virus infection, dissemination and transmission in the vector, Culicoides variipennis (Diptera: Ceratopogonidae). Arch Virol 144(4):747–761

Gale P, Brouwer A, Ramnial V, Kelly L, Kosmider R, Fooks AR, Snary EL (2010) Assessing the impact of climate change on vector-borne viruses in the EU through the elicitation of expert opinion. Epidemiol Infect 138(2):214–225

Gambles RM (1949) Bluetongue of sheep on Cyprus. J Comp Pathol 59:176–190

Gard GP, Shorthose JE, Weir RP, Erasmus BJ (1987) The isolation of a bluetongue serotype new to Australia. Aust Vet J 64(3):87–88

Gibbs EP, Lawman MJ (1977) Infection of British deer and farm animals with epizootic haemorrhagic disease of deer virus. J Comp Pathol 87(3):335–343

Gold S, Monaghan P, Mertens P, Jackson T (2010) A clathrin independent macropinocytosis-like entry mechanism used by bluetongue virus-1 during infection of BHK cells. PLoS One 5(6):e11360

Gonzalez HA, Knudson DL (1987) Genetic relatedness of corriparta serogroup viruses. J Gen Virol 68(Pt 3):661–672

Gonzalez HA, Knudson DL (1988) Intra- and inter-serogroup genetic relatedness of orbiviruses. I. Blot hybridization of viruses of Australian serogroups. J Gen Virol 69(Pt 1):125–134

Gorman BM (1978) Susceptibility of orbiviruses to low pH and to organic solvents. Aust J Exp Biol Med Sci 56(3):359–367

Gorman BM (1979) Variation in orbiviruses. J Gen Virol 44(1):1–15

Gorman BM (1986) Evolutionary relationships among orbiviruses. Rev Sci Tech Off Int Epiz 5(2):323–332
Gorman BM, Taylor J (1978) The RNA genome of Tilligerry virus. Aust J Exp Biol Med Sci 56(3):369–371
Gorman BM, Taylor J, Walker PJ, Davidson WL, Brown F (1981) Comparison of bluetongue type 20 with certain viruses of the bluetongue and Eubenangee serological groups of orbiviruses. J Gen Virol 57(Pt 2):251–261
Gorman BM, Taylor J, Walker PJ (1983) Orbiviruses. In: Joklik WK (ed) The Reoviridae. Plenum Press, New York, pp 287–357
Goto Y, Miura Y, Kono Y (1988a) Epidemiological survey of an epidemic of congenital abnormalities with hydranencephaly-cerebellar hypoplasia syndrome of calves occurring in 1985/86 and seroepidemiological investigations on Chuzan virus, a putative causal agent of the disease, in Japan. Nippon Juigaku Zasshi 50(2):405–413
Goto Y, Miura Y, Kono Y (1988b) Serologic evidence for the etiologic role of Chuzan virus in an epizootic of congenital abnormalities with hydranencephaly-cerebellar hypoplasia syndrome of calves in Japan. Am J Vet Res 49(12):2026–2029
Gouet P, Diprose JM, Grimes JM, Malby R, Burroughs JN, Zientara S, Stuart DI, Mertens PP (1999) The highly ordered double-stranded RNA genome of bluetongue virus revealed by crystallography. Cell 97(4):481–490
Gould AR (1988) The use of recombinant DNA probes to group and type orbiviruses. A comparison of Australian and South African isolates. Arch Virol 99(3–4):205–220
Gould AR, Eaton BT (1990) The amino acid sequence of the outer coat protein VP2 of neutralizing monoclonal antibody-resistant, virulent and attenuated bluetongue viruses. Virus Res 17(3):161–172
Gould AR, Hyatt AD (1994) The orbivirus genus. Diversity, structure, replication and phylogenetic relationships. Comp Immunol Microbiol Infect Dis 17(3–4):163–188
Gould AR, Pritchard LI (1991) Phylogenetic analyses of the complete nucleotide sequence of the capsid protein (VP3) of Australian epizootic haemorrhagic disease of deer virus (serotype 2) and cognate genes from other orbiviruses. Virus Res 21(1):1–18
Gould EA, de Lamballerie X, Zanotto PM, Holmes EC (2003) Origins, evolution, and vector/host coadaptations within the genus Flavivirus. Adv Virus Res 59:277–314
Gould E, Pettersson J, Higgs S, Charrel R, de Lamballerie X (2017) Emerging arboviruses: why today? One Health 4:1–13
Grimes J, Basak AK, Roy P, Stuart D (1995) The crystal structure of bluetongue virus VP7. Nature 373(6510):167–170
Grimes JM, Burroughs JN, Gouet P, Diprose JM, Malby R, Zientara S, Mertens PP, Stuart DI (1998) The atomic structure of the bluetongue virus core. Nature 395(6701):470–478
Guirakhoo F, Catalan JA, Monath TP (1995) Adaptation of bluetongue virus in mosquito cells results in overexpression of NS3 proteins and release of virus particles. Arch Virol 140(5):967–974
Gumm ID, Newman JF (1982) The preparation of purified bluetongue virus group antigen for use as a diagnostic reagent. Arch Virol 72(1–2):83–93
Guthrie AJ, Pardini AD, Howell PG (2009) Equine encephalosis. Equine Vet Edu Manual 8:145–150
Gutsche T (1979) There was a man. Timmins, Cape Town, p 4
Hamblin C, Graham SD, Anderson EC, Crowther JR (1990) A competitive ELISA for the detection of group-specific antibodies to African horse sickness virus. Epidemiol Infect 104(2):303–312
Hamblin C, Mellor PS, Boned J (1991) The use of ELISA for the detection of African horse sickness viruses in Culicoides midges. J Virol Methods 34(2):221–225
Han Z, Harty RN (2004) The NS3 protein of bluetongue virus exhibits viroporin-like properties. J Biol Chem 279(41):43092–43097
Hanada K, Suzuki Y, Gojobori T (2004) A large variation in the rates of synonymous substitution for RNA viruses and its relationship to a diversity of viral infection and transmission modes. Mol Biol Evol 21(6):1074–1080

Hassan SS, Roy P (1999) Expression and functional characterization of bluetongue virus VP2 protein: role in cell entry. J Virol 73(12):9832–9842
Hassan SH, Wirblich C, Forzan M, Roy P (2001) Expression and functional characterization of bluetongue virus VP5 protein: role in cellular permeabilization. J Virol 75(18):8356–8367
Hawkes RA, Kirkland PD, Sanders DA, Zhang F, Li Z, Davis RJ, Zhang N (2000) Laboratory and field studies of an antigen capture ELISA for bluetongue virus. J Virol Methods 85(1–2):137–149
He CQ, Ding NZ, He M, Li SN, Wang XM, He HB, Liu XF, Guo HS (2010) Intragenic recombination as a mechanism of genetic diversity in bluetongue virus. J Virol 84(21):11487–11495
Hemati B, Contreras V, Urien C, Bonneau M, Takamatsu HH, Mertens PP, Breard E, Sailleau C, Zientara S, Schwartz-Cornil I (2009) Bluetongue virus targets conventional dendritic cells in skin lymph. J Virol 83(17):8789–8799
Henning MM (1956) African horse sickness, perdesiekte, Pestis equorum. In: Animal diseases in South Africa, 3rd edn. Central News Agency, Johannesburg, pp 785–808
Hewat EA, Booth TF, Loudon PT, Roy P (1992a) Three-dimensional reconstruction of baculovirus expressed bluetongue virus core-like particles by cryo-electron microscopy. Virology 189(1):10–20
Hewat EA, Booth TF, Roy P (1992b) Structure of bluetongue virus particles by cryoelectron microscopy. J Struct Biol 109(1):61–69
Hoff GL, Trainer DO (1974) Observations on bluetongue and epizootic hemorrhagic disease viruses in white-tailed deer: (1) distribution of virus in the blood (2) cross-challenge. J Wildl Dis 10(1):25–31
Hoff GL, Griner LA, Trainer DO (1973) Bluetongue virus in exotic ruminants. J Am Vet Med Assoc 163(6):565–567
Hoffmann B, Eschbaumer M, Beer M (2009) Real-time quantitative reverse transcription-PCR assays specifically detecting bluetongue virus serotypes 1, 6, and 8. J Clin Microbiol 47(9):2992–2994
Hofmann MA, Renzullo S, Mader M, Chaignat V, Worwa G, Thuer B (2008) Genetic characterization of Toggenburg orbivirus, a new bluetongue virus, from goats, Switzerland. Emerg Infect Dis 14(12):1855–1861
Holland J, Spindler K, Horodyski F, Grabau E, Nichol S, VandePol S (1982) Rapid evolution of RNA genomes. Science 215(4540):1577–1585
Hooper PT, Lunt RA, Gould AR, Hyatt AD, Russell GM, Kattenbelt JA, Blacksell SD, Reddacliff LA, Kirkland PD, Davis RJ, Durham PJ, Bishop AL, Waddington J (1999) Epidemic of blindness in kangaroos—evidence of a viral aetiology. Aust Vet J 77(8):529–536
Hosseini M, Hawkes RA, Kirkland PD, Dixon RJ (1998) Rapid screening of embryonated chicken eggs for bluetongue virus infection with an antigen capture enzyme linked immunosorbent assay. J Virol Methods 75(1):39–46
House C, Mikiciuk PE, Berninger ML (1990) Laboratory diagnosis of African horse sickness: comparison of serological techniques and evaluation of storage methods of samples for virus isolation. J Vet Diagn Invest 2(1):44–50
Howell PG, Kumm NA, Botha MJ (1970) The application of improved techniques to the identification of strains of bluetongue virus. Onderstepoort J Vet Res 37(1):59–66
Huismans H (1979) Protein synthesis in bluetongue virus-infected cells. Virology 92(2):385–396
Huismans H, Cloete M (1987) A comparison of different cloned bluetongue virus genome segments as probes for the detection of virus-specified RNA. Virology 158(2):373–380
Huismans H, Els HJ (1979) Characterization of the tubules associated with the replication of three different orbiviruses. Virology 92(2):397–406
Huismans H, Erasmus BJ (1981) Identification of the serotype-specific and group-specific antigens of bluetongue virus. Onderstepoort J Vet Res 48(2):51–58
Huismans H, Bremer CW, Barber TL (1979) The nucleic acid and proteins of epizootic haemorrhagic disease virus. Onderstepoort J Vet Res 46(2):95–104

Huismans H, van der Walt NT, Cloete M, Erasmus BJ (1987) Isolation of a capsid protein of bluetongue virus that induces a protective immune response in sheep. Virology 157(1):172–179
Hundley F, McIntyre M, Clark B, Beards G, Wood D, Chrystie I, Desselberger U (1987) Heterogeneity of genome rearrangements in rotaviruses isolated from a chronically infected immunodeficient child. J Virol 61(11):3365–3372
Hutcheon D (1902) Malarial catarrhal fever of sheep. Vet Rec 14:629–633
Hyatt AD, Eaton BT, Brookes SM (1989) The release of bluetongue virus from infected cells and their superinfection by progeny virus. Virology 173(1):21–34
Hyatt AD, Zhao Y, Roy P (1993) Release of bluetongue virus-like particles from insect cells is mediated by BTV nonstructural protein NS3/NS3A. Virology 193(2):592–603
Inumaru S, Roy P (1987) Production and characterization of the neutralization antigen VP2 of bluetongue virus serotype 10 using a baculovirus expression vector. Virology 157(2):472–479
Janardhana V, Andrew ME, Lobato ZI, Coupar BE (1999) The ovine cytotoxic T lymphocyte responses to bluetongue virus. Res Vet Sci 67(3):213–221
Jansen CC, Prow NA, Webb CE, Hall RA, Pyke AT, Harrower BJ, Pritchard IL, Zborowski P, Ritchie SA, Russell RC, Van Den Hurk AF (2009) Arboviruses isolated from mosquitoes collected from urban and peri-urban areas of eastern Australia. J Am Mosq Control Assoc 25(3):272–278
Jeggo MH, Wardley RC (1982) Generation of cross-reactive cytotoxic T lymphocytes following immunization of mice with various bluetongue virus types. Immunology 45(4):629–635
Jeggo MH, Wardley RC (1985) Bluetongue vaccine: cells and/or antibodies. Vaccine 3(1):57–58
Jensen MJ, Wilson WC (1995) A model for the membrane topology of the NS3 protein as predicted from the sequence of segment 10 of epizootic haemorrhagic disease virus serotype 1. Arch Virol 140(4):799–805
Jimenez-Clavero MA, Aguero M, San Miguel E, Mayoral T, Lopez MC, Ruano MJ, Romero E, Monaco F, Polci A, Savini G, Gomez-Tejedor C (2006) High throughput detection of bluetongue virus by a new real-time fluorogenic reverse transcription-polymerase chain reaction: application on clinical samples from current Mediterranean outbreaks. J Vet Diagn Invest 18(1):7–17
Jochim MM, Chow TL (1969) Immunodiffusion of bluetongue virus. Am J Vet Res 30(1):33–41
Jones LD, Williams T, Bishop D, Roy P (1997) Baculovirus-expressed nonstructural protein NS2 of bluetongue virus induces a cytotoxic T-cell response in mice which affords partial protection. Clin Diagn Lab Immunol 4(3):297–301
Kar AK, Bhattacharya B, Roy P (2007) Bluetongue virus RNA binding protein NS2 is a modulator of viral replication and assembly. BMC Mol Biol 8:4
Karabatsos N (1985) International catalogue of arboviruses including certain other viruses of vertebrates, 3rd edn. American Society of Tropical Medicine and Hygiene, San Antonio, TX
Karstad L, Winter A, Trainer DO (1961) Pathology of epizootic hemorrhagic disease of deer. Am J Vet Res 22:227–235
Katz JB, Alstad AD, Gustafson GA, Moser KM (1993) Sensitive identification of bluetongue virus serogroup by a colorimetric dual oligonucleotide sorbent assay of amplified viral nucleic acid. J Clin Microbiol 31(11):3028–3030
Kirkland P (2005) Epidemic viral diseases of wildlife—sudden death in tammar wallabies, blind kangaroos, herpesviruses in pilchards-what next? Microbiol Aust 26:82–84
Koekemoer JJ, Dijk AA (2004) African horse sickness virus serotyping and identification of multiple co-infecting serotypes with a single genome segment 2 RT-PCR amplification and reverse line blot hybridization. J Virol Methods 122(1):49–56
Kusari J, Roy P (1986) Molecular and genetic comparisons of two serotypes of epizootic hemorrhagic disease of deer virus. Am J Vet Res 47(8):1713–1717
Laviada MD, Babin M, Dominguez J, Sanchez-Vizcaino JM (1992) Detection of African horse sickness virus in infected spleens by a sandwich ELISA using two monoclonal antibodies specific for VP7. J Virol Methods 38(2):229–242

Laviada MD, Roy P, Sanchez-Vizcaino JM, Casal JI (1995) The use of African horse sickness virus NS3 protein, expressed in bacteria, as a marker to differentiate infected from vaccinated horses. Virus Res 38(2–3):205–218

Libikova H, Buckley SM (1971) Serological characterization of Eurasian Kemerovo group viruses. II. Cross plaque neutralization tests. Acta Virol 15(1):79–86

Libikova H, Tesarova J, Rajcani J (1970) Experimental infection of monkeys with Kemerovo virus. Acta Virol 14(1):64–69

Libikova H, Heinz F, Ujhazyova D, Stunzner D (1978) Orbiviruses of the Kemerovo complex and neurological diseases. Med Microbiol Immunol 166(1–4):255–263

Luedke AJ, Jones RH, Walton TE (1977) Overwintering mechanism for bluetongue virus: biological recovery of latent virus from a bovine by bites of Culicoides variipennis. Am J Trop Med Hyg 26(2):313–325

Maan S (2004) Complete nucleotide sequence analysis of genome segment 2 from the twenty-four serotypes of bluetongue virus: development of nucleic acid based typing methods and molecular epidemiology. PhD thesis submitted in partial fulfilment for the degree of Doctor of Philosophy, University of London

Maan S, Maan NS, Samuel AR, Rao S, Attoui H, Mertens PP (2007) Analysis and phylogenetic comparisons of full-length VP2 genes of the 24 bluetongue virus serotypes. J Gen Virol 88(Pt 2):621–630

Maan S, Maan NS, Ross-smith N, Batten CA, Shaw AE, Anthony SJ, Samuel AR, Darpel KE, Veronesi E, Oura CA, Singh KP, Nomikou K, Potgieter AC, Attoui H, van Rooij E, van Rijn P, De Clercq K, Vandenbussche F, Zientara S, Breard E, Sailleau C, Beer M, Hoffman B, Mellor PS, Mertens PP (2008) Sequence analysis of bluetongue virus serotype 8 from the Netherlands 2006 and comparison to other European strains. Virology 377(2):308–318

Maan NS, Maan S, Nomikou K, Johnson DJ, El Harrak M, Madani H, Yadin H, Incoglu S, Yesilbag K, Allison AB, Stallknecht DE, Batten C, Anthony SJ, Mertens PP (2010a) RT-PCR assays for seven serotypes of epizootic haemorrhagic disease virus & their use to type strains from the Mediterranean region and North America. PLoS One 5(9):e12782

Maan S, Maan NS, van Rijn PA, van Gennip RG, Sanders A, Wright IM, Batten C, Hoffmann B, Eschbaumer M, Oura CA, Potgieter AC, Nomikou K, Mertens PP (2010b) Full genome characterisation of bluetongue virus serotype 6 from the Netherlands 2008 and comparison to other field and vaccine strains. PLoS One 5(4):e10323

Maan NS, Maan S, Nomikou K, Belaganahalli MN, Bachanek-Bankowska K, Mertens PP (2011a) Serotype specific primers and gel-based RT-PCR assays for 'typing' African horse sickness virus: identification of strains from Africa. PLoS One 6(10):e25686

Maan S, Maan NS, Nomikou K, Batten C, Antony F, Belaganahalli MN, Samy AM, Reda AA, Al-Rashid SA, El Batel M, Oura CA, Mertens PP (2011b) Novel bluetongue virus serotype from Kuwait. Emerg Infect Dis 17(5):886–889

Maan S, Maan NS, Nomikou K, Veronesi E, Bachanek-Bankowska K, Belaganahalli MN, Attoui H, Mertens PP (2011c) Complete genome characterisation of a novel 26th bluetongue virus serotype from Kuwait. PLoS One 6(10):e26147

Maan NS, Maan S, Belaganahalli MN, Ostlund EN, Johnson DJ, Nomikou K, Mertens PPC (2012) Identification and differentiation of the twenty six bluetongue virus serotypes by RT–PCR amplification of the serotype-specific genome segment 2. PLoS One 7(2):e32601

Maan S, Kumar A, Gupta AK, Dalal A, Chaudhary D, Gupta TK, Bansal N, Kumar V, Batra K, Sindhu N, Kumar A, Mahajan NK, Maan NS, Mertens PPC (2018) Concurrent infection of Bluetongue and Peste-des-petits-ruminants virus in small ruminants in Haryana State of India. Transbound Emerg Dis 65(1):235–239

MacLachlan NJ (1994) The pathogenesis and immunology of bluetongue virus infection of ruminants. Comp Immunol Microbiol Infect Dis 17(3–4):197–206

Maclachlan NJ, Guthrie AJ (2010) Re-emergence of bluetongue, African horse sickness, and other orbivirus diseases. Vet Res 41(6):35

MacLachlan NJ, Thompson J (1985) Bluetongue virus-induced interferon in cattle. Am J Vet Res 46(6):1238–1241
MacLachlan NJ, Nunamaker RA, Katz JB, Sawyer MM, Akita GY, Osburn BI, Tabachnick WJ (1994) Detection of bluetongue virus in the blood of inoculated calves: comparison of virus isolation, PCR assay, and in vitro feeding of Culicoides variipennis. Arch Virol 136:1–2), 1–8
Maclachlan NJ, Drew CP, Darpel KE, Worwa G (2009) The pathology and pathogenesis of bluetongue. J Comp Pathol 141(1):1–16
Marr JS, Malloy CD (1996) An epidemiologic analysis of the ten plagues of Egypt. Caduceus 12(1):7–24
Marshall ID, Woodroofe GM, Gard GP (1980) Arboviruses of coastal south-eastern Australia. Aust J Exp Biol Med Sci 58(1):91–102
Martin LA, Meyer AJ, O'Hara RS, Fu H, Mellor PS, Knowles NJ, Mertens PP (1998) Phylogenetic analysis of African horse sickness virus segment 10: sequence variation, virulence characteristics and cell exit. Arch Virol Suppl 14:281–293
Martinez-Torrecuadrada JL, Casal JI (1995) Identification of a linear neutralization domain in the protein VP2 of African horse sickness virus. Virology 210(2):391–399
Martinez-Torrecuadrada JL, Langeveld JP, Meloen RH, Casal JI (2001) Definition of neutralizing sites on African horse sickness virus serotype 4 VP2 at the level of peptides. J Gen Virol 82(Pt 10):2415–2424
Martins LC, Diniz JA, Silva EV, Barros VL, Monteiro HA, Azevedo RS, Quaresma JA, Vasconcelos PF (2007) Characterization of Minacu virus (Reoviridae: Orbivirus) and pathological changes in experimentally infected newborn mice. Int J Exp Pathol 88(1):63–73
McIntosh BM (1955) Horse sickness antibodies in the sera of dogs in enzootic areas. J S Afr Vet Med Assoc 26:269–272
McLaughlin BE, DeMaula CD, Wilson WC, Boyce WM, MacLachlan NJ (2003) Replication of bluetongue virus and epizootic hemorrhagic disease virus in pulmonary artery endothelial cells obtained from cattle, sheep, and deer. Am J Vet Res 64(7):860–865
Mecham JO, Dean VC (1988) Protein coding assignment for the genome of epizootic haemorrhagic disease virus. J Gen Virol 69(Pt 6):1255–1262
Mecham JO, Johnson DJ (2005) Persistence of bluetongue virus serotype 2 (BTV-2) in the southeast United States. Virus Res 113(2):116–122
Mellor PS (1994) Bluetongue. State Vet J 4:7–10
Mellor PS (2000) Replication of arboviruses in insect vectors. J Comp Pathol 123(4):231–247
Mellor PS, Boorman J (1995) The transmission and geographical spread of African horse sickness and bluetongue viruses. Ann Trop Med Parasitol 89(1):1–15
Mellor PS, Hamblin C (2004) African horse sickness. Vet Res 35(4):445–466
Mellor PS, Boorman J, Baylis M (2000) Culicoides biting midges: their role as arbovirus vectors. Annu Rev Entomol 45:307–340
Mellor P, Baylis M, Mertens P (2009) Bluetongue. Elsevier/Academic Press, London
Menzies FD, McCullough SJ, McKeown IM, Forster JL, Jess S, Batten C, Murchie AK, Gloster J, Fallows JG, Pelgrim W, Mellor PS, Oura CA (2008) Evidence for transplacental and contact transmission of bluetongue virus in cattle. Vet Rec 163(7):203–209
Mertens P (2001) Orbiviruses and Bluetongue virus. In: Encyclopedia of life sciences (ELS). John Wiley & Sons, Ltd., Chichester
Mertens P, Attoui H (2009) Orbiviruses and Bluetongue virus. In: Encyclopedia of life sciences (ELS). John Wiley & Sons, Ltd., Chichester
Mertens PP, Diprose J (2004) The bluetongue virus core: a nano-scale transcription machine. Virus Res 101(1):29–43
Mertens PP, Sangar DV (1985) Analysis of the terminal sequences of the genome segments of four orbiviruses. Virology 140(1):55–67
Mertens PP, Brown F, Sangar DV (1984) Assignment of the genome segments of bluetongue virus type 1 to the proteins which they encode. Virology 135(1):207–217

Mertens PP, Burroughs JN, Anderson J (1987) Purification and properties of virus particles, infectious subviral particles, and cores of bluetongue virus serotypes 1 and 4. Virology 157(2):375–386

Mertens PP, Pedley S, Cowley J, Burroughs JN, Corteyn AH, Jeggo MH, Jennings DM, Gorman BM (1989) Analysis of the roles of bluetongue virus outer capsid proteins VP2 and VP5 in determination of virus serotype. Virology 170(2):561–565

Mertens PP, Burroughs JN, Walton A, Wellby MP, Fu H, O'Hara RS, Brookes SM, Mellor PS (1996) Enhanced infectivity of modified bluetongue virus particles for two insect cell lines and for two Culicoides vector species. Virology 217(2):582–593

Mertens PPC, Arella M, Attoui H et al (2000) Family Reoviridae. In: van Regenmortel MHV, Fauquet CM, Bishop DHL et al (eds) Virus taxonomy. Seventh report of the international committee on taxonomy of viruses. Academic Press, New York, pp 395–480

Mertens PPC, Attoui H, Duncan R, Dermody TS (2005a) Reoviridae. In: Fauquet CM, Mayo MA, Maniloff J, Desselberger U, Ball LA (eds) Virus taxonomy. Eighth report of the international committee on taxonomy of viruses. Elsevier/Academic Press, London, pp 447–560

Mertens PPC, Maan S, Samuel A, Attoui H (2005b) Orbiviruses, Reoviridae. In: Fauquet CM, Mayo MA, Maniloff J, Desselberger U, Ball LA (eds) Virus taxonomy. Eighth report of the international committee on taxonomy of viruses. Elsevier/Academic Press, London, pp 466–483

Mertens PP, Maan NS, Prasad G, Samuel AR, Shaw AE, Potgieter AC, Anthony SJ, Maan S (2007) Design of primers and use of RT-PCR assays for typing European bluetongue virus isolates: differentiation of field and vaccine strains. J Gen Virol 88(Pt 10):2811–2823

Mildenberg Z, Westcott D, Bellaiche M, Dastjerdi A, Steinbach F, Drew T (2009) Equine encephalosis virus in Israel. Transbound Emerg Dis 56(8):291

Miura Y, Goto Y, Kubo M, Kono Y (1988a) Isolation of Chuzan virus, a new member of the Palyam subgroup of the genus Orbivirus, from cattle and Culicoides oxystoma in Japan. Am J Vet Res 49(12):2022–2025

Miura Y, Goto Y, Kubo M, Kono Y (1988b) Pathogenicity of Chuzan virus, a new member of the Palyam subgroup of genus Orbivirus for cattle. Nippon Juigaku Zasshi 50(3):632–637

Mohammed ME, Aradaib IE, Mukhtar MM, Ghalib HW, Riemann HP, Oyejide A, Osburn BI (1996) Application of molecular biological techniques for detection of epizootic hemorrhagic disease virus (EHDV-318) recovered from a sentinel calf in central Sudan. Vet Microbiol 52(3–4):201–208

Mohd Jaafar F, Goodwin AE, Belhouchet M, Merry G, Fang Q, Cantaloube JF, Biagini P, de Micco P, Mertens PP, Attoui H (2008) Complete characterisation of the American grass carp reovirus genome (genus Aquareovirus: family Reoviridae) reveals an evolutionary link between aquareoviruses and coltiviruses. Virology 373(2):310–321

Mohd Jaafar F, Belhouchet M, Belaganahalli M, Tesh RB, Mertens PP, Attoui H (2014) Full-genome characterisation of orungo, lebombo and changuinola viruses provides evidence for co-evolution of orbiviruses with their arthropod vectors. PLoS One 9(1):e86392

Monaco F, Polci A, Lelli R, Pinoni C, Di Mattia T, Mbulu RS, Scacchia M, Savini G (2011) A new duplex real-time RT-PCR assay for sensitive and specific detection of African horse sickness virus. Mol Cell Probes 25(2–3):87–93

Monath TP, Guirakhoo F (1996) Orbiviruses and coltiviruses. In: Fields BN, Knipe DM, Howley PM (eds) Fields virology. Lippincott-Raven, Philadelphia, PA, pp 1735–1766

Moore DL (1974) Bluetongue and related viruses in Ibadan, Nigeria: serologic comparison of bluetongue, epizootic hemorrhagic disease of deer, and Abadina (Palyam) viral isolates. Am J Vet Res 35(8):1109–1113

Moore DL, Lee VH (1972) Antigenic relationship between the virus of epizootic haemorrhagic disease of deer and bluetongue virus. Brief report. Arch Gesamte Virusforsch 37(2):282–284

Moss SR, Ayres CM, Nuttall PA (1987) Assignment of the genome segment coding for the neutralizing epitope(s) of orbiviruses in the Great Island subgroup (Kemerovo serogroup). Virology 157(1):137–144

Moss SR, Jones LD, Nuttall PA (1992) Comparison of the major structural core proteins of tick-borne and Culicoides-borne orbiviruses. J Gen Virol 73(Pt 10):2585–2590
Moule L (1896) Histoire de la Médecine Vétérinaire. Maulde, Paris, p 38
Moulton JE (1961) Pathology of bluetongue of sheep in California. J Am Vet Med Assoc 138:493–498
Murao K, Uyeda I, Ando Y, Kimura I, Cabauatan PQ, Koganezawa H (1996) Genomic rearrangement in genome segment 12 of rice dwarf phytoreovirus. Virology 216(1):238–240
Murphy FA, Borden EC, Shope RE, Harrison A (1971) Physicochemical and morphological relationships of some arthropod-borne viruses to bluetongue virus—a new taxonomic group. Electron microscopic studies. J Gen Virol 13(2):273–288
Myers RM, Carey DE, Reuben R, Jesudass ES, Shope RE (1971) Vellore virus: a recently recognized agent of the Palyam group of arboviruses. Indian J Med Res 59(8):1209–1213
Nason EL, Rothagel R, Mukherjee SK, Kar AK, Forzan M, Prasad BV, Roy P (2004) Interactions between the inner and outer capsids of bluetongue virus. J Virol 78(15):8059–8067
Nevill EM (1971) Cattle and Culicoides biting midges as possible overwintering hosts of bluetongue virus. Onderstepoort J Vet Res 38(2):65–71
Nomikou K, Dovas CI, Maan S, Anthony SJ, Samuel AR, Papanastassopoulou M, Maan NS, Mangana O, Mertens PP (2009) Evolution and phylogenetic analysis of full-length VP3 genes of Eastern Mediterranean bluetongue virus isolates. PLoS One 4(7):e6437
Noon TH, Wesche SL, Cagle D, Mead DG, Bicknell EJ, Bradley GA, Riplog-Peterson S, Edsall D, Reggiardo C (2002a) Hemorrhagic disease in bighorn sheep in Arizona. J Wildl Dis 38(1):172–176
Noon TH, Wesche SL, Heffelfinger J, Fuller A, Bradley GA, Reggiardo C (2002b) Hemorrhagic disease in deer in Arizona. J Wildl Dis 38(1):177–181
O'Hara RS, Meyer AJ, Burroughs JN, Pullen L, Martin LA, Mertens PP (1998) Development of a mouse model system, coding assignments and identification of the genome segments controlling virulence of African horse sickness virus serotypes 3 and 8. Arch Virol Suppl 14:259–279
Ohashi S, Yoshida K, Yanase T, Tsuda T (2002) Analysis of intratypic variation evident in an Ibaraki virus strain and its epizootic hemorrhagic disease virus serogroup. J Clin Microbiol 40(10):3684–3688
OIE (2007) Chapter 2.2.13. Bluetongue. Terrestrial animal health code, pp 131–136. Accessed 23 Dec 2018
OIE (2008) Report reference: OIE, Ref OIE: 6964, report date: 14/04/2008, Australia. http://www.oie.int/wahis/reports/en_imm_0000006964_20080415_200102.pdf
OIE (2010a) Chapter 8.3. Bluetongue. Terrestrial animal health code
OIE (2010b) Chapter 1.2. Criteria for listing diseases. Terrestrial animal health code. Accessed 23 Dec 2018
Okamoto K, Endo Y, Inoue S, Nabeshima T, Nga PT, Guillermo PH, Yu F, Loan do P, Trang BM, Natividad FF, Hasebe F, Morita K (2010) Development of a rapid and comprehensive proteomics-based arboviruses detection system. J Virol Methods 167(1):31–36
Omori T, Inaba Y, Morimoto T, Tanaka Y, Ishitani R (1969a) Ibaraki virus, an agent of epizootic disease of cattle resembling bluetongue. I. Epidemiologic, clinical and pathologic observations and experimental transmission to calves. Jpn J Microbiol 13(2):139–157
Omori T, Inaba Y, Morimoto T, Tanaka Y, Kono M (1969b) Ibaraki virus, an agent of epizootic disease of cattle resembling bluetongue. II. Isolation of the virus in bovine cell culture. Jpn J Microbiol 13(2):159–168
Orru G, Ferrando ML, Meloni M, Liciardi M, Savini G, De Santis P (2006) Rapid detection and quantitation of Bluetongue virus (BTV) using a Molecular Beacon fluorescent probe assay. J Virol Methods 137(1):34–42
Osburn BI (1994) The impact of bluetongue virus on reproduction. Comp Immunol Microbiol Infect Dis 17(3–4):189–196
Oura C (2011) Bluetongue: new insights and lessons learnt. Vet Rec 168(14):375–376

Oura CA, El Harrak M (2011) Midge-transmitted bluetongue in domestic dogs. Epidemiol Infect 139(9):1396–1400

Owens RJ, Limn C, Roy P (2004) Role of an arbovirus nonstructural protein in cellular pathogenesis and virus release. J Virol 78(12):6649–6656

Ozkul A, Erturk A, Caliskan E, Sarac F, Ceylan C, Mertens P, Kabakli O, Dincer E, Cizmeci SG (2009) Segment 10 based molecular epidemiology of bluetongue virus (BTV) isolates from Turkey: 1999–2001. Virus Res 142(1–2):134–139

Palacios G, Cowled C, Bussetti AV, Savji N, Weir R, Wick I, Travassos da Rosa A, Calisher CH, Tesh RB, Boyle D, Lipkin WI (2011) Rapid molecular strategy for orbivirus detection and characterization. J Clin Microbiol 49(6):2314–2317

Papadopoulos O, Mellor PS, Mertens PPC (2009) Bluetongue control strategies. In: Mellor PS, Baylis M, Mertens PPC (eds) Bluetongue monograph, 1st edn. Elsevier/Academic Press, London, pp 429–452

Parker J, Herniman KA, Gibbs EP, Sellers RF (1975) An experimental inactivated vaccine against bluetongue. Vet Rec 96(13):284–287

Parkes H, Gould AR (1996) Characterisation of Wongorr virus, an Australian orbivirus. Virus Res 44(2):111–122

Parsonson IM (1990) Pathology and pathogenesis of bluetongue infections. Curr Top Microbiol Immunol 162:119–141

Patton JT, Spencer E (2000) Genome replication and packaging of segmented double-stranded RNA viruses. Virology 277(2):217–225

Pearson JE, Jochim MM (1979) Protocol for the immunodiffusion test for bluetongue. Proc Am Assoc Vet Lab Diagn 22:463

Peiris JS, Amerasinghe PH, Amerasinghe FP, Calisher CH, Perera LP, Arunagiri CK, Munasingha NB, Karunaratne SH (1994) Viruses isolated from mosquitoes collected in Sri Lanka. Am J Trop Med Hyg 51(2):154–161

Polydorou K (ed) (1985) Bluetongue in Cyprus. Alan R Liss Inc., New York

Prasad BV, Rothnagel R, Zeng CQ, Jakana J, Lawton JA, Chiu W, Estes MK (1996) Visualization of ordered genomic RNA and localization of transcriptional complexes in rotavirus. Nature 382(6590):471–473

Prasad G, Minakshi, Malik YS, Maan S (1999) RT-PCR and its detection limit for cell culture grown bluetongue virus 1 using NS1 gene group specific primers. Indian J Exp Biol 37:1255–1258

Prasad G, Minakshi, Malik YS, Maan S (2000) Trends in bluetongue virus diagnostics: a review. Indian J Anim Sci 70:103–109

Prasad G, Minakshi, Malik YS (2007) Bluetongue. Publisher Indian Council of Agricultural Research, New Delhi, p 163

Pritchard LI, Gould AR, Wilson WC, Thompson L, Mertens PP, Wade-Evans AM (1995) Complete nucleotide sequence of RNA segment 3 of bluetongue virus serotype 2 (Ona-A). Phylogenetic analyses reveal the probable origin and relationship with other orbiviruses. Virus Res 35(3):247–261

Purse BV, Mellor PS, Rogers DJ, Samuel AR, Mertens PP, Baylis M (2005) Climate change and the recent emergence of bluetongue in Europe. Nat Rev Microbiol 3(2):171–181

Purse BV, Brown HE, Harrup L, Mertens PP, Rogers DJ (2008) Invasion of bluetongue and other orbivirus infections into Europe: the role of biological and climatic processes. Rev Sci Tech 27(2):427–442

Qiu W, Scholthof KB (2001) Defective interfering RNAs of a satellite virus. J Virol 75(11):5429–5432

Quan M, Lourens CW, MacLachlan NJ, Gardner IA, Guthrie AJ (2010) Development and optimisation of a duplex real-time reverse transcription quantitative PCR assay targeting the VP7 and NS2 genes of African horse sickness virus. J Virol Methods 167(1):45–52

Quist CF, Howerth EW, Bounous DI, Stallknecht DE (1997) Cell-mediated immune response and IL-2 production in white-tailed deer experimentally infected with hemorrhagic disease viruses. Vet Immunol Immunopathol 56(3–4):283–297

Rafyi A (1961) Horse sickness. Bull Off Int Epizoot 56:216–250

Ramadevi N, Roy P (1998) Bluetongue virus core protein VP4 has nucleoside triphosphate phosphohydrolase activity. J Gen Virol 79(Pt 10):2475–2480

Ramadevi N, Burroughs NJ, Mertens PP, Jones IM, Roy P (1998) Capping and methylation of mRNA by purified recombinant VP4 protein of bluetongue virus. Proc Natl Acad Sci U S A 95(23):13537–13542

Rao CD, Kiuchi A, Roy P (1983) Homologous terminal sequences of the genome double-stranded RNAs of bluetongue virus. J Virol 46(2):378–383

Ratinier M, Caporale M, Golder M, Franzoni G, Allan K, Nunes SF, Armezzani A, Bayoumy A, Rixon F, Shaw A, Palmarini M (2011) Identification and characterization of a novel nonstructural protein of bluetongue virus. PLoS Pathog 7(12):e1002477

Ratinier M, Shaw AE, Barry G, Gu Q, Di Gialleonardo L, Janowicz A, Varela M, Randall RE, Caporale M, Palmarini M (2016) Bluetongue virus NS4 protein is an interferon antagonist and a determinant of virus virulence. J Virol 90(11):5427–5439

Reddacliff L, Kirkland P, Philbey A, Davis R, Vogelnest L, Hulst F, Blyde D, Deykin A, Smith J, Hooper P, Gould A, Hyatt A (1999) Experimental reproduction of viral chorioretinitis in kangaroos. Aust Vet J 77(8):522–528

Reinisch KM, Nibert ML, Harrison SC (2000) Structure of the reovirus core at 3.6 A resolution. Nature 404(6781):960–967

Rodriguez-Sanchez B, Fernandez-Pinero J, Sailleau C, Zientara S, Belak S, Arias M, Sanchez-Vizcaino JM (2008) Novel gel-based and real-time PCR assays for the improved detection of African horse sickness virus. J Virol Methods 151(1):87–94

Rose KA, Kirkland PD, Davies RJ, Cooper DW, Blumstein DT (2000) An epizootic of sudden death in Tammar Wallabies (*Macropus eugenii*). In: Proceedings of the AAZV and IAAAM joint conference, pp 166–170

Rose K, Kirkland P, Davis R, Cooper D, Blumstein D, Pritchard L, Newberry K, Lunt R (2012) Epizootics of sudden death in tammar wallabies (Macropus eugenii) associated with an orbivirus infection. Aust Vet J 90(12):505–509

Roy P (1989) Bluetongue virus genetics and genome structure. Virus Res 13(3):179–206

Roy P (1996) Orbivirus structure and assembly. Virology 216(1):1–11

Roy P (2005) Bluetongue virus proteins and particles and their role in virus entry, assembly, and release. Adv Virus Res 64:69–123

Roy P (2008) Bluetongue virus: dissection of the polymerase complex. J Gen Virol 89(Pt 8):1789–1804

Roy P, Ritter GD Jr, Akashi H, Collisson E, Inaba Y (1985) A genetic probe for identifying bluetongue virus infections in vivo and in vitro. J Gen Virol 66(Pt 7):1613–1619

Sanchez-Vizcaino JM (2004) Control and eradication of African horse sickness with vaccine. Dev Biol (Basel) 119:255–258

Sangar DV, Mertens PPC (1983) Comparison of type-1 bluetongue virus protein synthesis in vivo and in vitro. In: Compans RW, Bishop DHL (eds) Double-stranded RNA viruses. Elsevier, New York, pp 183–191

Santman-Berends IM, Hage JJ, van Rijn PA, Stegeman JA, van Schaik G (2010a) Bluetongue virus serotype 8 (BTV-8) infection reduces fertility of Dutch dairy cattle and is vertically transmitted to offspring. Theriogenology 74(8):1377–1384

Santman-Berends IM, van Wuijckhuise L, Vellema P, van Rijn PA (2010b) Vertical transmission of bluetongue virus serotype 8 virus in Dutch dairy herds in 2007. Vet Microbiol 141(1–2):31–35

Savini G, Afonso A, Mellor P, Aradaib I, Yadin H, Sanaa M, Wilson W, Monaco F, Domingo M (2011) Epizootic haemorrhagic disease. Res Vet Sci 91(1):1–17

Schwartz-Cornil I, Mertens PP, Contreras V, Hemati B, Pascale F, Breard E, Mellor PS, MacLachlan NJ, Zientara S (2008) Bluetongue virus: virology, pathogenesis and immunity. Vet Res 39(5):46

Shaw AE, Monaghan P, Alpar HO, Anthony S, Darpel KE, Batten CA, Guercio A, Alimena G, Vitale M, Bankowska K, Carpenter S, Jones H, Oura CA, King DP, Elliott H, Mellor PS, Mertens PP (2007) Development and initial evaluation of a real-time RT-PCR assay to detect bluetongue virus genome segment 1. J Virol Methods 145(2):115–126

Shaw AE, Bruning-Richardson A, Morrison EE, Bond J, Simpson J, Ross-Smith N, Alpar O, Mertens PP, Monaghan P (2013) Bluetongue virus infection induces aberrant mitosis in mammalian cells. Virol J 10:319

Shope RE, MacNamara LG, Mangold R (1955) Epizootic hemorrhagic disease of deer. N J Outdoor 6:17

Spence L, Anderson CR, Aitken TH, Downs WG (1967) Bushbush, Ieri and Lukuni viruses, three unrelated new agents isolated from Trinidadian forest mosquitoes. Proc Soc Exp Biol Med 125(1):45–50

Spreull J (1902) Report from veterinary surgeon Spreull on the result of his experiments with the malarial catarrhal fever of sheep. Agric J Cape Good Hope 20:469–477

Spreull J (1905) Malarial catarrhal fever (bluetongue) of sheep in South Africa. J Comp Pathol 18:321–337

Squire KR, Chuang RY, Chuang LC, Doi RH, Osburn BI (1985a) Molecular cloning and hybridization studies on bluetongue virus serotype 17. Prog Clin Biol Res 178:355–361

Squire KR, Chuang RY, Chuang LF, Doi RH, Osburn BI (1985b) Detecting bluetongue virus RNA in cell culture by dot hybridization with a cloned genetic probe. J Virol Methods 10(1):59–68

Stauber N, Martinez-Costas J, Sutton G, Monastyrskaya K, Roy P (1997) Bluetongue virus VP6 protein binds ATP and exhibits an RNA-dependent ATPase function and a helicase activity that catalyze the unwinding of double-stranded RNA substrates. J Virol 71(10):7220–7226

Stewart ME, Roy P (2010) Role of cellular caspases, nuclear factor-kappa B and interferon regulatory factors in Bluetongue virus infection and cell fate. Virol J 7:362

Stewart M, Hardy A, Barry G, Pinto RM, Caporale M, Melzi E, Hughes J, Taggart A, Janowicz A, Varela M, Ratinier M, Palmarini M (2015) Characterization of a second open reading frame in genome segment 10 of bluetongue virus. J Gen Virol 96(11):3280–3293

Stuart DI, Grimes JM (2006) Structural studies on orbivirus proteins and particles. Curr Top Microbiol Immunol 309:221–244

Sutton G, Grimes JM, Stuart DI, Roy P (2007) Bluetongue virus VP4 is an RNA-capping assembly line. Nat Struct Mol Biol 14(5):449–451

Swanepoel R, Blackburn NK (1976) A new member of the Palyam serogroup of orbiviruses. Vet Rec 99(18):360

Tabachnick WJ, MacLachlan NJ, Thompson LH, Hunt GJ, Patton JF (1996) Susceptibility of Culicoides variipennis sonorensis to infection by polymerase chain reaction-detectable bluetongue virus in cattle blood. Am J Trop Med Hyg 54(5):481–485

Takamatsu H, Jeggo MH (1989) Cultivation of bluetongue virus-specific ovine T cells and their cross-reactivity with different serotype viruses. Immunology 66(2):258–263

Takamatsu H, Mellor PS, Mertens PP, Kirkham PA, Burroughs JN, Parkhouse RM (2003) A possible overwintering mechanism for bluetongue virus in the absence of the insect vector. J Gen Virol 84(Pt 1):227–235

Tamura K, Peterson D, Peterson N, Stecher G, Nei M, Kumar S (2011) MEGA5: molecular evolutionary genetics analysis using maximum likelihood, evolutionary distance, and maximum parsimony methods. Mol Biol Evol 28(10):2731–2739

Taraporewala ZF, Chen D, Patton JT (2001) Multimers of the bluetongue virus nonstructural protein, NS2, possess nucleotidyl phosphatase activity: similarities between NS2 and rotavirus NSP2. Virology 280(2):221–231

Theal GM (1899) Records of South-Eastern Africa collected in various libraries and archive departments in Europe, vol 3. Government of Cape Colony, Cape Town, p 224

Theiler A (1910) Notes on a fever in horses simulating horse-sickness. Transvaal Agric J 8:581–586

Theodoridis A, Nevill EM, Els HJ, Boshoff ST (1979) Viruses isolated from Culicoides midges in South Africa during unsuccessful attempts to isolate bovine ephemeral fever virus. Onderstepoort J Vet Res 46(4):191–198

Theron J, Huismans H, Nel LH (1996a) Identification of a short domain within the non-structural protein NS2 of epizootic haemorrhagic disease virus that is important for single strand RNA-binding activity. J Gen Virol 77(Pt 1):129–137

Theron J, Huismans H, Nel LH (1996b) Site-specific mutations in the NS2 protein of epizootic haemorrhagic disease virus markedly affect the formation of cytoplasmic inclusion bodies. Arch Virol 141(6):1143–1151

Thevasagayam JA, Mertens PP, Burroughs JN, Anderson J (1995) Competitive ELISA for the detection of antibodies against epizootic haemorrhagic disease of deer virus. J Virol Methods 55(3):417–425

Thevasagayam JA, Wellby MP, Mertens PP, Burroughs JN, Anderson J (1996) Detection and differentiation of epizootic haemorrhagic disease of deer and bluetongue viruses by serogroup-specific sandwich ELISA. J Virol Methods 56(1):49–57

Thomas CP, Booth TF, Roy P (1990) Synthesis of bluetongue virus-encoded phosphoprotein and formation of inclusion bodies by recombinant baculovirus in insect cells: it binds the single-stranded RNA species. J Gen Virol 71(Pt 9):2073–2083

Toussaint JF, Sailleau C, Breard E, Zientara S, De Clercq K (2007) Bluetongue virus detection by two real-time RT-qPCRs targeting two different genomic segments. J Virol Methods 140(1–2):115–123

Travassos da Rosa AP, Tesh RB, Pinheiro FP, Travassos da Rosa JF, Peralta PH, Knudson DL (1984) Characterization of the Changuinola serogroup viruses (Reoviridae: Orbivirus). Intervirology 21(1):38–49

Troupin C, Schnuriger A, Duponchel S, Deback C, Schnepf N, Dehee A, Garbarg-Chenon A (2011) Rotavirus rearranged genomic RNA segments are preferentially packaged into viruses despite not conferring selective growth advantage to viruses. PLoS One 6(5):e20080

Uitenweerde JM, Theron J, Stoltz MA, Huismans H (1995) The multimeric nonstructural NS2 proteins of bluetongue virus, African horse sickness virus, and epizootic hemorrhagic disease virus differ in their single-stranded RNA-binding ability. Virology 209(2):624–632

Urakawa T, Roy P (1988) Bluetongue virus tubules made in insect cells by recombinant baculoviruses: expression of the NS1 gene of bluetongue virus serotype 10. J Virol 62(11):3919–3927

Urakawa T, Ritter DG, Roy P (1989) Expression of largest RNA segment and synthesis of VP1 protein of bluetongue virus in insect cells by recombinant baculovirus: association of VP1 protein with RNA polymerase activity. Nucleic Acids Res 17(18):7395–7401

Uren MF (1986) Clinical and pathological responses of sheep and cattle to experimental infection with five different viruses of the epizootic hemorrhagic disease of deer serogroup. Aust Vet J 63(6):199–201

van Niekerk M, van Staden V, van Dijk AA, Huismans H (2001) Variation of African horse sickness virus nonstructural protein NS3 in southern Africa. J Gen Virol 82(Pt 1):149–158

van Regenmortel MH, Mahy BW (2004) Emerging issues in virus taxonomy. Emerg Infect Dis 10(1):8–13

Van Rensberg IB, De Clerk J, Groenewald HB, Botha WS (1981) An outbreak of African horse sickness in dogs. J S Afr Vet Assoc 52(4):323–325

Venter EH, Viljoen GJ, Nel LH, Huismans H, van Dijk AA (1991) A comparison of different genomic probes in the detection of virus-specified RNA in Orbivirus-infected cells. J Virol Methods 32(2–3):171–180

Verwoerd DW, Erasmus BJ (2004) Bluetongue. In: Coetzer JAW, Tustin RC (eds) Infectious diseases of livestock. Oxford University Press, Cape Town, pp 1201–1220

Verwoerd DW, Louw H, Oellermann RA (1970) Characterization of bluetongue virus ribonucleic acid. J Virol 5(1):1–7

Verwoerd DW, Els HJ, De Villiers EM, Huismans H (1972) Structure of the bluetongue virus capsid. J Virol 10(4):783–794

Verwoerd DW, Huismans H, Erasmus BJ (1979) Orbiviruses. In: Fraenkel-Conrat H, Wagner RR (eds) Comprehensive virology. Plenum Press, New York, pp 285–345

Vieira Cde M, Nunes MR, da Silva EV, Carvalho VL, Nunes Neto JP, Cruz AC, Casseb SM, Vasconcelos HB, Quaresma JA, Vasconcelos PF (2009) Full-length sequencing and genetic characterization of Breu Branco virus (Reoviridae, Orbivirus) and two related strains isolated from Anopheles mosquitoes. J Gen Virol 90(Pt 9):2183–2190

Wade-Evans AM (1990) The complete nucleotide sequence of genome segment 7 of bluetongue virus, serotype 1 from South Africa. Nucleic Acids Res 18(16):4919

Wade-Evans AM, Mertens PP, Bostock CJ (1990) Development of the polymerase chain reaction for the detection of bluetongue virus in tissue samples. J Virol Methods 30(1):15–24

Wade-Evans AM, Mertens PP, Belsham GJ (1992) Sequence of genome segment 9 of bluetongue virus (serotype 1, South Africa) and expression analysis demonstrating that different forms of VP6 are derived from initiation of protein synthesis at two distinct sites. J Gen Virol 73(Pt 11):3023–3026

Wang LF, Crameri G (2014) Emerging zoonotic viral diseases. Rev Sci Tech 33(2):569–581

Whistler T, Swanepoel R (1988) Characterization of potentially foetotropic Palyam serogroup orbiviruses isolated in Zimbabwe. J Gen Virol 69(Pt 9):2221–2227

Whitehead RH, Doderty RL, Domrow R, Standfast HA, Wetters EJ (1968) Studies of the epidemiology of arthropod-borne virus infections at Mitchell River Mission, Cape York Peninsula, North Queensland. 3. Virus studies of wild birns, 1964–1967. Trans R Soc Trop Med Hyg 62(3):439–445

Wilson A, Carpenter S, Gloster J, Mellor P (2007) Re-emergence of bluetongue in northern Europe in 2007. Vet Rec 161(14):487–489

Wilson A, Mellor PS, Szmaragd C, Mertens PP (2009a) Adaptive strategies of African horse sickness virus to facilitate vector transmission. Vet Res 40(2):16

Wilson WC, Hindson BJ, O'Hearn ES, Hall S, Tellgren-Roth C, Torres C, Naraghi-Arani P, Mecham JO, Lenhoff RJ (2009b) A multiplex real-time reverse transcription polymerase chain reaction assay for detection and differentiation of Bluetongue virus and Epizootic hemorrhagic disease virus serogroups. J Vet Diagn Invest 21(6):760–770

Wilson WC, O'Hearn ES, Tellgren-Roth C, Stallknecht DE, Mead DG, Mecham JO (2009c) Detection of all eight serotypes of Epizootic hemorrhagic disease virus by real-time reverse transcription polymerase chain reaction. J Vet Diagn Invest 21(2):220–225

Wu X, Chen SY, Iwata H, Compans RW, Roy P (1992) Multiple glycoproteins synthesized by the smallest RNA segment (S10) of bluetongue virus. J Virol 66(12):7104–7112

Yadav PD, Albarino CG, Nyayanit DA, Guerrero L, Jenks MH, Sarkale P, Nichol ST, Mourya DT (2018) Equine encephalosis virus in India, 2008. Emerg Infect Dis 24(5):898–901

Yamakawa M, Ohashi S, Kanno T, Yamazoe R, Yoshida K, Tsuda T, Sakamoto K (2000) Genetic diversity of RNA segments 5, 7 and 9 of the Palyam serogroup orbiviruses from Japan, Australia and Zimbabwe. Virus Res 68(2):145–153

Yang DK, Hwang IJ, Kim BH, Kweon CH, Lee KW, Kang MI, Lee CS, Cho KO (2008) Serosurveillance of viral diseases in Korean native goats (Capra hircus). J Vet Med Sci 70(9):977–979

Zanotto PM, Gao GF, Gritsun T, Marin MS, Jiang WR, Venugopal K, Reid HW, Gould EA (1995) An arbovirus cline across the northern hemisphere. Virology 210(1):152–159

Zanotto PM, Gould EA, Gao GF, Harvey PH, Holmes EC (1996) Population dynamics of flaviviruses revealed by molecular phylogenies. Proc Natl Acad Sci U S A 93(2):548–553

Zientara S, Sailleau C, Moulay S, Cruciere C, el-Harrak M, Laegreid WW, Hamblin C (1998) Use of reverse transcriptase-polymerase chain reaction (RT-PCR) and dot-blot hybridisation for the detection and identification of African horse sickness virus nucleic acids. Arch Virol Suppl 14:317–327

Zientara S, Sailleau C, Dauphin G, Roquier C, Remond EM, Lebreton F, Hammoumi S, Dubois E, Agier C, Merle G, Breard E (2002) Identification of bluetongue virus serotype 2 (Corsican strain) by reverse-transcriptase PCR reaction analysis of segment 2 of the genome. Vet Rec 150(19):598–601

Chapter 9
Equine Influenza Virus

Nitin Virmani, S. Pavulraj, B. C. Bera, Taruna Anand, R. K. Singh, and B. N. Tripathi

Abstract Equine influenza (EI), commonly known as 'horse flu,' is a major respiratory disease of equines caused by equine influenza virus (EIV) and has huge economic implications. The disease is characterised by pyrexia, anorexia, depression, dyspnoea, dry hacking cough and serous nasal discharge many a times followed by secondary bacterial pneumonia. Morbidity in the naive population may reach up to 60–90% although mortality may vary. Two subtypes of influenza A virus, viz. H7N7 and H3N8, were known to cause the disease in horses. However, outbreaks due to H7N7 have not been recorded in the last three decades. The H3N8 prototype was isolated from Miami in 1963 and circulated globally as a pre-divergent lineage till 1987. Further, the virus made two distinct lineages—European and American—out of which American lineage further diversified into Florida, Kentucky and Argentina, with current outbreaks occurring globally due to clade 1 and clade 2 viruses belonging to Florida sublineage. Equine influenza virus has jumped species and has been reported in canines, swine, camels and cats. One of the most effective ways to prevent equine influenza, is vaccination along with standstill in case of infection in premises/area. The control of the disease is mainly focused on vaccine strategies for generating robust antibody responses against the major surface glycoproteins, particularly HA. Despite vaccination, EIV is circulating and responsible for epizootics in recent years. This is attributed to continual antigenic variation in the surface glycoproteins. Antigenic and genetic drift (the accumulation of point mutations) in the gene encoding for the HA and NA epitopes helps the virus to evade the protective humoral immune system of the host and leads to failure of vaccine, thus warranting repeated harmonisation of vaccine strains. A BALB/c mice model was established by NRCE, India, for detail investigation of EIV pathogenicity and quick efficacy testing of inactivated vaccine. Currently, available EI vaccines include inactivated whole virus vaccines, subunit vaccines and modified live vaccines. A strong correlation

N. Virmani (✉) · B. C. Bera · T. Anand · B. N. Tripathi
ICAR-National Research Centre on Equines, Hisar, Haryana, India

S. Pavulraj
Institut für Virologie, Freie Universität, Berlin, Germany

R. K. Singh
ICAR-Indian Veterinary Research Institute, Bareilly, Uttar Pradesh, India

Y. S. Malik et al. (eds.), *Emerging and Transboundary Animal Viruses*, Livestock Diseases and Management, https://doi.org/10.1007/978-981-15-0402-0_9

has been observed between humoral immunity level and protective immunity against infection with antigenically homologous viruses in challenge studies. Along with vaccination, standstill with regard to the movement of horses and other biosecurity measures are a must to control the spread of equine influenza.

Keywords Equine influenza · Influenza A virus · EIV · Genome · Clinical signs · Pathology · Immunobiology · Detection · Vaccines

9.1 Preamble

Equine influenza (EI) is a highly contagious acute respiratory disease of equines caused by equine influenza virus (EIV). The EIV is responsible for two-third of viral respiratory infections in horses associated with huge economic impact (Mumford et al. 1998; Singh et al. 2018). The disease is called as 'horse flu' and 'Newmarket Cough' and characterised by pyrexia, anorexia, depression, dyspnoea, dry hacking cough and serous nasal discharge followed by secondary bacterial pneumonia (Gerber 1970). Morbidity in the naive population may reach up to 60–90%, and mortality with confirmed infection has ranged from 1% to 20%. Higher mortality rates are observed in foals and malnourished, immune-compromised, old and stressed equids and donkeys (Paillot et al. 2006; Waghmare et al. 2010). First recorded outbreak of EI was reported in 1872 in the USA and is known by 'great epizootic of 1872,' although causative agent was not isolated at that time (McClure, 1998). EIV was first isolated from horse with respiratory signs in Prague in 1956, which revealed a novel influenza virus causing infection only in equines (Sovinova et al. 1958) and the virus was subtyped as H7N7. The retrospective serological analysis suggested that the same subtype of the virus was also circulated among equine population in Europe before 1956 (Beveridge 1965). Later in 1963, horses in America suffered from respiratory illness caused by different subtype of influenza virus, which was antigenically dissimilar with H7N7, and the virus was designated as H3N8. The H3N8 prototype was isolated from Miami in 1963 (Waddell et al. 1963). Till the end of the 1970s, EI infections in horses were caused by both H7N7 and H3N8 subtype of viruses. However, the H7N7 virus is no longer in circulation (Webster, 1993) since the last 35 years, and it is H3N8 subtype that continues to circulate and causes disease world over. EI outbreaks were continuously being reported from different countries of the world in spite of routine vaccination since the 1960s (Bryans et al. 1966). The disease outbreaks predominantly have been recorded among naive and unvaccinated population globally and frequently reported in last few years from Argentina, Brazil, Canada, Chile, China, Dominican Republic, France, Germany, Ireland, India, Mongolia, Sweden, UAE, the UK, the USA and Uruguay (OIE 2019). Equine influenza virus was believed to restrict in equids only but recently species jumping of EIV was observed from equines to other novel hosts as reported in canines (Crawford et al. 2005; Yamanaka et al. 2009; Kirkland et al.

2010), swine (Tu et al. 2009) and camels (Yondon et al. 2014). Recent studies also showed that cats are also susceptible to EIV infection (Su et al. 2014). One of the most effective ways to prevent viral diseases, including equine influenza, is vaccination. The control of the disease is mainly focused on vaccine strategies for generating robust antibody responses against the surface glycoproteins, particularly HA (Daly et al. 2011). HA is the major surface protein of influenza A viruses and a recognised key antigen in the host response to influenza virus in both natural infection and vaccination. Despite vaccination, EIV has been circulating and responsible for epizootics in recent years. This is attributed to continual antigenic variation in the surface glycoproteins. Antigenic and genetic drift (the accumulation of point mutations) in the gene encoding for the HA and NA epitopes helps the virus to evade the protective humoral immune system of the host and leads to failure of vaccine. Thus, active research on EIV is the main focus of the scientists globally for deciphering the disease pathogenesis, evolution of the virus and development of the effective vaccines for generating both humoral and cellular immunity for control of the disease.

9.2 Biology of Influenza Virus

EIV belongs to the *Orthomyxoviridae* family containing five genera, viz. influenza A, B and C virus; thogotovirus; and isavirus (Cox and Subbarao 2000; Lamb and Krug 2001). Influenza viruses are broadly classified into type A, B and C based on the antigenic differences in nucleocapsid (NP) and matrix (M) proteins. Influenza A viruses cause infection in horses, humans, poultry, swine, rarely dogs, seals, mink and whales; influenza B infections are associated only in human with few case reports in seals and infections with influenza C virus have been reported in both humans and swine. All the above-said viruses are transmitted by aerosol and droplet infection. Thogotovirus is transmitted by ticks and produces diseases rarely in livestock as well as humans. Infectious salmon anaemia virus is the sole member of the genus and is a virus causing disease in Atlantic salmon fish (MacLachlan 2011). The genome of influenza A virus (IAV) is segmented RNA of about 13.6 kb containing eight separate single-stranded, negative-sense RNA segments enclosed in a lipid envelope. The IAV genome encodes for 12 distinct proteins, viz. haemagglutinin (HA), neuraminidase (NA), nucleoprotein (NP), viral polymerase proteins (PA, PA-X, PB1, PB1-F2 and PB2), matrix 1 (M1), matrix 2 (M2), nonstructural protein 1 (NS1) and nuclear export protein (NEP) (Chen et al., 2001, Jagger et al., 2012). IAVs are further classified into different subtypes based on antigenic relationships of their surface glycoproteins, namely haemagglutinin (H1-H18) which is rod-shaped homotrimers and neuraminidase (N1-N11) homotetramer with mushroom shape. The H18 and N11 subtypes were recently identified in a fruit bat (*Artibeus planirostris*) from Peru (Tong et al. 2013). The shape of the IAVs may vary from spherical (with a diameter of 80–120 nm) to filamentous. The outer structure of virus consists of a lipid bilayer with two glycoproteins—HA (responsible for binding with a receptor on host cells) and NA (directs release of virus from infected cells and spreading). The two matrix proteins (M1 and M2) support the envelope structure of the IAVs. Each segment of IAVs is packaged with polymerase proteins, and helically arranged multiple copies of nucleoprotein to form rod-shaped ribonucleoprotein (RNP) complex, which is the core structure of the virus. Three polymerase proteins, viz. PB1, PB2 and PA proteins, interact with each other to form the RNA-dependent RNA polymerase complex, which controls the replication of the virus. This polymerase protein complex is also associated with the nucleocapsid. The RNP is a transcriptionally active complex, which leads to transcription of messenger RNAs (mRNAs) and synthesis of complementary RNAs (cRNAs) which acts as a template for the production of nascent viral RNAs (vRNAs). The newly produced negative-sense viral RNAs are incorporated into progeny viruses. The NS1 protein plays a vital role in RNA transport, splicing, translation and subverting host immune response, whereas NEP is involved in the nuclear export of virion RNAs (Paillot et al. 2006). The lack of proofreading activity of the viral RNA-dependent RNA polymerases causes error-prone replication of virus which leads to the antigenic and genetic drift (the accumulation of point mutations) in the HA gene and thereby generation of new virus strains.

9.3 Epidemiology and Evolution of EIV

EI is endemic in Europe and North America and is of major economic significance (Cullinane and Newton 2013). EI has not yet been reported in island countries like New Zealand and Iceland. Epizootics of EI have been reported at frequent intervals in various parts of the world like South Africa in 1986 and 2003 (Guthrie 2006), India in 1987 and 2008 (Virmani et al. 2008; Singh et al. 2018), China in 1989 (Guo et al., 1992), Hong Kong in 1992 (Powell et al. 1995), China in 2007–2008, Mongolia in 2007 and 2011, Uruguay in 2012 (Acosta et al. 2012) and Pakistan in 2015–2016 (Khan et al. 2017), multifocal outbreaks in Europe in 2018–19. Australia had the EI-free status till 2007 when the country suffered from a huge outbreak of disease, and more than 50,000 horses were affected. The disease was, however, eradicated from the country within 4 months of first outbreak (Cowled et al. 2009). Disease control and eradication strategies do not include destroying of animals. In Australia EI outbreak was effectively controlled by movement control and following strict biosecurity measures.

All age group of horses are susceptible to infection, including newborn foals. Immunocompromised horses are at higher risk of developing diseases. The higher prevalence is noticed between the age group of 2 and 6 months (Nyaga et al. 1980). Outbreaks are commonly observed at the time of yearling sale, transport of horses for racing and other movements (Cullinane and Newton 2013; Daly et al. 2011). Transmission of EIV takes place through both direct and indirect ways. Naive infected horse sheds the virus in nasal secretions after 24 h following exposure, and it continues up to 10 days (Myers and Wilson 2006). An infected horse can spread the virus up to 32 metres through its forceful cough (Miller 1965). Aerosolised equine influenza can be taken by wind up to several kilometres (km) such as more than 8 km as reported in South Africa (Huntington 1990), 3.2 km in Jamaica (Dalglish 1992) and 5 km in Australia (Firestone et al. 2011). Indirectly, EIV spreads through fomites such as equipment, vehicles and personnel (Guthrie et al. 1999). International movement of horses for racing and breeding is important in the spread of influenza throughout the world (Powell et al. 1995; Mumford 1999; Wernery et al. 1999).

Influenza viruses are continuously evolving by processes like the accumulation of single-point mutations, reassortment and species jumping (Webster 2002). Due to the lack of proofreading function of RNA polymerase, influenza viruses continuously accumulate mutations in their genome. These mutations may be silent or may alter virulence or pathogenicity of the virus. Continuous accumulation of single-point mutations in surface glycoproteins over a long period produces structural and antigenic differences which affect vaccine efficacy adversely (Nakajima et al. 2005). Antigenic evolutions are mainly due to immune selection and other factors involved in virus adaptation in a new host. At least four antigenic changes in two different antigenic sites of HA gene are required for significant antigenic drift in case of human influenza A virus (Wilson and Cox 1990). One or more mutations in any of five major antigenic sites of the gene coding for HA1 glycoprotein of the virus lead to failure of circulating antibodies to recognise virus produced by vaccination or infection (Wiley and Skehel 1987; Daly et al. 2003). In influenza viruses, antigenic variations are very common in both HA and NA genes due to constant exposure to the immune system of the host (Webster and Laver 1975) leading to new isolated results in an influenza outbreak in every 5–7 years in a different part of the world.

Segmented natures of the influenza viruses allow genetic reassortment by random shuffling of the genome during superinfection, and progeny virus will receive genome segments from both parents and emerge as new subtypes (antigenic shifts). Influenza A virus is known for rapid evolution in aberrant hosts of different species even though they are in evolutionary stasis in the natural reservoir—aquatic birds. In H3N8 viruses, the rate of evolution is much slower than the human influenza virus. Some viruses have been circulating for more than 25 years without much antigenic and genetic variations (Guo et al. 1992); this phenomenon is known as 'frozen evolution' (Endo et al. 1992).

Two different subtypes of EIV, H7N7 and H3N8 were associated with acute respiratory disease in equines. H7N7 was first recognised in 1956 in Eastern Europe (Sovinova et al. 1958), but not reported after the 1970s, and is thought to be extinct from the equine population (Webster and Thomas 1993). Infection due to H3N8 subtype was first reported in 1963 in Florida (Waddell et al. 1963); since then, it caused outbreaks in every part of the world, and current EI outbreaks are caused only by virus belonging to H3N8. These viruses are more stable than human influenza viruses but still undergo antigenic drift periodically due to point mutations in HA and NA genes leading to a substantial change in the antigenicity of the virus (Cullinane and Newton 2013). A/equine/Miami/63 and Fontainebleau/76 are early prototypes of these viruses in the 1960s and 1970–1980s, respectively. Phylogenetic analysis of HA gene revealed that in 1980s H3N8 EIVs diverged into two distinct types of lineages named as American-like and European-like lineages based on the place of isolation (Daly et al. 1996). But later both viruses co-circulated without any geographical limitation. American- and European-like lineage prototype viruses differ from each other in nine residues of HA antigenic regions throughout the length A–D. No cross protection was noticed in between two lineages' virus infection/vaccination, so both lineages were included in vaccines (Daly et al. 1996, 2004; Yates and Mumford 2000; Mumford et al. 2003). American lineage further evolved into three distinct sublineages, namely Florida, Kentucky and Argentina (Lai et al. 2004), out of which Florida sublineage was predominant. Further, Florida sublineage diverged into two antigenically distinct clades (i.e. clade 1 and clade 2) (Bryant et al. 2009). Florida clade 1 virus is predominant in America, but has caused outbreaks in Asia (Japan), Australia, Africa and Europe (Cullinane and Newton 2013; Yamanaka et al. 2008). Clade 2 viruses predominate in Europe but have caused large outbreaks in Asian countries, namely Mongolia (2007), China (2007–2008) and India (2008–2009) (Virmani et al. 2010a; Qi et al. 2010; Yondon et al. 2013). From 2013 to 2015, UK, Germany and Italy had majority of the outbreaks due to Florida clade 2 viruses; however, in few of the outbreaks the viruses from the UK had changed in HA at position 144 (A144V) in amino acid sequence while viruses from Germany and Italy had changes at position 179 (I179V) (Rash et al. 2017). Chile had an outbreak due to Florida clade 1 virus in 2011–2012, which spread to many countries in South America (Perglione et al. 2016). During the year 2015, Malaysia had an outbreak with Florida clade 1 virus (Gahan et al. 2019). In the year 2018–2019, many countries in Europe including the UK, France, Germany, Ireland, the Netherlands, Denmark, Sweden and Belgium had major outbreaks of equine influenza with a virus belonging to Florida clade 1 origin (Animal Health Trust 2019. https://www.aht.org.uk/wp-content/uploads/2019/04/Equiflunet-outbreaks-to-15-April 2019.pdf).

The host adaptation of influenza viruses is influenced by the genomic composition as well as codon usage pattern because of complete viral replication dependency on the host cellular system. Few studies have highlighted the role of synonymous codon usage in the overall adaptation of EIV in hosts. The codon usage bias in overall EIV genome is lower and primarily influenced by nucleotide composition, mutation pressure and natural selection (Kumar et al. 2016). Further, codon usage also highlighted the underlying factor for the extinction of H7N7 subtype due to greater bias in codon usage, less mutation pressure and lower adaptation to tRNA pool specific to equine host in comparison to H3N8 subtype (Kumar et al. 2016). In another study, the influence of various evolutionary factors has been analysed for polymerase genes of equine influenza virus (Bera et al. 2017). The polymerase genes play an important role in host adaptation and pathogenesis due to their direct involvement in virus replication, transcription and host transmission. Similar to the codon usage pattern of the EIV genome, polymerase genes also have a lower bias in codon usage. The natural selection is the main factor for codon usage and evolution of these genes; however, other factors such as nucleotide composition, mutation pressure, hydrophobicity and aromaticity also influence the codon usage (Bera et al. 2017).

9.4 Species Involved

Host range determinant for influenza A virus is polygenic in nature and requires a balanced function of HA and NA proteins along with internal proteins. It also depends on receptor distribution on the host, especially type of linkage to galactose on the host cell surface. EIV preferentially binds to *N*-acetylneuraminic acid α-2,3-galactose (NeuAc α-2,3-Gal) linkage on sialyloligosaccharides on the host cell before entry (Lee and Saif 2009) which is predominant in equine upper respiratory tract, especially in epithelial cells, goblet cells and nasal glandular epithelium of nasal mucosa, tracheal and bronchiolar epithelial cells of lower respiratory tract (Scocco and Pedini 2008). Similar receptor present in birds and separate H3N8 strain are established, but no equine disease was reported by bird strain virus with the exception of China in 1989 where avian-like influenza A virus (A/eq/Jilin/1/89 [H3N8]) caused an outbreak in horses with 20% mortality (Guo et al. 1991). Horses, donkeys and mules are the regular target species for EIV and considered as an isolated or dead-end reservoir hosts for EI viruses. However, recently, in 2004, EIV has shown species jumping into canines (Crawford et al. 2005). Interspecies transmission of influenza A virus depends on many factors like the interaction between different host species and presence of appropriate receptors (α2,3-sialic acid linkage) on the respiratory tract of novel host for attachment of HA to respiratory tract epithelial cells.

As we know, cell line derived from canine renal tissue—Madin-Darby canine kidney cells (MDCK)—supports growth of EIV due to presence of cell surface α-2,3 sialic acid residues (Kovbasnjuk and Spring 2000) and has played a role of workhorse in EIV propagation and diagnosis, however, EIV infection was not observed in canine even though it has similar α2,3-sialic acid residues in the upper respiratory tract. At the beginning of 2004 virus crossed species barrier from equine to dogs in Florida, where 8 greyhounds suffered from fatal haemorrhagic-tracheitis, bronchopneumonia and pleuritis and 22 dogs experienced infection and showed seroconversion which was confirmed by virus isolation (Crawford et al. 2005). Further few more incidences were reported in greyhounds and pet dogs at the end of 2004–2007 in the USA (Payungporn et al. 2008). Phylogenetic analysis of HA gene of canine isolates revealed a close relationship with each other with common ancestor origin from Florida lineage of equine H3N8 strain circulated in the 1990s. Retrospective studies revealed that H3N8 virus was responsible for the infection in foxhound dogs which showed respiratory signs in the UK in 2002 (Daly et al. 2008) and initial transmission to dogs might have occured in 1998–2003 (Payungporn et al. 2008). Transmission of EIV to dogs has been reported in recent EI outbreak in Australia in 2007 (Kirkland et al. 2010). The predominance of α2,3 receptors on canine tracheal epithelial cells (Daly et al. 2008) may be responsible for infection in canines, but why infection in canines was not reported before 2004 is not clear. It may have been overlooked for kennel cough syndrome. However, EIVs circulating currently may be transferrable to dogs (Daly et al. 2011). Totally six isolates have been isolated from canines, viz. A/canine/Jacksonville/2005, A/canine/Miami/2005,

A/canine/Florida/242/2003, A/canine/Florida/43/2004, A/canine/Texas/2004 and A/canine/Iowa/2005 (Crawford et al. 2005; Yoon et al. 2005). All these six canine isolates had five conserved amino acid substitutions in H3 haemagglutinin differentiating them from equine influenza viruses. These conserved substitutions are N54K, N83S, W222L, I328T and N483T, which can be considered as a signature of the circulating CIV H3 haemagglutinin.

In addition to equids and dogs, two viruses isolated from pigs showing respiratory signs in China in 2005 and 2006 belonged to the EIV, subtype H3N8. Sequencing studies of HA revealed close resemblances of the current virus with European lineage virus circulated in the 1990s (Tu et al. 2009). Later in 2012, nasal swab collected from one camel was positive for H3N8 EIV in Mongolia. However, no clinical signs or subsequent spread among camels was reported (Yondon et al. 2014).

9.5 History of Equine Influenza

EI-like disease has been reported in history several times. First well-known outbreak of EI occurred in 1872 in North America, popularly known as 'The Great Epizootic of 1872,' that devastated equine population with 100% morbidity and mortality ranging from 1% to 10% (McClure, 1998). EIV was isolated from Sweden and Eastern Europe in 1955–1956 for the first time and belongs to subtype H7N7—A/eq/Prague/56 (Sovinova et al. 1958; Tumova and Sovinova-Fiserova 1959). The retrospective serological analysis revealed that these viruses circulated in America and Europe before its original virus detection (Doll 1961). Virus subtype belonging to H3N8 was first reported in Miami and the strain was A/eq/Miami/63 (Waddell et al. 1963). Till the end of the 1970s, both H7N7 and H3N8 strains were circulated in the equine population and produced diseases. EI caused by H7N7 strain was not reported in the last 35 years and considered to be extinct (Webster 1993). In recent times, Clade 1 and Clade 2 of Florida sublineage of EIV/H3N8 are circulating in horses globally. In Europe and North America, the disease is endemic (Cullinane and Newton 2013). The EI epizootics have been reported in various countries like South Africa in 1986 and in 2003 (Guthrie 2006); India in 1987 and 2008 (Virmani et al. 2008); China in 1989 (Guo et al., 1992); Hong Kong in 1992 (Powell et al. 1995); China in 2007–2008; Mongolia in 2007 and 2011; Uruguay in 2012 (Acosta et al. 2012); and Pakistan in 2015–2016 (Khan et al. 2017) that caused huge economic losses. Australia had a major outbreak in 2007, affecting more than 50,000 horses, and the disease had been eradicated from the country (Cowled et al. 2009).

9.6 EI in India

In India, there has been a report of EI-like disease in equines that originated from Kolkata after importation of horses from Australia and spread across the country in between the period of April and September 1915. The disease was suspected as EI because of its contagious nature, short incubation period, febrile illness, respiratory involvement and rapid recovery after rest. A total of 16,921 cases were reported with 893 deaths, but the actual causative agent could not be isolated (Williams 1924). Another incidence of respiratory disease in horses similar to EI was reported by Manjrekar et al. (1965) in which about 400 horses from Royal Western India Turf Club, Mumbai, were involved. The causative agent was not isolated from any of the infected animal. This coughing outbreak 'Newmarket Cough' started in the last week of October 1964 and continued till December 1964.

India had obnoxious experience with EI in 1987 when more than 83,000 equines suffered from the disease in northern states (Uppal and Yadav 1987). Equines imported from France during this period were found to be the source of entry of infection to India. Two virus isolates, namely A/eq/Bhiwani/87 and A/eq/Ludhiana/87, were confirmed in an outbreak (Uppal et al. 1989). Whole virus adjuvanted killed vaccine using A/equi-2/Ludhiana/87 was prepared indigenously and used (Gupta et al. 1993). After 1987, no other EI outbreak was observed till 2008 for 20 years (Virmani et al, 2010a) , except detection of antibodies against A/Equine-2 in some parts of the country which might be due to vaccine titres.

The disease again re-emerged in India in June 2008 and nidus of disease was the northernmost state of the country, Jammu and Kashmir. The disease further spread to 14 states of the country (Virmani et al. 2008, 2010a, b). The virus isolates belong to subtype H3N8 and are named as A/eq/Jammu-Katra/08, A/eq/Mysore/08 and A/eq/Ahamadabad/09 based on place of isolation. Phylogenetic analysis of the HA gene confirmed the virus related to clade 2 of the Florida sublineage in American lineage (H3N8) (Virmani et al. 2010a). Phylogenetic analysis of the M gene in all isolates shared 98.41% and 99.54% homology with other clade 2 viruses of Asian origin for M1 and M2 amino acid (aa) sequences, respectively. In addition, 3 and 4 unique aa residue changes, respectively, in M1 and M2 proteins in all Asian isolates along with Chinese and Mongolian isolates were noticed (Virmani et al. 2011). Continuous serosurveillance and monitoring and preventive measures carried out by the National Research Centre on Equines, Hisar, Haryana, in India led to control of the disease in the country, and the last case was reported in July 2009.

9.7 Pathogenesis

The incubation period for EI ranges between 18 h and 5 days as negatively correlated with the dose of virus challenge. EIV produces lesions in the entire respiratory tract, especially lower respiratory tract after binding to α2,3-sialic acid receptors with the help of HA (Muranaka et al. 2010). To have access to cellular receptors, EIV cleaves sialic acid residues in mucus with the help of NA protein which allows their access to cell surface receptors that leads to the increase in mobility of the virus and favours secondary bacterial infections (Patterson-Kane et al. 2008). Following virus attachment, internalisation takes place by receptor-mediated endocytosis (Eierhoff et al. 2010). Irreversible conformational changes in HA take place after acidification in endosome (Sieczkarski and Whittaker 2002). The fusion of viral and cellular membrane allows the release of viral RNPs into cell cytoplasm which is then transported to the nucleus for the synthesis of mRNA. Nucleus synthesis of complementary RNA and viral RNA undergoes association with the help of proteins like NS1 and NP (Lee and Saif 2009). Newly synthesised viral proteins, viz. M1, HA and NA, are produced in membrane-bound form and undergo various conformational changes during transport in the endoplasmic reticulum and Golgi bodies (Bosch et al. 1981). Budding of virions takes place by clustering of HA and NA on cell surface membrane (lipid raft domain) (Wang et al. 2007). M1 will bind to cytoplasmic tails of HA and NA which causes docking of vRNPs. The polymerisation of M1 causes elongation of budding virions with localisation of vRNPs to one corner. Membrane scission of complete virion takes place with the help of M2 protein. At last, fully formed infectious virion particles are released into the luminal surface of mucosa (Rossman et al. 2010; Rossman and Lamb 2011).

Viral infection damages ciliated epithelium and leads to reduced tracheal clearance (Willoughby et al. 1992). Interaction of NA in the clearing of sialic acid residues in mucosa favours viral attachment and secondary bacterial infection. It is a very good example of virus–bacteria synergism (Lopez 2007).

The molecular mechanism behind how the influenza virus kills the cells was little understood. Influenza virus infection in mammalian cells triggers apoptosis and causes a cytopathic effect (Takizawa et al. 1993). In vitro study of EIV in MDCK explored some basic mechanisms of pathogenesis. The infected MDCK cells cause cellular oxidative stress and induce stress-related transcription factors like c-jun/Ap-1, which is associated with apoptosis and cell death. EIV also induces cellular expression of cytokine TGF-β1, which can induce apoptosis upon higher expression of the cytokine. The growth inhibition by TGF-β is mediated by activation of JNK/SAPK pathway. TGF-β activates JNK and stimulates JNK phosphorylation of c-JUN, which induces apoptosis. Inhibition of TGF-B1 by neutralising antibodies attenuates apoptosis induced by EIV through attenuated c-JUN/AP1 upregulation. The role of c-JUN and TGF-β1 in the induction of apoptosis was evaluated by antioxidant NAC, which leads to overexpression of BCL2 (anti-apoptotic gene). Use of antioxidant drugs like carvedilol showed strong inhibition of EIV-infected MDCK cell death. Influenza virus surface protein HA stimulates transcription factor NF-κB and NA protein can directly activate TGF-β. C-Jun/AP-1 and NF-κB activation are both mediated by the concentration of reactive oxygen species (ROS). These factors target the genes for cytokines such as TNF-α and TGF-β. The higher level of the cytokines has a cytotoxic effect on cells. It has also been reported that influenza virus activates Fas gene, which is associated with apoptosis. Surprisingly, the NS gene product of influenza virus has around 50% similarity with Fas gene antigen over a 68-amino acid cytoplasmic domain region of Fas protein. It was thought that NS protein might have an important role in apoptosis induced by the influenza virus. EIV infection also stimulates multiple caspases in connection with JNK/SAPK activation and apoptosis. EIV might be playing a similar mechanism of activation of various signalling pathways and apoptosis for the pathogenesis of EIV infection in equines to that observed in EIV-infected MDCK cells (Chengbin et al. 2001).

9.8 Risk Factors

Few studies revealed that male is more susceptible to infection than female. It may be due to the immunosuppressive action of testosterone in males than oestrogens in females (Nuria et al. 2007). So females develop rapid protective humoral and cell-mediated immune response followed by infection than males. All age groups and breeds of equines are susceptible to infection. However, the disease is common among young horses of less than 2 years old. Though mortality is uncommon, death due to secondary bacterial infection following influenza virus infection is observed in old and immune-compromised horses.

Further, death of 50 EI-infected animals was reported during the outbreak in India in which animals were 14 km uphill in inclement weather condition, and the cause for the death was EI-associated secondary bacterial infections (Virmani et al. 2010a, b). Outbreaks of the disease can occur throughout the year; however, most of the outbreaks were associated with yearling sale, animal fair, racing/breeding movement, overcrowding and unrestricted movements (Wilson 1993; Virmani et al. 2010a, b). Upon infection, infected animals develop protective antibody titre against homologous strain, and the immunity persists for 1 year, which may confer partial protection against hetero-virus strain. Horses with partial protective titre may not suffer from a severe form of infection but may silently shed the viruses, pose the risk of infection to naive horses in the stud and also induce antigenic drift (Paillot, 2014).

9.9 Clinical Signs

Clinical signs and severity of the diseases in equines are dependent on the strain of the virus and the immune status of the host (Mumford et al. 1990). A severe form of clinical disease has been observed in naive unvaccinated horses (Daly et al. 2011). Clinical signs appear as early as 24 h following infection, characterised by a biphasic febrile illness, dyspnoea, persistent dry and harsh cough and serous nasal discharge (Muranaka et al. 2012). Subsequently, the initial serous nasal discharge becomes mucopurulent between 3 and 4 days. The second peak of pyrexia is observed around a week after a primary illness. Other variable clinical signs include anorexia, depression, myalgia and enlargement of respiratory-associated lymph nodes (leads to tachypnoea) and limb oedema. Few unusual cases exhibit signs of encephalitis and myocarditis. However, involvement of EIV as a cause needs to be elucidated (Daly et al. 2006). Haematological and biochemical changes are non-specific and include anaemia, leucopenia, lymphopenia and mild hypoalbuminaemia. Clinical signs may persist for 7–14 days in uncomplicated infections, but the cough may persist up to 21 days following infection (Newton et al. 2006). Secondary bacterial pneumonia commonly observed with EI is characterised by pulmonary consolidations. Morbidity may reach up to 100% in most of the outbreaks under favourable conditions; however, mortality is rare and commonly observed in foals, old horses and animals living in inclement climatic conditions due to bacterial bronchial and interstitial pneumonia. Naive infected animals may shed influenza viruses in nasal discharge for at least 4–6 days. No carrier state is observed for EI infection in equine. Vaccinated animals or animals exposed to EI during last season may suffer from less severe infection according to immune status. Partially protected animals still shed the viruses in nasal discharge without showing visible clinical signs and play a major role in virus persistence in the environment.

9.10 Pathological Findings

As death due to EI infection is rare, only limited information is available regarding pathologic findings of natural EI infection in equines. Further, few experimental infection studies were preformed in equine to uncover complete viral pathogenesis. Natural/experimental infection of EIV shows gross lesions restricted to respiratory organs and associated lymph nodes. Common findings are hyperaemia of the nasal mucosa and serous nasal secretions and acute phase with mild swelling of retropharyngeal and pulmonary lymph nodes. Petechial haemorrhage and mucopurulent exudate in trachea and bronchi with interstitial oedema can be seen in a portion of the affected lung after a week of infection. Hepatisation/consolidation of the lung due to secondary bacterial pneumonia appears 2 weeks after exposure.

Histopathological lesions are limited to the upper respiratory tract in early infection. Rhinitis and tracheitis are characterised by epithelial degeneration and necrosis with loss of ciliated epithelium, reduction in goblet cell numbers and moderate lymphocytic infiltration in lamina propria. More severe tracheal lesions can be seen in the terminal portion. Lesions are extended to bronchioles and alveoli at 7 days after infection. Marked epithelial hyperplasia, squamous metaplasia; proliferation of type 2 pneumocytes and suppurative bronchopneumonia with pulmonary oedema; and proliferation of neutrophils, macrophages and necrotic epithelial cells were found to be filling bronchiole and lung parenchyma (Begg et al. 2011). Severe tracheitis and bronchitis with epithelial hyperplasia, squamous metaplasia with short villi and small goblet cells are observed at 14 days after infection. Infiltration of neutrophils can be observed in lamina propria of bronchial and bronchiolar epithelium. Subsequently, bronchopneumonia develops in response to secondary bacterial infection mainly caused by *Streptococcus zooepidemicus*, *Streptococcus equi*,

Actinobacillus equuli, *Bordetella bronchiseptica*, etc., which causes damage to ciliated epithelium and goblet cells. Extensive infiltration of neutrophils and alveolar macrophages in alveoli, fibrin exudate, pulmonary oedema and marked proliferation of type 2 pneumocytes are also observed. Reactive lymphocytic hyperplasia is found in all respiratory-associated lymph node in all stages of infection (Muranaka et al. 2012).

A BALB/c mice model was established for detailed investigation of EIV pathogenicity and quick efficacy testing of inactivated vaccine (Pavulraj et al. 2015, 2017) at NRCE, India. For pathogenicity study, the BALB/c mice were inoculated intranasally with EIV (H3N8) belonging to clade 2 of Florida sublineage and monitored for the establishment of infection and associated parameters. EIV-infected mice showed clinical signs, viz. loss in body weights, lethargy, dyspnoea, etc., between 3 and 5 days which was further commensurate with lesions observed in the respiratory tract including rhinitis, tracheitis, bronchitis, bronchiolitis, alveolitis and diffuse interstitial pneumonia. The active viral infection in the upper and lower respiratory tract was demonstrated through transmission electron microscopy, immunohistochemistry, virus quantification by titration and qRT-PCR. Serological investigation showed a higher level of serum lactate dehydrogenase and seroconversion. The overall disease progression pattern, pathological lesions and virus recovery from nasal washings and lungs of infected mice were comparable to natural and experimental EIV infection in equines. Subsequently, established EIV mice model was utilized for efficacy testing of inactivated EIV vaccine through sequential evaluation of serology, clinical signs, gross and histopathology lesions with grading, immunohistochemistry and virus quantification. Immunisation of mice generated protective HAI and SRH antibody titre after two boosters and generated balanced Th1/Th2 responses. Upon EIV challenge, immunised mice were well protected as evident by a significant rise in serum antibody titre, with mild clinical signs, early recovery, lower gross and histopathological lesion score, less severe intensity of viral antigen distribution, restricted virus replication in the respiratory tract and less virus detection in nasal washes for short duration. Vaccinated mice showed only the congestion of lung parenchyma without consolidation. The findings have established that BALB/c mice could be used as small animal model for studying EIV (H3N8) infection and will have immense potential for dissecting viral pathogenesis, vaccine efficacy studies, preliminary screening of vaccine candidates and antiviral therapeutics against EIV.

9.11 Cell-Mediated Immunity

Stimulation of cell-mediated immune response following EIV infection and/or vaccination has not been investigated in detail. Complete protection against EI infection in ponies which did not have EIV-specific SRH antibody in serum at the time of infection correlates with supportive role of another arm of the immune system—

cell-mediated immunity (CMI). The previous study revealed that whole EIV vaccine adjuvanted with aluminium hydroxide or aluminium phosphate was unable to stimulate cytotoxic T lymphocytes (CTL) (Hannant et al. 1994). Adjuvant of choice in vaccine plays an essential role in the stimulation of immune response towards CMI and/or HMI. Aluminium hydroxide is well known to stimulate Th2 response, which results in inducing HMI rather than CMI (Lindblad 2004). ISCOM matrix-based adjuvanted EIV vaccine induces strong CMI response characterised by T-cell proliferation and an increase in the percent of EIV-specific CTL (Paillot and Prowse 2012). Stimulation of CMI responses following vaccination with Duvaxyn IE-T Plus has also been reported recently (Paillot et al. 2013), in which percent of cells positive for IFN-gamma synthesis was considered as CMI marker. Percentage of EIV-specific IFN-gamma-producing cells was measured prior to and 3 weeks following the first vaccination, and 2 weeks after each first and second booster vaccine. Increase in the percent of specific IFN-gamma-producing cells was observed in immunised ponies when compared with pre- and post-immunisation level and in control ponies. Also, stimulation of CMI response characterised by strong IFN-gamma response and HMI has been reported in mice following immunisation with Carbopol 974P-based vaccine. However, the protective role of CTL response during EIV infection is yet to be understood and is an essential component for the development of better vaccines.

9.12 Zoonotic Potential

EIV binds to α2,3-sialic acid receptor on the cell surface for entry into the cells before replication. In contrast, airway epithelial cells in a human contain α2,6-sialic acid receptor. Absence of specific receptors on the cell surface does not exclude the possibilities of cross-species transmission as observed during highly pathogenic avian influenza (HPAI) H5N1 outbreak in human beings. However, H5 of these HPAI has special multi-basic amino acids at the cleavage site of HA, which is not the case with HA protein of H3 viruses. Few EIV experimental challenge studies in human volunteers demonstrated the possibilities of human infection with the equine H3N8 virus. Though seroconversion has been observed in humans following virus challenge, no clinical signs were observed (Kasel et al. 1965, 1969; Alford et al. 1967). Recently, the serological response against EIV infection has been reported in persons in contact with equines in Iowa. However, no virus could be isolated (Larson et al. 2015). It is worthy of mentioning here that though observational studies support the evidence of human infection with EIV, the results need to be interpreted with caution. Serological response against H3N8 EIV should not be confused because of pre-existing antibodies against another subtype of human H3 viruses or vaccination in humans.

9.13 Treatment

There is no specific treatment available for EI, and the treatment is given as supportive care based on clinical signs which include adequate rest. Broad-spectrum antibiotics (to avoid secondary bacterial complications), anti-histamines and non-steroidal anti-inflammatory drugs (to a reduced inflammatory response in respiratory tract) are choices of treatment approach. Infected animal nutrition and hydration status may help in quick recovery. Adequate rest (for at least 3–4-week or 1-week rest per each day of sickness) is necessary for a speedy recovery. Preliminary studies performed regarding antiviral treatment of EI in equines with amantadine and rimantadine (targeting the transmembrane domain of the M2 ion channel protein) upon oral administration showed promising results; further, zanamivir and oseltamivir (NA inhibitors) showed antiviral activity against EIV in vitro. Only a few studies were carried out regarding antiviral therapies in equines. However, the results were promising. Difficulty in performing clinical trial or challenge experiments in equines due to cost, labour, need of special premises, etc. make progress slow.

9.14 Control of Outbreaks

Duration of virus shedding from an infected animal depends upon the strain of the virus and immune status of animal. Virus shedding may occur from 3 to 8 days (Ragni-Alunni and Zande 2009; Daly et al. 2007; Paillot et al. 2008) and rarely up to 12 days (Paillot et al. 2013). It is important to vaccinate all the animals irrespective of their previous vaccination schedule after the early diagnosis of influenza outbreak (blanket vaccination) to mitigate (Nuria et al. 2007). Basic outbreak containment strategies followed in Australia 2007 EI outbreak were movement restrictions, public awareness and communication, zoning of the areas based on risk, laboratory testing, disease tracing and surveillance, enhanced biosecurity measures to prevent spreading and emergency vaccination. Out of the above said, movement restriction played an immense role in disease containment and eradication of EI from Australia in 5 months. Role of vaccination in outbreak containment needs further extensive research (Cowled et al. 2009). EIV can be easily inactivated by sunlight, UV rays, heat, cold and common disinfectants (Yadav et al. 1993). So, routinely used disinfectants can be used in studs during EI outbreaks to prevent subsequent virus spread through fomites.

9.15 Prevention and Vaccination

Countries like the UK follow mandatory vaccination for race horses since 1981 after the outbreak in 1979 (Elton and Bryant 2011). Polyvalent inactivated whole virus vaccine against EI was developed early in 1966 by Bryans and co-workers (Bryans et al. 1966) based on the experience of human influenza vaccines and further by many other workers (Petermann et al. 1970; Burki and Sibalin 1973; Frerichs et al. 1973). Currently available EI vaccine contains inactivated whole virus with adjuvants like oil, alhydrogel, carbomer or subunit vaccines like ISCOMs or micelles combined with Quil A (Table 9.1). Vaccines with adjuvants like aluminium phosphate or hydroxide induce durable antibody response up to 16–20 weeks. The schedule consists of two doses of primary vaccine 4–6 weeks apart followed by repeating the vaccine annually or after checking the SRH/HI titres in serum. Homologous virus challenge in vaccinated animals always induces a high level of humoral immune response and confers the highest level of protective immunity, unlike with heterologous virus challenge (Mumford and Wood 1992). SRH antibody levels between >120 mm^2 and >154 mm^2 confer complete protection against experimental nebulised aerosol virus challenge in ponies (Mumford and Wood 1992), which was later confirmed through field studies during a natural disease outbreak. The animals with pre-infection SRH antibody levels of >140 mm^2 did not develop any clinical signs during infection or were seroconverted as detected by nucleoprotein-based antigen ELISA (Townsend et al. 1999; Newton et al. 2000). It was well documented that the SRH antibody levels of >80 mm^2 can confer protection against clinical infection (Mumford et al. 1988, 1990). Protection against EIV infection is in absolute correlation with antibody response/titre (Park et al. 2003; Paillot et al. 2013). Experimental EIV challenge studies on ponies revealed the highest levels of SRH antibody titre associated with fewer virus shedding (ELISA) and fewer copies of viral RNA shedding (qRT-PCR) through nostrils. On the other side, ponies, which had lower levels of SRH antibody titre, showed higher virus shedding and more copies of viral RNA shedding. Protection against antigenically different virus challenge in vaccinated/immunised animals may require much higher levels of SRH antibody titre to confer similar protection level against antigenically similar viruses (Nuria et al. 2007; Paillot et al. 2013).

Protective immunity following natural infection persists for a longer duration even after SRH antibodies vanish in the host (Hannant et al. 1988a, b). Natural virus infection in host induces long-lasting immunity, which is characterised by the production of IgA antibody at the local mucosa level along with IgGa and IgGb (Nelson et al. 1998). On the other hand, the immune response elicited following inactivated whole virus vaccine immunisation is characterised by IgGa, IgGb and IgG(T) antibodies with almost no IgA response at mucosa (Wilson et al. 2001). Taken together, productive virus infection is mandatory for the production of specific mucosal IgA antibody (Lunn et al. 1999). Besides, CMI characterised by CTL responses is generally observed only after natural virus infection except for recently developed vaccines as described earlier due to the fact of differences in antigen processing and presentation (Hannant et al. 1989).

Table 9.1 Currently available EI vaccines

Type	Vaccine	Company	Vaccine strain	Nature of vaccine	Adjuvant
Whole virus	Duvaxyn™ IE T Plus	Elanco Animal Health	A/eq/Suffolk/89 (H3N8) A/eq/Newmarket1/93 (H3N8) A/eq/Prague/56 (H7N7)	Inactivated	Carbopol 934P and Aluminium hydroxide
	Calvenza®-03 EIV	Boehringer Ingelheim Animal Health	A/eq/Newmarket/2/93 (H3N8), A/eq/ Kentucky/2/95 (H3N8), A/eq/Ohio/03 (H3N8)	Inactivated	Carbimmune
	Equilis Prequenza (updated 2013)	MSD Animal Health	A/eq/Newmarket/2/93 (H3N8), A/eq/South Africa/4/03 (H3N8)	Inactivated	ISCOM-Matrix
	Equip F	Zoetis	A/eq/Borlänge/91(H3N8), A/eq/Kentucky/98 (H3N8), A/eq/ Newmarket/77 (H7N7)	Inactivated	Quil A, Aluminium Phosphate
	Fluvac Innovator	Zoetis	A/eq/Kentucky/97 (H3N8)	Inactivated	MetaStim®
	Equip™ F	Pfizer Ltd.	A/eq/Newmarket/77 (H7N7), A/eq/ Borlänge/91 (H3N8), A/eq/Kentucky/98 (H3N8)	Inactivated	Quil A
Subunit	Equilis Prequenza Te	MSD Animal Health	A/eq/South Africa/4/03(H3N8), A/eq/ Newmarket/2/93(H3N8) and Tetanus toxoid	Subunit HA	Purified saponin
Viral-vector based	PROTEQ FLU™	Merial Animal Health Ltd.	A/eq/Ohio/03 (H3N8), A/eq/ Richmond/1/07 (H3N8)	Inactivated recombinant canary pox-HA subunit	Carbomer
Modified live EIV	Flu Avert® I.N.	Intervet/ Schering-Plough Animal Health (USA)	A/eq/Kentucky/91 (H3N8)	Attenuated, cold-adapted whole EIV	NA

Inadequate protection or suboptimal immune response is the well-known phenomenon in which majority of the immunised horses are unable to produce an optimal immune response, and the population remains vulnerable to the infection. Though the exact reasons behind suboptimal immune response in horses remain unknown, it was suggested to measure SRH antibody levels in horses following immunisation to achieve optimum protection at the population level (Gildea et al.

2011). Further, it may pave the way to identify poor responders and to optimise the vaccination schedules. Early vaccination in the presence of maternally derived antibody (MDA) adversely influences the risk of susceptibility to EI in the later part of life. Vaccination should be done beyond 6 months of age after the decline of MDA (Nuria et al. 2007). In spite of vaccination, several vaccine breakdowns have been noticed in the past like 2004 in Croatia (Barbic et al. 2009) and 2005 in Italy (Vito Martella et al. 2007), due to mismatch of vaccine and outbreak strain. The potency of a vaccine, adjuvants, schedule of vaccination and surveillance for antigenic drift are the current focus of study.

Acknowledgements All the authors of the manuscript thank and acknowledge their respective universities and institutes.

Conflict of interest: There is no conflict of interest.

References

Acosta R, Rimondi A, Mino S, Tordoya MS, Fernandez A, Betancor G, Larrauri H, Larrauri H, Nuñez A, Barrandeguy M (2012) Outbreak of equine influenza among Thoroughbred horses in Maronas, Montevideo (Uruguay) during March and April 2012. J Equine Vet 32:7

Alford RH, Kasel JA, Lehrich JR, Knight V (1967) Human responses to experimental infection with influenza a/equi 2 virus. Am J Epidemiol 86:185–192

Animal Health Trust, UK (2019) Equine influenza outbreaks reported in 2019. https://www.aht.org.uk/wp-content/uploads/2019/04/Equiflunet-outbreaks-to-15-April2019.pdf

Barbic L, Madic J, Turk N, Daly J (2009) Vaccine failure caused an outbreak of equine influenza in Croatia. Vet Microbiol 133:164–171

Begg AP, Reece RL, Hum S, Townsend W, Gordon A, Carrick J (2011) Pathological changes in horses dying with equine influenza in Australia. Aust Vet J 89:19–22

Bera BC, Virmani N, Kumar N, Anand T, Pavulraj S, Rash A, Elton D, Rash N, Bhatia S, Sood R, Singh RK, Tripathi BN (2017) Genetic and codon usage bias analyses of polymerase genes of equine influenza virus and its relation to evolution. BMC Genomics 18(1):652. https://doi.org/10.1186/s12864-017-4063-1

Beveridge WI (1965) Some topical comments on influenza in horses. Vet Rec 77:42

Bosch FX, Garten W, Klenk HD, Rott R (1981) Proteolytic cleavage of influenza virus hemagglutinins: primary structure of the connecting peptide between HA1 and HA2 determines proteolytic cleavability and pathogenicity of avian influenza viruses. Virology 113(2):725–735

Bryans JT, Doll ER, Wilson JC, McCollum WH (1966) Immunization for equine influenza. J Am Vet Med Assoc 148:413–417

Bryant NA, Rash AS, Russell CA, Ross J, Cooke A, Bowman S, MacRae S, Lewis NS, Paillot R, Zanoni R, Meier H, Griffiths LA, Daly JM, Tiwari A, Chambers TM, Newton JR, Elton DM (2009) Antigenic and genetic variations in European and North American equine influenza virus strains (H3N8) isolated from 2006 to 2007. Vet Microbiol 138:41–52

Burki F. Sibalin M (1973) Conclusions and questions arising from a study of serology and immunology of equine influenza. In: Proceedings of the 3rd International Conference on Equine Infectious Diseases, Eds: J.T. Bryans and H. Gerber, Karger Basel, New York, pp 510–526

Chen W, Calvo PA, Malide D, Gibbs J, Schubert U, Bacik I, Basta S, O'Neill R, Schickli J, Palese P, Henklein P, Bennink JR, Yewdell JW (2001) A novel influenza A virus mitochondrial protein that induces cell death. Nat Med 7:1306–1312

Chengbin L, Zimmer SG, Lu Z, Holland RE Jr, Dong Q, Chambers TM (2001) The involvement of a stress-activated pathway in equine influenza virus-mediated apoptosis. Virology 287:202–213

Cowled B, Ward MP, Hamilton S, Garner G (2009) The equine influenza epidemic in Australia: Spatial and temporal descriptive analyses of a large propagating epidemic. Prev Vet Med 92:60–70

Cox NJ, Subbarao K (2000) Global epidemiology of influenza: past and present. Annu Rev Med 51(1):407–421

Crawford PC, Dubovi EJ, Castleman WL, Stephenson I, Gibbs EPJ, Chen L, Smith C, Hill RC, Ferro P, Pompey J, Bright RA (2005) Transmission of equine influenza virus to dogs. Science 310(5747):482–485

Cullinane A, Newton JR (2013) Equine influenza—a global perspective. Vet Microbiol 167:205–214

Dalglish RA (1992) The international movement of horses—the current infectious disease situation. In: Short CR (ed) Proceedings of the 9th International Conference of Racing Analysts and Veterinarians. Louisiana State University, New Orleans, LA, pp 37–53

Daly JM, Lai AC, Binns MM, Chambers TM, Barrandeguy M, Mumford JA (1996) Antigenic and genetic evolution of equine H3N8 influenza A viruses. J Gen Virol 77:661–761

Daly JM, Yates PJ, Browse G, Swann Z, Newton JR, Jessett D, Davis-Poynter N, Mumford JA (2003) Comparison of hamster and pony challenge models for evaluation of effect of antigenic drift on cross-protection afforded by equine influenza vaccines. Equine Vet J 35:458–462

Daly JM, Yates PJ, Newton JR, Park A, Henley W, Wood JL, Davis-Poynter N, Mumford JA (2004) Evidence supporting the inclusion of strains from each of the two co-circulating lineages of H3N8 equine influenza virus in vaccines. Vaccine 22:4101–4109

Daly JM, Whitwell KE, Miller J, Dowd G, Cardwell JM, Smith KC (2006) Investigation of equine influenza cases exhibiting neurological disease: coincidence or association? J Comp Pathol 134(2):231–235

Daly JM, Sindle T, Tearle J, Barquero N, Newton JR, Corning S (2007) Equine influenza vaccine containing older H3N8 strains offers protection against A/eq/South Africa/4/03 (H3N8) strain in a short-term vaccine efficacy study. Equine Vet J 39:446–450

Daly JM, Blunden AS, Macrae S, Miller J, Bowman SJ, Kolodziejek J, Nowotny N, Smith KC (2008) Transmission of equine influenza virus to English foxhounds. Emerg Infect Dis 14:461–464

Daly JM, MacRae S, Newton JR, Wattrang E, Elton DM (2011) Equine influenza: a review of an unpredictable virus. Vet J 189(1):7–14

Doll ER (1961) Influenza of horses. Am Rev Respir Dis 83:48

Eierhoff T, Hrincius ER, Rescher U, Ludwig S, Ehrhardt C (2010) The epidermal growth factor receptor (EGFR) promotes uptake of influenza A viruses (IAV) into host cells. PLoS Pathog 6:e1001099

Elton D, Bryant N (2011) Facing the threat of equine influenza. Equine Vet J 43(3):250–258

Endo A, Pecoraro R, Sugita S, Nerome K (1992) Evolutionary pattern of the H3 haemagglutinin of equine influenza viruses: multiple evolutionary lineages and frozen replication. Arch Virol 123:73–87

Firestone SM, Schemann KA, Toribio Jenny-Ann LML, Ward MP, Dhand NK (2011) A case-control study of risk factors for equine influenza spread onto horse premises during the 2007 epidemic in Australia. Prev Vet Med 100:53–63

Frerichs GN, Burrows R, Frerichs CC (1973) Serological response of horses and laboratory animals to equine influenza vaccines. In: Bryans JT, Gerber H (eds) Proceedings of the 3rd International Conference on Equine Infectious Diseases. Karger Basel, New York, pp 503–509

Gahan J, Garvey M, Asmah Abd Samad R, Cullinane A (2019) Whole genome sequencing of the first H3N8 equine influenza virus identified in Malaysia. Pathogens 8(2):62. https://doi.org/10.3390/pathogens8020062

Gerber H (1970) Clinical features, sequelae and epidemiology of equine influenza. In: Proceedings of the Second International Conference on Equine Infectious Diseases, Paris, 1969. S. Karger, Basel, pp 63–80

Gildea S, Arkins S, Cullinane A (2011) Management and environmental factors involved in equine influenza outbreaks in Ireland 2007–2010. Equine Vet J 43:608–617

Guo Y, Wang M, Zheng S, Wang P, Ji W, Chen Q (1991) Aetiologic study on an influenza-like epidemic in horses in China. Acta Virol 35:190–195

Guo Y, Wang M, Kawaoka Y, Gorman O, Ito T, Saito T, Webster RG (1992) Characterization of a new avian-like influenza A virus from horses in China. Virology 188:245–255

Gupta AK et al (1993) Single radial immunodiffusion potency test for standardization of indigenous influenza vaccine. Indian J Exp Biol 31:944–947

Guthrie AJ (2006) Equine influenza in South Africa, 2003 outbreak. In: Proceedings of the 9th International Congress of World Equine Veterinary Association. International Veterinary Information Service, Ithaca, NY

Guthrie AJ, Stevens KB, Bosman PP (1999) The circumstances surrounding the outbreak and spread of equine influenza in South Africa. Rev Sci Tech 18:179–185

Hannant D, Jessett DM, O'Neill T, Sundquist B, Mumford JA, Powell DG (1988a) Nasopharyngeal, tracheobronchial, and systemic immune responses to vaccination and aerosol infection with equine-2 influenza A virus (H3N8). In: Equine infectious diseases. V: Proceedings of the Fifth International Conference, vol 5. University Press of Kentucky, Lexington, KY, pp 66–73

Hannant D, Mumford JA, Jessett DM (1988b) Duration of circulating antibody and immunity following infection with equine influenza virus. Vet Rec 122(6):125–128

Hannant D, Mumford JA (1989) Cell mediated immune responses in ponies following infection with equine influenza virus (H3N8): the influence of induction culture conditions on the properties of cytotoxic effector cells. Vet Immunol Immunopathol 21:327–337

Hannant D, Jessett DM, O'Neill T, Livesay GJ, Mumford JA (1994) Cellular immune responses stimulated by inactivated virus vaccines and infection with equine influenza virus (H3N8). In: Nakajima H, Plowright W (eds) Equine infectious diseases. VII. Proceedings of Seventh International Conference on Equine Infectious Diseases. R & W Publications Ltd, Newmarket, pp 169–174

Huntington PJ (1990) Equine influenza—the disease and its control. Technical Report Series No. 184. Department of Agriculture and Rural Affairs, Victoria

Jagger BW, Wise HM, Kash JC, Walters KA, Wills NM, Xiao YL, Dunfee RL, Schwartzman LM, Ozinsky A, Bell GL, Dalton RM, Lo A, Efstathiou S, Atkins JF, Firth AE, Taubenberger JK, Digard P (2012) An overlapping protein-coding region in influenza a virus segment 3 modulates the host response. Science 337:199–204

Kasel JA, Alford RH, Knight V, Waddell GH, Sigel MM (1965) Experimental infection of human volunteers with equine influenza virus. Nature 206:41–43

Kasel JA, Couch RB (1969) Experimental infection in man and horses with Influenza A viruses. Bull World Health Organ 41:447–452

Khan A, Mushtaq MH, Ahmad MUD, Nazir J, Farooqi SH, Khan A (2017) Molecular Epidemiology of a novel re-assorted epidemic strain of equine influenza virus in Pakistan in 2015–16. Virus Res 240:56–63

Kirkland PD, Finlaison DS, Crispe E, Hurt AC (2010) Influenza virus transmission from horses to dogs, Australia. Emerg Infect Dis 16:699–702

Kovbasnjuk ON, Spring KR (2000) The apical membrane glycocalyx of MDCK cells. J Membr Biol 176:19–29

Kumar N, Bera BC, Greenbaum BD, Bhatia S, Sood R, Selvaraj P, Anand T, Tripathi BN, Virmani N (2016) Revelation of influencing factors in overall codon usage bias of equine influenza viruses. PLoS One 11(4):e0154376. https://doi.org/10.1371/journal.pone.0154376

Lai AC, Rogers KM, Glaser A, Tudor L, Chambers T (2004) Alternate circulation of recent equine-2 influenza viruses (H3N8) from two distinct lineages in the United States. Virus Res 100:159–164

Lamb RA, Krug RM (2001) Orthomyxoviridae: the viruses and their replication. In: Knipe DM, Howle PM (eds) Fields virology. Lippincott Williams & Wilkins, Philadelphia, pp 1487–1532

Larson KR, Heil GL, Chambers TM, Capuano A, White SK, Gray GC (2015) Serological evidence of equine influenza infections among persons with horse exposure, Iowa. J Clin Virol 67:78–83
Lee CW, Saif YM (2009) Avian influenza virus. Comp Immunol Microbiol Infect Dis 32:301–310
Lindblad EB (2004) Aluminium adjuvants—In retrospect and prospect. Vaccine 22:3658–3668
Lopez A (2007) Respiratory system. In: McGavin D, Zachary JF (eds) Pathologic basis of veterinary disease, 4th edn. Mosby Elsevier, St Louis, MO, p 470
Lunn DP, Soboll G, Schram BR, Quass J, McGregor MW, Drape R, Macklin MD, McCabe DE, Swain WF, Olsen CW (1999) Antibody responses to DNA vaccination of horses using the influenza virus hemagglutinin gene. Vaccine 17:2245–2258
MacLachlan N, Dubovi EJ (eds) (2011) Fenner's veterinary virology, 4th edn. Academic, San Diego
Manjrekar SL, Gorhe DS, Paranjape VL (1965) Observation on the coughing outbreaks 'Newmarket Cough' in the race horses in Bombay. Indian Vet J 48:460–464
Martella V, Elia G, Decaro N, di Trani L, Lorusso E, Campolo M, Desario C, Parisi A, Cavaliere N, Buonavoglia C (2007) An outbreak of equine influenza virus in vaccinated horses in Italy is due to an H3N8 strain closely related to recent North American representatives of the Florida sub-lineage. Vet Microbiol 121:56–63
McClure JP (1998) The epizootic of 1872: horses and disease in a nation in motion. New York History 79:4–22
Miller WC (1965) Equine influenza—further observations on "coughing" outbreak 1965. Vet Rec 77:455
Mumford JA (1999) Control of influenza from an international perspective. In: Proceedings of the Eighth International Conference on Equine Infectious Diseases. R & W Publications, Newmarket, pp 11–24
Mumford JA, Wood J (1992) Establishing an acceptability threshold for equine influenza vaccines. Dev Biol Stand 79:137–146
Mumford JA, Wood JM, Folkers C, Schild GC (1988) Protection against experimental infection with influenza virus A/equine/Miami/63 (H3N8) provided by inactivated whole virus vaccines containing homologous virus. Epidemiol Infect 100(3):501–510
Mumford J, Hannant D, Jessett DM (1990) Experimental infection of ponies with equine influenza (H3N8) viruses by intranasal inoculation or exposure to aerosols. Equine Vet J 22:93
Mumford EL, Traub-Dargatz JL, Salman MD, Collins JK, Getzy DM, Carman J (1998) Monitoring and detection of acute viral respiratory tract disease in horses. J Am Vet Med Assoc 213:385–390
Mumford J, Cardwell J, Daly J, Newton R (2003) Efforts to preempt an equine influenza epidemic. Vet Rec 152:405–406
Muranaka M, Yamanaka T, Katayama Y, Hidari K, Kanazawa H, Suzuki T, Oku K, Oyamada T (2010) Distribution of influenza virus sialoreceptors on upper and lower respiratory tract in horses and dogs. J Vet Med Sci 73:125–127
Muranaka M, Yamanaka T, Katayama Y, Niwa H, Oku K, Matsumura T, Oyamada T (2012) Time-related pathological changes in horses experimentally inoculated with equine influenza A virus. J Equine Sci 23(2):17–26
Myers C, Wilson WD (2006) Equine influenza virus. Vet Clin North Am Equine Pract 5:187–196
Nakajima K, Nobusawa E, Nagy A, Nakajima S (2005) Accumulation of amino acid substitutions promotes irreversible structural changes in the hemagglutinin of human influenza A H3 virus during evolution. J Virol 79(10):6472–6477
Nelson KM, Schram BR, McGregor MW, Sheoran AS, Olsen CW, Lunn DP (1998) Local and systemic isotype-specific antibody responses to equine influenza virus infection versus conventional vaccination. Vaccine 16:1306–1313
Newton JR, Lakhani KH, Wood JLN, Baker DJ (2000) Risk factors for equine influenza serum antibody titres in young Thoroughbred racehorses given an inactivated vaccine. Prev Vet Med 46:129–141

Newton JR, Daly JM, Spencer L, Mumford JA (2006) Description of the outbreak of equine influenza (H3N8) in the United Kingdom in 2003, during which recently vaccinated horses in Newmarket developed respiratory disease. Vet Rec 158:185–192

Nuria B, Janet M, Daly J, Newton R (2007) Risk factors for influenza infection in vaccinated racehorses: lessons from an outbreak in Newmarket, UK in 2003. Vaccine 25:7520–7529

Nyaga PN et al (1980) Epidemiology of equine influenza, risk by age, breed and sex. Comp Immunol Microbiol Infect Dis 3:67

OIE, Equine influenza (2019). https://www.oie.int/fileadmin/Home/eng/Health_standards/tahm/3.05.07_EQ_INF.pdf

Paillot R (2014) A systematic review of recent advances in equine influenza. Vaccines (Basel) 2014(2):797–831. https://doi.org/10.3390/vaccines2040797

Paillot R, Prowse L (2012) ISCOM-matrix-based equine influenza (EIV) vaccine stimulates cell-mediated immunity in the horse. Vet Immunol Immunopathol 145(1–2):516–521

Paillot R, Hannant D, Kydd JH, Daly JM (2006) Vaccination against equine influenza: Quid novi? Vaccine 24:4047–4061

Paillot R, Grimmett H, Elton D, Daly JM (2008) Protection, systemic IFN gamma γ, and antibody responses induced by an ISCOM-based vaccine against a recent equine influenza virus in its natural host. Vet Res 39(3):1

Paillot R, Prowse L, Montesso F, Huang CM, Barnes H, Escala J (2013) Whole inactivated equine influenza vaccine: efficacy against a representative clade 2 equine influenza virus, IFN gamma synthesis and duration of humoral immunity. Vet Microbiol 162:396–407

Park AW, Wood JL, Newton JR, Daly J, Mumford JA, Grenfell BT (2003) Optimising vaccination strategies in equine influenza. Vaccine 21:2862–2870

Patterson-Kane JC, Carrick JB, Axon JE, Wilkie I, Begg AP (2008) The pathology of bronchointerstitial pneumonia in young foals associated with the first outbreak of equine influenza in Australia. EquineVet J 40:199–203

Pavulraj S, Bera BC, Joshi A, Anand T, Virmani M, Vaid RK, Shanmugasundaram K, Gulati BR, Rajukumar K, Singh R, Misri J (2015) Pathology of equine influenza virus (H3N8) in murine model. PLoS One 10(11):143094

Pavulraj S, Virmani N, Bera BC, Joshi A, Anand T, Virmani M, Singh R, Singh RK, Tripathi BN (2017) Immunogenicity and protective efficacy of inactivated equine influenza (H3N8) virus vaccine in murine model. Vet Microbiol 210:188–196

Payungporn S, Crawford PC, Kouo TS, Chen LM, Pompey J, Castleman WL, Dubovi EJ, Katz JM, Donis RO (2008) Influenza A virus (H3N8) in dogs with respiratory disease, Florida. Emerg Infect Dis 14:902–908

Perglione CO, Gildea S, Rimondi A, Miño S, Vissani A, Carossino M et al (2016) Epidemiological and virological findings during multiple outbreaks of equine influenza in South America in 2012. Influenza Other Respir Viruses 10:37–46

Petermann HG, Fayet MT, Fontaine M. Fontaine MP (1970) Vaccination against equine influenza. In: Bryans JT (ed) Proceedings of the 2nd International Conference On Equine Infectious Diseases. Karger Basel, New York, pp 105–110

Powell DG, Watkins KL, Li PH, Shortridge KF (1995) Outbreak of equine influenza among horses in Hong Kong during 1992. Vet Rec 136:531–536

Qi T, Guo W, Huang WQ, Li HM, Zhao LP, Dai LL, He N, Hao XF, Xiang WH (2010) Genetic evolution of equine influenza viruses isolated in China. Arch Virol 155:1425–1432

Ragni-Alunni R, van de Zande S (2009) Efficacy of Equilis Prequenza te against Recent Equine Influenza Isolates. Equitana; Essen, Germany: 2009. Animal studies with individual challenge method

Rash A, Morton R, Woodward A, Maes O, McCauley J, Bryant N, Elton D (2017) Evolution and divergence of H3N8 equine influenza viruses circulating in the United Kingdom from 2013 to 2015. Pathogens 6:6

Rossman JS, Lamb RA (2011) Influenza virus assembly and budding. Virology 411:229–236

Rossman JS, Jing X, Leser GP, Lamb RA (2010) Influenza virus M2 protein mediates ESCRT-independent membrane scission. Cell 142(6):902–913

Scocco P, Pedini V (2008) Localization of influenza virus sialoreceptors in equine respiratory tract. Histol Histopathol 23:973–978

Sieczkarski SB, Whittaker GR (2002) Influenza virus can enter and infect cells in the absence of clathrin-mediated endocytosis. J Virol 76:10455–10464

Singh RK, Dhama K, Karthik K, Khandia R, Munjal A, Khurana SK, Chakraborty S, Malik YS, Virmani N, Singh R, Tripathi BN, Munir M, van der Kolk JH (2018) A comprehensive review on equine influenza virus: etiology, epidemiology, pathobiology, advances in developing diagnostics, vaccines, and control strategies. Front Microbiol 9:1941. https://doi.org/10.3389/fmicb.2018.01941

Sovinova O, Tumova B, Pouska F, Nemec J (1958) Isolation of a virus causing respiratory disease in horses. Acta Virol 2:52

Su S, Wang L, Fu X, He S, Hong M, Zhou P, Lai A, Gray G, Li S (2014) Equine influenza A(H3N8) virus infection in cats. Emerg Infect Dis 20:2096–2099

Takizawa T, Matsukawa S, Higuchi Y, Nakamura S, Nakanishi Y, Fukuda R (1993) Induction of programmed cell death (apoptosis) by influenza virus infection in tissue culture cells. J Gen Virol 74:2347–2355

Tong S, Zhu X, Li Y, Shi M, Zhang J et al (2013) New world bats harbor diverse influenza A viruses. PLoS Pathog 9(10):e1003657. https://doi.org/10.1371/journal.ppat.1003657

Tu J, Zhou H, Jiang T, Li C, Zhang A, Guo X, Zou W, Chen H, Jin M (2009) Isolation and molecular characterization of equine H3N8 influenza viruses from pigs in China. Arch Virol 154:887–890

Tumova B, Sovinova-Fiserova O (1959) Properties of influenza viruses. A/Asia/57 and A-equi/Praha/56. Bull World Health Organ 20:445

Townsend HGG, Cook A, Watts TC (1999) Efficacy of a cold-adapted, modified-live virus influenza vaccine: a double-blind challenge trial. Proc Am Assoc Equine Pract 45:41–42

Uppal PK, Yadav MP (1987) Outbreak of equine influenza in India. Vet Rec 121:569–570

Uppal PK, Yadav MP, Oberoi MS (1989) Isolation of A/Equi-2 virus during 1987 equine influenza epidemic in India. Equine Vet J 21(5):364–366

Virmani N, Singh BK, Gulati BR, Kumar S (2008) Equine influenza outbreak in India. Vet Rec 163:607–608

Virmani N, Bera BC, Singh BK, Shanmugasundaram K, Gulati BR, Barua S, Vaid RK, Gupta AK, Singh RK (2010a) Equine influenza outbreak in India (2008–09): virus isolation, sero-epidemiology and phylogenetic analysis of HA gene. Vet Microbiol 143:224–237

Virmani N, Bera BC, Gulati BR, Karuppusamy S, Singh BK, Kumar Vaid R, Kumar S, Kumar R, Malik P, Khurana SK, Singh J, Manuja A, Dedar R, Gupta AK, Yadav SC, Chugh PK, Narwal PS, Thankur VL, Kaul R, Kanani A, Rautmare SS, Singh RK (2010b) Descriptive epidemiology of equine influenza in India (2008–2009): temporal and spatial trends. Vet Ital 46(4):449–458

Virmani N, Bera BC, Shanumugasundaram K, Singh BK, Gulati BR, Singh RK (2011) Genetic analysis of the matrix and non-structural genes of equine influenza virus (H3N8) from epizootic of 2008–2009 in India. Vet Microbiol 152:169–175

Waddell GH, Teigland MB, Sigel MM (1963) A new influenza virus associated with equine respiratory disease. J Am Vet Med Assoc 143:587–590

Waghmare SP, Mode SG, Kolte AY, Babhulkar N, Vyavahare SH, Patel A (2010) Equine influenza: an overview. Vet World 3:194–197

Wang X, Hinson ER, Cresswell P (2007) The interferon-inducible protein viperin inhibits influenza virus release by perturbing lipid rafts. Cell Host Microbe 2:96–105

Webster RG, Laver WG (1975) Antigenic variation of influenza viruses. In: Kilbourne ED (ed) The influenza viruses and influenza. Academic, New York, pp 269–314

Webster RG, Thomas TL (1993) Efficacy of equine influenza vaccines for protection against A/Equine/Jilin/89 (H3N8)—a new equine influenza virus. Vaccine 11:987–993

Webster RG (1993) Are equine 1 influenza viruses still present in horses? Equine Vet J 25:537–538

Webster RG (2002) The importance of animal influenza for human disease. Vaccine 20:16–20

Wernery R, Yates PJ, Wernery U, Mumford JA (1999) Equine influenza outbreak in a polo club in Dubai, United Arab Emirates in 1995/96. In: Wernery U, Wade JF, Mumford JA, Kaaden OR (eds) Proc. of the 8th Int. Conference on equine infectious diseases. Dubai, 1998, pp. 342–346.
Wiley DC, Skehel JJ (1987) The structure and function of the hemagglutinin membrane glycoprotein of influenza virus. Annu Rev Biochem 56:365–394
Williams AJ (1924) Analogies between influenza of horses and influenza of man. Section of epidemiology and state medicine. Proc R Soc Med 17:47–58
Willoughby R, Ecker G, McKee S, Riddolls L, Vernaillen C, Dubovi E, Lein D, Mahony JB, Chernesky M, Nagy E et al (1992) The effects of equine rhinovirus, influenza virus and herpesvirus infection on tracheal clearance rate in horses. Can J Vet Res 56:115–121
Wilson WD (1993) Equine influenza. Vet Clin North Am Equine Pract 9(2):257–282
Wilson IA, Cox NJ (1990) Structural basis of immune recognition of influenza virus hemagglutinin. Annu Rev Immunol 8:737–771
Wilson WD, Mihalyi JE, Hussey S, Lunn DP (2001) Passive transfer of maternal immunoglobulin isotype antibodies against tetanus and influenza and their effect on the response of foals to vaccination. Equine Vet J 33:644–650
Yadav MP, Uppal PK, Mumford JA (1993) Physico-chemical and biological characterization of A/Equl-2 virus isolated from 1987 equine influenza epidemic in India. Int J Anim Sci 8:99–99
Yamanaka T, Niwa H, Tsujimura K, Kondo T, Matsumura T (2008) Epidemic of equine influenza among vaccinated racehorses in Japan in 2007. J Vet Med Sci 70:623–625
Yamanaka T, Nemoto M, Tsujimura K, Kondo T, Matsumura T (2009) Interspecies transmission of equine influenza virus (H3N8) to dogs by close contact with experimentally infected horses. Vet Microbiol 139:351–355
Yates P, Mumford JA (2000) Equine influenza vaccine efficacy: the significance of antigenic variation. Vet Microbiol 74:173–177
Yondon M, Heil GL, Burks JP, Zayat B, Waltzek TB, Jamiyan BO, McKenzie PP, Krueger WS, Friary JA, Gray GC (2013) Isolation and characterization of H3N8 equine influenza A virus associated with the 2011 epizootic in Mongolia. Influenza Other Respir Viruses 7:659–665
Yondon M, Zayat B, Nelson MI, Heil GL, Anderson BD, Lin X, Halpin R, McKenzie PP, White SK, Wentworth DE, Gray GC (2014) Equine influenza A (H3N8) virus isolated from Bactrian camel, Mongolia. Emerg Infect Dis 20:2144–2147
Yoon KJ, Cooper VL, Schwartz KJ, Harmon KM, Kim WI, Janke BH, Strohbehn J, Butts D, Troutman J (2005) Influenza virus infection in racing greyhounds. Emerg Infect Dis 11:1974–1976

Chapter 10
Schmallenberg Virus

S. B. Sudhakar, P. N. Gandhale, F. Singh, A. A. Raut, A. Mishra, D. D. Kulkarni, and V. P. Singh

Abstract Schmallenberg virus (SBV) infection is an emerging infectious disease of ruminants first discovered in summer 2011 applying metagenomic tools in North-Western Europe and which caused an epidemic proportion and later in the other European countries. It is an enveloped, negative-sense, segmented, single-stranded RNA virus, of the Simbu serogroup, *Orthobunyavirus* genus and the *Bunyaviridae* family, and is arthropod-borne. SBV affects mostly wild and domestic ruminants but has got no zoonotic potential and is horizontally spread by various species of *Culicoides* biting midges. Transplacental transmission can occur during the early part of pregnancy in ruminants after placentomes have been formed and cause teratogenic effects. Schmallenberg virus has also been found to be shed in the semen of cattle and sheep. SBV infection is usually asymptomatic in adult cattle, sheep and goats. The disease is characterised by fever, reduced milk production and diarrhoea in cattle and abortions, stillbirths and foetal abnormalities in sheep and goats when infection of the dam occurs at a critical period of gestation. In response, various molecular and serological tests and inactivated vaccines have been developed rapidly to diagnose and monitor the disease. Schmallenberg virus infections can have an all-round effect on production and considerable economic impact. This chapter details the updated knowledge on the discovery, epidemiology, impact, clinical symptoms, molecular characteristics and diagnostic techniques and the possibilities for preventing infections.

Keywords Schmallenberg virus · Ruminants · Emerging infection · Vector · Risk · Congenital malformations · Abortion · Impact · Diagnosis · Prevention and control

---

S. B. Sudhakar (✉) · P. N. Gandhale · F. Singh · A. A. Raut · A. Mishra · D. D. Kulkarni · V. P. Singh
ICAR-National Institute of High Security Animal Diseases, Bhopal, MP, India

Y. S. Malik et al. (eds.), *Emerging and Transboundary Animal Viruses*, Livestock Diseases and Management, https://doi.org/10.1007/978-981-15-0402-0_10

10.1 Preamble

Schmallenberg virus infection is an emerging threat, first noticed in dairy cattle in the border region of Dutch and Germany where in summer and autumn 2011 a unique syndrome characterised by high body temperature, diarrhoea and fall in milk yield was detected. Investigation for common causes was tested negative, and the aetiology of the syndrome was subsequently identified using metagenomic tools and isolated by cell culture technique. The name Schmallenberg virus is given to this agent as it was first isolated from tissue samples from Schmallenberg in Germany (Hoffmann et al. 2012). Since then, this virus has transmitted to various countries in Europe. SBV is a novel vector-borne orthobunyavirus of Simbu serogroup, member of *Bunyaviridae* family, and transmitted through Culicoides biting midges and also crosses the placenta and causes teratogenic effects with extremely negligible risk for public health. Currently, viruses belonging to the Simbu serogroup reported from Asia, Africa and Australia and the World Organisation for Animal Health (OIE) have not been classified as notifiable. The virus mainly affects domestic ruminants and causes congenital malformations, stillbirth in lambs and goat kids as well as in calves, and abortion (Hoffmann et al. 2012; Bayrou et al. 2014; Peperkamp et al. 2015). The epizootic spread was affirmed in wild ruminants (EFSA 2014). In northern Europe, a significant economic loss was documented due to large-scale SBV outbreak. Consequently, restrictions were imposed on the trade of livestock and its products across the borders (Hoffmann et al. 2012).

Since the first identification of SBV in 2011 considerable information has been acquired about virus origin, emergence, epidemiology, molecular virology, clinical signs, pathogenesis, diagnosis, vaccine development, seroprevalence, potential for further outbreaks and re-emergence and the economic impact of this disease which has been compiled in this chapter and is based on the analysis of the research conducted and the already existing data reported so far on this in scientific journals and Web-based reporting tools.

10.2 History

An unknown disease syndrome in dairy cows was observed in Germany in the year 2011 where herd morbidities were 20–70% and recovery occurred in a few days. At the same time, similar cases with severe scouring were also identified in the Netherlands (Hoffmann et al. 2012; Tarlinton et al. 2012; Bilk et al. 2012; Elbers et al. 2012; EFSA 2012; Beer et al. 2013; Tarlinton and Daly 2013).

In the later part of the year, abortions and stillbirths among newborn lambs but also in goats and cattle along with some dystocia in mature animals were also observed in the Netherlands, Germany and Belgium (van den Brom et al. 2012). In the subsequent year 2012, during February, mid-March, May and August months various outbreaks had been reported from Belgium, Denmark, France, Germany,

Italy, Luxembourg, the Netherlands, Spain, Switzerland and the United Kingdom. Sheep farms were highly infected, followed by cattle and goat farms. The initial cases of congenital abnormalities were investigated in detail, and all suspected aetiological agents were ruled out, and by using metagenomic analysis a new virus was isolated and identified. By metagenomic and full-length sequence analysis, this new virus has shown resemblance to three viruses which were identified in cattle from Japan: Aino virus, Shamonda virus and Akabane virus of the genus Orthobunyavirus viruses and *Bunyaviridae* family (Hoffmann et al. 2012; van den Brom et al. 2012). This agent was subsequently named Schmallenberg virus based on the place of its origin.

In Germany, before this event, there was not a single evidence of SBV protein and RNA based on a retrospective study from 1961 to 2010 in ruminants (Gerhauser et al. 2014). In another study in Turkey, 1 buffalo in 2006 and 12 cows and 1 buffalo sample in 2007 were found positive by antibody ELISA test. However, the viral nucleic acid was only identified in June 2012 (Azkur et al. 2013). Furthermore, in the year 2012, in Mozambique, seropositive sheep, goats and cattle were detected (Blomstrom et al. 2014).

10.3 Disease

Disease in adult cattle causes inapparent or subclinical symptoms (Hoffmann et al. 2012; Schulz et al. 2014). The incubation period ranges from 1 to 4 days, and viraemic stage is very short (1–6 days) which is followed by decreased appetite, high body temperature during high vector activity (more than 40 °C), loss of body condition, reduced milk production (up to 50%) and diarrhoea, followed by recovery within a few days (Hoffmann et al. 2012; Laloy et al. 2015; Lechner et al. 2017). The virus affects equally both the genders (Wernike et al. 2013b).

There may not be any clinical symptoms in adult sheep and goats except for the increased risk of abortions and congenital malformations in offspring. However, some sheep may show very mild clinical signs such as diarrhoea, lethargy, depression, snotty nose and fever (Wernike et al. 2013c; Helmer et al. 2013).

It is a reproductive disorder where dams are capable of transmitting the virus to the foetus, if they get an infection during a certain period of gestation, and develop severe congenital abnormality categorised as arthrogryposis hydranencephaly syndrome (AHS). This includes premature birth, mummified foetuses, stillbirth, disproportionate metatarsus, bent limbs and fixed joints, severe torticollis, ankylosis, flattened skull and brachygnathia inferior (van den Brom et al. 2012; Gelagay et al. 2018). When more than one offspring are infected in utero, only one of them may show clinical symptoms or arthrogryposis may be shown in one and neurological disease in the other. In another situation, one of the twins can be malformed, and the other is viable or only shows delayed growth (van den Brom et al. 2012; Wernike et al. 2014).

The neurological form is manifested by amaurosis, ataxia and/or behavioural abnormalities, recumbency, an inability to suck and occasionally fits ("dummy syndrome"), tetany, paresis, swimming and circling movements. The affected newborn mostly shows multiple malformations of the vertebral column (torticollis, lordosis, kyphosis and scoliosis). Calves are most frequently affected by torticollis and lambs by scoliosis (Bayrou et al. 2014; Peperkamp et al. 2014). Besides the malformations, the body weight of the newborn calves is significantly less (Bayrou et al. 2014).

10.4 Post-mortem Findings

The gross lesions are characterised by arthrogryposis, brachygnathia inferior, and malformations of the vertebral column and central nervous system which include cerebellar hypoplasia hydranencephaly, hydrocephalus, porencephaly, micromyelia and subcutaneous oedema (calves) (van den Brom et al. 2012). The CNS lesions are characterised by cavity formation in white matter, loss of neurons in the cerebrum and cerebellum cortex, nuclei of brainstem and gray columns of the spinal cord. Lambs are more severely affected than calves (Peperkamp et al. 2014; Laloy et al. 2017). Additionally, in case of in utero-infected lambs and calves, the tendons of the affected joints appear smaller, and the related muscles show a change in colour and loss in mass (Bayrou et al. 2014).

10.5 Impacts

Various factors set the impact of the disease, for example, the number of congenitally abnormal lambs, lower milking periods and stage of gestation in which the infection took place (Wuthrich et al. 2016). The direct impact of the disease on adult animals incorporates a rise in body temperature, diarrhoea and decreased milk yield as well as non-pregnancy, repeat breeding, abortion and fatal dystocia. During certain gestational stages, there is a chance of complications featuring deformation and loss of newborn. The vertical transmission during the first half of gestation has the highest visible impact (Hoffmann et al. 2012; Elbers et al. 2012; Bilk et al. 2012; EFSA 2012; Tarlinton et al. 2012; Wernike et al. 2013a; Tarlinton and Daly 2013). During 2012, the observed direct impact was the consequence of the spread of the virus into a host population which was naive. Other consequences of SBV infection are the treatment costs or calving and lambing complications besides the cost of buying the replacement stock to compensate the reproductive losses along with unsold replacement animals, as well as restrictions on the movement (Alarcon et al. 2014).

Although the immediate financial impact of this infection in the ruminants is limited, the appearance of this virus has a major economic impact on international trade and has caused considerable damage to export due to imposed restrictions on the import of livestock products, such as embryos, semen and live animals from affected countries. Such restrictions on trades have caused considerable financial losses in Europe. During 2011 and 2012, the purebred animals' export value declined by 20%, and in 2012 the bovine semen trade declined by 11–26%, corresponding to 8.9 million doses (EFSA 2014).

At the individual farm level, the impact ranges from minimal to more than 50% losses of young animals (Helmer et al. 2013). In general, the effect of SBV is more on sheep farms experiencing increased numbers of abortions, lamb mortality, dystocia, malformations and lower fertility rate.

In cattle farms, the financial losses owing to the fall in milk yield and return to service are bigger than the congenitally malformed calves (EFSA 2012; Beer et al. 2013). Effect of the virus on domestic goats is lesser than sheep. However, heavy financial loss up to 50% was registered in affected goat farms due to kid's mortality and fall in milk yield (Helmer et al. 2013). Wuthrich et al. (2016) conducted a study in an SBV-infected standardised farm and calculated that the mean loss was 1338 EUR, which at the national scale may be low, but the high fluctuations in losses were observed between farms so that particular farms might have experienced considerable losses (8333 EUR).

10.6 Virus

Within the family *Bunyaviridae*, the largest genus *Orthobunyavirus* is divided into 18 serogroups and the Simbu serogroup holds more than 25 viruses including SBV and is classified into seven species (Simbu virus, Akabane virus, Oropouche virus, Shamonda virus, Sathuperi virus, Shuni virus and Manzanilla virus) on the basis of cross-neutralisation tests and cross-haemagglutination inhibition tests (Goller et al. 2012; Yanase et al. 2012; Hoffmann et al. 2012; Plyusnin et al. 2012).

SBV is an enveloped, spherical (with a diameter ranging 80–120 nm having short surface projections), three-segmented, single-stranded, negative-sense RNA virus. The envelope holds the large (L), medium (M) and small (S) genome segments and forms a panhandle structure using complementary non-coding bases at the end of the segment. These complexes are consociated to a few copies of L polymerase and many copies of the N protein to make ribonucleoproteins (RNPs), the infectious viral particles (Tilston et al. 2017). The L segment codes for an essential protein the viral RNA-dependent RNA polymerase (RdRp) L protein which is accountable for the replication and transcription of the viruses (Kraatz et al. 2018).

The M segment codes for a precursor polyprotein which is co-translationally cleaved by cellular proteases into two surface glycoproteins (Gn and Gc) and a non-structural protein (NSm). Together, Gn and Gc form heterodimeric complexes responsible for virus entry into cells and function as antigenic determinants and are identified by neutralising antibodies whereas NSm protein's role is not fully known, but this protein appears to be involved in the assembly of virus particles. In general, among the S, M and L segments, the M RNA segment is highly variable. Natural genetic reassortment is responsible for the emergence of new virus strains with a prospective change in their host range, virulence and antigenicity.

The S segment codes for the nucleocapsid protein (N) as well as another small non-structural protein (NSs) in an overlapping ORF. The primary function of this protein is to encapsidate the viral genome and protect its disintegration in the cells and it is also necessary for viral RNA transcription and replication. The N protein is the highly available protein in the virion and infected host cells. Therefore, it is mostly used for molecular and serological identification of SBV (Bilk et al. 2012) and is the major SBV antigen responsible for complement fixation (Goller et al. 2012; Yanase et al. 2012) and also modulates the host innate immune response (Elliott et al. 2013).

10.7 Resistance to Physical and Chemical Action

The virus loses its infectivity at 50–60 °C temperatures in 30 min. Exposure to common disinfectants such as 2% glutaraldehyde, 1% sodium hypochlorite, formaldehyde and 70% ethanol affects the viral virulence and virus does not sustain outside the vector or host for long term (OIE 2017).

10.8 Phylogeny

After the discovery of this virus, the full genome was sequenced by Hoffmann et al. (2012) and the S (830 nucleotides), M (4415 nucleotides) and L (6865 nucleotides) segments were compared with other Orthobunyaviruses and they observed that the small segment was 97% similar to Shamonda virus; the medium segment was 71% similar to Aino virus; and the large segment was 69% similar to Akabane virus, all detected in cattle of Japan. Rooting upon such observations, Schmallenberg virus was placed in the Simbu serogroup as Shamonda-like virus.

Later, Yanase et al. (2012) advocated that SBV emerged as a result of reassortant phenomena between Sathuperi and Shamonda viruses with the small and large segments emerging from Shamonda virus and the medium segment emerging from Douglas and Sathuperi virus. Afterwards, almost complete genome sequences were ascertained for nine viruses of the five species in the Simbu serogroups (i.e. species Shamonda virus (Sango virus, Peaton virus and Shamonda virus), species Sathuperi virus (Douglas virus and Sathuperi virus), species Shuni virus (Shuni virus and Aino virus), species Akabane virus (Sabo virus) and species Simbu virus (Simbu virus)). Upon phylogenetic analysis, it was observed that Schmallenberg virus belongs to the species Sathuperi virus and possibly is not a reassortant virus and as a fact may be an ancestor to Shamonda virus, which itself is a reassortant with small and large segments from Schmallenberg virus and medium segment from an unspecified virus. This conclusion is also assisted by a serological examination wherein Douglas and Sathuperi viruses are neutralised by anti-SBV serum, but not the Shamonda virus (Goller et al. 2012).

10.9 Epidemiology of Disease

10.9.1 Incidence and Prevalence

Since its first appearance in north-western Europe (Hoffmann et al. 2012), SBV has extended over considerable parts of Europe. SBV infections have been spotted in Germany, the Netherlands and Belgium (seroprevalence up to 99.8%), the United Kingdom, France (seroprevalence up to 90%), Italy, Luxembourg, Spain, Italy, Denmark, Estonia, Ethiopia (seroprevalence up to 56.6%), Northern Ireland, Switzerland, Norway, Austria, Sweden, Finland, Poland and Turkey (Elbers et al. 2012; Azkur et al. 2013; Afonso et al. 2014; Gelagay et al. 2018). After the initial epidemic in 2011–2013, possible recirculation of virus has been reported from several countries, including Germany (Wernike et al. 2015a), Belgium (Delooz et al. 2016) and England and Wales (APHA 2017). Orthobunyaviruses in the Simbu serogroup are identified in Africa, Asia, Australia and the Middle East. Serological studies have shown SBV antibody-positive results from African countries. Since viruses of the Simbu serogroup are spotted in many locations of Africa and due to the paramount issue of cross-reactivity, Mathew et al. (2015) inferred that the seropositivity in ELISA might be due to other viruses of the Simbu serogroup instead of SBV. Likewise, Sathuperi virus (Simbu serogroup) was in the first place to be isolated in India from a pool of *Culex vishnui* mosquitoes and later in Nigeria from dairy cattle and pools of *Culicoides* spp. (Dandawate et al. 1969; Causey et al. 1972). Another member of Simbu serogroup, Kaikalur virus, was isolated from a pool of *Culex tritaeniorhynchus* mosquitoes collected from Krishna district, India (Rodrigues et al. 1977). Because of very close two-way cross-reaction, Kaikalur and Aino viruses are considered identical or varieties of a single virus.

Among cattle, younger animals had lower prevalence than adult (Gelagay et al. 2018). However, Elbers et al. (2012) have not found any age-related difference in seroprevalence in the Netherlands.

10.9.2 Risk Factors

Hitherto, not a single human case has been reported from any country. Thus, the public health risk of SBV should be deemed to be negligible. Simbu serogroup does not have zoonotic implication except Oropouche virus, which causes severe flulike symptoms in humans.

Nevertheless, SBV was expanding and overcame the European boundary. Availability of vectors, reservoirs and susceptible host populations will facilitate it to spread further and/or become endemic as well as determines its persistence. Every fresh batch of animals significantly introduces fresh susceptible hosts. The rate of mixing of a new susceptible host with the existing population will determine the duration and amplitude of inter-annual epidemic cycles. This also depends on herd replacement rates, level of vaccination and durations of immunity. The rate of restocking may vary based on farm management and production systems. In farm enterprises, the regular replacement rate is 20 or 25%, which gives rise to a substantial number of susceptible hosts in a herd annually.

Furthermore, before the introduction of the breeding male(s), if naive females are infected, no unfavourable effects are to be expected. But infection during early pregnancy results in early embryonic death and dams become repeat breeders. Nonetheless, in the next pregnancy, normal results may be expected. When animal health status and meteorological settings become favourable for the vectors and virus, also when a substantial number of the hosts become susceptible, especially at the boundary of the endemic area, the virus can re-emerge.

10.10 Transmission

10.10.1 Susceptible Species

Since the first detection of SBV, the existence of viral RNA and/or antibodies has been investigated in a variety of animals and found that domestic ruminants (cattle, sheep and goats) and multiple wild (alpacas, Anatolian water buffalo, elk, bison, red deer, fallow deer, roe deer, sika deer, muntjac, chamois, moufflons and wild boar) and zoo species (bongo, babirusa, banteng, Congo buffalo, European bison, gaur, gemsbok, greater kudu, Grevy's zebra, moose, Nile lechwe, Nubian goat, onager, reindeer, roan antelope, scimitar-horned oryx, sitatunga and yak) are susceptible to SBV (EFSA 2014). Among domestic ruminants, it is observed that susceptibility of goats is lesser than cattle and sheep.

There is a report indicating that dogs may be infected with SBV and pregnant females show teratogenic effects, but this is possibly a rare phenomenon (Sailleau et al. 2013). Virus infection in pigs has also been reported where it induces seroconversion but does not involve its epizootiology (Poskin et al. 2014). Infection in camelids is also reported (Wernike et al. 2012).

10.10.2 Horizontal Transmission

The virus does not spread through direct contact. The oral route is also not the probable way of spread. Experimental injection by subcutaneous route in cattle (Hoffmann et al. 2012; Wernike et al. 2013b), sheep (Wernike et al. 2013c; Martinelle et al. 2017) and goats (Laloy et al. 2015) and intradermal route in sheep (Martinelle et al. 2017) results in viraemia.

10.10.3 Vector

The SBV is also transmitted by Culicoides biting midges, principally members of the *Culicoides obsoletus* complex, but other *Culicoides* spp. are also capable of spreading, (*Culicoides chiopterus*, *Culicoides dewulfi*, *Culicoides scoticus*, *C. pulicaris*); they are active 1 h before sunrise and sunset (Hoffmann et al. 2012; Bilk et al. 2012; Elbers et al. 2012; Tarlinton et al. 2012; Beer et al. 2013; Wernike et al. 2013a; Tarlinton and Daly 2013).

10.10.4 Vertical Transmission

Vertical transmission is of great significance as SBV gets across the placenta. Transplacental infection occurs when the first placentome appears till the foetus becomes immunocompetent, days 30–150 post-conception in bovine (Bayrou et al. 2014) and days 28–56 post-conception in ovine and caprine (Helmer et al. 2013; Laloy et al. 2017). The clinical manifestation is highly dependent on the age of the foetus (Bayrou et al. 2014). When the dam is infected during the early stage of pregnancy, it can result in foetal death, lower fertility and stillbirth. But if infected later in pregnancy, developed immune system of the foetus is capable of resisting the virus; however mummification, stillbirth and abortion can also occur (Helmer et al. 2013). Infected offspring does not show viraemia, and there is no evidence of virus transmission from the infected progeny to vectors (EFSA 2014).

10.10.5 Semen

The virus can be secreted in the semen of infected animals, and viral nucleic acid has been identified both in the plasma and cell fraction of semen (Kesik and Larska 2016). Semen containing SBV is not likely to infect embryo, but if the dam gets viraemic, vector transmission may occur (Schulz et al. 2014).

10.11 Immunopathobiology

There is not much information on the immune response and duration of immunity following the infection. An innate immune response takes place immediately after the infection in cattle (Wernike et al. 2013b). Nevertheless, SBV alters the innate immune response of the host by suppressing interferon production at the transcription level, particularly mediated by NSs protein. This protein behaves like a virulence factor and antagonises IFN likely by suppressing cellular metabolism. Thus, the innate immune response of the host is vanquished, and effective replication of the virus takes place (Elliott et al. 2013). Seroconversion in cattle occurs after 8–14 days of infection and remains for more than 3 years in naturally infected cows; therefore, a long-standing immunity can be anticipated, but in experimentally infected cattle immunity lasted for at least 8 weeks which was able to prevent reinfection (Wernike et al. 2013b; Elbers et al. 2014; Schulz et al. 2014; Wernike et al. 2015b). Seroconversion in sheep takes place after 6–22 days of infection and lasts for at least 15 months and in goats between 7 and 14 days after infection (Wernike et al. 2013c; Poskin et al. 2015; Laloy et al. 2015). Colostrum feeding is the only way of transferring maternal antibodies to calves, which remains for 5–6 months as the transplacental transfer does not take place in ruminants (Elbers et al. 2014). Furthermore, foetuses are capable of producing neutralising antibodies and are also identified in stillborn or aborted calves and lambs. However, if the foetus is immunocompetent, clinical signs are not visible at birth in case of in utero infection.

10.12 Diagnostics

Clinical diagnosis is usually done through observing clinical evidence of the disease, which may be different in different species. Various laboratory procedures have been described to detect the SBV infections which include (i) real-time reverse transcriptase PCR, (ii) neutralisation and indirect immunofluorescence assays, (iii) ELISA and (iv) isolation and identification of the virus by cell culture technique.

10.12.1 Sample Collection

Schmallenberg virus remains present for a longer time in infected foetuses than adult live animals (4–6 days) and can be detected in malformed newborns (Laloy et al. 2017). From an acutely infected animal, viral nucleic acid could be identified in mesenteric lymph nodes, spleen as well as semen for a long period. Placenta and amniotic fluid are also materials of choice for diagnosis (Bilk et al. 2012).

From acutely infected adults, samples of whole blood in EDTA and serum should be collected and properly packaged and shipped to the designated laboratory under the cold chain. Samples from aborted foetuses or newborn animals may be collected for histopathology (fixed central nervous system, including spinal cord), serology and virological investigations. Brain sample, ideally cerebrum and cerebellum as well as central nervous system, lymphatic organs and blood from dead animals and pre-colostrum blood, serum and meconium samples from live animals should be collected and properly packaged and shipped to the designated laboratory under the cold chain (OIE 2017).

10.12.2 Nucleic Acid Detection

Real-time quantitative reverse transcription PCR is a reliable detection method for viral nucleic acids (L or S segment of the viral genome) in clinical samples. For this purpose, RT-qPCR assay is developed by the FLI (Bilk et al. 2012; Hoffmann et al. 2012).

10.12.3 Virus Isolation in Cell Culture

Virus isolation can be performed from blood collected at the height of temperature from diseased adult animals and from the dead foetus and its brain (Laloy et al. 2017). The virus can be propagated in various cell lines originated from different animal species and humans like Vero cells, sheep choroid plexus cells, bovine foetal aorta endothelial cells, human 293T, Madin-Darby canine kidney cells and baby hamster kidney-21 cells and BSR cells or insect KC (*Culicoides variipennis* larvae) and it induces cytopathic effect (CPE) in most of these cell lines. Among these cell lines, sheep choroid plexus cells were found most sensitive for SBV propagation (Hoffmann et al. 2012; Wernike et al. 2013c; Ilchmann et al. 2017).

10.12.4 Serological Test

The SBV seroprevalence and also the serological status of the individual animal may be determined using virus neutralisation, immunofluorescence assays and ELISA techniques (Loeffen et al. 2012). The gold standard test for SBV diagnosis is virus neutralisation test (VNT) having almost 100% sensitivity and specificity (Loeffen et al. 2012). In spite of the fact that cross-reactions within Simbu serogroup of viruses have been reported, the enzyme-linked immunosorbent assay is a most sensitive, specific, sturdy and approved technique for anti-SBV antibody detection and can be employed for surveillance studies. Bulk milk antibody tests are available and can be used for surveillance as they indicate herd-level exposure (Hoffmann et al. 2012; Tarlinton et al. 2012; EFSA 2012; Elbers et al. 2012; Bilk et al. 2012; Beer et al. 2013; Tarlinton and Daly 2013). Both ELISA and RT-qPCR kits are commercially available. For multispecies serum or plasma viral antibody detection, competitive ELISA kits are commercially available (ID Screen® Schmallenberg virus Competition Multi-species, IDvet Laboratories, Montpellier, France)

10.12.5 Differential Diagnosis

Differential diagnosis of the disease is important as SBV infection does not show conclusive clinical symptoms in adults; therefore, all potential causes of high body temperature, liquid faecal discharge, lower milk yield stillbirth and abortion must be considered such as bovine viral diarrhoea (BVD), border disease (BD) and other pestiviruses, bovine herpesvirus 1 and other herpesviruses, bovine ephemeral fever, epizootic haemorrhagic disease (EHD), foot and mouth disease (FMD), bluetongue, Rift Valley fever (RVF), and toxic substances, e.g. *Veratrum californicum* and *Lupinus* spp. However, Cache Valley virus, Orthobunyavirus infections (Akabane), genetic factors (e.g. spider lamb syndrome), toxic substances and nutritional deficiencies (e.g. gestational protein deficiency, manganese) must also be taken into consideration while investigating the congenital malformations.

10.13 Prevention and Control

As it is a viral infection, presently, there is no therapy available for this disease; hence, supportive care is an only reliable option for intervention. The direct impact of this disease can be mitigated by the use of a potent vaccine, or by avoiding the risk of pregnancy when vector activity is high. Different inactivated vaccines have been formulated and successfully tested (Wernike et al. 2013c; Hechinger et al. 2014), and two of them, SBVvax (Merial 2013) and Bovilis SBV (MSD Animal Health n.d.), have been given provisional permission to the market in the United Kingdom and France. As per the manufacturer's instruction, the vaccination regime for the large animal is two injections at 28-day interval, and for sheep only one injection is adequate where immunity develops within 3 weeks. Vaccination before the breeding season is the most efficient measure to prevent infection. After vaccination, anti-SBV antibodies develop in a previously uninfected host which protect the foetus from teratogenic effects, once the animal is pregnant, by inhibiting transplacental infection.

Furthermore, the vaccine also protects susceptible animals from sickness and purportedly prevents further transmission of the vector (Tarlinton et al. 2012). DNA immunisation can raise multiple Th response and antibody response. So far, no DNA vaccine is available for this disease. For the development of a prospective DNA vaccine, both the nucleoprotein and the putative GC ectodomain gene have been targeted (Boshra et al. 2017).

One feasible alternative is to control the midge vectors by implementing the procedure like use of insecticides/larvicide and pathogens to natural dwelling where they grow and removal of larva breeding enclosures through environmental interference; adult midges may be controlled by treating either resting sites such as livestock house or its body with insecticides and repellents, e.g. pyrethroids and host kairomones (Carpenter et al. 2008).

Besides, the naive animals or herds with low within-herd seroprevalence can be protected by the improved breeding system. The breeding period may be planned in such a fashion where it falls in late autumn and *Culicoides* spp. are scarce; the animals may be kept mostly inside house, all of the sunsets to sunrise. To protect animals during pregnancy, a susceptible individual may also be shifted to the endemic zone soon enough to raise acquired immunity (Helmer et al. 2013). In calves, maternal antibodies disappear in 5–6 months but in adults specific antibodies persist for a minimum of 2 years (Elbers et al. 2014). Movement restrictions are not imposed as this is not a notifiable disease.

Acknowledgements All the authors of the manuscript thank and acknowledge their institutes.

Conflict of interest: There is no conflict of interest.

References

Afonso A, Abrahantes JC, Conraths F et al (2014) The Schmallenberg virus epidemic in Europe 2011–2013. Prev Vet Med 116:391–403

Alarcon P, Hasler B, Raboisson D et al (2014) Application of integrated production and economic models to estimate the impact of Schmallenberg virus for various sheep production types in the UK and France. Vet Rec Open 1:e000036

Animal & Plant Health Agency (APHA) (2017) Disease surveillance in England and Wales, February 2017. Vet Rec 180:243–246

Azkur AK, Albayrak H, Risvanli A et al (2013) Antibodies to Schmallenberg virus in domestic livestock in Turkey. Trop Anim Health Prod 45:1825–1828

Bayrou C, Garigliany MM, Sarlet M et al (2014) Natural intrauterine infection with Schmallenberg virus in malformed newborn calves. Emerg Infect Dis 20:1327–1330

Beer M, Conraths FJ, van der Poel WH (2013) Schmallenberg virus—a novel orthobunyavirus emerging in Europe. Epidemiol Infect 141:1–8

Bilk S, Schulze C, Fischer M et al (2012) Organ distribution of Schmallenberg virus RNA in malformed newborns. Vet Microbiol 159:236–238

Blomstrom AL, Stenberg H, al SI (2014) Serological screening suggests presence of Schmallenberg virus in cattle, sheep and goat in the Zambezia Province, Mozambique. Transbound Emerg Dis 61:289–292

Boshra HY, Charro D, Lorenzo G et al (2017) DNA vaccination regimes against Schmallenberg virus infection in IFNAR−/− mice suggest two targets for immunization. Antivir Res 141:107–115

van den Brom R, Luttikholt SJ, Lievaart-Peterson K et al (2012) Epizootic of ovine congenital malformations associated with Schmallenberg virus infection. Tijdschr Diergeneeskd 137:106–111

Carpenter S, Mellor PS, Torr SJ (2008) Control techniques for Culicoides biting midges and their application in the UK and northwestern Palaearctic. Med Vet Entomol 22:175–187

Causey OR, Kemp GE, Causey CE (1972) Isolations of Simbu-group viruses in Ibadan, Nigeria 1964–69, including the new types Sango Shamonda, Sabo and Shuni. Ann Trop Med Parasitol 66:357–362

Dandawate CN, Rajagopalan PK, Pavri KM et al (1969) Virus isolations from mosquitoes collected in North Arcot district, Madras state and Chittoor district, Andhra Pradesh between November 1955 and October 1957. Indian J Med Res 57(8):1420–1426

Delooz L, Saegerman C, Quinet C et al (2016) Resurgence of Schmallenberg virus in Belgium after 3 years of epidemiological silence. Transbound Emerg Dis 64(5):1641–1642

EFSA (2012) Scientific report of European Food Standards Agency: Schmallenberg virus: analysis of the epidemiological data and assessment of impact. EFSA J 10:2768

EFSA (2014) Schmallenberg virus: state of art. EFSA J 12(5):3681

Elbers AR, Loeffen WL, Quak S et al (2012) Seroprevalence of Schmallenberg virus antibodies among dairy cattle, the Netherlands, winter 2011–2012. Emerg Infect Dis 18:1065–1071

Elbers AR, Stockhofe-Zurwieden N, van der Poel et al (2014) Schmallenberg virus antibody persistence in adult cattle after natural infection and decay of maternal antibodies in calves. BMC Vet Res 10:103

Elliott RM, Blakqori G, van Knippenberg IC (2013) Establishment of a reverse genetics system for Schmallenberg virus, a newly emerged orthobunyavirus in Europe. J Gen Virol 94:851–859

Gelagay A, Endrias Z, Gebremedhinc E et al (2018) Seroprevalence of Schmallenberg virus in dairy cattle in Ethiopia. Acta Trop 178:61–67

Gerhauser I, Weigand M, Hahn K et al (2014) Lack of Schmallenberg virus in ruminant brain tissues archived from 1961 to 2010 in Germany. J Comp Pathol 150:151–154

Goller KV, Hoper D, Schirrmeier H et al (2012) Schmallenberg virus as possible ancestor of Shamonda virus. Emerg Infect Dis 18:1644–1646

Hechinger S, Wernike K, Beer M (2014) Single immunization with an inactivated vaccine protects sheep from Schmallenberg virus infection. Vet Res 45:79

Helmer C, Eibach R, Tegtmeyer PC et al (2013) Survey of Schmallenberg virus (SBV) infection in German goat flocks. Epidemiol Infect 141:2335–2345

Hoffmann B, Scheuch M, Hoper D et al (2012) Novel orthobunyavirus in cattle, Europe. Emerg Infect Dis 18:469–472

Ilchmann A, Armstrong AA, Clayton RF et al (2017) Schmallenberg virus, an emerging viral pathogen of cattle and sheep and a potential contaminant of raw materials, is detectable by classical in-vitro adventitious virus assays. Biologicals 49:28–32

Kesik MJ, Larska M (2016) Detection of Schmallenberg virus RNA in bull semen in Poland. Pol J Vet Sci 19(3):655–657

Kraatz F, Wernike K, Reiche S et al (2018) Schmallenberg virus non-structural protein NSm: intracellular distribution and role of non-hydrophobic domains. Virology 516:46–54

Laloy E, Breard E, Trapp S et al (2017) Fetopathic effects of experimental Schmallenberg virus infection in pregnant goats. Vet Microbiol 211:141–149

Laloy E, Riou M, Barc C et al (2015) Schmallenberg virus: experimental infection in goats and bucks. BMC Vet Res 11:221

Lechner I, Wuthrich M, Meylan M et al (2017) Association of clinical signs after acute Schmallenberg virus infection with milk production and fertility in Swiss dairy cows. Prev Vet Med 146:121–129

Loeffen W, Quak S, de Boer-Luijtze E et al (2012) Development of a virus neutralisation test to detect antibodies against Schmallenberg virus and serological results in suspect and infected herds. Acta Vet Scand 54:44

Martinelle L, Poskin A, Dal Pozzo F et al (2017) Three different routes of inoculation for experimental infection with Schmallenberg virus in sheep. Transbound Emerg Dis 64(1):305–308

Mathew C, Klevar S, Elbers AR (2015) Detection of serum neutralizing antibodies to Simbu serogroup viruses in cattle in Tanzania. BMC Vet Res 11:208

Merial (2013) Merial receives approval for new vaccine to prevent Schmallenberg disease in livestock. http://www.merial.com/EN/PressRoom/PressRelease/Pages/MerialApprovalSchmallenbergVaccine.aspx. Accessed 5 Nov 2014

Merck Animal Health (2013) Veterinary medicines directorate grants provisional marketing authorisation to MSD animal health for first vaccine targeting Schmallenberg virus. http://www.merck-animal-health.com/news/2013-12-18.aspx. Accessed 5 Nov 2014

OIE (2017): Schmallenberg virus: Technical Fact Sheet

Peperkamp K, Van Schaik G, Vellema P (2014) Risk factors for malformations and impact on reproductive performance and mortality rates of Schmallenberg virus in sheep flocks in the Netherlands. PLoS One 9:e100135

Peperkamp NH, Luttikholt SJ, Dijkman R et al (2015) Ovine and bovine congenital abnormalities associated with intrauterine infection with Schmallenberg virus. Vet Pathol 52:1057–1066

Plyusnin A, Beaty BJ, Elliott RM (2012) Virus taxonomy: ninth report of the International Committee on taxonomy of viruses. Elsevier Academic Press, London, pp 725–741

Poskin A, Van Campe W, Mostin L (2014) Experimental Schmallenberg virus infection of pigs. Vet Microbiol 170:398–402

Poskin A, Verite S, Comtet L et al (2015) Persistence of the protective immunity and kinetics of the isotype specific antibody response against the viral nucleocapsid protein after experimental Schmallenberg virus infection of sheep. Vet Res 46:119

Rodrigues FM, Singh PB, Dandawate CN et al (1977) Kaikalur virus a new arthropod borne virus belonging to the Simbu group isolated in India from *Culex tritaeniorhynchus* (Giles). Indian J Med Res 66(5):719–725

Sailleau C, Boogaerts C, Meyrueix A (2013) Schmallenberg virus infection in dogs, France, 2012. Emerg Infect Dis 19(11):1896–1898

Schulz C, Wernike K, Beer M et al (2014) Infectious Schmallenberg virus from bovine semen, Germany. Emerg Infect Dis 20:338–340

Tarlinton R, Daly J (2013) Testing for Schmallenberg virus. Vet Rec 172:190

Tarlinton R, Daly J, Dunham S et al (2012) The challenge of Schmallenberg virus emergence in Europe. Vet J 194:10–18

Tilston LNL, Shi X, Elliott RM, Acrani GO (2017) The Potential for Reassortment between Oropouche and Schmallenberg Orthobunyaviruses. Viruses 9(8):220

Wernike K, Breithaupt A, Keller M (2012) Schmallenberg virus infection of adult type I interferon receptor knockout mice. PLoS One 7:e40380

Wernike K, Eschbaumer M, Schirrmeier H et al (2013b) Oral exposure, reinfection and cellular immunity to Schmallenberg virus in cattle. Vet Microbiol 165:155–159

Wernike K, Hoffman B, Conraths FJ et al (2015a) Schmallenberg virus recurrence, Germany, 2014. Emerg Infect Dis 21:1202–1204

Wernike K, Hoffmann B, Bréard E et al (2013a) Schmallenberg virus experimental infection of sheep. Vet Microbiol 166:461–466

Wernike K, Holsteg M, Saßerath M et al (2015b) Schmallenberg virus antibody development and decline in a naturally infected dairy cattle herd in Germany, 2011–2014. Vet Microbiol 181:294–287

Wernike K, Holsteg M, Schirrmeier H et al (2014) Natural infection of pregnant cows with Schmallenberg virus a follow up study. PLoS One 9:e98223

Wernike K, Nikolin VM, Hechinger S et al (2013c) Inactivated Schmallenberg virus prototype vaccines. Vaccine 31:3558–3563

Wuthrich M, Lechner I, Aebi M et al (2016) A case control study to estimate the effects of acute clinical infection with the Schmallenberg virus on milk yield, fertility and veterinary costs in Swiss dairy herds. Prev Vet Med 126:54–65

Yanase T, Kato T, Aizawa M et al (2012) Genetic reassortment between Sathuperi and Shamonda viruses of the genus orthobunyavirus in nature: implications for their genetic relationship to Schmallenberg virus. Arch Virol 157:1611–1616

Chapter 11
Crimean-Congo Haemorrhagic Fever Virus

Ashwin Ashok Raut, Pradeep N. Gandhale, D. Senthil Kumar, Naveen Kumar, Atul Pateriya, Anamika Mishra, and Diwakar D. Kulkarni

Abstract Crimean-Congo haemorrhagic fever virus belongs to genus *Nairovirus*, of family *Bunyaviridae*. It is a tick-borne virus, classified as a BSL4 agent. The genome is a tripartite, single-stranded RNA. The virus has been reported from 45 countries in the world. However, the viruses isolated from different regions of the world show considerable genetic diversity. The geographic distribution of CCHF concurs with *Ixodid* ticks mostly of genus *Hyalomma*. The ticks transmit the virus to mammals, causing transient viremia. However, ticks remain infected throughout their lifetime. The infection in livestock is asymptomatic but plays a crucial role in the transmission, amplification and perpetuation of the virus. It causes severe haemorrhagic fever in humans. High mortality has been observed in healthcare workers due to nosocomial infections arising as a result of contact with infected blood or body secretions from patients. Pathogenesis includes endothelial damage and immune response impairment, leading to multiple-organ failure and shock. Delayed type I interferon induction, inhibition of maturation of antigen-presenting cells and weak antibody response cause uncontrolled replication of the virus. The laboratory confirmation of the CCHFV is done by virus isolation, genome detection is done by RT-PCR and, RT-qPCR, virus detection is done by antigen capture ELISA while antibody detection is done by ELISA. Currently, there are no licensed vaccines or approved therapeutics for CCHF. Continuous surveillance of animals, with regular control of tick infestation in livestock, in conjunction with prophylactic measures in high-risk human populations, is required for control of CCHF infection.

Keywords Crimean-Congo haemorrhagic fever · Tick fever · Haemorrhagic fever · Nairovirus Bunyaviridae · Genome organization and evolution · Zoonotic virus · Ruminant · Hyalomma · Epidemiology · Diagnosis

---

A. A. Raut (✉) · P. N. Gandhale · D. S. Kumar · N. Kumar · A. Pateriya · A. Mishra · D. D. Kulkarni
ICAR-National Institute of High Security Animal Diseases, Bhopal, Madhya Pradesh, India

Y. S. Malik et al. (eds.), *Emerging and Transboundary Animal Viruses*, Livestock Diseases and Management, https://doi.org/10.1007/978-981-15-0402-0_11

11.1 Preamble

Crimean-Congo haemorrhagic fever virus (CCHFV) is a tick-borne, zoonotic virus that causes acute, haemorrhagic fever with case fatality up to 40% in humans. The infection in wild and domestic animals is asymptomatic. The disease was first observed between 1944 and 1945, in Russian soldiers who assisted the farmers in Crimea at the end of World War II. The disease was named Crimean haemorrhagic fever. Epidemics continued to occur in Astrakhan between 1953 and 1968, in Rostov Oblast between 1963 and 1971 of former USSR and in Bulgaria between 1953 and 1973. The virus was isolated by intra-cerebral inoculations in suckling mice in 1967 by the technique known then as the newborn white mice technique (Hoogstraal 1979). On the other hand, Congo fever virus was isolated from the blood of a patient in the Belgian Congo of Africa in 1959. Later in 1969 a collaborative study between the Soviet and American groups confirmed that this etiological agent was antigenically identical with that of the Congo fever virus (Casals 1969). The virus was renamed by linking the two names, as Crimean-Congo haemorrhagic fever virus in 1973.

The geographic range of CCHFV is known to be the most extensive of the tick-borne viruses relevant to human health and is now identified as a public health problem in Eastern Europe, Asia, the Middle East and Africa. The virus is maintained in nature in *Ixodid* ticks mostly of genus *Hyalomma* by trans-ovarial and trans-stadial transmission. Apart from tick bites, the disease can be transmitted from animals to humans through contact with blood and meat of viremic animals. High case fatality in humans, along with symptomless infection in animals, makes difficult to manage this zoonotic disease. The gravity of Crimean-Congo haemorrhagic fever (CCHF) in humans is further enhanced by the occurrence of nosocomial transmission and in-house transmission from human to human (Watts et al. 2005). The increase in the tick population and their global spread as a result of climate change, expanding farming and trading practices, are amongst the major causes for the spread of CCHF to newer areas of the world like Turkey, Greece and India. CCHFV is recognised as a possible agent of bioterrorism (Sidwell and Smee 2003). Its potential as a biological weapon has been studied in Iraq, and aerosolisation has been done for potential dissemination of the virus (Zilinskas 1997; Bronze et al. 2002).

Presently, only supportive and preventative strategies are used to control the spread of the infection as there are no approved vaccines for CCHF (Papa et al. 2015). Under such situations and owing to the high case fatality rates, CCHFV is classified as a biosafety level 4 (BSL4) agent. Reliable and robust diagnostics, which can promptly and accurately diagnose CCHF during the first few days of infection, are very critical for successful management of disease as well as for preventing nosocomial transmissions. Being a BSL4 agent, CCHFV isolation cannot be done in routine clinical laboratories. Molecular diagnostics are increasingly being used for rapid and reliable detection of the virus in clinical samples. The enormous genetic variability, along with the potential for recombination and reassortment in CCHFV, is one of the major bottlenecks in the development of vaccines and diagnostics.

11.2 Genome Organisation

CCHFV belongs to the genus *Nairovirus*, which is one of the five genera belonging to the family *Bunyaviridae* (Schmaljohn and Nichol 2007). There are 32 members in *Nairovirus* genus, transmitted by *Ixodid* ticks, and CCHFV is the most important and widespread virus amongst them (World Health Organization 2011). The CCHF virions are enveloped and spherical measuring about 80–120 nanometres (nm) in diameter, though pleomorphic forms are commonly observed.

The CCHFV genome is a tripartite, single-stranded RNA with 17,100–22,800 nucleotides. The three segments of the genome are small (S), medium (M) and large (L) (Fig. 11.1). The complementary base pairing in the ends of three segments forms three individual non-covalently closed circular stable RNAs. Complementary non-coding regions (NCRs) which serve as viral promoter regions are present at 5′ and 3′ ends in all the three segments; S, M and L are conserved within the genera. These NCRs have nine terminal nucleotides which are conserved between nairoviruses (5′-UCUCAAAGA and 3′-AGAGUUUCU). The complete NCR sequence is required for initiation of replication, transcription, encapsidation and packaging of virions. The complete NCR sequences differ between the viral gene segments and are necessary for binding of viral RNA-dependent-RNA polymerase (L protein) and initiate replication and transcription of viral genomes (Zivcec et al. 2016).

The S segment is approximately 1.6 kb in size, which is comparable to other bunyaviruses and has one ORF in a negative orientation that codes for the viral nucleoprotein (NP). This nucleoprotein encapsidates the viral RNA genome into ribonucleoprotein particles forming helical structures (Fig. 11.1). The second ORF in the S segment is in opposite orientation with respect to nucleoprotein ORF and codes for a non-structural protein (NS_S) indicating an ambisense nature of this segment (Barnwal et al. 2015). There is ambisense coding by S segment in other bunyaviruses too, but in CCHF it is by overlapping coding regions, whereas in other bunyaviruses it is separated by intergenic region which acts as a transcription termination signal (Albariño et al. 2007). The M segment is approximately 5.4 kb in length, which is considerably larger than other members of the *Bunyavirus* genera. It encodes for envelope glycoproteins G_N and G_C as well as non-structural protein (NS_m) (Bergeron et al. 2007). The entry of the virus into the cell is facilitated by the viral glycoproteins G_N and G_C (Fig. 11.1). These virally encoded membrane proteins are targeted by neutralising antibodies. The L segment encodes the 4000-amino acid-long, 448 kDa RNA-dependent RNA polymerase RdRp. This segment has a single ORF and is 12 kb long, which is larger than the L segment of other bunyaviruses (Fig. 11.1) (Honig et al. 2004).

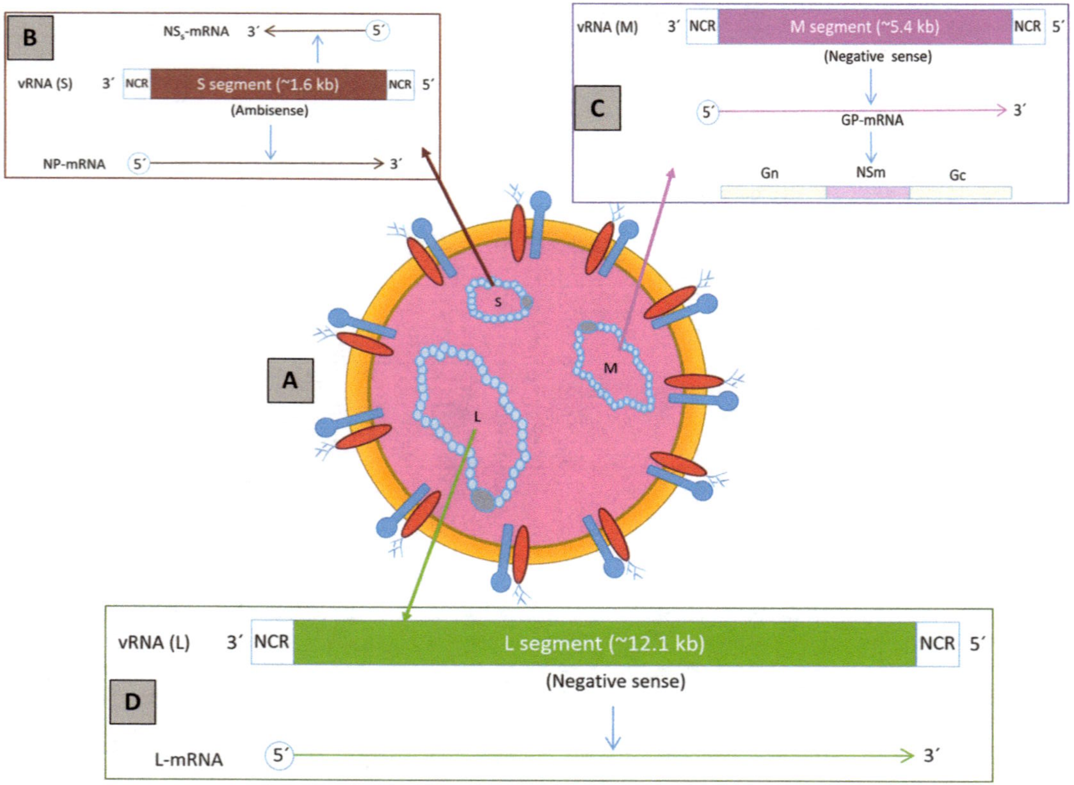

Fig. 11.1 Crimean-Congo haemorrhagic fever virion structure and genome organisation. (**a**) *Virion structure*: CCHFV has a spherical structure, surrounded by a lipid bilayer containing two viral glycoproteins Gn (37 kDa) and Gc (75 kDa). The entry of the virus into the cell is facilitated by the fusogenic activity of viral glycoproteins Gn and Gc through unknown receptors. Underneath is the core of the virus made of the three vRNA segments (S, M and L) that are encapsidated

11.3 Genetic Diversity

Like all other viruses with segmented genomes, CCHFV also has ample potential of segment reassortment (Hewson et al. 2004). Reassortments are more common in the M compared to the S and L segments. Large-scale genome sequence analysis has confirmed the occurrence of recombination in the S segment. However, the evidence of recombination in M and L segments is ambiguous (Lukashev 2005). Consequently, the CCHF viruses isolated from different regions of the world show considerable genetic diversity. Across the globe, the genetic variation in the nucleotide sequence of S, M and L segment is 20%, 31% and 22%, respectively. The variation at the amino acid levels of the N protein, GPC and L protein is 8%, 27% and 10%, respectively (Deyde et al. 2006). However, the overall lengths of RNA segments, ORF lengths and important motifs like the RNA polymerase core domains are well conserved. The rate of evolution of S, M and L segments has been estimated to be 1.09×10^{-4}, 1.52×10^{-4} and 0.58×10^{-4} substitutions/site/year (Carroll et al. 2010).

The S segment has been extensively sequenced and more than 62 full-length S segment sequences are available in NCBI database. The virus is grouped into seven clades based on the phylogenetic analysis of this segment. Clade I includes isolates from West Africa, clade II includes Central African isolates, Clade III has isolates of South Africa and West Africa, Clade IV includes the isolates from the Middle East and Asia, Clade V includes the isolates of Europe as well as Turkey and clade VI includes the isolates of Greece (Hewson et al. 2004). Clade IV is segregated into Asia 1 and Asia 2 subgroups with the Indian isolates forming a separate cluster under Asia 2 which has the isolates of Far-East Asia including China, Uzbekistan and Tajikistan. The Indian isolates share 98.5% nucleotide and 99.5% amino acid identity with the Tajikistan virus. The isolates from neighbouring Pakistan are clustered under Asia 1 subgroup (Yadav et al. 2013) (Fig. 11.2).

The analysis concerning the L segment sequence gives a phylogenetic tree similar to that of S segment. The genetic diversity regarding the M segment RNA is of immense importance. Though limited numbers of full-length M segment sequences are available, there are evident differences in the phylogenetic tree as compared to that of S segment. The Chinese strains 79121 and 7001 cluster into West Africa clade I and some South African strains SPU415/85 and SPU97/85 move into clade IV. An additional clade VII with Mauritania is detected only with the M segment.

Fig. 11.1 (continued) by the viral nucleoprotein (NP). Associated with each vRNP complex is the viral RNA-dependent RNA polymerase (RdRp, L protein). Together, vRNA, NP and RdRp form three non-covalently bonded circular genomic ribonucleoprotein complexes (RNP) of three vRNA segments in each virion particle. (**b**) *S segment*: It is a single-stranded, ambisense vRNA, of approximately 1.6 kb, having conserved 3′ and 5′ NCRs indicated as hatched boxes. In negative sense it codes for structural nucleoprotein whereas in positive sense codes for non-structural S protein (NS_S). (**c**) *M segment*: It is a single-stranded, negative-sense vRNA, of approximately 5.4 kb, having conserved 3′ and 5′ NCRs indicated as hatched boxes. It codes for single precursor glycoprotein, which is processed by proteolytic cleavage to form two mature viral glycoproteins Gn and Gc. (**d**) *L segment*: It is a single-stranded, negative-sense viral RNA, of approximately 12.1 kb, having conserved 3′ and 5′ NCRs indicated as hatched boxes. It codes for single large protein, viral RNA-dependent RNA polymerase (RdRp)

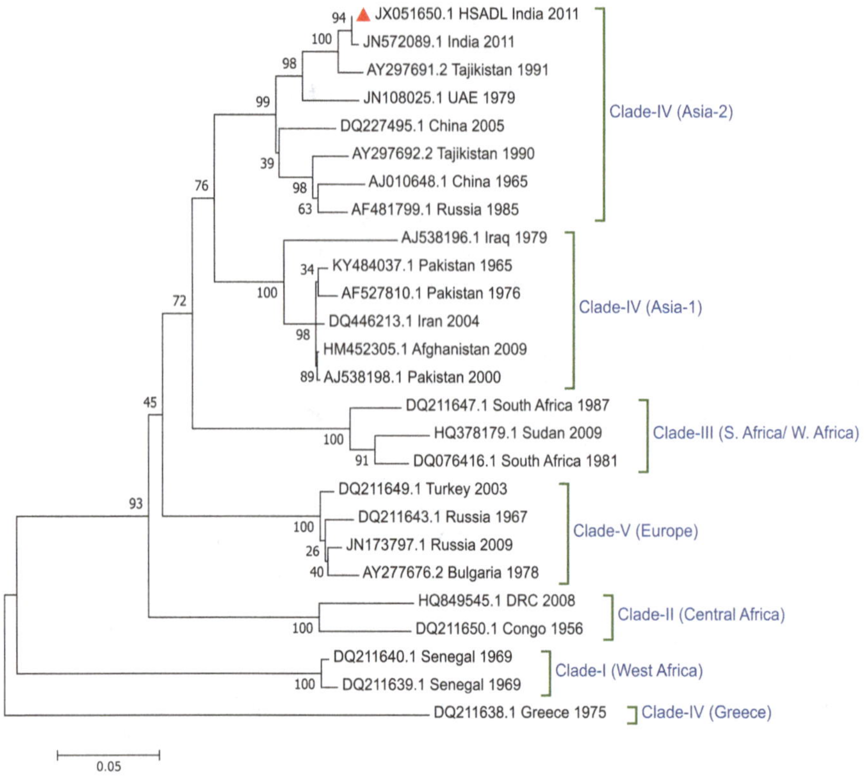

Fig. 11.2 Phylogenetic relationship of CCHF viruses with reference to viruses circulating in India based on full-length S segment sequences. Maximum-likelihood trees with 1000 bootstrap replicates were inferred using MEGA5. Phylogenetic tree constructed from 26 complete S segment nucleotides. Brackets indicate the virus genetic clades and their geographic distribution

There are also changes in clustering patterns within clades V and III. These studies are a clear indication that reassortments are more frequent in the M segment as compared to the S and L segments (Hewson et al. 2004; Lukashev 2005).

Concerning the timeline of CCHFV evolution, the genetic variability of CCHFV is suitable for analysis by the relaxed clock model rather than the more conventional constant molecular clock model. A large amount of sequence variation indicates that though CCHFV was identified in the 1940s it is an ancient virus. The most recent common ancestor (MRCA) of these viruses is estimated to have existed around 1599–1100 BC. The MRCA of the virus dates to 3198 years, 3560 years and 7358 years before present based on S, M and L segments, respectively. Amongst the lineages with multiple sequence representatives, clade IV is the farthest in time, and clade VI is one of the earliest offshoots of CCHFV (Carroll et al. 2010).

Overall the genetic diversity study when projected onto the geographic distribution of the virus reveals that related viruses are found in geographically apart territories like those found in South Africa and West Africa or Iraq and China. On the other hand, multiple genetic lineages are also found in the same geographic area as clade I and III viruses in West Africa. This has been explained with the movement of livestock by trade, movement of infected ticks by migratory birds or movement of infected migratory birds themselves. It is also suggested that probably CCHFV is an older virus than what it is known to be under modern virology (Chen 2013).

The kind of genetic diversity seen in case of CCHFV is unusual for any tick-borne virus because tick-borne viruses have to maintain fitness both in the arthropod vector and the vertebrate amplifying host; this is referred to as a double filter. However, in case of CCHFV, the double filter may not be effective because the vertebrate host is not very critical for the maintenance of the virus as the virus shows effective trans-ovarian and trans-stadial transmission in the ticks. The ability of the virus to maintain high fitness in the tick-vertebrate-tick cycle is supported by the presence of three M segment reassortants from a very wide region namely Iran, Tajikistan and Afghanistan throughout 33 years, i.e. from 1976 to 2006 (Chen 2013). Irrespective of the driving factors for the genetic diversity of CCHFV in nature, owing to its vast genetic diversity, the complete genome sequence including all three segments should be used for drawing any molecular epidemiological inference at least for the key strains. Inferences drawn with only a single segment sequence or incomplete genome sequence may lead to development as well as application of inefficient serodiagnostics and vaccines.

11.4 Epidemiology

The CCHF infection in animals is asymptomatic, but it causes severe hemorrhagic fever in humans. Viremia in livestock is transient and of low intensity but plays a crucial role in the life cycle of ticks, transmission, amplification and perpetuation of the virus and hence is important from an epidemiology point of view. The CCHFV covers the greatest geographic range of any tick-borne virus and virus isolation and/or disease has been reported from more than 45 countries corresponding with the south of the 50° parallel north (World Health Organization, Regional Office for South-East Asia 2014; Centers for Disease Control and Prevention 2018). The range of vertebrate and invertebrate animals involved in the viral-tick life cycle, numerous environmental factors playing a role in tick survival and abundance, spread of the virus to previously unaffected areas due to the legal or illegal animal trade or through bird migration, more than one route of transmission to humans and various educational, cultural, and behavioural parameters make the epidemiology of CCHF complex and unique (Papa et al. 2015). Based on sero-epidemiological studies in enzootic countries, various domestic animals could be asymptomatically infected with the virus. The enzootic tick-vertebrate-tick cycle of CCHFV maintenance mostly coincides with tick abundance and seroprevalence in animals, and thus animals heavily infested with ticks were more likely to be CCHFV seropositive (Adam et al. 2013; Ibrahim et al. 2015). As cattle, in general, are highly infested with *Hyalomma* ticks than those found on small ruminants, they are most sensitive indicators of low-level CCHFV circulation (Camicas et al. 1990). Detection of CCHFV antibodies in domestic animals has been important in identifying hotspots for the localisation of CCHFV foci and increased risk for human infection and initial evidence of circulating virus (Spengler et al. 2016).

11.4.1 India

India is surrounded by CCHF-affected countries and thus at high risk for CCHF. Because of the complex border with adjoining countries coupled with animal/human movement across the border, the risk of CCHFV being passed to the Indian subcontinent was evident. Even though the first confirmed case of the disease was reported in 2011 (Mishra et al. 2011), CCHFV could be demonstrated in retrospective human samples collected in February 2010 (Mourya et al. 2012). Since then regular sporadic CCHF outbreaks/cases transmitted either by ticks via livestock or nosocomial infections were reported from Gujarat, Rajasthan and Uttar Pradesh.

A couple of studies documented the presence of anti-CCHFV antibodies in different animal species, well before the first laboratory confirmed an outbreak of CCHF in 2010/2011. A study by Shanmugam et al. (1976) analysed 656 sera from goat (n = 186), sheep (n = 149), horse (n = 282), cow (n = 25), bulls (n = 12) and buffaloes (n = 02) by AGPT indicating percent seropositivity of 16.1, 1.1 and 0.7 in goat, horses and sheep, respectively. Rodrigues et al. (1986) also analysed sheep (n = 38), goat (n = 75), cattle (n = 66), buffalo (n = 23), horse (n = 16), donkey (n = 06), mule (n = 64) and camel (n = 03) sera from Jammu and Kashmir but did not find evidence of CCHFV activity. In recent past, livestock sera collected from abattoirs of Rajasthan (n = 34), Maharashtra (n = 132) and North West Bengal (n = 73) were screened for the presence of antibodies wherein buffalo (n = 1), goat (n = 3) and sheep (n = 6) from Rajasthan were positive (Mourya et al. 2012). Moreover, none of the 61 sera samples of 10 migratory bird species at the Keoladeo Ghana National Park Bharatpur, Rajasthan, was positive for anti-CCHFV antibodies (Ghosh et al. 1978).

After the first report, multiple serological studies documented the presence of anti-CCHFV antibodies in India. While screening for the CCHFV in ticks and animals in Gujarat, 66 of 305 serum samples of domestic animals were positive for anti-CCHFV antibodies (Mourya et al. 2012). In continuation, Mourya et al. (2014) tested a total of 1226 sera of bovines (n = 711), goat (n = 279) and sheep (n = 236) from 15 districts of Gujarat reporting 12.09%, 41.21% and 33.62% seropositivity in bovine, goat and sheep, respectively. The cross-sectional serosurvey spanning 22 states and 1 union territory comprising 5636 sera from bovine, goat and sheep revealed seropositivity of 5.43% and 10.99% in bovines and sheep/goats, respectively (Mourya et al. 2015).

11.4.2 World

Across the globe, many domestic animal species including buffalo, camels, cattle, sheep, goats, horses, pigs, dogs, chickens, ostriches and wild species have been investigated for the anti-CCHFV antibodies. However, cattle are the most often studied species, followed by sheep and goats. South Africa and Zimbabwe together have reported the largest sample size of cattle so far, comprising almost 9430 sera with percent positivity of 28 and 45, respectively (Swanepoel et al. 1987). Few countries have also reported high seroprevalence in animals with seropositivity of 79.1% in cattle (Afghanistan), 75.0% sheep (Afghanistan), 66.0% goats (Turkey), 58.8% horses (Iraq), 39.5% donkeys (Tajikistan) and 26% camels (Spengler et al. 2016).

Variation in abiotic factors such as season, increased mean temperature, normalised difference vegetation index (NDVI) and habitat fragmentation may also increase the incidence of CCHF in humans and CCHFV seroprevalence in animals. Vescio et al. (2012) found that warmer temperatures in the autumn before the case-reporting year had an increased probability of reporting zero CCHF cases.

About companion animals, studies are very limited and hence could not be interpreted. However, antibodies to CCHFV were reported in 5.9% of dogs (Shepherd et al. 1987). The association between human CCHF cases and ticks collected from livestock and dogs was also studied, wherein *R. evertsievertsi* ticks from sheep were CCHFV positive, but none of the *R. sanguineus* ticks collected from dogs was positive (Nabeth et al. 2004). However, additional data on companion animals and associated vector species will aid in more clearly evaluating the role of companion animals in the ecology of CCHFV.

11.4.3 Wild Animals

The consistent and high seroprevalence was observed in hares (3–22%), buffalo (10–20%) and rhinoceroses (40–68%). The seropositivity was also detected in several hedgehogs, viz. *Erinaceus europaeus* and *Hemiechinus auritus*, with an infestation of larval and nymphal *H. marginatum* ticks during the peak season of immature tick activity. Though the role of hedgehogs in enzootic maintenance depends on species, *H. auritus* could develop viremia upon experimental infection and might be a natural CCHFV reservoir while serving as a source of CCHFV for ticks. However, European hedgehog, *E. europaeus*, did not produce viremia after experimental infection maybe due to reduced susceptibility to infection or efficient viral clearance (Spengler et al. 2016). Few studies have also reported antibodies to CCHFV or related Nairoviruses in the mammalian order *Chiroptera*. Earlier Tkachenko et al. (1969) and Saidi et al. (1975) have documented evidence of anti-CCHFV antibodies in sera of two bat species in France and greater mouse-eared bat and the common noctule in Iran. Recently, 1135 sera from 16 different bat species collected from Congo, Gabon, Ghana, Germany and Panama were analysed using a CCHFV glycoprotein-based indirect immunofluorescence test (IIFT) and antibodies were identified in 10.0% of bat sera. The seropositivity of 3.6–42.9% in cave-dwelling bats and 0.6–7.1% in foliage-living bats suggests that they might be playing a role in the life cycle and geographic dispersal of CCHFV (Müller et al. 2016).

11.4.4 Birds

The birds infested with CCHFV-infected ticks are a major concern regarding the regional spread of the virus as many birds are important hosts for *Hyalomma* ticks and can efficiently transport ticks over long distances. Although an increase in CCHF cases was also correlated with increased rook populations, CCHFV infection and presence or absence of antibodies in avian species remain unclear. In spite of investigation of numerous species, the majority of serosurveys of wild avian species did not show any serological evidence of CCHFV infection. This absence of viremia is contrasting, as many species support CCHFV-infected ticks whereas CCHFV could replicate in red-billed hornbill (*T. erythrorhynchus*) without viremia and act as a source of infection to other ticks. Though domestic chickens were refractory to CCHFV infection, 0.2% CCHFV seroprevalence was reported in chickens and ducks. This absence of detectable anti-CCHFV antibodies in birds may be due to assay sensitivity as most of the serological surveys on birds were based on the AGDP test, and the AGDP test is less sensitive than the RPHI or IFA tests. Nevertheless, the absence of antibody response in birds is not general as ostriches appear to be an exception amongst birds in harbouring CCHFV and might be transmitting to humans (Spengler et al. 2016).

11.5 Transmission

The geographic distribution of CCHF concurs with *Ixodid* ticks, particularly those of the genus *Hyalomma*. Although CCHFV has been isolated from more than 30 tick species, few are competent as vector and the global distribution of the disease is solely associated with *Hyalomma* ticks (Estrada-Pena et al. 2007). *Hyalomma* ticks favour dry climates and arid types of vegetation coupled with abundant mammals supporting the life cycle. Given the wide distribution of the animals which act as a host for the *Hyalomma* ticks, and the favourable climatic and ecologic conditions across many countries, the incidences of CCHF may increase geographically over time.

The ticks transmit the virus to mammals and the resulting viremia in mammals is transient, but ticks remain infected throughout their lifetimes; they are therefore the true natural reservoirs. Because humans are not a source of infection for ticks, they are only accidental, dead-end hosts for the virus (Bente et al. 2013). In the enzootic cycle of transmission, variable seroprevalence is often associated with competent vector dispersion, their host preference and tick load on a particular animal species. With Hyalomma being the most competent vector, the endemic transmission is found only where these are present, and epizootic transmission occurs during their periods of increased abundance and activity (Spengler et al. 2016).

Humans acquire infection by the bite of or by crushing an infected tick against bare skin. The infection can also be acquired by percutaneous and permucosal routes, and contact with infected animal blood or tissues. Few studies have also reported possible aerosol and sexual transmission, but further research is warranted (Al-Abri et al. 2017). While investigating human CCHF cases, domestic and peri-domestic animal species are often implicated as a source of CCHFV. Several instances have recognised sheep as very important CCHFV reservoirs in certain enzootic regions and often epidemiologically linked to human cases. Livestock and abattoir workers and persons involved in backyard slaughtering in enzootic areas remain at an increased risk of infection. As such, increased CCHFV seropositivity in livestock often corresponds to the CCHF cases in humans with exposure to livestock, especially those who handle blood and organs from infected livestock (Spengler et al. 2016).

High mortality has also been observed in healthcare workers due to nosocomial infections arising as a result of contact with infected blood or body secretions from patients. The lack of rapid and sensitive early diagnostics poses the highest risk of nosocomial transmission to healthcare personnel, where transmission due to splash and needle-stick injuries has occurred in the absence of adequate personal protective equipment (Al-Abri et al. 2017).

11.6 Immunopathobiology

11.6.1 Incubation Period

The incubation period of the disease ranges from 3 days to 1 week, depending on the route of transmission and amount of the inoculum. The length is shorter for infections by tick bite or livestock contact compared to nosocomial infections (Vorou et al. 2007). Amount of virus and route of entry may affect the severity of the disease. Infections acquired through tick bites are comparatively less lethal than nosocomially transmitted infections; this could be due to the lower amount of inoculum through tick bites (Chinikar et al. 2010; Whitehouse 2004).

11.6.2 Clinical Signs in Animals

The disease in animals is asymptomatic. Replication of CCHF virus in experimentally inoculated sheep and cattle with a mild increase in body temperature and no evidence of clinical disease was reported. The viremic phase is relatively short, and antibodies are detectable shortly after the viremic phase. Antibody prevalence in adult livestock (cattle, sheep and goat) species in endemic regions can be more than 50%.

11.6.3 Clinical Signs in Humans

Initial symptoms are vague and usually characterised by sudden onset of fever, headache, myalgia, weakness, nausea and vomiting lasting for up to 3 days or more. Wide range of clinical symptoms from asymptomatic or mild infections to severe disease and death were reported (Bodur et al. 2012). Haemorrhagic manifestations develop 3–6 days after the onset of the illness in severe cases. Other signs like splenomegaly and hepatomegaly were observed in one-third of the infected patients. Leukopenia, thrombocytopenia, elevated liver enzymes and prolonged bleeding times are the main laboratory features of CCHF (Vorou et al. 2007; Cevik et al. 2008; Hatipoglu et al. 2010). Affected patients recover by 2–3 weeks after the onset of disease; symptoms like lethargy, dizziness and weakness may persist more than a year (Whitehouse 2004). The case fatality rate of CCHF ranges between 5% and 40%, depending on the geographic region and route of virus entry (Yilmaz et al. 2009). Multiple organ failure due to severe anaemia, dehydration and shock is considered as one of the important causes of death. The prognosis or outcome of the disease depends on the amount of the virus in the blood, platelet counts, liver enzyme levels, bleeding time, fibrinogen levels and gastrointestinal bleeding (Cevik et al. 2008; Hatipoglu et al. 2010).

11.6.4 Pathogenesis

CCHF pathogenesis is poorly understood due to the necessity of handling the virus in BSL-4 laboratories and lack of proper animal models. It involves virus entry, dissemination, endothelial damage and vascular leakage. The main contributors of pathogenesis are endothelial cells (ECs) and immune cells.

11.6.5 Entry of Virus

The primary barrier for viral entry is epithelium which may be overcome by a tick bite. After a tick bite, the virus attaches to the attachment proteins in the basolateral compartment of endothelial cells and virus is released into the bloodstream for systemic dissemination (Connolly-Andersen 2010). The entry of the virus into host cells is mediated by the interaction between envelope glycoprotein and cell-surface receptors probably nucleolin in humans (Xiao et al. 2011).

11.6.6 Dissemination

After virus entry through a tick bite or other routes, virus replication takes place in resident tissue macrophages and dendritic cells. Release into the blood facilitates the dissemination of the virus to regional lymph nodes, spleen and other organs leading to systemic spread of the disease (Connolly-Andersen 2010). The liver is considered as an important target organ; lack of basement membrane could help in easier viral entry into the hepatocytes (Connolly-Andersen 2010). Involvement of the brain is also reported at a later stage of the infection (Bente et al. 2010). Transmission of the virus to the cerebrospinal fluid (CSF) may be due to increased vascular permeability by cytokine release and subsequently disruption of the blood–CSF barrier (Kang and McGavern 2010).

11.6.7 Endothelial Cell (EC) Damage and Vascular Leakage

Endothelium appears to be a major target of CCHFV, which is indicated by pathological characteristics such as haemorrhage and increased vascular permeability (Bodur et al. 2010b; Burt et al. 1997; Ozturk et al. 2010). ECs play a very important role in initiating inflammatory response such as chemotaxis, leucocyte attachment, vascular permeability, diapedesis into inflamed areas and initiation of the innate and adaptive immune response to an infection via the release of chemical mediators (Connolly-Andersen et al. 2011). The intrinsic coagulation cascade is activated either by the virus directly or indirectly by virus-induced endothelial damage or through host-derived soluble mediators (Connolly-Andersen et al. 2011). Upregulation of soluble molecules like E-selectin, vascular cell adhesion molecule-1 (VCAM-1) and intercellular adhesion molecule-1 (ICAM-1) in CCHFV-infected ECs favours the increased adhesion of leukocytes in response to inflammatory mediators (Connolly-Andersen et al. 2011). These soluble adhesion molecules can be used as biomarkers to assess the endothelial activation, vascular damage and severity of disease (Bodur et al. 2010b; Ozturk et al. 2010). One of the main contributors of CCHF pathogenesis and disease progression is cytokine release. The proinflammatory cytokines, interleukin (IL) IL-1, IL-6, IL-8 and IL-10 and tumour necrosis factor-α (TNF-α) play an important role in mediating the pathogenesis and immune response of the disease. Although macrophages and dendritic cells are major sources of cytokines, ECs are also found to release molecules upon infection (Connolly-Andersen et al. 2011). Expression of ICAM-1 in ECs is in turn mediated by the release of cytokines and other soluble mediators from CCHFV-infected DCs (Schnittler and Feldmann, 2003; Connolly-Andersen et al. 2011). TNF-α was found to be a crucial cytokine in the activation of ECs (Connolly-Andersen 2010). Increased serum proinflammatory cytokines IL-6, IL-8 and IL-10 and TNF-α levels are implicated as important prognostic factors in CCHF patients (Bente et al. 2010; Saksida et al. 2010). In severe cases, immune dysregulation and excessive release of the cytokines (cytokinemia) lead to increased vascular permeability, vasodilatation, multiple organ failure and shock (Connolly-Andersen 2010). Thus, CCHF pathogenesis largely involves the events related to immune dysregulation and cytokine storm leading to a systemic vascular collapse. The positive correlation between the viral load, endothelial activation and cytokine release is frequently observed in CCHF-affected patients.

11.6.8 Impairment of Immune Response

The non-specific immune response, i.e. innate immune response, is the first defence against viruses. CCHFV paralyses the innate immune system and delays the specific immune response which targets the virus replication and clearance (Bente et al. 2010; Saksida et al. 2010; Peyrefitte et al. 2010). The CCHFV utilises different strategies to tackle the immune response. Interferons (IFNs) (IFN-α and -β) play an important role in limiting the spread of viral infection by their antiviral response. Activation of the innate immune system by the molecules produced during viral replication leads to the induction of interferon regulatory factor-3 (IRF-3) which in turn induces type I IFN secretion (Andersson et al. 2008). Delayed induction of interferons is one of the important strategies used by CCHFV to manipulate the host defence mechanism. INF-induced MxA protein expression has been reported to interact with nucleocapsid protein and reduces the replication (Andersson et al. 2008). The mechanism of paralysing IFN response seems to be human specific as CCHFV infection did not develop in immune-competent mice, but rapid development of the disease leading to death was observed in immunodeficient mice (Bente et al. 2010). The specific adaptive immune response is mainly elicited by the production of a specific antibody for the antigen. In CCHFV infection, antibody response is usually undetectable and sufficient antibody against the virus is an important indicator of survival of the infected patients. As in the case of other viral diseases, the reverse relationship is observed between specific antibody level and viral load (Saksida et al. 2010). It is speculated that insufficient antibody response in CCHFV infection is due to blockage of complete maturation of antigen-presenting cells, namely dendritic cells and macrophages (Bente et al. 2010; Peyrefitte et al. 2010). Lymphocytes and natural killer (NK) cells are an important part of both innate and adaptive immune response against viruses in the detection of infected cell and lysis. Depletion of immune cells like NK cells. T and B lymphocytes is a common feature reported in CCHFV-infected patients (Yilmaz et al. 2009). The studies on CCHF in a mouse model showed an increase in NK and T and B lymphocytes on the first day of infection due to increased production of cytokines in an early phase, but later the significant depletion and lymphopenia due to excessive apoptosis of lymphocytes (Bente et al. 2010; Geisbert et al. 2000). Thus CCHFV dysregulates both innate and adaptive immune response for its replication.

11.6.9 Haemophagocytosis

Hyper-activation of monocytes and macrophages leads to excessive phagocytosis, which results in severe cytopenia associated with fever, splenomegaly and haemophagocytosis in bone marrow, also called as a haemophagocytic syndrome. It is reported in 50% of the CCHFV-infected patients with severe bleeding (Tasdelen Fisgin et al. 2008). Hence, it is also considered as one of the mechanisms of pathogenesis in CCHFV infection in humans.

11.6.10 Impairment of Haemostasis and Coagulation

Extensive bleeding is one of the important signs of CCHF, which is effected primarily by impairment of haemostasis through thrombocytes and disruption of ECs and secondly by immune-mediated cell injury (Chen and Cosgriff 2000). Hence, disseminated intravascular coagulation (van Gorp et al. 1999), thrombocytopenia and prolonged prothrombin time (Cevik et al. 2008) are considered as important features in assessing the prognosis of the disease.

11.6.11 Histopathology

Histopathological lesions are mainly observed in the liver as a varying degree of hepatocellular necrosis in multiple foci, Kupffer cell hyperplasia, fatty change and periportal mononuclear infiltration. The necrotic areas are usually marked by haemorrhage and associated with the Councilman bodies in the liver (Burt et al. 1997). The damage in liver correlates with an elevation of liver enzymes, namely alanine aminotransferase (ALT) and aspartate aminotransferase (AST) (Cevik et al. 2008; Hatipoglu et al. 2010). In the spleen, lymphoid depletion accompanied by necrosis and karyorrhexis of lymphocytes is reported (Bente et al. 2010). Apoptosis in liver and spleen is frequently reported in severe CCHFV-infected patients (Rodrigues et al. 2012). Other lesions include alveolar damage; intra-alveolar and intestinal haemorrhage is also reported (Burt et al. 1997). CCHFV antigen is usually demonstrated in hepatocytes, Kupffer cells and endothelial cells of liver and spleen (Bente et al. 2010).

In summary, the pathogenesis of the disease is orchestrated by endothelial damage and immune response impairment (Fig. 11.3). The sequence of events that occur during the pathogenesis viz. increased the release of cytokines, activation of endothelial cells, and vasodilation leading to multiple organ failure and shock. Delayed type I interferon induction, inhibition of maturation of antigen-presenting cells and weak antibody response cause uncontrolled replication of the virus.

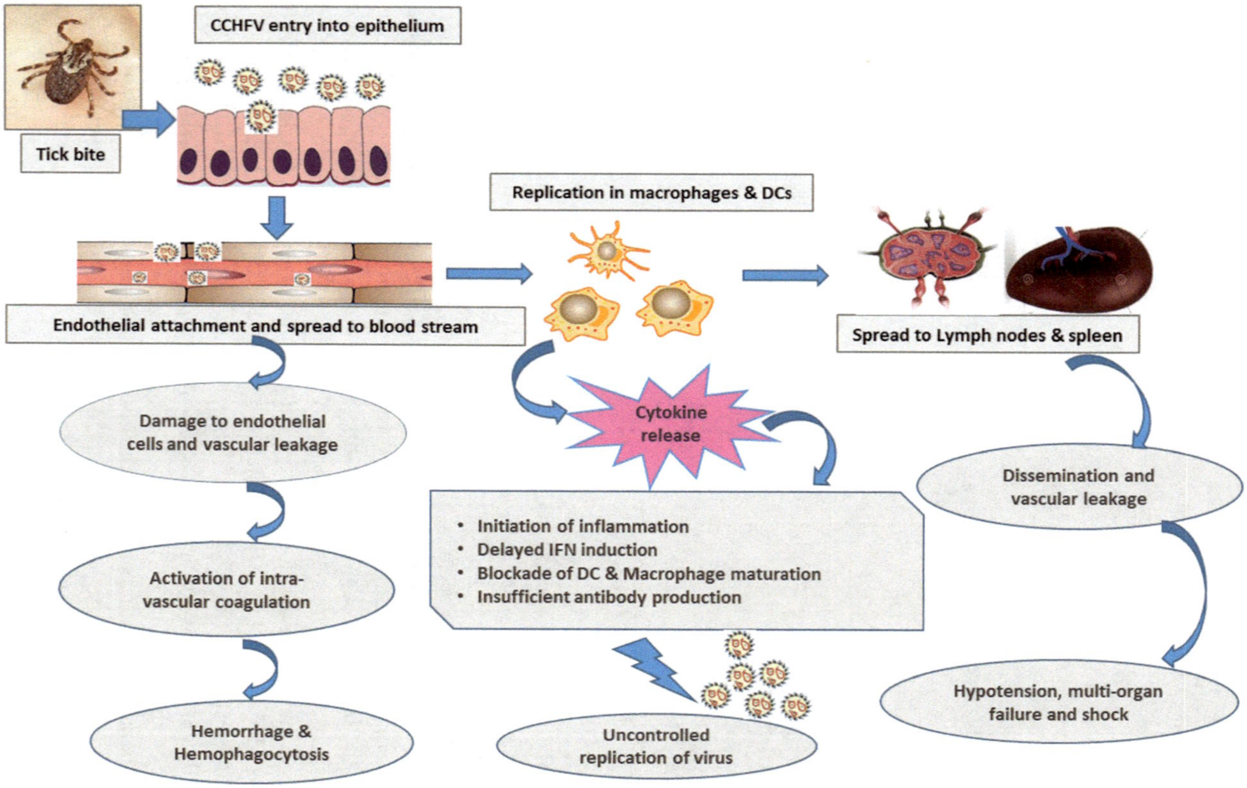

Fig. 11.3 Sequence of events during CCHFV pathogenesis

11.7 Diagnostics

The non-specific clinical symptoms in the early phase of CCHF are similar to other viral haemorrhagic fevers (VHFs); therefore, confirmatory diagnosis relies mainly on laboratory testing. CCHFV being BSL-4 agent, isolation of virus is limited to the laboratories with the highest level of biocontainment facilities. The samples can also be tested in non-BSL-4 laboratories using molecular tests only after the inactivation of the biological specimen.

The laboratory confirmation of the CCHFV is done by virus isolation, genome detection is done by RT-PCR and, RT-qPCR, virus detection is done by antigen capture ELISA and while antibody detection is done by ELISA. Mostly serum and blood plasma are collected as clinical materials for virus isolation and molecular detection. Preferably collection of blood is done in EDTA to increase the PCR efficiency (Drosten et al. 2003). Tissue, saliva and urine samples are also collected as clinical materials for molecular investigations. The infectivity of the virus remains for a few days in serum samples at ambient temperature, and is therefore considered as the primary material for investigation (Bodur et al. 2010a).

During viremia, isolation of virus is possible for up to 12 days after onset of infection, where chances are more in the first week (highest titre 1–6 days) (Shepherd et al. 1988). However, genome detection is possible for up to 18 days in serum samples. The viral RNA has been observed to increase with time in fatal cases, whereas it decreases in nonfatal cases (Saksida et al. 2010).

11.7.1 Virus Isolation

The virus isolation is considered as the gold standard method for CCHF diagnosis, but due to the requirement of BSL-4 containment facility, sensitivity and risk involved make it a poor choice for routine diagnosis. Mostly, clinical materials, including blood and organ suspensions, are inoculated intra-cerebrally in suckling mice or cell culture (Vero, LLCMK2, SW-13, CER and BHK-21 cells). CCHFV does not produce any definite cytopathic effect. Hence, confirmation is done by immunofluorescence test or molecular tests. The virus can be isolated in 3.3 days (mean) in cell culture comparison to mean 7.7 days in mice. The inoculation in suckling mice is a more sensitive method than the cell culture (Shepherd et al. 1986).

11.7.2 Antigen Detection

The antigen capture ELISA is an alternative to the virus isolation for the detection of CCHV antigen. The assay provides results within a few hours and does not require containment setup if pre-inactivated samples are used. However, its sensitivity is comparatively lower than the virus isolation and RT-PCR test (Shepherd et al. 1988; Burt 2011). The monoclonal antibodies developed against the nucleoprotein (NP) gene are taken as a capture antibody for the ELISA. The antigen capture ELISA is a rapid method and an alternative for the detection of illness in acute infection. The fatal cases of patient with higher viremia can easily be detected by this test. The test is unable to detect antibody-positive samples (nonfatal cases) which is one of its limitations (Saijo et al. 2005).

11.7.3 Molecular Diagnostics

Over the conventional tests, the detection of CCHFV by molecular tests, including RT-PCR and real-time reverse transcriptase PCR (RT-qPCR), provides the rapidity, specificity and sensitivity without the requirement of BSL-4 biocontainment facility. The well-timed availability of test results within a few hours also helps in taking the immediate control measures and controlling the infection. However, the molecular diagnosis is dependent on the status of the sample including the collection time, material/organ collected, transportation, cold chain maintenance, etc. The circulation of many different strains of CCHFV worldwide with the variable sequence is one of the major challenges for PCR-based diagnosis. The fact was also supported by an international quality assessment where all the 44 participating laboratories were found to be suboptimal in the molecular detection of CCHFV highlighting the need for improvement (Escadafal et al. 2012).

In 1990, the first reported conventional nested PCR assay was based on visualisation of results either on the agarose gel or by southern blotting. The assay was prone to cross-contamination and time taking. However, the nested PCR developed using two sets of primer increased the sensitivity of the assay (Burt et al. 1998).

The development of RT-qPCR assays offered increased sensitivity and specificity, rapidity and quantification with minimal to no cross-contamination (Table 11.1). The oligonucleotide primers and probes used were targeted to conserved S segment of CCHFV to increase the diagnostic scope of the assays for detection of all the circulating CCHFV strains. Amongst the many RT-qPCR-based chemistries, SYBR Green, TaqMan and Molecular Beacon-based RT-qPCR were developed targeting the NP gene of CCHFV. The first SYBR Green-based assay reported the sensitivity of 2779 virus genome equivalents per millilitre of plasma (Drosten et al. 2003). The oligo for the first TaqMan RT-qPCR assay was designed using the worldwide distributed 19 strains of CCHFV (Yapar et al. 2005). Further, qPCR assay specific for the detection of strains of the Balkan region was designed (Duh et al. 2006). Later on, TaqMan-minor groove-binding protein-based qPCR assay with more specific probe was developed (Garrison et al. 2007). Multiple probe-based qPCR assays were also developed to detect any of the strains of CCHFV (Wolfel et al. 2007). Simple probe technology under the real-time PCR format was used for detection

Table 11.1 Real-time PCR assays for detection of CCHFV

Sr No.	Primer sequences	Probe	Chemistry	Reference
1	CCS-ATGCAGGAACCATTAARTCTTGGGA CCAs-CTAATCATAATCTGACAACATTTC plus CTAATCATGTCTGACAGCATCTC	—	SYBR Green	Drosten et al. (2002)
2	CCHF L1-GCTTGGGTCAGCTCTACTGG CCHF D1-TGCATTGACACGGAAACCTA	CCHF S1 FAM-AGAAGGGGCTTGAGTGGTT-DABCYL	TaqMan	Duh et al. (2006)
3	Forward-CAAGGGGTACCAAGAAAATGAAGAAGGC Reverse-GCCACAGGGATTGTTCCAAAGCAGAC	FAM-ATCTACATGCACCCTGCTGTGTTGACA-TMARA	TaqMan	Wolfel et al. (2007)
4	Forward-GGAGTGGTGCAGGGAATTTG Reverse- CAGGGCGGGTTGAAAGC	6FAM-CAAGGCAAGGTACATCAT-MGBNFQ	TaqMan	Garrison et al. (2007)
6	Forward-AGTGTTCTCTTGAGTGCTA Reverse- CCACAAGTCCATTTCCTT	6-FAM-CGCGATCATCTCATCTTTGTTGTTCACCTCGATC GCG-BHQ-1	Molecular Beacon	Kamboj et al. (2014)

and genotyping of southern African isolates of CCHFV (Kondiah et al. 2010). Since most of the assays failed to detect the highly divergent AP92 genotype of CCHFV, Atkinson et al. 2012 demonstrated a novel single-probe-based RT-qPCR assay capable of detecting all known strains of CCHFV, including AP92. Also, a novel molecular beacon-based real-time RT-PCR assay was designed based on conserved N gene sequences of CCHFV strains circulating in India for specific detection of Indian CCHFV strains (Fig. 11.4, adapted from Kamboj et al. 2014). The molecular beacon-based assay was comparatively tenfold more sensitive to the previously introduced TaqMan assays. This assay is proposed as a cost-effective, rapid, specific and sensitive alternative test for the diagnosis of CCHFV infection (Kamboj et al. 2014) (Table 11.1).

For surveillance and primary genotyping purpose, one of the WHO reference laboratories successfully developed a low-density microarray using 20 different CCHF virus-specific capture probes and validated over confirmed clinical cases of CCHF from past 20 years (Wölfel et al. 2009). A high-throughput DNA sequencing microarray has been developed for simultaneous detection and differential diagnosis of various VHF viruses including CCHFV (Filippone et al. 2013). For field use, a non-PCR-based genome detection method was also developed using circle-to-circle amplification (C2CA) technology for easy and sensitive diagnosis of CCHFV especially for developing countries (Ke et al. 2011).

11.8 Prevention and Control

Currently, there are no licensed vaccines or approved therapeutics for CCHF. This has been primarily due to non-availability of suitable animal models for years, BSL-4 requirements and a relatively small research community focusing on CCHF. However, progress made in the area of prevention and control of CCHF is discussed here.

11.8.1 Supportive Treatment

The most important thing in the effective management of confirmed CCHF patients is timely administration of supportive treatment. Close monitoring of hospitalised patients using various laboratory tests such as white blood cell count, renal function tests, serum electrolytes and transaminases, and blood clotting parameters is essential. Supportive therapies in the form of transfusion of blood, plasma or platelet are effective when started as soon as possible. In the case of CCHF patients with refractory thrombocytopenia, intravenous immunoglobulin is advised.

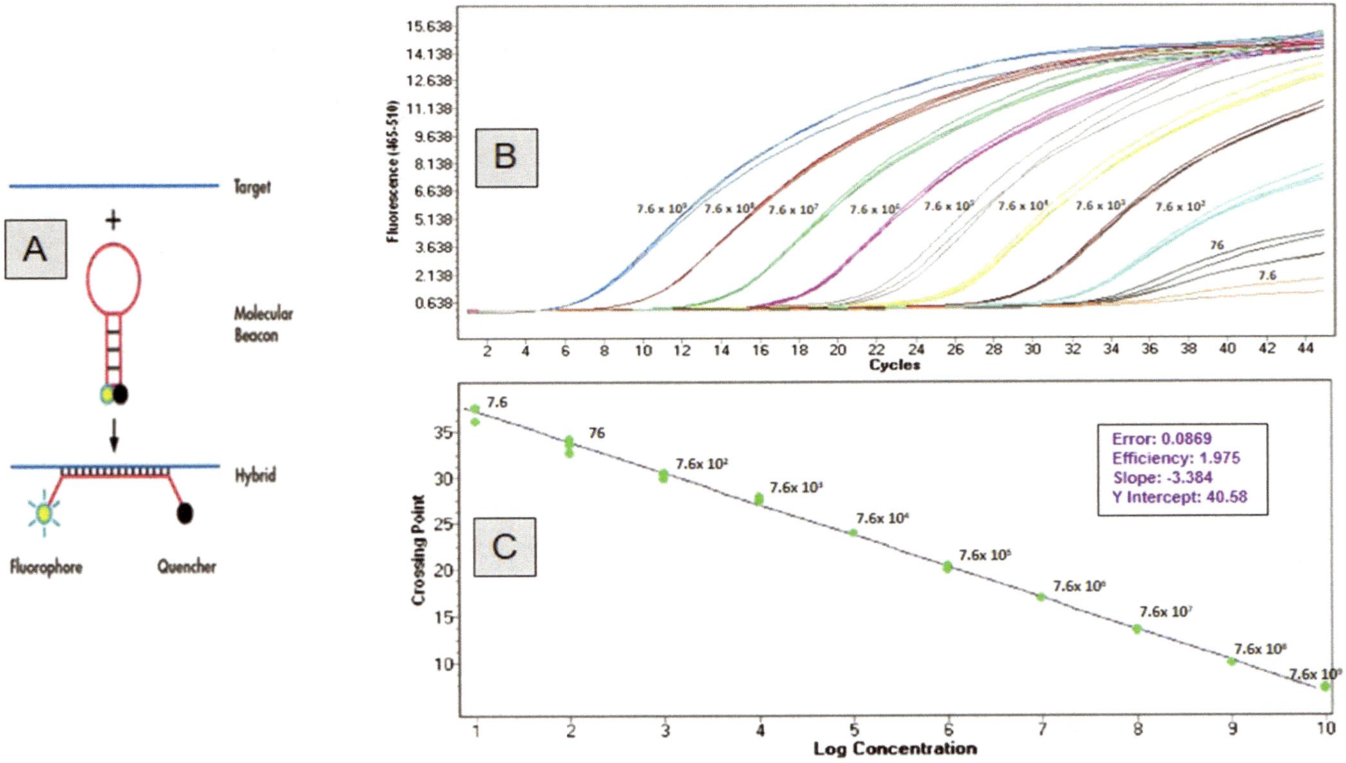

Fig. 11.4 Molecular beacon-based RT-qPCR test for rapid diagnosis of CCHF. (**a**) Schematic representation of molecular beacon chemistry. (**b**) Amplification curves and (**c**) standard curve for molecular beacon probe-based RT-qPCR test using tenfold serial dilutions of CCHFV S segment IVT-RNA. The assay was linear from 7.6×10^9 to 7.6 copies of IVT-RNA. Sensitivity RT-qPCR test is 7.6–76 copies (adapted from Kamboj et al. 2014)

11.8.2 Antiviral Therapy

There is no approved specific antiviral drug available for the treatment of CCHF. However, ribavirin is the only drug that is being used for the treatment of CCHF. It is a broad-spectrum synthetic purine nucleoside analogue whose exact mechanism of action is not fully understood. However, ribavirin has shown to inhibit CCHFV replication both in vivo and in vitro in a concentration-dependent manner. In cases of post-exposure prophylaxis, oral ribavirin has sometimes been used; however its efficacy still remains uncertain. Of note, meta-analysis of these studies suggested that ribavirin is not effective in treated patients as it did not improve the survival rate and shorten the hospitalisation compared to untreated patients (Ascioglu et al. 2011). The combination therapy (ribavirin and corticosteroids) might be beneficial especially when given to patients in early stage of CCHFV infection (Jabbari et al. 2006). Recently, Favipiravir (T-705), a new drug licensed for the treatment of influenza, has shown to have comparatively better efficacy in vivo than that of the current standard treatment of CCHF using ribavirin (Oestereich et al. 2014). Besides, specific immunoglobulin therapy has also been used, particularly in Bulgaria, wherein plasma of convalescent CCHF patients was given intramuscularly for prophylactic and therapeutic purposes (Christova et al. 2009). In one study performed in Turkey, CCHFV hyper-immunoglobulin derived from healthy donors was effective as a therapeutic approach, particularly for patients carrying high virus load (Kubar et al. 2011). Moreover, to decide and adopt effective antiviral therapy for CCHF patients, there is an urgent need for the large placebo-controlled, prospective randomised controlled trials.

11.8.3 Vaccines

The quest for CCHF vaccine began around 1960s, and the first vaccine (classic Russian/Bulgarian vaccine) was developed from the brain tissue of CCHFV-infected sucking mice in 1970 (Papa et al. 2011). Though this vaccine was approved by the Soviet Ministry of Health in 1970 and Bulgaria in 1974, it did not gain widespread approval from the international regulatory authority because of poor immune response and possibility of autoimmune responses induced by myelin basic protein (Hemachudha et al. 1987; Mousavi-Jazi et al. 2012). For decades, the progress towards the development of vaccines was severely hampered due to the lack of a suitable animal model. Recent approaches for CCHF vaccine candidate development include a DNA-based vaccine carrying the glycoprotein-encoding region of the CCHFV. This DNA vaccine induced neutralising antibodies in approximately 50% of vaccinated mice (Spik et al. 2006). Another vaccine candidate utilised transgenic tobacco leaves expressing the CCHF viral glycoproteins that induced both IgG and IgA in mice (Ghiasi et al. 2011). However, protection against lethal disease was not tested for these two vaccines' approach due to non-availability of a suitable laboratory animal model. In the year 2010, two mouse models, namely A129 (deficient in type I interferon receptor) and STAT-1 knockout mouse, have made it possible to study pathogenesis and test vaccine efficacy through a lethal challenge. Afterwards, a vaccine based on human serotype 5 adenoviruses (AdHu5) vector carrying the entire M segment of CCHFV was developed that induced both humoral and cellular immune response but failed to confer protection in STAT-1 knockout mouse model (Sahib 2010). Recently, adjuvanted subunit vaccine carrying insect cells expressing Gn or Gc ectodomains was shown to induce neutralising antibodies, but this vaccine failed to protect STAT-1 knockout mice in a subsequent lethal virus challenge (Kortekaas et al. 2015). The possible reason for the failure of this vaccine could be non-matching glycosylation in insect cells compared to mammalian cells. Another vaccine based on transcriptionally competent virus-like particles has also been developed, but it could protect only 40% of A129 mice despite the induction of strong neutralising antibody titres (Hinkula et al. 2017). Till date, the most promising CCHF vaccine candidate is a Modified Vaccinia Ankara (MVA) vector expressing the glycoprotein-encoding M segment ORF (MVA-GP). This vaccine induced both humoral and cellular immunity, and also provided 100% protection against CCHF virus infection in mice model (Buttigieg et al. 2014). Further, vaccine using the same MVA vector but carrying NP protein fails to confer protection against the lethal challenge (Dowall et al. 2016). Nevertheless, none of the vaccines has reached human clinical trials so far.

11.8.4 Modes of Prevention

Effective methods to prevent CCHFV infection can be divided into two broad categories: (a) controlling CCHF in animals and ticks and (b) reducing the risk of infection in human. The tick-animal-tick cycle usually goes undetected, and infection in domestic animals produces non-apparent symptoms. Additionally, the tick vectors are widespread, and the vaccine intended to use in animals is not available. Therefore, the use of acaricides to control ticks is a single effective way to control the infection in animals.

In the case of non-availability of approved vaccine for use in human, the most important factor is creating awareness and educating people, especially animal handlers, slaughterhouse workers, agriculture farmers and healthcare workers, about the measures to reduce the exposure to the virus. In general, to minimise tick exposure, use of approved tick repellent, use of light-coloured protective clothing and avoiding visiting the endemic areas are recommended. The healthcare workers are prone to frequent exposure to infected blood or tissue from CCHF-infected patients, and therefore it is recommended to wear gloves, gowns and face masks and avoid close physical contact with CCHF-infected people to reduce the risk of exposure. Besides, proper measures to prevent illegal transportation of animals across borders should be in place to limit the geographical extension of CCHF.

CCHFV infection occurrences are reported from a wide geographic area, and this is due to the widespread presence of the tick vectors. Owing to the impact of climate change, tick populations are anticipated to flourish and thus could facilitate the spread and occurrence of CCHFV. Intensive urban and suburban animal farming close to human populations along with suboptimal management practices facilitates the animal-to-human transmission by tick bites and exposure to infected animal blood and meat. One health approach including surveillance of animals, with prevention and control of tick infestation in livestock along with appropriate prophylactic measures in high-risk human populations, can effectively control CCHFV infections.

Acknowledgements All the authors of the manuscript thank and acknowledge their respective universities and institutes.

Conflict of interest: There is no conflict of interest.

References

Adam I, Mahmoud MA, Aradaib IE (2013) A seroepidemiological survey of Crimean Congo hemorrhagic fever among cattle in North Kordofan State, Sudan. Virol J 10:178

Al-Abri SS, Abaidani A, Fazlalipour M, Mostafavi E, Leblebicioglu H, Pshenichnayaf N, Memish ZA, Hewson R, Peterseni E, Mala P, Nguyenj TMN, Malik MR, Formenty P, Jeffries R (2017) Current status of Crimean-Congo haemorrhagic fever in the World Health Organization Eastern Mediterranean Region: issues, challenges, and future directions. Intl J Infect Dis 58:82–89

Albariño CG, Bird BH, Nichol ST (2007) A shared transcription termination signal on negative and ambisense RNA genome segments of Rift Valley fever, sandfly fever Sicilian, and Toscana viruses. J Virol 81:5246–5256

Andersson I, Karlberg H, Mousavi-Jazi M, Martı'nez-Sobrido L et al (2008) Crimean-Congo hemorrhagic fever virus delays activation of the innate immune response. J Med Virol 80(8):1397–1404

Ascioglu S, Leblebicioglu H, Vahaboglu H, Chan KA (2011) Ribavirin for patients with Crimean-Congo haemorrhagic fever: a systematic review and meta-analysis. J Antimicrob Chemother 66:1215–1222

Atkinson B, Chamberlain J, Logue CH, Cook N, Bruce C, Dowall SD, Hewson R (2012) Development of a real-time RT-PCR assay for the detection of Crimean-Congo hemorrhagic fever virus. Vector Borne Zoonotic Dis 12:786–793

Barnwal B, Karlberg H, Mirazimi A, Tan YJ (2015) Non-structural protein of Crimean-Congo hemorrhagic fever virus disrupts mitochondrial membrane potential and induces apoptosis. J Biol Chem 291:582–592

Bente DA, Alimonti JB, Shieh WJ, Camus G et al (2010) Pathogenesis and immune response of Crimean-Congo hemorrhagic fever virus in a STAT-1 knockout mouse model. J Virol 84:11089–11100

Bente DA, Forrester NL, Watts DM, McAuley AJ, Whitehouse CA, Bray M (2013) Crimean-Congo hemorrhagic fever: history, epidemiology, pathogenesis, clinical syndrome and genetic diversity. Antivir Res 100(1):59–189

Bergeron E, Vincent MJ, Nichol ST (2007) Crimean–Congo hemorrhagic fever virus glycoprotein processing by the endoprotease SKI-1/S1P is critical for virus infectivity. J Virol 81(23):13271–13276

Bodur H, Akinci E, Ascioglu S, Ongürü P, Uyar Y (2012) Subclinical infections with Crimean-Congo hemorrhagic fever virus, Turkey. Emerg Infect Dis 18:640–642
Bodur H, Akinci E, Ongürü P, Carhan A, Uyar Y, Tanrici A, Cataloluk O, Kubar A (2010a) Detection of Crimean–Congo hemorrhagic fever virus genome in saliva and urine. Int J Infect Dis 14:247–249
Bodur H, Akinci E, Ongürü P et al (2010b) Evidence of vascular endothelial damage in Crimean-Congo hemorrhagic fever. Int J Infect Dis 14:e704–e707
Bronze MS, Huycke MM, Machado LJ, Voskuhl GW, Greenfield RA (2002) Viral agents as biological weapons and agents of bioterrorism. Am J Med Sci 323(6):316–325
Burt FJ (2011) Laboratory diagnosis of Crimean–Congo hemorrhagic fever virus infections. Future Virol 6:831–841. Review on laboratory diagnosis of Crimean-Congo hemorrhagic fever virus
Burt FJ, Leman PA, Smith JF, Swanepoel R (1998) The use of a reverse transcription-polymerase chain reaction for the detection of viral nucleic acid in the diagnosis of Crimean–Congo haemorrhagic fever. J Virol Methods 70:129–137
Burt FJ, Swanepoel R, Shieh WJ, Smith JF et al (1997) Immunohistochemical and in situ localization of Crimean-Congo hemorrhagic fever (CCHF) virus in human tissues and implications for CCHF pathogenesis. Arch Pathol Lab Med 121:839–846
Buttigieg KR, Dowall SD, Findlay-Wilson S, Miloszewska A, Rayner E, Hewson R et al (2014) A novel vaccine against Crimean-Congo haemorrhagic fever protects 100% of animals against lethal challenge in a mouse model. PLoS One 9:e91516
Camicas JL, Wilson ML, Cornet J, Digoutte J, Calvo M, Adam F, Gonzalez JP (1990) Ecology of ticks as potential vectors of Crimean-Congo hemorrhagic fever virus in Senegal: epidemiological implications. Arch Virol 1:303–322
Carroll SA, Bird BH, Rollin PE, Nichol ST (2010) Ancient common ancestry of Crimean-Congo hemorrhagic fever virus. Mol Phylogenet Evol 55:1103–1110
Casals J (1969) Antigenic similarity between the virus causing Crimean hemorrhagic fever and Congo virus. Proc Soc Exp Biol Med 131(1):233–236
Centers for Disease Control and Prevention 2018, Crimean-Congo Hemorrhagic Fever (CCHF) distribution map. https://www.cdc.gov/vhf/crimean-congo/outbreaks/distribution-map.html. Accessed 16 Mar 2018
Cevik MA, Erbay A, Bodur H, Gülderen E et al (2008) Clinical and laboratory features of Crimean-Congo hemorrhagic fever: predictors of fatality. Int J Infect Dis 12:374–379
Chen JP, Cosgriff TM (2000) Hemorrhagic fever virus-induced changes in hemostasis and vascular biology. Blood Coagul Fibrinolysis 11:461–483
Chen S (2013) Molecular evolution of Crimean-Congo hemorrhagic fever virus based on complete genomes. J Gen Virol 94:843–850
Chinikar S, Ghiasi SM, Moradi M, Goya MM et al (2010) Phylogenetic analysis in a recent controlled outbreak of Crimean-Congo haemorrhagic fever in the south of Iran, December 2008. Euro Surveill 15(47):19720
Christova I, Di Caro A, Papa A, Castilletti C, Andonova L, Kalvatchev N et al (2009) Crimean-Congo hemorrhagic fever, southwestern Bulgaria. Emerg Infect Dis 15:983–985
Connolly-Andersen AM (2010) Pathogenesis of an emerging pathogen Crimean-Congo hemorrhagic fever. Thesis, Stockholm
Connolly-Andersen AM, Moll G, Andersson C, Akerström S et al (2011) Crimean-Congo hemorrhagic fever virus activates endothelial cells. J Virol 85:7766–7774
Deyde VM, Khristova ML, Rollin PE, Ksiazek TG, Nichol ST (2006) Crimean-Congo hemorrhagic fever virus genomics and global diversity. J Virol 80:8834–8842
Dowall SD, Buttigieg KR, Findlay-Wilson SJ, Rayner E, Pearson G, Miloszewska A et al (2016) A Crimean-Congo hemorrhagic fever (CCHF) viral vaccine expressing nucleoprotein is immunogenic but fails to confer protection against lethal disease. Hum Vaccin Immunother 12(2):519–517
Drosten C, Göttig S, Schilling S, Asper M, Panning M, Schmitz H, Günther S (2002) Rapid detection and quantification of RNA of Ebola and Marburg viruses, Lassa virus, Crimean-Congo

hemorrhagic fever virus, Rift Valley fever virus, dengue virus, and yellow fever virus by real-time reverse transcription-PCR. J Clin Microbiol 40:2323–2330. The first one-step real-time-RT-PCR for detection of Crimean-Congo hemorrhagic fever virus

Drosten C, Kummerer BM, Schmitz H, Gunther S (2003) Molecular diagnostics of viral haemorrhagic fevers. Antivir Res 57:61–87

Duh D, Saksida A, Petrovec M, Dedushaj I, Avsic-Zupanc T (2006) Novel one-step real-time RT-PCR assay for rapid and specific diagnosis of Crimean–Congo hemorrhagic fever encountered in the Balkans. J Virol Methods 133:175–179

Escadafal C, Olschläger S, Avšič-Županc T, Papa A, Vanhomwegen J, Wolfel R, Mirazimi A, Teichmann A, Donoso-Mantke O, Niedrig M (2012) First international external quality assessment of molecular detection of Crimean–Congo hemorrhagic fever virus. PLoS Negl Trop Dis 6:1706–1714

Estrada-Pena A, Zatansever Z, Gargili A, Aktas M, Uzun R, Ergonul O, Jongejan F (2007) Modeling the spatial distribution of Crimean-Congo hemorrhagic fever outbreaks in Turkey. Vector Borne Zoonotic Dis 7(4):667–678

Filippone C, Marianneau P, Murri S, Mollard N, Avsic-Zupanc T, Chinikar S, Desprès P, Caro V, Gessain A, Berthet N, Tordo N (2013) Molecular diagnostic and genetic characterization of highly pathogenic viruses: application during Crimean-Congo haemorrhagic fever virus outbreaks in Eastern Europe and the Middle East. Clin Microbiol Infect 19:118–128

Garrison AR, Alakbarova S, Kulesh DA, Shezmukhamedova D, Khodjaev S, Endy TP, Paragas J (2007) Development of a TaqMan minor groove binding protein assay for the detection and quantification of Crimean–Congo hemorrhagic fever virus. Am J Trop Med Hyg 77:514–520

Geisbert TW, Hensley LE, Gibb TR, Steele KE et al (2000) Apoptosis induced in vitro and in vivo during infection by Ebola and Marburg viruses. Lab Investig 80:171–186

Ghiasi SM, Salmanian AH, Chinikar S, Zakeri S (2011) Mice orally immunized with a transgenic plant expressing the glycoprotein of Crimean-Congo hemorrhagic fever virus. Clin Vaccine Immunol 18(12):2031–2037

Ghosh SN, Sokhey J, Dandawate CN, Gupta NP, Obukhova VR, Gaidamovich SY (1978) Arthropod-borne virus activity in migratory birds, Ghana Bird Sanctuary, Rajasthan, state, 1973. Indian J Med Res 68:192–196

van Gorp EC, Suharti C, ten Cate H, Dolmans WM et al (1999) Review: infectious diseases and coagulation disorders. J Infect Dis 180:176–186

Hatipoglu CA, Bulut C, Yetkin MA, Ertem GT et al (2010) Evaluation of clinical and laboratory predictors of fatality in patients with Crimean-Congo haemorrhagic fever in a tertiary care hospital in Turkey. Scand J Infect Dis 42:516–521

Hemachudha T, Griffin DE, Giffels JJ, Johnson RT, Moser AB, Phanuphak P (1987) Myelin basic protein as an encephalitogen in encephalomyelitis and polyneuritis following rabies vaccination. N Engl J Med 316:369–374

Hewson R, Gmyl A, Gmyl L, Smirnova SE, Karganova G, Jamil B, Hasan R, Chamberlain J, Clegg C (2004) Evidence of segment reassortment in Crimean-Congo haemorrhagic fever virus. J Gen Virol 85:3059–3070

Hinkula J, Devignot S, Akerstrom S, Karlberg H, Wattrang E, Bereczky S et al (2017) Immunization with DNA plasmids coding for Crimean-Congo hemorrhagic fever virus capsid and envelope proteins and/or virus-like particles induces protection and survival in challenged mice. J Virol 91(10):e02076–6

Honig JE, Osborne JC, Nichol ST (2004) Crimean-Congo hemorrhagic fever virus genome L RNA segment and encoded protein. Virology 321:29–35

Hoogstraal H (1979) The epidemiology of tick borne Crimean-Congo hemorrhagic fever in Asia, Europe and Africa. J Med Entomol 15:307–417

Ibrahim AM, Adam I, Osman BT, Aradaib IE (2015) Epidemiological survey of Crimean Congo hemorrhagic fever virus in cattle in East Darfur State, Sudan. Ticks Tick Borne Dis 6(4):439–444

Jabbari A, Besharat S, Abbasi A, Moradi A, Kalavi K (2006) Crimean-Congo hemorrhagic fever: case series from a medical center in Golestan province, Northeast of Iran (2004). Indian J Med Sci 60:327–329

Kamboj A, Pateriya AK, Mishra A, Ranaware P, Kulkarni DD, Raut AA (2014) Novel molecular beacon probe-based real-time RT-PCR assay for diagnosis of Crimean-Congo hemorrhagic fever encountered in India. Biomed Res Int 2014:496219

Kang SS, McGavern DB (2010) Microbial induction of vascular pathology in the CNS. J Neuroimmune Pharmacol 5:370–386

Ke R, Zorzet A, Göransson J, Lindegren G, Sharifi-Mood B, Chinikar S, Mardani M, Mirazimi A, Nilsson M (2011) Colorimetric nucleic acid testing assay for RNA virus detection based on circle-to-circle amplification of padlock probes. J Clin Microbiol 49:4279–4285

Kondiah K, Swanepoel R, Paweska JT, Burt FJ (2010) A simple-probe real-time PCR assay for genotyping reassorted and non-reassorted isolates of Crimean–Congo hemorrhagic fever virus in southern Africa. J Virol Methods 169:34–38

Kortekaas J, Vloet RP, McAuley AJ, Shen X, Bosch BJ, de Vries L et al (2015) Crimean-Congo hemorrhagic fever virus subunit vaccines induce high levels of neutralizing antibodies but no protection in STAT1 knockout mice. Vector Borne Zoonotic Dis 15:759–754

Kubar A, Haciomeroglu M, Ozkul A, Bagriacik U, Akinci E, Sener K et al (2011) Prompt administration of Crimean-Congo hemorrhagic fever (CCHF) virus hyperimmunoglobulin in patients diagnosed with CCHF and viral load monitorization by reverse transcriptase-PCR. Jpn J Infect Dis 64:439–433

Lukashev AN (2005) Evidence for recombination in Crimean-Congo hemorrhagic fever virus. J Gen Virol 86:2333–2338

Mishra AC, Mehta M, Mourya DT, Gandhi S (2011) Crimean-Congo haemorrhagic fever in India. Lancet 378(9788):372

Mourya DT, Yadav PD, Shete A, Majumdar TD, Kanani A, Kapadia D, Chandra V, Kachhiapatel AJ, Joshi PT, Upadhyay KJ, Dave P, Raval D (2014) Serosurvey of Crimean-Congo hemorrhagic fever virus in domestic animals, Gujarat, India, 2013. Vector Borne Zoonotic Dis 14(9):690–692

Mourya DT, Yadav PD, Shete AM, Gurav YK, Raut CG, Jadi RS, Pawar SD, Nichol ST, Mishra AC (2012) Detection, isolation and confirmation of Crimean-Congo hemorrhagic fever virus in human, ticks and animals in Ahmadabad, India, 2010–2011. PLoS Negl Trop Dis 6(5):e1653

Mourya DT, Yadav PD, Shete AM, Sathe PS, Sarkale P, Pattnaik B, Sharma G, Upadhyay KJ, Gosavi S, Patil DY, Chaubal GY, Majumdar TD, Katoch VM (2015) Cross-sectional serosurvey of Crimean-Congo hemorrhagic fever virus IgG in livestock, India, 2013–2014. Emerg Infect Dis 21(10):1837–1839

Mousavi-Jazi M, Karlberg H, Papa A, Christova I, Mirazimi A (2012) Healthy individuals' immune response to the Bulgarian Crimean-Congo hemorrhagic fever virus vaccine. Vaccine 30:6225–6229

Müller MA, Devignot S, Lattwein E, Corman VM, Maganga GD, Gloza-Rausch F, Binger T, Vallo P, Emmerich P, Cottontail VM, Tschapka M, Oppong S, Drexler JF, Weber F, Leroy EM, Drosten C (2016) Evidence for widespread infection of African bats with Crimean-Congo hemorrhagic fever-like viruses. Sci Rep 6:26637

Nabeth P, Cheikh DO, Lo B, Faye O, Vall IOM, Niang M, Wague B, Diop D, Diallo M, Diallo B, Diop OM, Simon F (2004) Crimean-Congo hemorrhagic fever, Mauritania. Emerg Infect Dis 10(12):2143–2149

Oestereich L, Rieger T, Neumann M, Bernreuther C, Lehmann M, Krasemann S et al (2014) Evaluation of antiviral efficacy of ribavirin, arbidol, and T-705 (favipiravir) in a mouse model for Crimean-Congo hemorrhagic fever. PLoS Negl Trop Dis 8(5):e2804

Ozturk B, Kuscu F, Tutuncu E, Sencan I et al (2010) Evaluation of the association of serum levels of hyaluronic acid, sICAM-1, sVCAM-1, and VEGF-A with mortality and prognosis in patients with Crimean-Congo hemorrhagic fever. J Clin Virol 47:115–119

Papa A, Mirazimi A, Köksal I, Estrada-Pena A, Feldmann H (2015 Mar) Recent advances in research on Crimean-Congo hemorrhagic fever. J Clin Virol 64:137–143

Papa A, Papadimitriou E, Christova I (2011) The Bulgarian vaccine Crimean-Congo haemorrhagic fever virus strain. Scand J Infect Dis 43:225–229

Peyrefitte CN, Perret M, Garcia S, Rodrigues R et al (2010) Differential activation profiles of Crimean-Congo hemorrhagic fever virus and Dugbe virus-infected antigen-presenting cells. J Gen Virol 91:189–198

Rodrigues FM, Padbidri VS, Ghalsasi GR, Gupta NP, Mandke VB, Pinto BD, Hoon RS, Bapat MB, Mohan Rao CV (1986) Prevalence of Crimean haemorrhagic-Congo virus in Jammu and Kashmir state. Indian J Med Res 84:134–138

Rodrigues R, Paranhos-Baccala G, Vernet G, Peyrefitte CN (2012) Crimean-Congo hemorrhagic fever virus-infected hepatocytes induce ER-stress and apoptosis crosstalk. PLoS One 7:e29712

Sahib MM (2010) Rapid development of optimized recombinant adenoviral vaccines biosafety level 4 viruses. University of Manitoba, Winnipeg

Saidi S, Casals J, Faghih MA, Faghih AA (1975) Crimean hemorrhagic fever-Congo (CHF-C) virus antibodies in man, and in domestic and small mammals, in Iran. Am J Trop Med Hyg 24:353–357

Saijo M, Tang Q, Shimayi B, Han L, Zhang Y, Asiguma M, Tianshu D, Maeda A, Kurane I, Morikawa S (2005) Antigen capture enzyme-linked immunosorbent assay for the diagnosis of Crimean–Congo hemorrhagic fever using a novel monoclonal antibody. J Med Virol 77:83–88

Saksida A, Duh D, Wraber B, Dedushaj I et al (2010) Interacting roles of immune mechanisms and viral load in the pathogenesis of Crimean-Congo hemorrhagic fever. Clin Vaccine Immunol 17:1086–1093

Schmaljohn CS, Nichol ST (2007) Bunyaviruses. In: Knipe DM, Howley PM (eds) Fields virology, 5th edn. Lippincott Williams & Wilkins, Philadelphia, pp 1741–1790

Schnittler HJ, Feldmann H (2003) Viral hemorrhagic fever—a vascular disease? Thromb Haemost 89:967–972

Shanmugam J, Smirnova SE, Chumakov MP (1976) Presence of antibody to arboviruses of the Crimean Haemorrhagic Fever-Congo (CHF-Congo) group in human beings and domestic animals in India. Indian J Med Res 64:1403–1413

Shepherd AJ, Swanepoel R, Gill DE (1988) Evaluation of enzyme-linked immunosorbent assay and reversed passive hemagglutination for detection of Crimean–Congo hemorrhagic fever virus antigen. J Clin Microbiol 26:347–353

Shepherd AJ, Swanepoel R, Leman PA, Shepherd SP (1986) Comparison of methods for isolation and titration of Crimean–Congo hemorrhagic fever virus. J Clin Microbiol 24:654–656

Shepherd AJ, Swanepoel R, Shepherd SP, McGillivray GM, Searle LA (1987) Antibody to Crimean-Congo hemorrhagic fever virus in wild mammals from southern Africa. Am J Trop Med Hyg 36(1):133–142

Sidwell RW, Smee DF (2003) Viruses of the Bunya- and Togaviridae families: potential as bioterrorism agents and means of control. Antivir Res 57(1–2):101–111

Spengler JR, Bergeron É, Rollin PE (2016) Seroepidemiological studies of Crimean-Congo Hemorrhagic fever virus in domestic and wild animals. PLoS Negl Trop Dis 10(1):e0004210

Spik K, Shurtleff A, McElroy AK, Guttieri MC, Hooper JW, SchmalJohn C (2006) Immunogenicity of combination DNA vaccines for Rift Valley fever virus, tick-borne encephalitis virus, Hantaan virus, and Crimean Congo hemorrhagic fever virus. Vaccine 24(21):4657–4656

Swanepoel R, Shepherd AJ, Leman PA, Shepherd SP, McGillivray GM, Erasmus MJ, Searle LA, Gill DE (1987) Epidemiologic and clinical features of Crimean-Congo hemorrhagic fever in Southern Africa. Am J Trop Med Hyg 36:120–132

Tasdelen Fisgin N, Fisgin T, Tanyel E, Doganci L et al (2008) Crimean-Congo hemorrhagic fever: five patients with hemophagocytic syndrome. Am J Hematol 83:73–76

Tkachenko E, Khanun K, Berezin V (1969) Serological investigation of human and animal sera in agar gel diffusion and precipitation (AGDP) test for the presence of antibodies of Crimean hemorrhagic fever and Grand Arbaud viruses. Mater Nauch Sess Inst Polio Virus Entsef 2:265

Vescio FM, Busani L, Mughini-Gras L, Khoury C, Avellis L, Taseva E, Rezza G, Christova I (2012) Environmental correlates of Crimean-Congo haemorrhagic fever incidence in Bulgaria. BMC Public Health 12:1116

Vorou R, Pierroutsakos IN, Maltezou HC (2007) Crimean-Congo hemorrhagic fever. Curr Opin Infect Dis 20:495–500

Watts DM, Flick R, Peters CJ, Shope R (2005) Bunyaviral fevers: Rift Valley fever and Crimean-Congo hemorrhagic fever. In: Guerrant RL, Walker DH, Weller PF (eds) Tropical infectious diseases: principles, pathogens, and practice. Elsevier Churchill Livingstone, Philadelphia, PA, pp 756–760

Whitehouse CA (2004) Crimean-Congo hemorrhagic fever. Antivir Res 64:145–160

Wolfel R, Paweska JT, Petersen N, Grobbelaar AA, Leman PA, Hewson R, Courbot MCG, Papa A, Günther S, Drosten C (2007) Virus detection and monitoring of viral load in Crimean–Congo hemorrhagic fever virus patients. Emerg Infect Dis 13:1097–1100

Wölfel R, Paweska JT, Petersen N, Grobbelaar AA, Leman PA, Hewson R, Georges-Courbot MC, Papa A, Heiser V, Panning M, Günther S, Drosten C (2009) Low density macroarray for rapid detection and identification of Crimean–Congo hemorrhagic fever virus. J Clin Microbiol 47:1025–1030

World Health Organization (2011) WHO Fact Sheet No. 208. WHO, Geneva. http://www.who.int/mediacentre/factsheets/fs208/en/

World Health Organization, Regional Office for South-East Asia (2014) A brief guide to emerging infectious diseases and zoonoses. WHO, New Delhi

Xiao X, Feng Y, Zhu Z, Dimitrov DS (2011) Identification of a putative Crimean-Congo hemorrhagic fever virus entry factor. Biochem Biophys Res Commun 411(2):253–258

Yadav PD, Cherian SS, Zawar D, Kokate P, Gunjikar R, Jadhav S, Mishra AC, Mourya DT (2013) Genetic characterization and molecular clock analyses of the Crimean-Congo hemorrhagic fever virus from human and ticks in India, 2010–2011. Infect Genet Evol 14:223–231

Yapar M, Aydogan H, Pahsa A, Besirbellioglu BA, Bodur H, Basustaoglu AC, Guney C, Avci IY, Sener K, Setteh MH, Kubar A (2005) Rapid and quantitative detection of Crimean–Congo hemorrhagic fever virus by one-step real-time reverse transcriptase-PCR. Jpn J Infect Dis 58:358–362

Yilmaz GR, Buzgan T, Irmak H, Safran A, Uzun R, Cevik MA, Torunoglu MA (2009) The epidemiology of Crimean-Congo hemorrhagic fever in Turkey, 2002–2007. Int J Infect Dis 13(3):380–386

Zilinskas RA (1997) Iraq's biological weapons. The past as future? JAMA 278(5):418–424

Zivcec M, Scholte FE, Spiropoulou CF, Spengler JR, Bergeron E (2016) Molecular insights into Crimean-Congo hemorrhagic fever virus. Viruses 8(4):106

Chapter 12
Porcine Reproductive and Respiratory Syndrome Virus

Tridib Kumar Rajkhowa

Abstract Porcine reproductive and respiratory syndrome (PRRS) is an economically important swine disease worldwide. There are two distinct antigenically different genotypes of porcine reproductive and respiratory syndrome virus (PRRSV): the European (type I) and North American (type II), which cause the same disease symptoms. In the last three decades, the disease has evolved from classical form to most devastating highly pathogenic PRRS. Although tremendous progress has been made in understanding the disease, there are many unanswered questions regarding pathogenesis and immunobiology of PRRSV. The virus appears to use several evasion strategies to circumvent both innate and acquired immunity, including interference with antigen presentation, antibody-mediated enhancement, reduced cell surface expression of viral proteins and shielding of neutralising epitopes. Presently available vaccines are ineffective because they suffer both from the immune evasion strategies of the virus and the antigenic heterogeneity of field strains. Recent signs of progress made on viral reverse genetics, genomics and host transcriptomics, vaccinology and adjuvant technologies have identified new areas for control and prevention of PRRS.

Keywords Porcine reproductive and respiratory syndrome virus · PRRS · Pathology · Immunobiology · Diagnosis · Epidemiology · Vaccines

12.1 Preamble

The porcine reproductive and respiratory syndrome (PRRS), an acute respiratory disease in pigs of all ages and reproductive failure in sows, was recorded first in the late 1980s, in pig farms of the United States and shortly after that in Europe (Botner

T. K. Rajkhowa (✉)
Department of Veterinary Pathology, College of Veterinary Sciences and Animal Husbandry, Central Agricultural University, Aizawl, Mizoram, India

Y. S. Malik et al. (eds.), *Emerging and Transboundary Animal Viruses*, Livestock Diseases and Management, https://doi.org/10.1007/978-981-15-0402-0_12

et al. 1997; Done et al. 1996). This disease of unknown aetiology was initially known as 'Mystery Swine Disease' (MSD). Pigs that suffered from the disease first manifest acute illness including anorexia, lethargy, pyrexia, dyspnoea, cyanosis of extremities and often red-blue discoloration of the ears. Hence the disease was also named as 'blue-eared pig disease'. Reproductive failure in sows, which includes stillbirths, mummifications, weak-born piglets and high pre-weaning mortality, occurs concurrently or soon after the acute illness (Meulenberg 2000; Rowland 2007). Therefore the disease was also referred to as 'porcine epidemic abortion and respiratory syndrome' (PEARS) or 'swine infertility and respiratory syndrome' (SIRS) (Christianson et al. 1992; Collins et al. 1992; Wensvoort et al. 1991).

The aetiology of the disease, porcine reproductive and respiratory syndrome virus (PRRSV), was first isolated from the Netherlands, in 1991, and was named as Lelystad virus (LV) (Terpstra et al. 1991; Wensvoort et al. 1991). Shortly after that, the aetiology of PRRS was confirmed by isolation of the virus and experimental reproduction of the disease in the United States (Benfield et al. 1992; Collins et al. 1992). The origin and independent emergence of PRRSV in two separate continents during the same time period are not clearly understood. The lactate dehydrogenase-elevating virus (LDV) in mice is generally agreed as the common ancestor for PRRSV (Carman et al. 1995; Hanada et al. 2005; Plagemann 2003). However, there is no consensus about the divergence time of PRRSV and the history of PRRSV before its emergence in the swine industry remains largely unknown (Forsberg 2005; Hanada et al. 2005).

In subsequent years the PRRSV has been detected in the majority of swine-producing countries worldwide. The disease was recorded as subclinical infections to severe respiratory disease in piglets and reproductive failure in sows (Baron et al. 1992; Dea et al. 1992; Hopper et al. 1992; Jiang et al. 2000; Plana-Duran et al. 1992; Shimizu et al. 1994). Some exceptionally serious PRRS outbreaks with a high frequency of abortions and sow mortality have been recorded in the United States around 1996. Recently, China and surrounding countries have suffered from an extremely severe epidemic that was attributed to the emergence of a novel highly virulent PRRSV strain (Halbur and Bush 1997; Hurd et al. 2001; Normile 2007; Tian et al. 2007). Since its emergence, PRRS has resulted in huge economic losses in the swine industry. In the United States, the economic losses due to PRRSV have been roughly estimated at half a billion dollar per year (Neumann et al. 2005; Holtkamp et al. 2013). Despite all the efforts to prevent and control the disease, it will continue to impact the global swine industry.

12.2 PRRS Virion and Genome Structure

The porcine reproductive and respiratory syndrome virus (PRRSV) belongs to the family *Arteriviridae* in the order of the *Nidovirales* (Gorbalenya et al. 2006). *Arteriviridae* comprises three other viruses: equine arteritis virus (EAV), simian haemorrhagic fever virus (SHFV) and lactate dehydrogenase-elevating virus (LDV).

All these viruses have distinct host range for each one of them: pigs for PRRSV, horses and donkeys for EAV, monkeys for SHFV and mice for LDV (Snijder and Meulenberg 1998). There are two distinct antigenically different genotypes of PRRSV: the European (type I) and North American (type II) (Meng et al. 1995). Both the genotypes cause similar disease symptoms in susceptible pigs (Kapur et al. 1996; Meng et al. 1995; Meng 2000; Meulenberg 2000; Nelsen et al. 1999; Ropp et al. 2004; Stadejek et al. 2006).

The PRRS virion is roughly spherical and of 50–60 nm in diameter with a relatively smooth external surface (Fig. 12.1). The viral genome is packed with nucleocapsid proteins and surrounded by a lipid-bilayer envelope, on which surface glycoproteins (GPs) and membrane proteins are inserted to form the virion particles. PRRSV genome is approximately 15 kb in length; has a positive-stranded, 3′-polyadenylated molecule; and contains 11 known open reading frames (ORFs) (Table 12.1). The 5′-end of the polycistronic genome contains ORFs 1a and 1b, which encode two large non-structural polyproteins, pp1a and pp1ab, with the expression of the latter depending on a −1 ribosomal frameshift signal in the ORF1a/ORF1b overlap region. Following their synthesis, the pp1a and pp1ab replicase polyproteins are processed into at least 14 non-structural proteins (NSPs) by 4 ORF1a-encoded proteinases residing in nsp1α, nsp1β, nsp2 and nsp4. The size of ORF1a is quite variable due to the hypervariability in the central region of non-structural protein 2 (nsp2). A new ORF (TF) and −1/−2 programmed ribosomal frameshift signal were discovered recently in the central region of ORF1a, which expresses two novel proteins, nsp2TF and nsp2N (Fang et al. 2012; Li et al. 2014). The 3′ end of the viral genome contains eight relatively small genes. These genes have both 5′ and 3′ terminal sequences

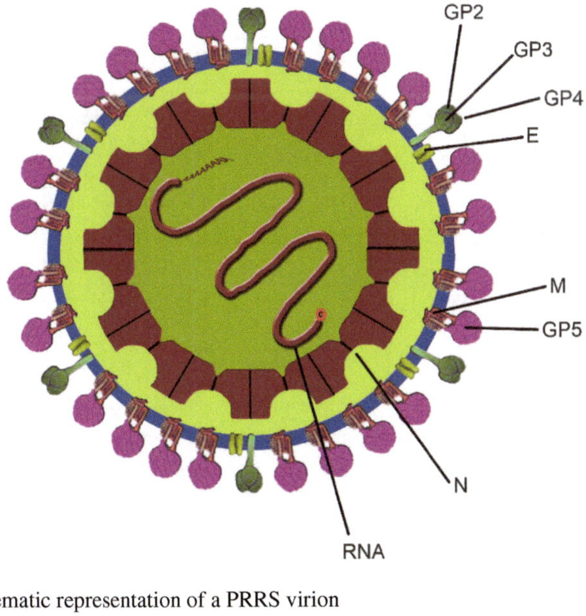

Fig. 12.1 Schematic representation of a PRRS virion

Table 12.1 Porcine reproductive and respiratory syndrome virus (PRRSV) proteins and their functions

Sl. No.	Gene and expressed protein		Protein length		Known or predicted function
	Gene	Protein	Genotype I (LV, GenBank accession # M96262)	Genotype II (VR-2332, GenBank accession # AY150564)	
1	ORF1a	nsp1α	180	180	Contains papain-like cysteine protease (PLPα); zinc-finger protein; regulator of subgenomic (sg)mRNA synthesis; potential interferon (IFN) antagonist
		nsp1β	205	203	Contains protease PLPβ; potential IFN antagonist
		nsp2	1078	1196	Contains protease PLP2; deubiquitinating enzyme; potential IFN antagonist; transmembrane protein involved in membrane modification forming replication complex
2	ORF1a′ -TF	nsp2TFc	902	1019	Contains PLP2 domain
		nsp2N∗	733	850	Contains PLP2 domain
3	ORF1a	nsp3	230	230	Transmembrane (TM) domain protein involved in membrane modification, forming replication complex
		nsp4	203	204	Main serine protease (SP); apoptosis inducer; potential IFN antagonist
		nsp5	170	170	TM protein possibly involved in membrane modification
		nsp6	16	16	?
		nsp7α	149	149	Recombinant nsp7 is highly antigenic
		nsp7β	120	110	
		nsp8	45	45	N-terminal domain of nsp9
4	ORF1b∗	nsp9	685	685	RNA-dependent RNA polymerase
		nsp10	442	441	RNA NTPase/helicase; contains putative zinc-binding domain
		nsp11	224	223	Uridylate-specific endoribonuclease (NendoU)
		nsp12	152	153	?
5	ORF2a	GP2a	249	256	Minor glycosylated structural protein; essential for virus infectivity; incorporated into virion as a multimeric complex with GP3-4; viral attachment protein

(continued)

Table 12.1 (continued)

Sl. No.	Gene and expressed protein: Gene	Protein	Protein length: Genotype I (LV, GenBank accession # M96262)	Genotype II (VR-2332, GenBank accession # AY150564)	Known or predicted function
6	ORF2b	E	70	73	Minor unglycosylated and myristoylated structural protein; essential for virus infectivity; incorporated into virion as a multimeric complex; possesses ion channel-like properties and may function as a viroporin in the envelope
7	ORF3	GP3	265	254	Minor glycosylated structural protein; essential for virus infectivity; highly antigenic and may be involved in viral neutralisation; incorporated into virions as a multimeric complex with GP2a and GP4; a subset of GP3 could be secreted as a non-virion-associated soluble protein
8	ORF4	GP4	183	178	Minor glycosylated structural protein; essential for virus infectivity; forms GP2a-3-4 complex to incorporate into virion; viral attachment protein and may involve in viral neutralisation
9	ORF5	GP5	201	200	Major glycosylated structural protein; TM protein with a variable number of potential N-glycosylation sites; the most variable structural protein; forms a disulphide-linked heterodimer with M protein
10	ORF5a	ORF5a	43	41	Minor unglycosylated, hydrophobic structural protein; essential for virus viability; incorporated into virion as a multimeric complex
11	ORF6	M	173	174	Highly conserved major unglycosylated structural protein; GP5-M hetero-dimerisation is crucial for virus infectivity; plays a key role in virus assembly and budding
12	ORF7	N	128	123	Unglycosylated and phosphorylated structural protein; component of the viral capsid; highly antigenic; potential IFN antagonist

overlapping with neighbouring genes, except ORF4/ORF5 of type 2 PRRSV. They encode four membrane-associated glycoproteins (GP2a, GP3, GP4 and GP5), three non-glycosylated membrane proteins (E, ORF5a and M) and a nucleocapsid protein (N) (Fig. 12.2).

12.3 Molecular Epidemiology of PRRSV

PRRSV emerged almost simultaneously in North America (genotype 2) and Western Europe (genotype 1) in the late 1980s and early 1990s, respectively (Keffaber 1989; Wensvoort et al. 1991). The PRRSV strains isolated from both the continents are strikingly different with only 55–70% nucleotide identity (Allende et al. 1999; Nelsen et al. 1999). The evolutionary distance between the two lineages has led to the hypothesis that these two lineages have evolved separately from a very distant common ancestor (Plagemann 2003). Since LDV was identified as the closest relative of PRRSV within the *Arteriviridae* family, it was hypothesised that PRRSV had an LDV-like ancestor circulating in rodents (Plagemann 2003). The ancestral PRRSV in rodents was later adapted to Eurasian wild boars, which served as an intermediate host that brought the virus from rodents to the domestic swine population. It was postulated further that the divergence of the two types of PRRSV started from the introduction of wild boars from Europe to the United States (Plagemann 2003; Mayer and Brisbin 1991). Following the introduction, PRRSV evolved separately in the wild boar populations in Eurasia and North American continents before their independent introduction and establishment to the domestic pig population (Plagemann 2003).

12.3.1 Diversity and Evolution of Type 1 PRRSV

The first type 1 PRRSV epidemic was documented in the early 1990s in Western Europe (Wensvoort et al. 1991). Lelystad virus, the prototype of type 1 PRRSV, which was isolated first in the Netherlands from the epidemic tends to form a highly

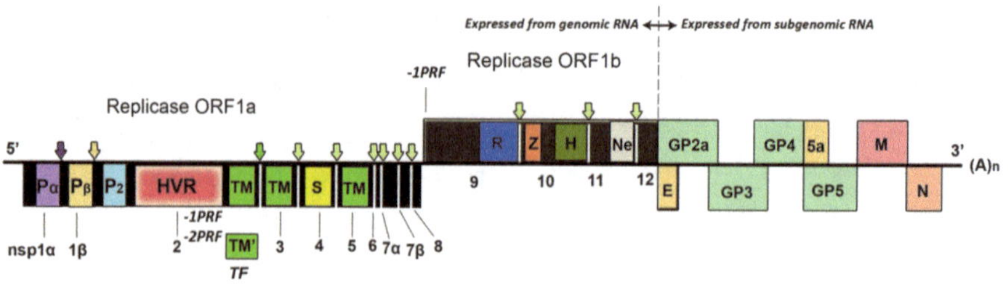

Fig. 12.2 PRRS virus genome organisation

homologous phylogenetic cluster (Lelystad-like) with contemporaneous samples from Belgium, France, Germany, Great Britain, the Netherlands and Spain (Drew et al. 1997; Forsberg et al. 2002; Le Gall et al. 1998; Suarez et al. 1996). However, Lelystad-like cluster could barely represent the diversity of type 1 PRRSV, nor could it be viewed as the ancestral virus type from which most of the current type 1 PRRSV derived. The increasing sample range which was obtained after the isolation of Lelystad-like viruses showed well-supported diverged clades in the phylogenetic analysis indicating that they were not likely to be 'progenies' of Lelystad-like PRRSV. Therefore, the evidence suggested that type 1 PRRSV was present in the field long before the 'original' epidemic in Western Europe.

Based on the ORF5 sequence divergence and phylogenetic grouping, type 1 PRRSV was classified into four subtypes. RecentlyORF7 size polymorphism has been recommended as a stable and independent indicator for subtyping (Stadejek et al. 2008), which has also confirmed three previously identified subtypes. This includes a pan-Europe subtype I; a subtype II with samples from Belarus, Lithuania and Russia; and a subtype III mainly constituted of Belarus samples with nucleoprotein sizes of 128, 125 and 124 amino acids, respectively (Stadejek et al. 2008). Although the majority of the genotype I diversity is in Europe, type 1 PRRSVs have been introduced now to five non-European countries, including the United States (Fang et al. 2007; Ropp et al. 2004), Canada (Dewey et al. 2000), South Korea (Lee et al. 2010), China and Thailand (Thanawongnuwech et al. 2004).

12.3.2 Diversity and Evolution of Type 2 PRRSV

VR-2332, the prototype strain of type 2 PRRSV, was isolated from the epidemic of the late 1980s in North America (Collins et al. 1992; Hill 1990; Keffaber 1989). A significant portion of US isolates from 1989 to the early 1990s were found to be closely related to VR-2332 (Kapur et al. 1996; Meng et al. 1995; Wesley et al. 1998; Shi et al. 2010). During the late summers of 1996, there were increased numbers of outbreaks of severe PRRS characterised by an elevated abortion rate and sow mortality in southeast Iowa as well as in many other states of the United States. These outbreaks occurred even in vaccinated herd either with Ingelvac PRRS MLV alone or both Ingelvac PRRS MLV and PrimePac PRRS vaccines (Bush et al. 1999; Halbur and Bush 1997; Key et al. 2001). The apparent protection failure by both live attenuated vaccines was considered because PRRSV isolates in this epidemic were genetically distinct from available live attenuated vaccines at that time (Key et al. 2001). Late in 2001, a virulent variant of PRRSV was identified in the US state of Minnesota (Han et al. 2006). These isolates (MN184 A, B and C) were characterised by three discontinuous deletions of 11, 1 and 19 amino acids in the nsp2 region of the replicase ORF, corresponding to VR-2332 positions 324–434, 486 and 505–523 (Han et al. 2006). The sudden emergence of the virus is still a mystery.

Because of its original identification and major distribution in North America, type 2 PRRSV was also referred to as North American type PRRSV. Recent studies based on sequence analysis and phylogenetic reconstruction have identified three clades of type 2 PRRSV, which constituted mainly of North American sequences (one clade contained PrimePac PRRS vaccine, one contained Canadian and Thai sequences, and one was associated with VR-2332 and Ingelvac PRRS MLV vaccine) and several highly diverse Asian clusters (Wang et al. 2008). Recent studies using Bayesian phylogenies, based on the entire collection ($n \cong 8500$) of currently available high-quality ORF5 sequences (Shi et al. 2010), have divided type 2 PRRSV isolates into nine well-supported lineages, including five large clusters ($n > 1000$) and four small groups, each separated by larger than 10% genetic distance from neighbouring lineages. Of the nine lineages defined, seven have their diversity maintained by North American samples, while the remaining two are found only in Asian countries (Shi et al. 2010). Within the seven North American lineages, introductions to Asian and European countries were frequently observed, some resulting in local outbreaks (An et al. 2007; Cha et al. 2004; Chen et al. 2006; Kang et al. 2004; Kim et al. 2009; Madsen et al. 1998; Shi et al. 2010; Thanawongnuwech et al. 2004). Among all, lineage 5 containing Ingelvac PRRS MLV vaccine is found as most cosmopolitan, with viruses introduced to more than eight countries outside of the North American continent.

12.3.3 Emergence of Highly Pathogenic PRRSV in China

Highly pathogenic PRRS (HP-PRRS), characterised by prolonged high fever, cutaneous red colorations and high mortality rate in pigs of all age group (An et al. 2007; Tian et al. 2007; Li et al. 2007), was first detected in the summer of 2006 in China. The disease quickly spread to more than 20 provinces in China, causing huge losses to the country's swine industry. The causative agent was later identified as a novel variant of PRRSV (Tian et al. 2007; Tong et al. 2007; Zhou et al. 2008) with the genetic marker of discontinuous deletions of 1 and 29 amino acids in the nsp2 region of PRRSV genome. However, the genetic basis of this unusually highly virulent PRRSV remains unknown. Phylogenetic reconstructions had indicated that the highly pathogenic PRRSV variant actually evolved from the local diversity of PRRSV in China (An et al. 2009; Hu et al. 2009; Zhou et al. 2008), which may be traced to a single introduction from North America (Shi et al. 2010). HP-PRRSV has subsequently spread to other Southeast Asian countries including Laos, Viet Nam, Cambodia, Bhutan and Myanmar, resulting in huge loss to the local pig husbandries. The first outbreak of PRRS in India in the pig population of Mizoram was recorded in 2013 (Rajkhowa et al. 2015a). Subsequent outbreaks were reported in 2015, 2016 and 2017. All these outbreaks were later characterised as HP-PRRS (Rajkhowa et al. 2015a, b, 2016, 2018; Gogoi et al. 2017).

12.4 Transmission, Infectivity and Pathogenesis of PRRSV

Transmission of PRRSV may occur via distance-dependent (area spread) and distance-independent routes. In distance-dependent route, the transmission from one herd to a nearby herd occurs even without active communication between the two farms. Various mechanisms, including aerosol (Brockmeier and Lager 2002), animal and insect vectors (Otake et al. 2003) and contaminated fomites (Dee et al. 2003; Otake et al. 2003), have been proposed for area spread. Whereas in independent distance routes, communication between the two places is often required. In this route, transmission can occur via the transportation of pigs, introduction of semen (Christopher-Hennings et al. 2008) and even vaccine usage if vaccine-related viruses are involved. Besides, indiscriminate needle use during drug and vaccine administration can facilitate iatrogenic transmission (Otake et al. 2003).

The PRRSV has a very restricted tropism for cells of the monocytic lineage, and the fully differentiated porcine alveolar macrophage (PAM) is the primary target cell (Duan et al. 1997). Also, dendritic cells were also reported to be able to support PRRSV replication (Loving et al. 2007). For in vitro replication, only the African green monkey kidney cell line MA-104 and derivatives such as MARC-145 are found to be fully permissive. PRRSV enters host cells through standard clathrin-mediated endocytosis. The viral genome is released into the cytosol following endosome acidification and membrane fusion (Nauwynck et al. 1999). The major receptor that mediates viral internalisation and disassembly has been identified as CD163 (Welch and Calvert 2010). The minor structural protein, GP2a, and GP4 mediate virus entry into susceptible host cells by interacting with CD163 (Das et al. 2010; Welch and Calvert 2010). Following entry by receptor-mediated endocytosis and disassembly, genome translation yields replicase polyproteins pp1a-nsp2TF, pp1a-nsp2N, pp1a and pp1ab. These polyproteins are cleaved by viral internal proteinases to generate at least 14 non-structural proteins, which are assembled into a replication and transcription complex (RTC). The RTC first engages in minus-strand RNA synthesis to produce both single-strand full-length and subgenomic length minus-strand RNAs. Subsequently, the subgenomic mRNAs serve as templates for the synthesis of plus-strand subgenomic mRNAs required to express the structural protein genes that reside in the 3′-proximal quarter of the genome. Newly generated RNA genomes are packaged into a nucleocapsid that becomes enveloped by budding from smooth intracellular membranes. The new virions are released from the cell using the exocytic pathway.

PRRSV infection can be divided into three distinct stages: acute infection, persistence and extinction. Each one of this stage is unique in terms of immunology, virology and clinical disease. The acute infection is characterised by an acute respiratory infection, during which the lung serves as a preferential site of infection. PRRSV replicates mainly in macrophages and dendritic cells in the lungs and the upper respiratory tract, resulting in viremia by 6–12 h postinfection (pi). Despite the presence of circulating antibodies serum viremia may last for several weeks. The second stage is characterised by localisation and replication of virus primarily in lymphoid organs, including tonsil and lymph nodes but not in the spleen, resulting in persistent infection (Wills et al. 1997; Allende et al. 2000; Rowland et al. 2003). During this stage, the virus is no longer detected in blood and lungs, and pigs no longer exhibit overt signs of clinical disease. However, virus replication in regional lymph nodes continues, which accounts for the efficient transmission of the virus to naive pigs via oral-nasal secretions and semen. Virus replication gradually decays during the final stage of infection until the virus becomes extinct in the host. The replication of virus can be maintained for as long as 250 days after infection (Wills et al. 2003) and thereby PRRSV establishes a 'lifelong' infection.

12.5 Immunopathobiology of PRRSV Infection

Pigs with infection of PRRSV mount a rapid antibody response by 7–9 days postinfection (dpi) directed mainly to the N protein and a lesser extent to the M protein and the antibodies are non-neutralising (Loemba et al. 1996). The neutralising antibodies which appear typically >28 dpi are mainly directed against GP5. The neutralising antibodies are usually specific for the homologous strain, with lower or no titres of cross-neutralising antibodies against heterologous strain (Vu et al. 2011; Zhou et al. 2012).

12.5.1 Innate Immune Responses to PRRSV Infection

The innate immune system is the first line of host defence against viral infections. Adequate activation of the host innate immune system in response to viral infection is critical to prevent viral replication and invasion into mucosal tissues and importantly in the initiation of the strong adaptive immune response to fight against intracellular pathogens (Koyama et al. 2008). Infection with PRRSV elicits poor innate and adaptive immune responses through immune modulation and results in incomplete viral clearance (Albina et al. 1998a, b; Renukaradhya et al. 2010).The NK cell is the innate lymphocyte subset that helps in non-specific clearance of any virus-infected cell from the body. PRRSV infection induces significant suppression of NK cell cytotoxic activity, which was noticed as early as 2dpi and continued for 3–4 weeks (Renukaradhya et al. 2010). One of the most remarkable features of PRRSV infection is that it fails to elicit any significant inflammatory cytokine expression, in particular of type I interferons, interleukin (IL)-1 and TNFα (Thanawongnuwech et al. 2001; Van Reeth et al. 1999). PRRSV elicits only a minimal TNFα production, both in alveolar macrophages in vitro and in vivo at the site of infection (Genini et al. 2008; Loving et al. 2007; Lee et al. 2004). The downregulation of type I IFN, in particular of TNFα, is considered as a crucial step in the pathogenesis because TNFα has been shown to inhibit PRRSV replication (Albina et al. 1998a, b; Le Bon et al. 2001).

PRRSV can infect both mature and immature porcine DCs (Loving et al. 2007; Flores-Mendoza et al. 2008; Chang et al. 2008; Charerntantanakul et al. 2006b; Wang et al. 2007; Park et al. 2008), resulting in apoptosis; downregulation of the expression of CD11b/c, CD14 and CD80/86 and major histocompatibility complex (MHC) class I and II molecules; reduced allogeneic stimulation of T cells; and upregulated expression of IL-10, IL-12 and TNF-α (Genini et al. 2008; Flores-Mendoza et al. 2008; Wang et al. 2007; Park et al. 2008). In total, it was observed that PRRSV might dampen both innate and specific immune responses by altering the cytokine production of macrophages and DCs and by modifying the expression of molecules involved in antigen presentation. As a consequence, natural killer (NK) cell activation and mobilisation of cells from the acquired arm of the immune system are delayed, resulting in delayed neutralising antibody production, lymphoproliferative response and IFN-γ responses to PRRSV (Flores-Mendoza et al. 2008; Chang et al. 2008; Butler et al. 2008). This weak initial innate immune response to PRRSV contributes to the long survival of the virus in infected pigs.

12.5.2 Acquired Immunity

The level of protective immunity following PRRSV infection is a debated issue. Development of immunity after natural infection needs at least 3 months' time to reach peak levels and does not appear to be solid enough to prevent reinfection, especially caused by antigenically heterologous strains (Murtaugh et al. 2002; Zuckermann et al. 2007). Due to the broad genetic and antigenic variation of simultaneously circulating PRRSV field strains, the efficacy of humoral and cell-mediated immune responses is insufficient to protect under field circumstances.

Following initial PRRSV infection, antibodies are generally detected from 5 to 14 dpi, and antibody titres increase rapidly to a maximum level around 4 weeks postinfection (Albina et al. 1998b; Batista et al. 2004; Diaz et al. 2005; Labarque et al. 2000). Immunoglobulin M (IgM) titres in serum peak around 14–21 dpi and then rapidly decrease to undetectable levels. Virus-specific IgG (both IgG1 and IgG2) reaches maximum values around the third or fourth week postinfection and remains at high levels for months (Labarque et al. 2000; Loemba et al. 1996; Mulupuri et al. 2008). However, even during persistent infection, antibody titres and plasma cell counts ultimately decline, and secondary contact with the virus often does not result in an anamnestic antibody response. While the PRRSV-specific antibody response upon infection in naive animals shows a normal course, a different picture arises when talking in terms of protection. PRRSV-infected macrophages are refractory to ADCML and most likely also to ADCC since viral proteins are not expressed on the cell surface, or at least not in a way rendering them detectable for antibodies (Costers et al. 2006).

Conversely, the interaction of antibodies with free virions likely does occur, since in vitro virus neutralisation can be obtained with antisera and purified antibodies, both in MARC-145 and in PAM. Virus-neutralising antibodies as measured by in vitro seroneutralisation (SN) test however only appear very slowly and in low amounts upon PRRSV infection in naive animals. This delayed response of neutralising antibody against PRRSV has been postulated due to the presence of an immunodominant 'decoy' epitope in GP5 (aa 27–30), which may evoke a robust, early and non-protective immune response that masks and/or impairs the response to the major neutralising epitope (Ostrowski et al. 2002). An alternative explanation is identified as glycan-shielding effects of N-linked glycosylation in GPs as proposed for the human and simian immunodeficiency viruses. 'Glycan shielding' may be a primary mechanism to explain evasion from neutralising antibodies, ensuring in vivo persistence of these viruses (Wei et al. 2003). Additionally, antibody-dependent enhancement of viral entry into target cells (Cancel-Tirado et al. 2004), suppression of innate immune responses (Sang et al. 2011) and prevention of normal B-cell repertoire development (Butler et al. 2014) are also identified as contributing towards poor acquired immunity against the PRRSV infection.

12.6 Clinical Manifestations of PRRS

The clinical disease of PRRS in a herd is primarily due to the consequence of acute viremia in individuals and transplacental transmission, resulting in reproductive failure. Clinical epidemics occur when PRRSV enters in an immunologically naive herd. Endemic PRRS occurs in herds that have homologous immunity to the infecting strain of PRRSV and in which clinical disease is observed only in a susceptible subpopulation (Zimmerman et al. 2006).

The first phase of PRRS epidemics lasts for 2 or more weeks and is characterised by anorexia and lethargy in animals of all ages, as a result of acute viremia. Clinically affected animals show the high rise of body temperature (105–107 °F) and respiratory distress, and may have transient cutaneous hyperaemia or cyanosis on extremities. The second phase is characterised by reproductive failure in sows and high pre-weaning mortality in piglets, which may begin before the first one is completed and continues for 1–4 months.

Affected adult animals suffer from anorexia, fever and lethargy (Keffaber 1989; Loula 1991). Occasionally subcutaneous and hind limb oedema, nervous signs and skin lesions, such as purplish discoloration of the ears and vulva, are observed (Hopper et al. 1992; Rossow et al. 1998). PRRSV infection in adults can be fatal, especially in late-term gestating females. Reproductive failure in sows includes low conception and farrowing rates, decreasing number of live-born piglets and high incidence of weak-born, stillborn and mummified piglets (Zimmerman et al. 2006). The impact on the health status of lactating sows is similar to gestating sows. Loss of appetite and fever lead to agalactia, which results in starvation of the piglets or development of diseases that are controlled by maternal immunity such as colibacillosis. Pre-weaning mortality rates can exceed up to 80% in severe cases (Christianson et al. 1992). In boars, in addition to acute respiratory signs, there may be lack of libido and variable reduction in semen quality, which includes reduced motility and defects in sperm (Prieto and Castro 2005; Zimmerman et al. 2006).

High pre-weaning mortality is observed in pigs born prematurely and at term associated with listlessness, diarrhoea, emaciation, splay-legged posture, hyperpnoea and dyspnoea. Less common are tremors and paddling, anaemia and bacterial polyarthritis, and meningitis (Zimmerman et al. 2006). Other clinical signs reported in neonates include oedema around the eyes, conjunctivitis, blue discoloration of ears, bruising of the skin, diarrhoea, rough hair coats and profuse bleeding post-injection (Rossow 1998). The impact of PRRS virus infections in pigs after weaning can be highly variable (Rossow et al. 1994). Acute PRRSV infection in post-weaned pigs is characterised by anorexia, lethargy, cutaneous hyperaemia, hyperpnoea and dyspnoea without coughing, rough hair coats and elevated mortality (Zimmerman et al. 2006).

Once introduced into a farm, PRRSV usually becomes endemic, leading to regular or occasional outbreaks of typical acute PRRS in susceptible subpopulations of animals, such as nursery or grower pigs and replacement gilts and boars (Stevenson et al. 1993; Zimmerman et al. 2006). In nursery pigs, elevations in mortality mostly occur in winter months, suggesting that environmental factors, such as lowered ambient temperature, increased range in the fluctuation of ambient temperature, lowered ventilation rates and elevated relative humidity, probably contribute to recirculation of virus and decrease in immune responses (Stevenson et al. 1993). The reproductive consequences depend on the number of animals infected and the stage of their reproductive cycle when infected. If gilts are infected, there may be scattered abortions, irregular returns to oestrus, non-pregnant gilts and late-term reproductive failure (Zimmerman et al. 2006).

12.7 Pathology of PRRS

The characteristic lesions of PRRSV are seen in young pigs affected by the respiratory form of the syndrome. Gross and microscopic lesions are consistently observed from 4 to 28 or more dpi in lung and lymph nodes, where most viral replication takes place. The virulence of the strain of PRRSV to which pigs are exposed determines the severity and the distribution of lesions. Gross lesions are suggestive of PRRSV, but are not diagnostic since a variety of other viral and bacterial diseases can cause similar lesions. Although the presence of typical gross and microscopic lesions of PRRSV allows a strong presumptive diagnosis, definitive diagnosis always requires demonstration of PRRSV (Mengeling and Lager 2000; Zimmerman et al. 2006).

Interstitial pneumonia develops in the lungs of affected pigs from 3 to 28 dpi, with most severe lesions within 10–14 dpi. Affected lung parenchyma is resilient, non-collapsing, slightly firm, mottled or diffusely red-tan, rubbery and moist (Zimmerman et al. 2006). Microscopic thickening of alveolar septa due to infiltration of macrophages, lymphocytes, plasma cells, oedema, congestion and hyperplasia of pneumocyte II is observed in lungs. Alveoli may contain necrotic cell debris, macrophages and serous fluid, while lymphocytes and plasma cells form cuffs around airways and blood vessels. Macroscopic lesions in lymph nodes are presented mainly as tan to grey, enlarged and oedematous, which are often several times the size of normal nodes (Mengeling and Lager 2000). Microscopic lesions are predominantly observed in germinal centres that appear necrotic and depleted. The cortices may contain small cystic spaces that are variably lined by endothelium and contain proteinaceous fluid, lymphocytes and multinucleate prokaryocytes. Mild-to-moderate multifocal lymphohistiocytic vasculitis and perivascular myocarditis may develop in the heart. Lymphohistiocytic leukoencephalitis or encephalitis involving cerebellum, cerebrum and brainstem may develop in the brain. Segmental cuffing of blood vessels by lymphocytes and macrophages and multifocal gliosis may also occur. Mild periglomerular and peritubular lymphohistiocytic aggregation is seen in the kidney. Affected vessels have swollen endothelium, pooled subendothelial proteinaceous fluid, and intramural and perivascular aggregates of lymphocytes and macrophages. PRRSV-infected litters contain variable numbers of normal pigs, small weak pigs and dead pigs that are either fresh stillborn, autolytic stillborn or mummified foetuses. Gross foetal lesions include perirenal oedema, oedema of the splenic ligament, mesenteric oedema, ascites, hydrothorax and hydro-peritoneum. Segmental haemorrhagic enlargement of the umbilical cord of weak live-born and stillborn pigs is sometimes present (Zimmerman et al. 2006). Uterus of affected sows shows oedematous myometrium and endometrium with lymphohistiocytic perivascular cuffs. There may also be segmental lymphohistiocytic vasculitis in small vessels and microseparations between endometrial epithelium and placental trophoblasts. In boar atrophy seminiferous tubules of the testicle are observed 7–25 dpi, associated with apoptosis and depletion of germ cells (Zimmerman et al. 2006).

12.8 Diagnosis of PRRS

A tentative diagnosis of PRRS virus infection is suggested by clinical signs of reproductive problems in breeding stock or respiratory disease in pigs of any age. Infection with PRRS virus often does not induce unique gross or microscopic lesions and may resemble or be obscured by lesions caused by other infectious agents. Therefore, definitive diagnosis of PRRS needs differential diagnosis with porcine parvovirus, pseudorabies virus, haemagglutinating encephalomyelitis virus, porcine circovirus type 2, porcine enterovirus, swine influenza virus, classical swine fever (hog cholera) virus, porcine cytomegalovirus and leptospirosis (Keffaber 1989; Halbur et al. 1993, 1995; Paton et al. 1992a, b). When the clinical history and pathology are suggestive of PRRS, detection of viral antigens, viral genomic material or isolation of virus from clinical specimens must be used to confirm the tentative diagnosis. Alternatively, documentation of rising serum antibodies in a time frame compatible with the clinical episode may support the diagnosis.

12.8.1 Detection of Virus or Viral Antigen

The fluorescent antibody test (FA) and immunohistochemistry (IHC) test on frozen tissue section can be used for detecting PRRS virus antigen in tissues. Although the FA test is specific, it may greatly be affected by sample quality and is not always very sensitive. In contrast, IHC is useful for detecting the virus in formalin-fixed tissues. IHC is more sensitive than direct FA examination of frozen tissues. Thus, a definitive diagnosis can be accomplished by the detection of microscopic lesions characteristic of PRRS virus in conjunction with IHC or FA tests. Preferred tissues for these tests include lungs, lymph nodes, spleen, thymus and tonsil (Halbur et al. 1996; Rossow et al. 1998).

Reverse transcriptase-polymerase chain reaction (RT-PCR)-based tests have been developed for detecting PRRSV genome in clinical specimens and are believed to be highly sensitive and highly specific (Benson et al. 2002; Horter et al. 2002). Since the virus does not need to be isolated in cell culture to detect the viral RNA, RT-PCR can be performed in a shorter duration of time than virus isolation. Several types of RT-PCR-based assays have been developed, most being designed to detect the ORF7, ORF6, ORF5 or ORF1b gene of PRRSV, and can be run directly on diagnostic specimens (Benson et al. 2002; Chen and Plagemann 1995; Christopher-Hennings et al. 1995a; Cook and Spatz 1998; Egli et al. 2001; Gilbert et al. 1997; Legeay et al. 1997; Mardassi et al. 1994; Oleksiewicz et al. 1998; Shin et al. 1997; Spagnuolo Weaver et al. 1998; Suarez et al. 1994; Rajkhowa et al. 2016). For added sensitivity, some PCR assays use a 'nested' procedure (RT-nPCR). Recently, automated fluorogenic PCR-based tests, such as the TaqMan™ PCR (Egli et al. 2001; Spagnuolo Weaver et al. 2000) or 'molecular beacon' PCR (Carlson et al. 2002), have been developed for detecting PRRS virus in clinical specimens. However, the use of PCR assays has become more common both for the diagnosis of PRRS and to aid in herd monitoring (Bierk et al. 2001; Dee et al. 2001; Horter et al. 2002; Kleiboeker et al. 2002).

12.8.2 Assays for Detection of Serum Antibodies

The enzyme-linked immunosorbent assay (ELISA), indirect fluorescent antibody (IFA) test, serum virus neutralisation (SVN) test and immunoperoxidase monolayer assay (IPMA) have all been used for the detection of antibodies specific for PRRS virus. The ELISA has the advantages of detecting specific antibody against North American and European PRRS virus strains and a fast turnaround time and is reported to be sensitive and specific (Albina et al. 1992; Edwards et al. 1994; Nodelijk et al. 1996; O'Connor et al. 2002; Takikawa et al. 1996). Several ELISA formats like an indirect ELISA using a sample-to-positive (S/P) ratio system (Yoon et al. 1995), an indirect ELISA using direct OD values (Albina et al. 1992; Cho et al. 1996; Takikawa et al. 1996) and a blocking ELISA (Ferrin et al. 2002; Houben et al. 1995; Zhou et al. 2001) have been described.

The IFA has high specificity (99.5%) but unknown sensitivity for individual animals (Yoon et al. 1992a, b). An advantage of the IFA test compared to ELISA is that the magnitude of the antibody titre can be determined. It can detect specific antibodies for 2–3 months after infection (Frey et al. 1992; Yoon et al. 1995; Bautista et al. 1993). The IPMA is also considered to be a highly specific and sensitive test, which can detect the PRRS-specific antibodies from 7 to 15 days after infection to 2–3 months (Frey et al. 1992; Ohlinger et al. 1992; Wensvoort et al. 1991; Yoon et al. 1995). The relatedness of the virus strain used in the assay and the virus strain infecting the pig will likely affect the performance of the IPMA test (Wensvoort et al. 1992; Yoon et al. 1997).

The SVN test is considered as less sensitive than IFA and ELISA (Benson et al. 2002), which is due to the fact that neutralising antibodies (the type of antibody detected by the SVN test) against PRRS virus develop as late as 1–2 months after infection (Frey et al. 1992; Goyal and Collins 1992; Minehart et al. 1992; Nelson et al. 1994; Yoon et al. 1995a). Therefore, SVN is best considered as a research tool rather than a routine diagnostic test because of its laborious nature.

12.9 Isolation of Virus

PRRS virus replicates only in two types of cells: porcine alveolar macrophages (PAMs) and certain African monkey kidney cell lines (Bautista et al. 1993; Dea et al. 1992; Paton et al. 1992a, b; Wensvoort et al. 1991; Yoon et al. 1992b). PAMs are suitable for isolation of European-like PRRS virus, while North American PRRSV grows well in African monkey kidney cell lines like MA-104/MAC145 (Dewey et al. 2000; Wensvoort et al. 1991). PRRSV can be isolated from a variety of clinical specimens (Done et al. 1996; Goyal 1993; Paton et al. 1992b; Rossow 1998), including serum, plasma, peripheral blood mononuclear cells (i.e. buffy coat layer), bone marrow, tonsil, lungs, lymph nodes, thymus, spleen, heart, brain, liver, testis, epididymis, ductus deferens, bulbourethral gland, penile tissue, oropharyngeal scraping, nasal turbinate, nasal swabs, placenta, saliva, urine, faeces and semen (Baron et al. 1992; Christianson et al. 1992; Christopher Hennings et al. 1995b, 1998, 2001; Dea et al. 1992; Done et al. 1992; Goyal and Collins 1992; Horter et al. 2002; Ohlinger et al. 1992; Paton et al. 1992a, b; Rossow et al. 1994, 1998). Serum, lung and bronchio-alveolar lavage fluid are samples of choice for isolation of PRRS virus in acutely infected animals. In older animals, viremia is of short duration, and PRRSV may be found in tissues longer than those in serum (Christopher-Hennings et al. 2001). For virus isolation from persistently infected animals, tonsil, oropharyngeal scraping and bronchio-alveolar lavage fluid are better samples than serum and lung (Christopher-Hennings et al. 2001; Horter et al. 2002). In cases of late-term abortion and early farrowing, samples should be collected from weak-born, pre-suckle pigs, rather than mummified, aborted or stillborn pigs (Done et al. 1992).

12.10 PRRSV Vaccines

Vaccines against PRRSV were developed, registered and administered in the field, soon after the emergence of PRRSV. However, despite the efficacy of certain vaccines under experimental as well as field conditions, eradication of the virus was never obtained. This was attributed most likely due to lack of vaccine efficacy against genetically divergent field strains. While the currently registered vaccines are still broadly applied, global efforts are on to develop new and better PRRSV vaccines.

Presently two types of PRRSV vaccines are available for field use: modified live virus (MLV) or attenuated vaccines, and killed virus (KV) or inactivated vaccines. The MLV vaccines are generated by in vitro cell culture passage of virulent virus until an attenuated phenotype is achieved, while the KV vaccines are generated by chemically or physically inactivating the virulent virus and are administered in combination with an adjuvant, while most MLV vaccines do not use adjuvants. Both type I- and type II-based MLV and KV vaccines have been developed and used globally.

MLV vaccines are the most potent in inducing protection and protect pigs to a large extent against viremia and virus replication in lungs and lymphoid tissues upon homologous challenge (Sattler et al. 2018). Hence, the vaccinated animals are well protected against respiratory disease, reproductive failure and transplacental infection, caused by homologous virus (Labarque et al. 2004; Labarque et al. 2003; Martelli et al. 2007; Nielsen et al. 1997; Scortti et al. 2006a; Zuckermann et al. 2007). However, the efficacy of MLV vaccines against heterologous challenge virus is poor. The amino acid or nucleotide identity is also not found to be a good predictive parameter for cross-protection against different PRRSV strains (Diaz et al. 2006; Labarque et al. 2003, 2004; Mengeling et al. 2003; Nodelijk et al. 2001; Okuda et al. 2008; van Woensel et al. 1998; Zuckermann et al. 2007). The immune response following MLV vaccination suffers from the same weaknesses as infection-induced immunity and is characterised by slow development of cell-mediated immune mechanisms with virus-neutralising antibodies appearing very weakly. Virus-neutralising antibodies induced by MLV vaccination are often detected only in neutralisation tests using the vaccine virus as antigen, which explains to a certain extent the strain-specific protective efficacy of these vaccines (Charerntantanakul et al. 2006a; Meier et al. 2003; Okuda et al. 2008; Scortti et al. 2006b; Zuckermann et al. 2007).

MLV vaccines are attenuated to such extent that they do not grow efficiently in macrophages anymore, and do not cause disease. However, vaccination with MLV vaccines still results in a certain level of virus replication in vivo and is often detected in blood and lungs of (mostly with type II PRRSV) MLV-vaccinated animals (Beilage et al. 2009; Diaz et al. 2006; Labarque et al. 2004; Scortti et al. 2006b). Also, infectious virus in pregnant sows may spread to the foetuses, and in boars to the semen (Christopher Hennings et al. 1997; Nielsen et al. 1997; Scortti et al. 2006a). This causes a major safety risk, since mutant virus variants with increased virulence may emerge during vaccine virus replication, and may spread to negative animals. The risk of reversion to virulence was dramatically illustrated in Denmark during the late 1990s, where huge outbreaks of PRRSV-associated reproductive disorders occurred as a result of a vaccination campaign in a PRRS-negative swine population (Nielsen et al. 2001; Nielsen et al. 2002). PRRSV attenuation and reversion to virulence are associated with changes in non-structural proteins, but the specific virulence determinants are not yet identified, which compromises the targeted development of new and safe attenuated PRRSV vaccines (Kim and Yoon 2008; Kwon et al. 2008, 2009; Nielsen et al. 2001). Taken together, MLV vaccination induces adaptive immunity, highly resembling the immune response induced by natural infection, which leads to protection only against the homologous virus.

Concerning the safety aspects, KV vaccines against PRRSV may be a good alternative to MLV vaccines. However, the use of KV vaccines in the field revealed that they induce only a moderate anamnestic antibody response upon the challenge of vaccinated animals and do not induce virus-specific neutralising antibodies. A beneficial effect of KV vaccination on reproductive performance has been reported in case of challenge with the virulent vaccine virus, but KV vaccines generally do not influence viremia, virus replication in tissues and shedding, even not in nearly homologous conditions (Nielsen et al. 1997; Nilubol et al. 2004; Plana-Duran et al. 1997; Scortti et al. 2007; Zuckermann et al. 2007). KV vaccination does not induce the virus-specific IFN-γ-producing cells, indicating that this type of vaccines does not stimulate an adaptive cell-mediated immune response (Meier et al. 2003; Nilubol et al. 2004; Piras et al. 2005; Zuckermann et al. 2007).

12.11 Conclusion and Prospects

Porcine reproductive and respiratory syndrome (PRRS) has devastated the swine industry worldwide since its emergence in the late 1980s. In spite of exhaustive research in the last four decades, pathogenesis and immunobiology of PRRSV are not clearly understood. PRRSV replicates in macrophages, can induce prolonged viremia and may cause persistent infections that last for months after infection. During this prolonged period of persistent infection, novel phenotypic variants of PRRSV emerge. The virus uses several evasion strategies to escape both innate and acquired immunity. In particular the downregulation of type I interferon-α production appears to interfere with the induction of acquired immunity. Because of the genetic diversity observed among PRRSV strains, it is unlikely that a vaccine based on a single strain will confer effective protection against the antigenically and genetically diversified field strains currently circulating in swine herds worldwide. Therefore, the development of PRRS vaccine that possesses high immunogenicity confers broad protection and is safe, but will continue to be a challenge for PRRS researchers in the years to come.

Acknowledgements All the authors of the manuscript thank and acknowledge their respective universities and institutes.

Conflict of interest: There is no conflict of interest.

References

Albina E, Carrat C, Charley B (1998a) Interferon-alpha response to swine arterivirus (PoAV), the porcine reproductive and respiratory syndrome virus. J Interf Cytokine Res 1(18):485–490

Albina E, Leforban Y, Baron T, Plana Duran J, Vannier P (1992) An enzyme linked immunosorbent assay (ELISA) for the detection of antibodies to the porcine reproductive and respiratory syndrome (PRRS) virus. Ann Rech Vet 23:167–176

Albina E, Piriou L, Hutet E, Cariolet R, L'Hospitalier R (1998b) Immune responses in pigs infected with porcine reproductive and respiratory syndrome virus (PRRSV). Vet Immunol Immunopathol 61:49–66

Allende R, Laegreid WW, Kutish GF, Galeota JA, Wills RW, Osorio FA (2000) Porcine reproductive and respiratory syndrome virus: description of persistence in individual pigs upon experimental infection. J Virol 74:10834–10837

Allende R, Lewis TL, Lu Z, Rock DL, Kutish GF, Ali A, Doster AR, Osorio FA (1999) North American and European porcine reproductive and respiratory syndrome viruses differ in nonstructural protein coding regions. J Gen Virol 80:307–315

An TQ, Tian ZJ, Xiao Y, Li R, Peng JM, Wei TC, Zhang Y, Zhou YJ, Tong GZ (2009) Origin of highly pathogenic porcine reproductive and respiratory syndrome virus, China. Emerg Infect Dis 16(2):365–367

An TQ, Zhou YJ, Liu GQ, Tian ZJ, Li J, Qiu HJ, Tong GZ (2007) Genetic diversity and phylogenetic analysis of glycoprotein 5 of PRRSV isolates in mainland China from 1996 to 2006: coexistence of two NA-subgenotypes with great diversity. Vet Microbiol 123(1–3):43–52

Baron T, Albina E, Leforban Y, Madec F, Guilmoto H, Plana Duran J, Vannier P (1992) Report on the first outbreaks of the porcine reproductive and respiratory syndrome (PRRS) in France. Diagnosis and viral isolation. Ann Rech Vet 23:161–166

Batista L, Pijoan C, Dee S, Michael O, Molitor T, Joo HS, Xiao Z, Murtaugh M (2004) Virological and immunological responses to porcine reproductive and respiratory syndrome virus in a large population of gilts. Can J Vet Res 68:267–273

Bautista EM, Goyal SM, Yoon IJ, Joo HS, Collins JE (1993) Comparison of porcine alveolar macrophages and CL2621 for the detection of porcine reproductive and respiratory syndrome (PRRS) virus and anti-PRRS antibody. J Vet Diagn Invest 5:163–165

Beilage EG, Nathues H, Meemken D, Harder TC, Doherr MG, Grotha I, Greiser Wilke I (2009) Frequency of PRRS live vaccine virus (European and North American genotype) in vaccinated and non-vaccinated pigs submitted for respiratory tract diagnostics in North-Western Germany. Prev Vet Med 92:31–37

Benfield DA, Nelson E, Collins JE, Harris L, Goya SM, Robison D, Christianson WT, Morrison RB, Gorcyca DE, Chladek DW (1992) Characterization of swine infertility and respiratory syndrome (SIRS) virus (isolate ATCC-VR2332). J Vet Diagn Investig 4:127–133

Benson J, Yaeger M, Christopher-Hennings J, Lager K, Yoon KJ (2002) A comparison of virus isolation, immunohistochemistry, fetal serology and reverse-transcription polymerase chain reaction assay for the identification of porcine reproductive and respiratory syndrome virus transplacental infection in the fetus. J Vet Diagn Investig 14:8–14

Bierk MD, Dee SA, Rossow KD, Otake S, Collins JE, Molitor TW (2001) Transmission of porcine reproductive and respiratory syndrome virus from persistently infected sows to contact controls. Can J Vet Res 65:261–266

Botner A, Strandbygaard B, Sorensen KJ, Have P, Madsen KG, Madsen ES, Alexandersen S (1997) Appearance of acute PRRS-like symptoms in sow herds after vaccination with a modified live PRRS vaccine. Vet Rec 141(19):497–499
Brockmeier SL, Lager KM (2002) Experimental airborne transmission of porcine reproductive and respiratory syndrome virus and Bordetella bronchiseptica. Vet Microbiol 89(4):267–275
Bush E, Corso B, Zimmerman J, Swenson S, Pyburn D, Burkgren T (1999) Update on the acute PRRS investigative study. Swine Health Prod 7:179–180
Butler JE, Lager KM, Golde W, Faaberg KS, Sinkora M, Loving C, Zhang YI (2014) Porcine reproductive and respiratory syndrome (PRRS): an immune dysregulatory pandemic. Immunol Res 59:81–108
Butler JE, Wertz N, Weber P, Lager KM (2008) Porcine reproductive and respiratory syndrome virus subverts repertoire development by proliferation of germline encoded B cells of all isotypes bearing hydrophobic heavy chain CDR3. J Immunol 180:2347–2356
Cancel-Tirado SM, Evans RB, Yoon KJ (2004) Monoclonal antibody analysis of porcine reproductive and respiratory syndrome virus epitopes associated with antibody-dependent enhancement and neutralization of virus infection. Vet Immunol Immunopathol 102:249–262
Carlson DL, Fang Y, Nelson EA, Christopher-Hennings J. (2002) Discriminating between PRRSV isolates and vaccine with quantitative, real-time RT-PCR. Proceedings of the American Association of Veterinary Laboratory Diagnosticians, pp 57–67
Carman S, Sanford SE, Dea S (1995) Assessment of seropositivity to porcine reproductive and respiratory syndrome (PRRS) virus in swine herds in Ontario—1978 to 1982. Can Vet J 36:776–777
Cha SH, Chang CC, Yoon KJ (2004) Instability of the restriction fragment length polymorphism pattern of open reading frame 5 of porcine reproductive and respiratory syndrome virus during sequential pig-to-pig passages. J Clin Microbiol 42(10):4462–4467
Chang HC, Peng YT, Chang HL, Chaung HC, Chung WB (2008) Phenotypic and functional modulation of bone marrow-derived dendritic cells by porcine reproductive and respiratory syndrome virus. Vet Microbiol 129:281–293
Charerntantanakul W, Platt R, Johnson W, Roof M, Vaughn E, Roth JA (2006a) Immune responses and protection by vaccine and various vaccine adjuvant candidates to virulent porcine reproductive and respiratory syndrome virus. Vet Immunol Immunopathol 109:99–115
Charerntantanakul W, Platt R, Roth JA (2006b) Effects of porcine reproductive and respiratory syndrome virus-infected antigen-presenting cells on T cell activation and antiviral cytokine production. Viral Immunol 19:646–661
Chen J, Liu T, Zhu CG, Jin YF, Zhang YZ (2006) Genetic variation of Chinese PRRSV strains based on ORF5 sequence. Biochem Genet 44(9–10):425–435
Chen Z, Plagemann PG (1995) Detection of related positive-strand RNA virus genomes by reverse transcription/polymerase chain reaction using degenerate primers for common replicase sequences. Virus Res 39:365–375
Cho HJ, Deregt D, Joo HS (1996) An ELISA for porcine reproductive and respiratory syndrome: production of antigen of high quality. Can J Vet Res 60:89–93
Christianson WT, Collins JE, Benfield DA, Harris L, Gorcyca DE, Chladek DW, Morrison RB, Joo HS (1992) Experimental reproduction of swine infertility and respiratory syndrome in pregnant sows. Am J Vet Res 53:485–488
Christopher-Hennings J, Holler LD, Benfield DA, Nelson EA (2001) Detection and duration of porcine reproductive and respiratory syndrome virus in semen, serum, peripheral blood mononuclear cells, and tissues from Yorkshire, Hampshire and Landrace boars. J Vet Diagn Investig 13:133–142
Christopher-Hennings J, Nelson EA, Althouse GC, Lunney J (2008) Comparative antiviral and proviral factors in semen and vaccines for preventing viral dissemination from the male reproductive tract and semen. Anim Health Res Rev 9:59–69

Christopher-Hennings J, Nelson EA, Hines RJ, Nelson JK, Swenson SL, Zimmerman JJ, Chase CCL, Yaeger MJ, Benfield DA (1995a) Persistence of porcine reproductive and respiratory syndrome virus in serum and semen of adult boars. J Vet Diagn Investig 7:456–464

Christopher-Hennings J, Nelson EA, Nelson JA, Hines RJ, Swenson SL, Hill HT, Zimmermann JJ, Katz JB, Yaeger MJ, Chase CCL, Benfield DA (1995b) Detection of porcine reproductive and respiratory syndrome virus in boar semen by PCR. J Clin Microbiol 33:1730–1734

Christopher-Hennings J, Nelson EA, Nelson JK, Benfield DA (1997) Effects of a modified-live virus vaccine against porcine reproductive and respiratory syndrome in boars. Am J Vet Res 58:40–45

Christopher-Hennings J, Nelson EA, Nelson JK et al (1998) Identification of porcine reproductive and respiratory syndrome virus in semen and tissues from vasectomized and nonvasectomized boars. Vet Pathol 35:260–267

Collins JE, Benfield DA, Christianson WT, Harris L, Hennings JC, Shaw DP, Goyal SM, McCullough S, Morrison RB, Joo HS et al (1992) Isolation of swine infertility and respiratory syndrome virus (isolateATCCVR-2332) in North America and experimental reproduction of the disease in gnotobiotic pigs. J Vet Diagn Investig 4(2):117–126

Cook CA, Spatz SJ (1998) Development of a commercial diagnostic assay for the detection of PRRSV in body fluids. Proceedings of the American Association of Swine Practitioners, pp 267–274

Costers S, Delputte PL, Nauwynck HJ (2006) Porcine reproductive and respiratory syndrome virus-infected alveolar macrophages contain no detectable levels of viral proteins in their plasma membrane and are protected against antibody-dependent, complement-mediated cell lysis. J Gen Virol 87:2341–2351

Das PB, Dinh PX, Ansari IH, de Lima M, Osorio FA, Pattnaik AK (2010) The minor envelope glycoproteins GP2a and GP4 of porcine reproductive and respiratory syndrome virus interact with the receptor CD163. J Virol 84:1731–1740

Dea SA, Bilodeau R, Athanassious R (1992) PRRS in Quebec: virology and pathologic aspects. Am Assoc Swine Pract Newsl 4:2–7

Dee S, Deen J, Rossow K, Weise C, Eliason R, Otake S, Joo HS, Pijoan C (2003) Mechanical transmission of porcine reproductive and respiratory syndrome virus throughout a coordinated sequence of events during warm weather. Can J Vet Res 67(1):12–19

Dee SA, Bierk MD, Deen J, Molitor TW (2001) An evaluation of test and removal for the elimination of porcine reproductive and respiratory syndrome virus from 5 swine farms. Can J Vet Res 65:22–27

Dewey C, Charbonneau G, Carman S, Hamel A, Nayar G, Friendship R, Eernisse K, Swenson S (2000) Lelystad-like strain of porcine reproductive and respiratory syndrome virus (PRRSV) identified in Canadian swine. Can Vet J 41(6):493–494

Diaz I, Darwich L, Pappaterra G, Pujols J, Mateu E (2005) Immune responses of pigs after experimental infection with a European strain of porcine reproductive and respiratory syndrome virus. J Gen Virol 86:1943–1951

Diaz I, Darwich L, Pappaterra G, Pujols J, Mateu E (2006) Different European-type vaccines against porcine reproductive and respiratory syndrome virus have different immunological properties and confer different protection to pigs. Virology 351:249–259

Done SH, Paton DJ, White ME (1996) Porcine reproductive and respiratory syndrome (PRRS): a review, with emphasis on pathological, virological and diagnostic aspects. Br Vet J 152:153–174

Done SH, Paton DS, Edwards S (1992) Porcine reproductive and respiratory syndrome ("blue-eared" pig disease). Pig Vet J 28:9–23

Drew TW, Lowings JP, Yapp F (1997) Variation in open reading frames 3, 4 and 7 among porcine reproductive and respiratory syndrome virus isolates in the UK. Vet Microbiol 55(1–4):209–221

Duan X, Nauwynck HJ, Pensaert MB (1997) Effects of origin and state of differentiation and activation of monocytes/macrophages on their susceptibility to porcine reproductive and respiratory syndrome virus (PRRSV). Arch Virol 142:2483–2497

Edwards S, Robertson IB, Wilesmith JW (1994) PRRS ("blue-eared pig disease") in Great Britain. Am Assoc Swine Pract Newsl 4:32–36

Egli C, Thur B, Liu L (2001) Quantitative TaqMan RT-PCR for the detection and differentiation of European and North American strains of porcine reproductive and respiratory syndrome virus. J Virol Methods 98:63–75

Fang Y, Schneider P, Zhang WP, Faaberg KS, Nelson EA, Rowland RR (2007) Diversity and evolution of a newly emerged North American Type 1 porcine arterivirus: analysis of isolates collected between 1999 and 2004. Arch Virol 152(5):1009–1017

Fang Y, Treffers EE, Li Y, Tas A, Sun Z, van der Meer Y, de Ru AH, van Veelen PA, Atkins JF, Snijder EJ, Firth AE (2012) Efficient−2 frameshifting by mammalian ribosomes to synthesize an additional arterivirus protein. PNAS 109:E2920–E2928

Ferrin N, Fang Y, Carroll J (2002) Validation of a blocking ELISA for antibodies against PRRSV. Proceedings of the International Pig Veterinary Society Congress, p 365

Flores-Mendoza L, Silva-Campa E, Reséndiz M, Osorio FA, Hernández J (2008) Porcine reproductive and respiratory syndrome virus infects mature porcine dendritic cells and up-regulates interleukin-10 production. Clin Vaccine Immunol 15:720–725

Forsberg R (2005) Divergence time of porcine reproductive and respiratory syndrome virus subtypes. Mol Biol Evol 22:2131–2134

Forsberg R, Storgaard T, Nielsen HS, Oleksiewicz MB, Cordioli P, Sala G, Hein J, Botner A (2002) The genetic diversity of European type PRRSV is similar to that of the North American type but is geographically skewed within Europe. Virology 299(1):38–47

Frey ML, Eernisse KA, Landgraf JG (1992) Diagnostic testing for SIRS virus at the National Veterinary Services Laboratories (NVSL). Am Assoc Swine Pract Newsl 4:31

Genini S, Delputte PL, Malinverni R, Cecere M, Stella A, Nauwynck HJ (2008) Genome-wide transcriptional response of primary alveolar macrophages following infection with porcine reproductive and respiratory syndrome virus. J Gen Virol 89:2550–2564

Gilbert SA, Larochelle R, Magar R (1997) Typing of porcine reproductive and respiratory syndrome viruses by a multiplex PCR assay. J Clin Microbiol 35:264–267

Gogoi A, Rajkhowa TK, Singh YD, Ravindran R, Arya RS, Hauhnar L (2017) Epidemiology of porcine reproductive and respiratory syndrome (PRRS) outbreak in India. Indian J Vet Pathol 41(1):31–37

Gorbalenya AE, Enjuanes L, Ziebuhr J, Snijder EJ (2006) Nidovirales: evolving the largest RNA virus genome. Virus Res 117:17–37

Goyal SM (1993) Porcine reproductive and respiratory syndrome. J Vet Diagn Investig 5:656–664

Goyal SM, Collins J (1992) SIRS serology and virus isolation. Minnesota Board An Health Newsl 37:2

Halbur PG, Bush EJ (1997) Update on abortion storms and sow mortality. Swine Health Prod 5:73

Halbur PG, Paul PS, Janke BH. (1993) Viral contributors to the porcine respiratory disease complex. Proceedings of the American Association of Swine Practitioners. pp 343–350

Halbur P, Andrew J, Paul P (1995) Strain variation of PRRS virus: field and research experiences. Proceedings of the American Association of Swine Practitioners, pp 200–204

Halbur PG, Paul PS, Frey ML, Landgraf J, Eernisse K, Meng XJ, Andrews JJ, Lum MA, Rathje JA (1996) Comparison of the antigen distribution of two US porcine reproductive and respiratory syndrome virus isolates with that of the Lelystad virus. Vet Pathol 33:159–170

Han J, Wang Y, Faaberg KS (2006) Complete genome analysis of RFLP 184 isolates of porcine reproductive and respiratory syndrome virus. Virus Res 122(1–2):175–182

Hanada K, Suzuki Y, Nakane T, Hirose O, Gojobori T (2005) The origin and evolution of porcine reproductive and respiratory syndrome viruses. Mol Biol Evol 22:1024–1031

Hill H 1990 Overview and history of mystery swine disease (swine infertility respiratory syndrome). Proceedings of the Mystery Swine Disease Committee Meeting, October 6, Denver, CO, pp 29–30

Holtkamp DJ, Kliebenstein JB, Neumann EJ, Zimmerman J, Rotto HF, Yoder TK, Wang C, Yeske P, Mowrer CL, Haley CA (2013) Assessment of the economic impact of porcine reproductive and respiratory syndrome virus on United States pork producers. J Swine Health Prod 21:72–84

Hopper SA, White MEC, Twiddy N (1992) An outbreak of blue-eared pig-disease (porcine reproductive and respiratory syndrome) in 4 pig herds in Great Britain. Vet Rec 131:140–144

Horter DC, Pogranichniy RC, Chang CC, Evans RB, Yoon KJ, Zimmerman JJ (2002) Characterization of the carrier state in porcine reproductive and respiratory syndrome virus infection. Vet Microbiol 86:213–218

Houben S, Callebaut P, Pensaert MB (1995) Comparative study of a blocking enzyme-linked immunosorbent assay and the immunoperoxidase monolayer assay for the detection of antibodies to the porcine reproductive and respiratory syndrome virus in pigs. J Virol Methods 51:125–128

Hu H, Li X, Zhang Z, Shuai J, Chen N, Liu G, Fang W (2009) Porcine reproductive and respiratory syndrome viruses predominant in southeastern China from 2004 to 2007 were from a common source and underwent further divergence. Arch Virol 154(3):391–398

Hurd HS, Bush EJ, Losinger W, Corso B, Zimmerman JJ, Wills R, Swenson S, Pyburn D, Yeske P, Burkgren T (2001) Outbreaks of porcine reproductive failure: report on a collaborative field investigation. J Swine Health Prod 9:103–108

Jiang P, Chen PY, Dong YY, Cai JL, Cai BX, Jiang ZH (2000) Isolation and genome characterization of porcine reproductive and respiratory syndrome virus in P. R. China. J Vet Diagn Investig 12:156–158

Kang SY, Yun SI, Park HS, Park CK, Choi HS, Lee YM (2004) Molecular characterization of PL97-1, the first Korean isolate of the porcine reproductive and respiratory syndrome virus. Virus Res 104(2):165–179

Kapur V, Elam MR, Pawlovich TM, Murtaugh MP (1996) Genetic variation in porcine reproductive and respiratory syndrome virus isolates in the midwestern United States. J Gen Virol 77(6):1271–1276

Keffaber K (1989) Reproductive failure of unknown etiology. Am Assoc Swine Pract Newsl 1(2):1–9

Key KF, Haqshenas G, Guenette DK, Swenson SL, Toth TE, Meng XJ (2001) Genetic variation and phylogenetic analyses of the ORF5 gene of acute porcine reproductive and respiratory syndrome virus isolates. Vet Microbiol 83(3):249–263

Kim HK, Yang JS, Moon HJ, Park SJ, Luo Y, Lee CS, Song DS, Kang BK, Ann SK, Jun CH, Park BK (2009) Genetic analysis of ORF5 of recent Korean porcine reproductive and respiratory syndrome viruses (PRRSVs) in viremic sera collected from MLV-vaccinating or non-vaccinating farms. J Vet Sci 10(2):121–130

Kim WI, Yoon KJ (2008) Molecular assessment of the role of envelope associated structural proteins in cross neutralization among different PRRS viruses. Virus Genes 37:380–391

Kleiboeker SB, Lehman JR, Fangman TJ (2002) Concurrent use of reverse transcription-polymerase chain reaction testing of oropharyngeal scrapings and paired serological testing for detection of porcine reproductive and respiratory syndrome virus infection in sows. J Swine Health Prod 10(6):251–258

Koyama S, Ishii KJ, Coban C, Akira S (2008) Innate immune response to viral infection. Cytokine 43:336–341

Kwon B, Ansari IH, Pattnaik AK, Osorio FA (2008) Identification of virulence determinants of porcine reproductive and respiratory syndrome virus through construction of chimeric clones. Virology 380:371–378

Labarque G, Reeth KV, Nauwynck H, Drexler C, Van Gucht S, Pensaert M (2004) Impact of genetic diversity of European-type porcine reproductive and respiratory syndrome virus strains on vaccine efficacy. Vaccine 22:4183–4190

Labarque G, Van Gucht S, Van Reeth K, Nauwynck H, Pensaert M (2003) Respiratory tract protection upon challenge of pigs vaccinated with attenuated porcine reproductive and respiratory syndrome virus vaccines. Vet Microbiol 95:187–197

Labarque GG, Nauwynck HJ, Van Reeth K, Pensaert MB (2000) Effect of cellular changes and onset of humoral immunity on the replication of porcine reproductive and respiratory syndrome virus in the lungs of pigs. J Gen Virol 81:1327–1334

Le Bon A, Schiavoni G, D'Agostino G, Gresser I, Belardelli F, Tough DF (2001) Type I interferons potently enhance humoral immunity and can promote isotype switching by stimulating dendritic cells in vivo. Immunity 14:461–470

Le Gall A, Legeay O, Bourhy H, Arnauld C, Albina E, Jestin A (1998) Molecular variation in the nucleoprotein gene (ORF7) of the porcine reproductive and respiratory syndrome virus (PRRSV). Virus Res 54(1):9–21

Lee C, Kim H, Kang B, Yeom M, Han S, Moon H, Park S, Song D, Park B (2010) Prevalence and phylogenetic analysis of the isolated type I porcine reproductive and respiratory syndrome virus from 2007 to 2008 in Korea. Virus Genes 40(2):225–230

Lee SM, Schommer SK, Kleiboeker SB (2004) Porcine reproductive and respiratory syndrome virus field isolates differ in in vitro interferon phenotypes. Vet Immunol Immunopathol 102:217–231

Legeay O, Bounaix S, Denis M (1997) Development of a RT-PCR test coupled with a microplate colorimetric assay for the detection of a swine Arterivirus (PRRSV) in boar semen. J Virol Methods 68:65–80

Li Y, Treffers EE, Napthine S, Tas A, Zhu L, Sun Z, Bell S, Mark BL, van Veelen PA, van Hemert MJ, Firth AE, Brierley I, Snijder EJ, Fang Y (2014) Transactivation of programmed ribosomal frameshifting by a viral protein. PNAS 111:E2172–E2181

Li Y, Wang X, Bo K, Wang X, Tang B, Yang B, Jiang W, Jiang P (2007) Emergence of a highly pathogenic porcine reproductive and respiratory syndrome virus in the mid-eastern region of China. Vet J 174:577–584

Loemba HD, Mounir S, Mardassi H, Archambault D, Dea S (1996) Kinetics of humoral immune response to the major structural proteins of the porcine reproductive and respiratory syndrome virus. Arch Virol 141:751–761

Loula T (1991) Mystery pig disease. Agri Prac 12:23–34

Loving CL, Brockmeier SL, Sacco RE (2007) Differential type I interferon activation and susceptibility of dendritic cell populations to porcine arterivirus. Immunology 120:217–229

Madsen KG, Hansen CM, Madsen ES, Strandbygaard B, Botner A, Sorensen KJ (1998) Sequence analysis of porcine reproductive and respiratory syndrome virus of the American type collected from Danish swine herds. Arch Virol 143(9):1683–1700

Mardassi H, Wilson L, Mounir S, Dea SA (1994) Detection of porcine reproductive and respiratory syndrome virus and efficient differentiation between Canadian and European strains by reverse transcription and PCR amplification. J Clin Microbiol 32:2197–2203

Martelli P, Cordioli P, Alborali LG, Gozio S, De Angelis E, Ferrari L, Lombardi G, Borghetti P (2007) Protection and immune response in pigs intradermally vaccinated against porcine reproductive and respiratory syndrome (PRRS) and subsequently exposed to a heterologous European (Italian cluster) field strain. Vaccine 25:3400–3408

Mayer J, Brisbin I (1991) Wild pigs of the United States: their biology, history, morphology and current status. University of Georgia Press, Athens, GA

Meier WA, Galeota J, Osorio FA, Husmann RJ, Schnitzlein WM, Zuckermann FA (2003) Gradual development of the interferon-gamma response of swine to porcine reproductive and respiratory syndrome virus infection or vaccination. Virology 309:18–31

Meng XJ (2000) Heterogeneity of porcine reproductive and respiratory syndrome virus: implications for current vaccine efficacy and future vaccine development. Vet Microbiol 74:309–329

Meng XJ, Paul PS, Halbur PG, Morozov I (1995) Sequence comparison of open reading frames 2 to 5 of low and high virulence United States isolates of porcine reproductive and respiratory syndrome virus. J Gen Virol 76(12):3181–3188

Mengeling WL, Lager KM (2000) A brief review of procedures and potential problems associated with the diagnosis of porcine reproductive and respiratory syndrome. Vet Res 3:61–69

Mengeling WL, Lager KM, Vorwald AC, Koehler KJ (2003) Strain specificity of the immune response of pigs following vaccination with various strains of porcine reproductive and respiratory syndrome virus. Vet Microbiol 93:13–24

Meulenberg JJ (2000) PRRSV, the virus. Vet Res 31:11–21

Minehart MJ, Nelson EA, Harley DJ (1992) Comparison of virus neutralization (VN) and indirect fluorescent antibody (IFA) for the detection of antibodies to porcine reproductive and respiratory syndrome (PRRS) virus. Proceedings of the Conference for Research Workers in Animal Disease, p 66

Mulupuri P, Zimmerman JJ, Hermann J, Johnson CR, Cano JP, Yu W, Dee SA, Murtaugh MP (2008) Antigen-specific B-cell responses to porcine reproductive and respiratory syndrome virus infection. J Virol 82:358–370

Murtaugh MP, Xiao Z, Zuckermann F (2002) Immunological responses of swine to porcine reproductive and respiratory syndrome virus infection. Viral Immunol 15:533–547

Nauwynck HJ, Duan X, Favoreel HW, Van Oostveldt P, Pensaert MB (1999) Entry of porcine reproductive and respiratory syndrome virus into porcine alveolar macrophages via receptor-mediated endocytosis. J Gen Virol 80:297–305

Nelsen CJ, Murtaugh MP, Faaberg KS (1999) Porcine reproductive and respiratory syndrome virus comparison: divergent evolution on two continents. J Virol 73:270–280

Nelson EA, Christopher-Hennings J, Benfield DA (1994) Serum immune responses to the proteins of porcine reproductive and respiratory syndrome (PRRS) virus. J Vet Diagn Investig 6:410–415

Neumann EJ, Kliebenstein JB, Johnson CD, Mabry JW, Bush EJ, Seitzinger AH, Green AL, Zimmerman JJ (2005) Assessment of the economic impact of porcine reproductive and respiratory syndrome on swine production in the United States. J Am Vet Med Assoc 227:385–392

Nielsen HS, Oleksiewicz MB, Forsberg R, Stadejek T, Bøtner A, Storgaard T (2001) Reversion of a live porcine reproductive and respiratory syndrome virus vaccine investigated by parallel mutations. J Gen Virol 82:1263–1272

Nielsen J, Botner A, Bille-Hansen V, Oleksiewicz MB, Storgaard T (2002) Experimental inoculation of late term pregnant sows with a field isolate of porcine reproductive and respiratory syndrome vaccine-derived virus. Vet Microbiol 84:1–13

Nielsen TL, Nielsen J, Have P, Baekbo P, Hoff-Jorgensen R, Botner A (1997) Examination of virus shedding in semen from vaccinated and from previously infected boars after experimental challenge with porcine reproductive and respiratory syndrome virus. Vet Microbiol 54:101–112

Nilubol D, Platt KB, Halbur PG, Torremorell M, Harris DL (2004) The effect of a killed porcine reproductive and respiratory syndrome virus (PRRSV) vaccine treatment on virus shedding in previously PRRSV infected pigs. Vet Microbiol 102:11–18

Nodelijk G, de Jong MC, van Leengoed LA, Wensvoort G, Pol JM, Steverink PJ, Verheijden JH (2001) A quantitative assessment of the effectiveness of PRRSV vaccination in pigs under experimental conditions. Vaccine 19:3636–3644

Nodelijk G, Wensvoort G, Kroese B (1996) Comparison of a commercial ELISA and an immunoperoxidase monolayer assay to detect antibodies against porcine respiratory and reproductive syndrome virus. Vet Microbiol 49:285–295

Normile D (2007) China, Vietnam grapple with 'rapidly evolving' pig virus. Science 317:1017

O'Connor M, Fallon M, O'Reilly PJ (2002) Detection of antibody to porcine reproductive and respiratory syndrome (PRRS) virus: reduction of cut-off value of an ELISA, with confirmation by immunoperoxidase monolayers assay. Irish Vet J 55:73–75

Ohlinger VF, Haas B, Sallmüller A (1992) In vivo and in vitro studies on the immunobiology of PRRS. Am Assoc Swine Pract Newsl 4:24

Okuda Y, Kuroda M, Ono M, Chikata S, Shibata I (2008) Efficacy of vaccination with porcine reproductive and respiratory syndrome virus following challenges with field isolates in Japan. J Vet Med Sci 70:1017–1025

Oleksiewicz MB, Botner A, Madsen KG, Storgaard T (1998) Sensitive detection and typing of porcine reproductive and respiratory syndrome virus by RT-PCR amplification of whole viral genes. Vet Microbiol 64:7–22

Ostrowski M, Galeota JA, Jar AM, Platt KB, Osorio FA, Lopez OJ (2002) Identification of neutralizing and nonneutralizing epitopes in the porcine reproductive and respiratory syndrome virus GP5 ectodomain. J Virol 76:4241–4250

Otake S, Dee SA, Moon RD, Rossow KD, Trincado C, Pijoan C (2003) Evaluation of mosquitoes, Aedes vexans, as biological vectors of porcine reproductive and respiratory syndrome virus. Can J Vet Res 67(4):265–270

Park JY, Kim HS, Seo SH (2008) Characterization of interaction between porcine reproductive and respiratory syndrome virus and porcine dendritic cells. J Microbiol Biotechnol 18:1709–1716

Paton DJ, Brown IH, Scott AC (1992a) Isolation of a Lelystad virus-like agent from British pigs and scanning electron microscopy of infected macrophages. Vet Microbiol 33:195–201

Paton DJ, Drew TW, Brown IH (1992b) Laboratory diagnosis of porcine reproductive and respiratory syndrome. Pig Vet J 29:188–192

Piras F, Bollard S, Laval F, Joisel F, Reynaud G, Charreyre C, Andreoni C, Juillard V (2005) Porcine reproductive and respiratory syndrome (PRRS) virus-specific interferon-gamma(+) T-cell responses after PRRS virus infection or vaccination with an inactivated PRRS vaccine. Viral Immunol 18:381–389

Plagemann PG (2003) Porcine reproductive and respiratory syndrome virus: origin hypothesis. Emerg Infect Dis 9:903–908

Plana-Duran J, Bastons M, Urniza A, Vayreda M, Vila X, Mane H (1997) Efficacy of an inactivated vaccine for prevention of reproductive failure induced by porcine reproductive and respiratory syndrome virus. Vet Microbiol 55:361–370

Plana-Duran J, Vayreda M, Vilarrasa J, Bastons M, Rosell R, Martinez M, San Gabriel A, Pujols J, Badiola JL, Ramos JA, Mariano D (1992) Porcine epidemic abortion and respiratory syndrome (mystery swine disease). Isolation in Spain of the causative agent and experimental reproduction of the disease. Vet Microbiol 33:203–211

Prieto C, Castro JM (2005) Porcine reproductive and respiratory syndrome virus infection in the boar: a review. Theriogenology 63:1–16

Rajkhowa TK, Catherine V, Singh YD, Ravindran R, Arya RS (2018) Genetic variation of Indian highly pathogenic porcine reproductive and respiratory syndrome viruses after introduction in 2013. Indian J Anim Sci 88(10):1118–1126

Rajkhowa TK, Gogoi A, Hauhnar L, Lalrohlua I (2015a) Molecular detection, epidemiology and clinico-pathological studies on first outbreak of porcine reproductive and respiratory syndrome (PRRS) in pig population of Mizoram, India. Indian J Anim Sci 85(4):343–347

Rajkhowa TK, Mohan Rao GJ, Gogoi A, Hauhnar L (2016) Indian porcine reproductive and respiratory syndrome virus bears discontinuous deletion of 30 amino acids in nonstructural protein 2. Virusdisease 27:287–293. https://doi.org/10.1007/s13337-016-0341-9

Rajkhowa TK, Mohan Rao GJ, Gogoi A, Hauhnar L, Lalrohlua I (2015b) Porcine reproductive and respiratory virus (PRRSV) from first outbreak of India shows close relationship with the highly pathogenic variant of China. Vet Q 35(4):186–193. https://doi.org/10.1080/01652176.2015.1066043

Renukaradhya GJ, Alekseev K, Jung K, Fang Y, Saif LJ (2010) Porcine reproductive and respiratory syndrome virus-induced immunosuppression exacerbates the inflammatory response to porcine respiratory coronavirus in pigs. Viral Immunol 23:457–466

Ropp SL, Wees CE, Fang Y, Nelson EA, Rossow KD, Bien M, Arndt B, Preszler S, Steen P, Christopher-Hennings J, Collins JE, Benfield DA, Faaberg KS (2004) Characterization of emerging European-like porcine reproductive and respiratory syndrome virus isolates in the United States. J Virol 78(7):3684–3703

Rossow KD (1998) Porcine reproductive and respiratory syndrome. Vet Pathol 35:1–20

Rossow KD, Bautista EM, Goyal SM, Molitor TW, Murtaugh MP, Morrison RB, Benfield DA, Collins JE (1994) Experimental porcine reproductive and respiratory syndrome virus infection in one-, four-, and 10-week-old pigs. J Vet Diagn Investig 6:3–12

Rowland RR (2007) The stealthy nature of PRRSV infection: the dangers posed by that ever-changing mystery swine disease. Vet J 174(3):451

Rowland RRR, Lawson S, Rossow K, Benfield DA (2003) Lymphoid tissue tropism of porcine reproductive and respiratory syndrome virus replication during persistent infection of pigs originally exposed to virus in utero. Vet Microbiol 96:219–235

Sang Y, Rowland RR, Blecha F (2011) Interaction between innate immunity and porcine reproductive and respiratory syndrome virus. Anim Health Res Rev 12:149–167

Sattler T, Pikalo J, Wodak E, Revilla-Fernández S, Steinrigl A, Bagó Z, Entenfellner F, Claude JB, Pez F, Francillette M, Schmoll F (2018) Efficacy of live attenuated porcine reproductive and respiratory syndrome virus 2 strains to protect pigs from challenge with a heterologous Vietnamese PRRSV 2 field strain. Vet Res 14:133

Scortti M, Prieto C, Alvarez E, Simarro I, Castro JM (2007) Failure of an inactivated vaccine against porcine reproductive and respiratory syndrome to protect gilts against a heterologous challenge with PRRSV. Vet Rec 1(61):809–813

Scortti M, Prieto C, Martinez-Lobo FJ, Simarro I, Castro JM (2006a) Effects of two commercial European modified-live vaccines against porcine reproductive and respiratory syndrome viruses in pregnant gilts. Vet J 172:506–514

Scortti M, Prieto C, Simarro I, Castro JM (2006b) Reproductive performance of gilts following vaccination and subsequent heterologous challenge with European strains of porcine reproductive and respiratory syndrome virus. Theriogenology 66:1884–1893

Shi M, Lam TT, Hon C-C, Murtaugh MP, Davies PR, Hui RK, Li J, Wong LT, Yip CW, Jiang JW, Leung FC-C (2010) A phylogeny-based evolutionary, demographical and geographical dissection of North American type 2 porcine reproductive and respiratory syndrome viruses. J Virol 84(17):8700–8711

Shimizu M, Yamada S, Murakami Y, Morozumi T, Kobayashi H, Mitani K, Ito N, Kubo M, Kimura K, Kobayashi M, Yamamoto K, Miura Y, Yamamoto T, Watanabe K (1994) Isolation of porcine reproductive and respiratory syndrome (Prrs) virus from Heko-Heko disease of pigs. J Vet Med Sci 56:389–391

Shin J, Torrison J, Choi CS (1997) Monitoring of porcine reproductive and respiratory syndrome virus infection in boars. Vet Microbiol 55:337–346

Snijder EJ, Meulenberg JJ (1998) The molecular biology of arteriviruses. J Gen Virol 79(5):961–979

Spagnuolo Weaver M, Walker IW, Campbell ST (2000) Rapid detection of porcine reproductive and respiratory syndrome viral nucleic acid in blood using a fluorimeter based PCR method. Vet Microbiol 76:15–23

Spagnuolo Weaver M, Walker IW, McNeilly F (1998) The reverse transcription polymerase chain reaction for the diagnosis of porcine reproductive and respiratory syndrome: comparison with virus isolation and serology. Vet Microbiol 62:207–215

Stadejek T, Oleksiewicz MB, Potapchuk D, Podgorska K (2006) Porcine reproductive and respiratory syndrome virus strains of exceptional diversity in Eastern Europe support the definition of new genetic subtypes. J Gen Virol 87(Pt 7):1835–1841

Stadejek T, Oleksiewicz MB, Scherbakov AV, Timina AM, Krabbe JS, Chabros K, Potapchuk D (2008) Definition of subtypes in the European genotype of porcine reproductive and respiratory syndrome virus: nucleocapsid characteristics and geographical distribution in Europe. Arch Virol 153:1479–1488

Stevenson GW, Vanalstine WG, Kanitz CL, Keffaber KK (1993) Endemic porcine reproductive and respiratory syndrome virus-infection of nursery pigs in 2 swine herds without current reproductive failure. J Vet Diagn Investig 5:432–434

Suarez P, Zardoya R, Martin MJ, Prieto C, Dopazo J, Solana A, Castro JM (1996) Phylogenetic relationships of European strains of porcine reproductive and respiratory syndrome virus

(PRRSV) inferred from DNA sequences of putative ORF-5 and ORF-7 genes. Virus Res 42(1–2):159–165

Suarez P, Zardoya R, Prieto C (1994) Direct detection of the porcine reproductive and respiratory syndrome (PRRS) virus by reverse polymerase chain reaction (RT-PCR). Arch Virol 135:89–99

Takikawa N, Kobayashi S, Ide S (1996) Detection of antibodies against porcine reproductive and respiratory syndrome (PRRS) virus in swine sera by enzyme-linked immunosorbent assay. J Vet Med Sci 58:355–357

Terpstra C, Wensvoort G, Pol JMA (1991) Experimental reproduction of porcine epidemic abortion and respiratory syndrome (mystery swine disease) by infection with Lelystad virus: Koch's postulates fulfilled. Vet Q 13:131–136

Thanawongnuwech R, Amonsin A, Tatsanakit A, Damrongwatanapokin S (2004) Genetics and geographical variation of porcine reproductive and respiratory syndrome virus (PRRSV) in Thailand. Vet Microbiol 101(1):9–21

Thanawongnuwech R, Young TF, Thacker BJ, Thacker EL (2001) Differential production of proinflammatory cytokines: in vitro PRRSV and mycoplasma hyopneumoniae co-infection model. Vet Immunol Immunopathol 79:115–127

Tian K, Yu X, Zhao T, Feng Y, Cao Z, Wang C, Hu Y, Chen X, Hu D, Tian X, Liu D, Zhang S, Deng X, Ding Y, Yang L, Zhang Y, Xiao H, Qiao M, Wang B, Hou L, Wang X, Yang X, Kang L, Sun M, Jin P, Wang S, Kitamura Y, Yan J, Gao GF (2007) Emergence of fatal PRRSV variants: unparalleled outbreaks of atypical PRRS in China and molecular dissection of the unique hallmark. PLoS One 2(6):e526

Tong GZ, Zhou YJ, Hao XF, Tian ZJ, An TQ, Qiu HJ (2007) Highly pathogenic porcine reproductive and respiratory syndrome, China. Emerg Infect Dis 13:1434–1436

Van Reeth K, Labarque G, Nauwynck H (1999) Differential production of proinflammatory cytokines in the pig lung during different respiratory virus infections: correlations with pathogenicity. Res Vet Sci 67:47–52

Van Woensel PA, Liefkens K, Demaret S (1998) Effect on viremia of an American and a European serotype PRRSV vaccine after challenge with European wild-type strain of the virus. Vet Rec 142:510–512

Vu HL, Kwon B, Yoon KJ, Laegreid WW, Pattnaik AK, Osorio FA (2011) Immune evasion of porcine reproductive and respiratory syndrome virus through glycan shielding involves both glycoprotein 5 as well as glycoprotein 3. J Virol 85:5555–5564

Wang X, Eaton M, Mayer M, Li H, He D, Nelson E (2007) Porcine reproductive and respiratory syndrome virus productively infects monocyte-derived dendritic cells and compromises their antigen-presenting ability. Arch Virol 152:289–303

Wang Y, Liang Y, Han J, Burkhart KM, Vaughn EM, Roof MB, Faaberg KS (2008) Attenuation of porcine reproductive and respiratory syndrome virus strain MN184 using chimeric construction with vaccine sequence. Virology 371(2):418–429

Wei X, Decker JM, Wang S, Hui H, Kappes JC, Wu X (2003) Antibody neutralization and escape by HIV-1. Nature 422:307–312

Welch SK, Calvert JG (2010) A brief review of CD163 and its role in PRRSV infection. Virus Res 154:98–103

Wensvoort G, de Kluyver EP, Luijtze EA (1992) Antigenic comparison of Lelystad virus and swine infertility and respiratory syndrome (SIRS) virus. J Vet Diagn Investig 4:134–138

Wensvoort G, Terpstra C, Pol JM, terLaak EA, Bloemraad M, de Kluyver EP, Kragten C, van Buiten L, den Besten A, Wagenaar F, Broekhuijsen JM, Moonen PLJM, Zetstra T, de Boer EA, Tibben HJ, de Jong MF, van 't Veld P, GJR G, van Gennep JA, Voets M, JHM V, Braamskamp J (1991) Mystery swine disease in The Netherlands: the isolation of Lelystad virus. Vet Q 13(3):121–130

Wesley RD, Mengeling WL, Lager KM, Clouser DF, Landgraf JG, Frey ML (1998) Differentiation of a porcine reproductive and respiratory syndrome virus vaccine strain from North American field strains by restriction fragment length polymorphism analysis of ORF 5. J Vet Diagn Investig 10(2):140–144

Wills RW, Doster AR, Galeota JA, Sur JH, Osorio FA (2003) Duration of infection and proportion of pigs persistently infected with porcine reproductive and respiratory syndrome virus. J Clin Microbiol 41:58–62

Wills RW, Zimmerman JJ, Yoon KJ, Swenson SL, McGinley MJ et al (1997) Porcine reproductive and respiratory syndrome virus: a persistent infection. Vet Microbiol 55:231–240

Yoon IJ, Joo HS, Christianson WT (1992a) An indirect fluorescent antibody test for the detection of antibody to swine infertility and respiratory syndrome virus in swine sera. J Vet Diagn Investig 4:144–147

Yoon IJ, Joo HS, Christianson WT (1992b) Isolation of a cytopathic virus from weak pigs on farms with a history of swine infertility and respiratory syndrome. J Vet Diagn Investig 4:139–143

Yoon KJ, Wu LL, Zimmerman JJ, Platt KB (1997) Field isolates of porcine reproductive and respiratory syndrome virus (PRRSV) vary in their susceptibility to antibody dependent enhancement (ADE) of infection. Vet Microbiol 55:277–287

Yoon KJ, Zimmerman JJ, McGinley MJ (1995a) Failure to consider the antigenic diversity of porcine reproductive and respiratory syndrome (PRRS) virus isolates may lead to misdiagnosis. J Vet Diagn Invest 7:386–387

Yoon KJ, Zimmerman JJ, Swenson SL, McGinley MJ, Eernisse KA, Brevik A, Rhinehart LL, Frey ML, Hill HT, Platt KB (1995) Characterization of the humoral immune response to porcine reproductive and respiratory syndrome (PRRS) virus infection. J Vet Diagn Investig 7:305–312

Zhou E, Zimmerman JJ, Zhou KX .(2001) Development of a blocking ELISA for detection of swine antibodies to PRRSV. Proceedings of the American Association of Veterinary Laboratory Diagnosticians, p 60

Zhou L, Ni YY, Pineyro P, Sanford BJ, Cossaboom CM, Dryman BA, Huang YW, Cao DJ, Meng XJ (2012) DNA shuffling of the GP3 genes of porcine reproductive and respiratory syndrome virus (PRRSV) produces a chimeric virus with an improved cross-neutralizing ability against a heterologous PRRSV strain. Virology 434:96–109

Zhou YJ, Hao XF, Tian ZJ, Tong GZ, Yoo D, An TQ, Zhou T, Li GX, Qiu HJ, Wei TC, Yuan XF (2008) Highly virulent porcine reproductive and respiratory syndrome virus emerged in China. Transbound Emerg Dis 55(3–4):152–164

Zimmerman JJ, Benfield DA, Murtaugh MP, Osorio FA, Stevenson GW, Torremorell M (2006) Porcine reproductive and respiratory syndrome virus (porcine arterivirus). In: Straw BE, Zimmerman JJ, D'Allaire S, Taylor DJ (eds) Diseases of swine, 9th edn. Blackwell Publishing, Ames, IA, pp 387–417

Zuckermann FA, Garcia EA, Luque ID, Christopher-Hennings J, Doster A, Brito M (2007) Assessment of the efficacy of commercial porcine reproductive and respiratory syndrome virus (PRRSV) vaccines based on measurement of serologic response, frequency of gamma-IFN-producing cells and virological parameters of protection upon challenge. Vet Microbiol 123:69–85

Chapter 13
Peste Des Petits Ruminant Virus

Balamurugan Vinayagamurthy, Govindaraj Gurrappa Naidu, and Parimal Roy

Abstract The *peste des petits ruminant* (PPR) virus is a small ruminant morbillivirus (SRMV), formerly known as PPR virus, belonging to the family *Paramyxoviridae*, and the genotype of the virus is clustered into four lineages, based on the nucleoprotein (N) and fusion (F) gene-based sequence and phylogenetic analysis. The PPR virus causes highly contagious, OIE notifiable and economically important transboundary animal viral disease of domestic and wild small ruminants, known as "plague of small ruminant". PPR is clinically manifested by high fever, nasal and ocular discharges, oral necrotising and erosive ulcers, gastroenteritis, and diarrhoea with respiratory distress followed by bronchopneumonia and death. The most important epizootiological risk parameters of PPR in sheep and goats are the introduction of animals to the flock from the unknown source, along with other management factors and rearing patterns. Though natural transmission occurs in cattle, buffaloes and camels, the clinical form of the disease is generally not observed. PPR can be diagnosed based on the clinical signs, pathological lesions and specific detection of the virus antigen or antibodies or genomic nucleic acid by various serological or molecular assays. With the availability of effective and safe vaccine and enabling of institutional mechanism, many countries implemented the vaccination programme to control the disease. PPR is one of the foremost constraints in enhancing the productivity of sheep and goats in developing countries, particularly affecting the economy of the poor landless, and small or marginal farmers. Considering the importance of sheep and goats in food security and socio-economic growth, Food and Agricultural Organization (FAO) and the World Organisation of Animal Health (OIE) jointly launched an international plan for control and eradication of PPR by 2030. This chapter delivers structures of the PPR virus and its genome organisation and functions, disease transmission, epidemiology and risk factors of the disease, diagnosis, prevention and control measures with perspectives.

B. Vinayagamurthy (✉) · G. Gurrappa Naidu
ICAR-National Institute of Veterinary Epidemiology and Disease Information (ICAR-NIVEDI), Bengaluru, Karnataka, India
e-mail: b.vinayagamurthy@icar.gov.in

P. Roy
Indian Council of Agricultural Research-National Institute of Veterinary Epidemiology and Disease Informatics (ICAR-NIVEDI), Bengaluru, Karnataka, India

Y. S. Malik et al. (eds.), *Emerging and Transboundary Animal Viruses*, Livestock Diseases and Management, https://doi.org/10.1007/978-981-15-0402-0_13

Keywords PPR virus · History · Prevalence · Transmission · Diagnosis · Prevention · Control

13.1 Preamble

The husbandry of small ruminants plays an important role in maintaining the sustainable livelihood of the small, marginal and landless rural farmers in underdeveloped and developing countries in the world. Sheep and goats are the "Any Time Money—ATM" of the poor landless farmers, and it generates self-employment and income for their livelihood throughout the year. Peste des petits ruminants (PPR), otherwise called as "plague of small ruminants" or "goat plague", is concomitant with considerable morbidity and mortality in sheep and goats and causes a serious threat to the economy of the endemic countries. PPR is an acute, extremely contagious, transboundary animal viral disease (TAD) of domestic sheep and goats and small wild ruminants. It causes severe socio-economic implications to livestock owners and is reported to cause an economic loss of USD 1.45–2.1 billion annually (OIE and FAO 2015). The disease was first reported in Côte d'Ivoire, West Africa, during 1942 (Gargadennec and Lalanne 1942) and then spread to other parts of Africa, the Arabian Peninsula, the Middle East and Asia (Balamurugan et al. 2014a; Muthuchelvan et al. 2015; Parida et al. 2015b; Baron et al. 2016). Clinically, PPR is manifested by depression, high fever, nasal and ocular discharges, oral necrotic ulcers, necrotising gastroenteritis, and diarrhoea with respiratory distress followed by bronchopneumonia.

The small ruminant morbillivirus (SRMV), otherwise known as PPR virus (PPRV) (Gibbs et al. 1979) (http://ictvonline.org/virusTaxonomy.asp), causes disease, and the virus is a single negative-stranded RNA genome of single genotype (Adombi et al. 2017) and clustered into four lineages based on the analyses of the nucleocapsid (N) and fusion (F) gene sequences (Shaila et al. 1996; Balamurugan et al. 2010b). Generally, lineages I to III circulate in Africa, while lineage IV (Asian lineage) circulates in Asian continents. However, a recent emergence of Asian lineage IV virus in African countries (Morocco, Marmara Region of Turkey) (Banyard et al. 2010; Kwiatek et al. 2011) and African lineage to China (Zhou et al. 2018) challenges the risk of PPR being introduced to Europe as well (Kwiatek et al. 2011; Baazizi et al. 2017), as lately PPR outbreaks were reported in Bulgaria (Altan et al. 2019). Therefore, the disease spread to various countries in African, Asian and European continents with an association of different lineages is a cause of worldwide concern (Kwiatek et al. 2011; Balamurugan et al. 2014a; Kumar et al. 2014; Parida et al. 2015b; Shatar et al. 2017; Niyokwishimira et al. 2019).

Thus, the transboundary nature of SRMV is expected to accentuate the disease spread to wider geographical region affecting the livestock productivity, especially in developing and underdeveloped countries. As of today, a total of 76 countries with ~1.7 billion of sheep and goats in Africa and Asian continents are in threat of the disease (http://www.fao.org/ppr/en/). Considering the sheep and goats in food security and socio-economic growth in the underdeveloped and developing countries of the globe, Food and Agricultural Organization (FAO) and the OIE jointly initiated a worldwide programme for control and eradication of PPR by 2030 (OIE and FAO 2015; http://www.fao.org/news/story/en/item/282397/icode; Parida et al. 2019).

13.2 History

The first report of "small ruminants plague" was described in Ivory Coast, West Africa, during 1942 (Gargadennec and Lalanne 1942). They reported an epidemic disease in small ruminants, which was the resemblance to rinderpest (RP), without the involvement of in-contact cattle. After that, the disease spread to Nigeria, Senegal, Ghana, Sudan and other countries in Africa until 1984. Further, disease spread to various countries in the African and Asian continents. In India, the first outbreak was reported from South India during 1987 (Shaila et al. 1989) and thereafter from other parts of northern India, and as of now the disease is enzootic, and outbreaks are regularly occurring throughout the year (Balamurugan et al. 2014a). Based on the analyses of the reports of the outbreak in the National Animal Disease Referral Expert System (NADRES) (http://www.nadres.res.in) from 1991 to 2017, PPR is the topmost disease reported in sheep and goats in India (Balamurugan et al. 2016). PPR is of snowballing status and extends its topographical spreading across the enzootic country. Further, evidence of infection has been reported from several other countries from time to time. Spread of disease to different countries in Africa, Asia and Southern Europe with an association of different lineages of the virus is a cause of animal health alarm (Kwiatek et al. 2011; Balamurugan et al. 2014a; Banyard et al. 2010), as outbreaks were reported in Bulgaria during June 2018. As of today, PPR is enzootic in various countries with the expansion of the range and several outbreaks are being reported recurrently from diverse countries in Africa, Asia and Europe with the involvement of different lineages of virus (Banyard et al. 2015).

13.3 Peste Des Petits Ruminants Virus

PPR virus is an envelope small ruminant morbillivirus (SRMV), which belongs to the genus *Morbillivirus* in the family *Paramyxoviridae* and the order *Mononegavirales* (Gibbs et al. 1979; http://ictvonline.org/virusTaxonomy.asp). On account of its genetic similarity with other members of the genus, PPRV is a relatively recently identified virus. Therefore, most of our understanding of virus structure and molecular biology is based on the comparison with other morbilliviruses (Baron 2015). PPRV is immunologically cross-reactive with other *Morbillivirus*es, namely measles virus (MV), rinderpest virus (RPV), canine distemper virus (CDV), dolphin and porpoise morbillivirus (DMV) and phocine distemper virus (PDV) (Barrett et al. 1993). The PPRV is pleomorphic with an envelope lipoprotein membrane and a ribonucleoprotein core, which contains non-segmented single-stranded negative RNA genome (Haffar et al. 1999) of ≈16 kb length (Chard et al. 2008). It has an envelope of about 8–15 nm thickness, with spikes of 8.5–14.5 nm length. The herringbone-like structured ribonucleoprotein strand measures ≈4–23 nm. The genome follows the *"rule of six"* and the order of coding genes and their arrangements are as follows: 3′ N–P/C/V–M–F–H–L 5′ (Bailey et al. 2005). While studying the evolutionary dynamics of PPRV, an evolutionary rate of 2.61×10^6 nucleotide substitutions per site per day was estimated (Bao et al. 2017). Each transcriptional unit is separated by conserved intergenic trinucleotides. The viral genome is separated into six transcriptional elements encoding six structural proteins (fusion (F) and haemagglutinin (H), matrix protein (M), nucleoprotein (N), large (L) and phosphoprotein (P)) and two non-structural (V, C) proteins. The N and P proteins form the enzyme complex in connotation with large (L) protein. Generally, morbilliviruses are classified into two groups: one group comprises CDV and PDV, and another group consists of RPV, MV and PPRV, based on nucleotide (nt) and amino acid (aa) sequences of N protein (Diallo 1990). The close antigenic relationship between the three (N, V and H) proteins of PPRV with DMV has also been reported (Bailey et al. 2005). The conserved 3′ and 5′ termini sequences of the virus are complementary and play an important role as regulatory elements in replication, transcription and packaging of the genome (Banyard et al. 2010). The PPRV leader sequence with 3′ UTR of N gene produces the virus genome promoter (GP) and similarly trailer sequence with 5′ UTR of L gene constitutes the anti-genome promoter (AGP) (Muniraju et al. 2014).

13.3.1 Nucleocapsid (N) Protein

The N protein of virus is the most abundant, antigenically well-conserved and foremost constituent of the nucleocapsid core of virion and plays an important role in transcription and replication. The nucleocapsid is the much conserved immunogenic core protein and expressed at a high level in the infected cells as compared to the other viral proteins. This protein is a noble target candidate for diagnosis assay (Munir et al. 2013) due to its antigenic stability and contains both type-specific and cross-reactive epitopes. It is highly immunogenic, but nucleocapsid antibodies are not protective due to its internal location. The genetic diversity of the N gene sequences was used for classification of SRMV into different lineages by phylogenic analysis. The N gene is located at 3′ end of the genome and encodes for the most transcribed N protein among all the genes of SRMV. The N protein has 525 amino acids (aa) and is possible to interact with other N protein (N–N interaction), P protein (N–P interaction) and L polymerase protein (P–L interaction) for involving in the replication of the virus. Even though sequences essential for self-assembly of N proteins are mapped, evidence of two domains {N-terminus (at 1–120aa) and central region (at 146–241aa)} was established for N–N self-assembly with a short peptide (at 120–145aa) essential for the stability of nucleoprotein. Recently, Ma et al. (2018) identified PPRV viral synonymous codon usage at the two transition boundaries in PPRV N protein and suggested that tRNA abundance of host variation might have potential effects on the formation of the secondary structure of PPRV N protein. Further, the amino acids in the region (at 106–210aa) are essential for suppressing the nuclear translocation of IRF3 (IFN regulatory factor 3) and IFN-β production, and the region (at 140–400aa) of IRF3 was identified as the crucial region for the N protein-IRF3 interaction (Zhu et al. 2019). Therefore, PPRV N protein as an extremely important antagonistic viral factor that plays a vital role in the suppression of IRF3 function and type I IFN production by interacting with IRF3 and abrogating the phosphorylation of IRF3, and interfering with the TBK1-IRF3 complex formation and thereby inhibiting IFN production (Zhu et al. 2019). Protein plays a vital role in the replication, as silencing of the mRNA can block the production of transcripts and further it right away hinders the synthesis of M protein, which results in inhibition of virus progeny.

13.3.2 Matrix (M) Protein

It is the most conserved and one of the smallest proteins, possesses an inherent ability to bind to lipid membranes and is situated inside the viral envelope (Haffar et al. 1999). It plays a key role in the assembly of virus and in the formation of new virion and release from the cells by budding as well as is adequate for assembly and release of virus-like particles (Wang et al. 2017). M protein has 335aa with an MW of 37.8 kDa. The highly conserved nature of the protein plays a critical role in the progeny virus formation and interacts with glycoproteins in the cell membrane. Further, it mediates the virus budding process favourably at specialised areas of the host membrane. It constitutes the internal coat of the virus envelope and acts as a link to connect F and H proteins with RNP core. For example, MV budding occurs at the apical microvilli due to greatly concentrated actin filaments, which are required for cellular transport. Moreover, the motif (at FMYL 50–53aa) is required for the localisation of the protein to enable the budding in Nipah virus as that of SRMV. Nevertheless, it is not well known whether these viruses share functional homologies. The defected M protein of MV and CDV prevents the formation of infectious virus, which leads to persistent infection, and Liu et al. (2015) showed inhibition of virus replication in vitro and increased virus-mediated fusogenicity by targeting mRNA of M protein by SiRNA approaches.

13.3.3 Surface Glycoproteins

The surface glycoproteins (H and F) of the virus facilitate attachment and penetration. The 137 kDa protein is coded by a conserved F gene and the highly immunogenic F protein have an MW of 59 kDa (Bailey et al. 2005). F protein is embedded in the lipid bilayer envelope as spikes (Diallo et al. 2007). The virulence of the virus depends on the cleavage of F protein, which is synthesised as a precursor (F0), and cleaved into F1 and F2 subunits, linked by a disulphide bond. Cleavage is mandatory for the virus to become fusogenic and infective and F2 subunits of morbillivirus contain a highly conserved glycosylation NXS/T site. Though the mechanism is not well understood, SRMV has RRTRR (at 104–108aa) (Chard et al. 2008), which is recognisable by the trans-Golgi-associated furin endopeptidase reliable to the cleavage RRX1X2R (X1-any aa, X2—either arginine or lysine) site of morbilliviruses. The F1 subunit of SRMV has four conserved motifs, namely N-terminus fusion peptide, heptad repeat 1 and 2 and a transmembrane domain. The identically structured heptad repeats have a fusion mechanism, which is likely to play a role in budding. Further, anchoring the fusion peptide domain in the membrane and dimerisation of heptad repeat lead to fusion (Rahaman et al. 2003). A conserved leucine zipper motif (at 459–480aa in PPRV) is accountable for enabling the oligomerisation and fusion of the protein through the mysterious process. The F protein of virus is needed for penetration and cell-to-cell spread and plays a crucial role in virus-induced cytopathology, haemolysis (Devireddy et al. 1999) and fusion and hemifusion activities (Seth and Shaila 2001).

The H protein facilitates the virus to bind to the cellar receptor, CD 150 or SLAM (Tatsuo et al. 2001). H and F are equally involved in the fusion activity of the F protein. The ORF of H protein gene (7376–9152 nucleotides) encodes at least conserved 67 kDa H protein (haemagglutinin–neuraminidase). H protein of PPRV and RPV shares only 50% amino acid identity, even though both viruses have 609aa residues that reflects the virus specificity for cell tropism and host range. The H protein of the virus plays a fundamental role in the progression of viral infection and specific binding to the host cell membrane and generation of virus-neutralising antibodies in the host. Moreover, it is the vital antigenic determinant and a major determinant of cell tropism in MV and is the main cause of cross-species pathogenesis. Further, H protein of PPRV requires a homologous F protein for its exact function during replication, and it possesses haemadsorption and neuraminidase activities (Seth and Shaila 2001) whereas in some paramyxoviruses surface proteins can cause haemagglutination and can carry neuraminidase activities. In PPRV, it agglutinates the erythrocytes from diverse mammalian and avian hosts, including the PPRV-infected cell cultures that readily haemadsorb chicken erythrocytes. Further, identification of epitopes and their motifs advances our understanding of the antigenic features of H protein of SRMV and provides a foundation for the progress of epitope-based diagnostic assays and multiple epitope vaccines (Yu et al. 2017).

13.3.4 Large (L) Protein

The RNA-dependent RNA polymerase is the largest, conserved and least abundant protein among morbilliviruses. L protein of SRMV is 2183aa long with an MW of 247.3 kDa identical to that of rinderpest virus (Bailey et al. 2005). L protein is a multifunctional catalytic protein that executes transcription and replication of the genomic RNA, besides capping, methylation and polyadenylation of viral mRNA including RNA triphosphatase (RTPase), guanylyltransferase (GTase) and methyltransferase activities (Ansari et al. 2019). Three functional motif sequences have been identified, though the direct actions of this protein are not investigated for PPRV (Munir et al. 2013). The binding site sequences (at 9–21aa) for P and L proteins for P–L interaction are conserved in paramyxoviruses, except first valine amino acid in PPRV (Chard et al. 2008). Further, though significant functions of the protein are not yet defined, it will be possible to determine the multifunctional activities of L protein of PPRV using the reconstituted system (Yunus and Shaila 2012). Further, Ansari et al. (2019) identified RTPase domain in the C-terminal region (at 1640–1840aa) of the L protein, which exhibits RTPase activity as well as RTPase-associated nucleotide triphosphatase activity (NTPase).

13.3.5 Phosphoprotein (P) Protein

The P, V and C proteins of morbillivirus are encoded by the overlapping ORF sequences of P gene (Mahapatra et al. 2003). The P protein of PPRV is least conserved and is acidic due to the abundance of serine and threonine and undergoes intensive post-translational phosphorylation. The P protein has a varied length of amino acids from 506 to 509 among morbilliviruses, and PPRV has the longest and plays a role at multiple levels in both viral replication and transcription and immune regulation (Mahapatra et al. 2003). C-terminus of the P protein is more conserved than the N-terminus and is involved in the N–P interaction, which is required for cell cycle control, regulation of transcription and translation. The interactive motifs in the P protein of PPRV are conserved and required for the P–N interaction and are also a vital element of the polymerase complex and a key determinant of cross-species pathogenicity.

13.3.6 Accessory Proteins

The accessory non-structural C and V proteins are synthesised from the ORF of the P gene through the utilisation of alternative start codons and RNA editing mechanism, respectively, in virus-infected cells. The C-terminus of C protein of morbillivirus is highly conserved (Mahapatra et al. 2003) and plays a vital role in viral replication and also inhibits interferon-beta (IFN-β) production by blocking the transcription factors of IFN-β activation. This molecular machinery of C protein of RPV inhibition still needs further investigation. The biological function of the C protein is not known in PPRV, whereas in MV infection it is a virulence factor. The V protein of the virus is generated from mRNA of the phosphoprotein gene by adding one or more non-template G residues by RNA editing. The amino acid length of V protein is highly variable among morbilliviruses, and the N-terminus of V protein shares amino acid residues to the P protein with different cysteine-rich residues at C-terminus. Like P protein, the V protein undertakes phosphorylation and has 60%

of the serine residues. The V protein plays a regulatory role in the transcription and replication of the virus, is a potent inhibitor of IFN and blocks IFN signalling through interference and suppression of STAT-mediated signalling (Ma et al. 2015).

13.3.7 Molecular Biology of Virus

The foremost elements of the host range and tissue tropism of PPRV are the cellular receptors. The virus uses the cellular SLAM or Ovine Nectin-4 molecule (Birch et al. 2013). The continuous cell line such as Vero (African green monkey kidney) (Lefevre and Diallo 1990), marmoset B-lymphoblastoid-B95a (Sreenivasa et al. 2006) and VeroNectin-4 (Fakri et al. 2016) cells can be used for isolation and propagation and titration of the viruses in vitro. Pawar et al. (2008) observed that association exists among replication of PPRV in PBMC of livestock species and the level of SLAM mRNA, and they inferred that the virus uses SLAM receptors of the unnatural host at abridged efficiencies. Generally, the virus exhibits classical cytopathic effect (CPE) in infected cells, like the rounding of the cells or grape-bunch-clusters, vacuolation, granulation and syncytia within 3–5 days of infection. Further, the CPE of virus-infected PBMCs was characterised by degeneration, ballooning, rounding and clumping of cells without syncytia (Mondal et al. 2001), but syncytia were also observed in B95a cells 2–3 days postinfection (pi) (Sreenivasa et al. 2006). Generally, the virus is fragile and sensitive to the ether, but is comparatively stable at pH 5.8–10.0 and inactivated at 50 °C for 60 min (OIE 2013).

The first step in the replication cycle of the virus is the attachment on the surface of the cell membrane and fusion to release genetic materials into the cytoplasm. After the binding and entry of the virus in the endometrial epithelial cells, it intensely affects early cellular gene expression (Yang et al. 2018a, b). The H protein of virus is accountable for attachment to the sialic acid on the cell membrane (Munir et al. 2013) surface through either SLAM or CD 150 (Pawar et al. 2008) or nectin-4 (Birch et al. 2013) receptor. Attachment activates the fusion of the F protein, enabling a fusion of the envelope with cell membrane and release of genetic materials into the cell. Replication of the morbilliviruses occurs solely in the cytoplasm of host cells. The viral genome is never found as naked RNA and is encapsidated by the N protein to form the helical ribonucleoproteins (RNP). This complex contains the encapsidated RNA, in conjunction with the P and L proteins, forming the mini replicative unit (Parida et al., 2015b). The polymerase complex acts on the viral genome, binds to the genome promoter and starts transcribing short leader RNAs and subsequently transcribing each gene, which leads to the building up of a transcriptional gradient. At certain points, the polymerase complex switches its action from mRNA production to full-length positive-sense RNA, which is thought to be linked to the accumulation of viral proteins, although the precise mechanism remains unclear (Parida et al. 2015a, b). Following the production of a full-length anti-genome RNA, the polymerase binds to the anti-genome RNA at the anti-genome promoter (3′) and generates nascent full-length sense genome. The synthesised viral components finally lead to viral egress from the host cell. The M protein of virus plays a significant role in bringing the nascent RNPs and viral glycoproteins to the host cell membrane, which results in the packaging, budding and release of nascent virions (Parida et al. 2015a, b).

Further, Chaudhary et al. (2015) observed that receptor tyrosine kinase (RTK) signalling regulates virus replication and PVRL4 (nectin-4), tumour-associated marker, plays a vital role in pathogenesis and promises its use in cancer therapies (Delpeut et al. 2014). Moreover, Balamurugan et al. (2008) proved the antiviral effect of aqueous extract of *Acacia arabica* on in vitro virus multiplication, whereas Khandelwal et al. (2015) reported that SNPs proficiently inhibited the multiplication of the virus. Further, Kumar et al. (2019) demonstrated that sarco/endoplasmic reticulum calcium-ATPase (SERCA) could regulate virus replication by using SERCA-specific inhibitor (Thapsigargin) to block viral entry into the target cells as well as synthesis of viral proteins. Further, Qi et al. (2018) investigated the roles of host microRNA in PPR virus replication and pathogenesis. Yang et al. (2018a, b) reported that virus infection in caprine endometrial epithelial cells activates an autophagy response, mediated by C and N proteins of the PPRV, and induced autophagy inhibits caspase-dependent apoptosis and thus contributes to the enhancement of viral replication and maturity in host cells.

Molecular characterisation and phylogenetic analyses of different PPRV isolates/strains from various countries in African and Asian continents (Padhi and Ma 2014; Baron 2015; Banyard and Parida 2015) defined the prevalence of four lineages of the virus (Shaila et al. 1996; Dhar et al. 2002). During the 1970s, lineage I virus was initially isolated from Nigeria and Senegal countries in West Africa and later from Cote d'Ivoire, Guinea, Senegal and Burkina Faso. Subsequently, lineage II viruses were isolated in the 1980s from different countries in West Africa (Dundon et al. 2018) and as of now this African lineage did not cross the Red Sea to Asian continents. Further, lineage III viruses have been reported from Meilig, Ethiopia, Yemen, Sudan, Oman and Burundi (Niyokwishimira et al. 2019), whereas the Asian lineage IV viruses have been reported from Arabian Peninsula, the Middle East and southeast part of Asia. Moreover, recently, lineage IV was also reported in African countries and Southern Europe (Altan et al. 2019). Further, co-circulation of different lineages of the virus in Nigeria (Woma et al. 2016) and China (Liu 2018) with the emergence of lineage IV virus in Ethiopia (Muniraju et al. 2014) has also been reported. In India, since the first PPR report so far involves only PPRV lineage IV strains/isolates (Balamurugan et al. 2010b; Muthuchelvan et al. 2014), and one isolate of lineage III (PPR TN/92) (Shaila et al. 1996).

13.4 Prevalence of Disease

The prevalence of virus antibodies and detection of virus antigen/genome in small ruminants are reported from different countries in the world (Singh et al. 2004a; Balamurugan et al. 2011, 2012b, 2015; Khan et al. 2008; Mbyuzi et al. 2014; Gari et al. 2017; Balamurugan 2017; Ali et al. 2019) including cattle, buffaloes and camels (Govindarajan et al. 1997; Abraham et al. 2005; Balamurugan et al. 2012a, 2014b; Sen et al. 2014; Omani et al. 2019). The presence of PPRV antibodies in small and large ruminants (Abubakar et al. 2017; Woma et al. 2016) infers subclinical/inapparent/non-lethal infection, as vaccination is irregularly practiced or restricted in the endemic settings of underdeveloped and developing countries in the world. Moreover, antibodies in small adult ruminant animals are not constantly indicative of PPR. Cattle and swine are the seroconverters to the virus, but they neither exhibit clinical symptoms nor transmit the infection. However, recently, Schulz et al. (2018) reported that suids are the likely source of infection. Moreover, cattle are dead-end hosts and do not play an epizootiological role in the maintenance and spread of the virus (Couacy-Hymann et al. 2019). All these studies revealed that PPR could also be transmitted from small ruminants to large ruminants and provided machinery for the survival of the virus in the environment outside the host. In unnatural hosts, PPRV may have a vital role to restrict the disease spread in certain topographical niche (Balamurugan et al. 2012a, 2014b), where small and large ruminants are reared together in the farming systems (Balamurugan et al. 2014b), which might be due to the fortuitous of adaption and alteration in the virulence of the virus. However, it needs further confirmation, based on the virus survival, host susceptibility of the host, virus mutation, variation in the disease severity, etc.

The role of wildlife in the epizootiology of the disease has been emphasised in literature (Taylor 1984), especially in wildlife species (e.g. in vulnerable wild small ruminants (Marashi et al. 2017), wild ungulates (Rahman et al. 2018), wild and domestic animals (Li et al. 2017)) which could be of substantial significance for the spread of the virus. There is a likelihood of wildlife-domestic interactions for feed and water reservoirs through grazing pastureland (Mahapatra et al. 2015), resulting in the spillover of the infection to wildlife (Rahman et al. 2016). However, the definite role of the wildlife on the epizootiology of PPR remains uncertain at present and to be investigated (Banyard et al. 2010; Balamurugan et al. 2015). Additionally, the presence of PPRV antibodies in the bovines and wild ruminants besides sheep and goats divulges the natural transmission (Abraham et al. 2005; Balamurugan et al. 2015). Moreover, the detection of PPRV in wildlife has not yet been described except in gazelles, camels (Khalafallaa et al. 2010; Zakian et al. 2016), wild bharal and chousingha (*Tetracerus quadricornis*), a member of the subfamily Bovinae (Jaisree et al. 2018). Further, PPRV in the nasal swabs of dogs by microarray screening (Ratta et al. 2016) and detection of virus genomes in the tissue of lion (Balamurugan et al. 2012c) provided new insight into the possible crossing of the species barrier. Because of the risk of PPR, vaccination should be adopted to circumvent the virus, and the circulation between wild and domestic small ruminants.

13.5 Risk Factors

The virus primarily affects domestic sheep and goats and rarely camels and other wildlife. Many researchers stated that though PPRV infects small ruminants severity is more in goats (Tripathi et al. 1996; Singh et al. 2004a; Balamurugan et al.

2015). Even though there is no evidence of carrier state (Furley et al. 1987), the infected animals excrete the PPRV in the secretions and excretions from the 3rd day to 26th day postinfection or even up to 16 weeks after infection (Balamurugan et al. 2006, 2010a; Liu et al. 2014; Wasee Ullah et al. 2016), and thus they play a vital role as a source of silent infection in the epizootiology of PPR. Toll-like receptors and cytokines have played a role in the differential susceptibility of species/breed to PPRV (Dhanasekaran et al. 2015) and analysing these factors with host genetic factors might provide further understanding on polymorphisms in the host and its susceptibility. Breed of the animals may also have a different outcome of disease (Lefèvre and Diallo 1990). Besides breeds, there exists a relationship between disease and age of the animals (6 months to 1 year), i.e. between the severity of the disease and the young animals. Young animals below 1 year of age are the significant predictors in rural systems or pastoral settings (Huyam et al. 2014; Gitonga 2015) and the passive immunity in offspring disappears after 4–5 months in endemic settings (Balamurugan et al. 2012d). Gowane et al. (2016) reported that the presence of maternal antibodies affects the vaccination response and inferred that the moderate estimate of heritability for vaccine response predicts the selection of goats for the greater vaccine response.

Moreover, response to the PPR vaccine in sheep and goats is extremely variable and influenced by environmental and genetic constituents, including MHC class II (Gowane et al. 2016, 2017, 2018). In general, the species, age and sex of the animals and the herd size have been identified as risk factors for PPR (Teshale et al. 2018). Further, the seasonal incidence of PPR was associated with the movement of the small ruminants and climate factors. However, in enzootic settings, PPR occurs around the year with more frequency in the lean period (Balamurugan et al. 2011, 2012b, 2016). Generally, small ruminants are raised by the small or marginal farmers on free-range grassland, shrubs and forest areas, and these animals frequently travel for long distances in the lean period for the search of fodder and water, thus causing a considerable upsurge of PPR incidence (Balamurugan et al. 2016). Most researchers have related the outbreaks with the introduction of new animals from the strange sources to the flocks as a major risk factor. Further, Kardjadj et al. (2015) reported abortion and associated risk factors in small ruminants for PPR. In general, the epidemiological risk parameters of PPR in sheep and goats are the young animals aged between 4 months and 1 year that are being severely affected. This warrants the need for public awareness of biosecurity management in the flocks (Almeshay et al. 2017).

Feeding pattern including stall feeding and grazing was not found to be a significant risk factor as reported (Rahman et al. 2016), although this variable has been considered as a risk factor for acquiring infectious diseases. However, stall feeding might have other management practices like the purchase of animals that upsurge the risk of transmission. Even though grazing was not a substantial risk factor, there is a quiet likelihood of domestic animal interfaces in pastures and aquatic sources for feed and water reservoirs for spillover of the disease. Rony et al. (2017) identified the determinants and space-time clusters of PPR while conducting hospital-based case-control study design and stated that vaccination should focus prior to the onset of winter and monsoon on the disease hotspot areas, especially on high-risk groups (young animals aged 4–24 months), to raise their immunity level in the lean periods. Differences in PPRV seropositivity depending on species, sex, age, season and geographical location have previously been described (Abubakar et al. 2008). However, Khan et al. (2008) stated that significantly greater proportions of sero-positive female animals were observed compared to male in their study. This may be related to the physiological differences between female and male, where females reveal some degree of infection resulting from stress due to milk production and pregnancies. Due to the significance of productivity potential, females sustained for a longer period as compared to males, thus increasing the likelihood of female animals being exposed to PPRV over time. PPRV actively circulated in the endemic regions and the migration of animals was the main source of spreading the disease in these regions (Almeshay et al. 2017).

13.6 Pathobiology

The disease occurs by the entry of the PPRV through the upper respiratory tract (URT) epithelial cells (Parida et al. 2015a). Pope et al. (2013) demonstrated that the initial site for the virus replication is within the tonsillar tissue and lymph nodes draining the site of inoculation. They proposed that immune cells transport the PPRV to lymphoid tissues where primary virus replication occurs, and then virus enters circulation (primary viraemia), and subsequently clinical signs appear according to the multiplication of the virus in the target cells/tissues (Truong et al. 2015). The virus proliferation and its pathogenicity are relational to those of many epizootiological determinants of the host (Munir et al. 2013; Balamurugan et al. 2015) including genetics and non-genetic determinants (Gowane et al. 2016, 2017). Kumar et al. (2004) reported the detection of the PPRV antigen in the respiratory epithelial syncytial cells. An affinity of the virus to the lung parenchymatous cells and lymphocytolysis was observed in Peyer's patches and mesenteric lymph nodes and further the virus was also detected in the ileal epithelial cells as dark brown granules. Nevertheless, various researchers have demonstrated the viruses in various tissues and organs, including epithelial crypts in the intestine. The details of the pathogenesis and involvement of immune cells during PPRV infection have been described (Balamurugan 2017). The transient leucocytosis was considered as a response to stress and immune activation and speculated the activation of B cells during secondary viraemia. Moreover, PPRV infection induces IFN-β weakly and transiently, and the virus can actively block the induction of IFN-β for protection (Sanz Bernardo et al. 2017). Kinetics of immune response has also been deliberated in vaccinated sheep and goats (Singh et al. 2004b, 2004c; Rajak et al. 2005) and protective antibodies persist for 3–6 years after a single vaccination (Saravanan et al. 2010; Zahur et al. 2014). The immunological study conducted by Rojas et al. (2019) indicated that F and H viral antigens are natural antibody-dependent cell-mediated cytotoxicity targets during infection and they stated a novel effector immune mechanism against PPRV in the host that could contribute to virus clearance. Both humoral and cell-mediated immune responses are involved in the immune suppression (Rajak et al. 2005). However, the precise mechanism of suppression by the activation of the host RIG-I-like receptor (RLR) pathway has yet to be elucidated. Moreover, Zhu et al. (2019) demonstrated an antagonistic mechanism of PPRV that significantly suppressed the RLR pathway activation and type I IFN production by suppressing IFN-β- and IFN-stimulated gene (ISG) expression. Jagtap et al. (2012) reported that immunosuppressed animals had a short period of viraemia, and more extensive and severe disease progression with high mortality than the non-immunosuppressed one. Further, Mondal et al. (2001) demonstrated virus-induced apoptosis in PBMCs whereas Kumar et al. (2002) reported apoptotic lymphoid cells in PBMCs from infected animals. Nevertheless, genomic and transcriptomic analysis of host and virus revealed 985 differently expressed genes and transcription factors modulating immune-regulatory pathways with involvement of genes in regulation, spliceosomal and identified apoptotic pathways that are dysregulated (Manjunath et al. 2015). They identified PPRV-induced miR-21-3p, miR-320a and miR-363 transcriptome, which might act cooperatively to increase viral pathogenesis by downregulating several immune response genes. Manjunath et al. (2017) predicted induction of ISGs by IRFs in an interferon-independent manner to activate a strong immune response based on the transcriptome analysis and qRT-PCR validation. Further, Baron et al. (2015) specified constant decrease in CD4+ T cells during pathogenesis in goats, while Truong et al. (2015) reported that the predominant sites of virus replication are the lymph nodes, lymphoid tissue and GI tract organs.

13.7 Transmission

For the spread of infection, the virus requires nearby contact of the infected and susceptible host due to the low resistance or labile of the virus outside the environment (Balamurugan 2017). During infection, substantial quantities of viruses have been excreted in the discharges, secretion and excretion of the infected animals that are the important source of infection (Balamurugan 2017). Transmission occurs primarily between closely contacted animals especially by direct contact with neighbouring animals by inhalation of the fine infectious aerosol droplets. These virus particles indirectly spread through contaminated water and feed troughs, bedding materials, fomites, etc. This method appears to be less significant since the virus does not survive outside the host for a long period (Lefèvre and Diallo 1990). Further, trade at local marketplaces where animals from multiple sources are brought into close contact provides better likelihood for spread. Besides sheep and goats, virus transmission to wild ruminants and unnatural hosts, namely cattle, buffaloes, camels, swine, etc., provides outside mechanism for the survival of the virus (Abraham et al. 2005) and the presence of PPRV antibodies in these unnatural hosts and wild ruminants suggests the natural transmission (Abraham et al. 2005;

Balamurugan et al. 2012a). Recently, Şevik and Oz (2019) investigated the possible role of the *Culicoides* spp. in the transmission of the virus. Even though seroconversion has been reported, clinical disease/signs in these unnatural hosts have not been reported (Sen et al. 2014). Migrating small ruminants will often come into close contact with other infected animals (Singh et al. 2004a) and spread the infection. Kivaria et al. (2013) reported that the spread and persistence of PPRV infection with epizootiological patterns and predictions in developing and underdeveloped countries are required for a better understanding of transmission. Hence, animal movement plays a pivotal part in the disease spread and the conservation of PPRV in the environment. Recently, Fournié et al. (2018) estimated the required immune population to be more than 37% in at least 71% of rural village populations in an endemic setting to prevent the virus spread by fitting a metapopulation-simulating model.

13.8 Clinical Manifestations

Clinical symptoms of the disease in small ruminants are manifested by high fever, discharges from the eye and nasal orifices, oral necrotic ulcers, gastroenteritis, diarrhoea and respiratory distress and bronchopneumonia (Balamurugan 2017). In camels, clinical symptoms are similar with additional manifestations of keratoconjunctivitis and oedema of the abdomen (Omani et al. 2019). PPR manifests as either peracute or acute or mild forms relying on the disease severity (OIE 2013; Balamurugan 2017), which depends on the virulence of lineage of virus (Couacy-Hymann et al. 2007), host, breed, age and immune status of the species, etc. PPR exhibits 4–5 different phases/stages during the infection, viz. (i) short incubation, which ranged from 3 to 10 days; (ii) prodromal feverish/febrile stage; (iii) mucosal stage with high fever, conjunctivitis, discharges from eye and nasal orifices (Fig. 13.1a, b), and erosions of the oral mucosa especially on the tongue, palate, lips and other parts (Fig. 13.1c); (iv) diarrhoeal phase (Fig. 13.1d), with pneumonia, dehydration and death; and (v) recovery phase, especially for non-fatal stage, in which animals develop lifelong immunity. Studies on the clinical symptoms and its complications in the manifestation of the disease have been discussed (Tripathi et al. 1996).

13.9 Diagnosis

PPR is diagnosed by specific clinical signs and characteristic pathological lesions in the affected animals, which include consolidation of lungs; enlarged spleen and intestinal mesenteric lymph nodes; streaks of haemorrhages on the mucosa of large intestines, etc. (Fig. 13.1e, f); and confirmative diagnosis by an array of laboratory techniques. A surfeit of diagnostic assays is available for the detection of either viral antigen or antibodies or genome. In general, the confirmatory conventional assays/tests including the gold standard virus isolation are not rapid, are less sensitive, are tedious, require technical expertise and culture facilities, and may not be useful for

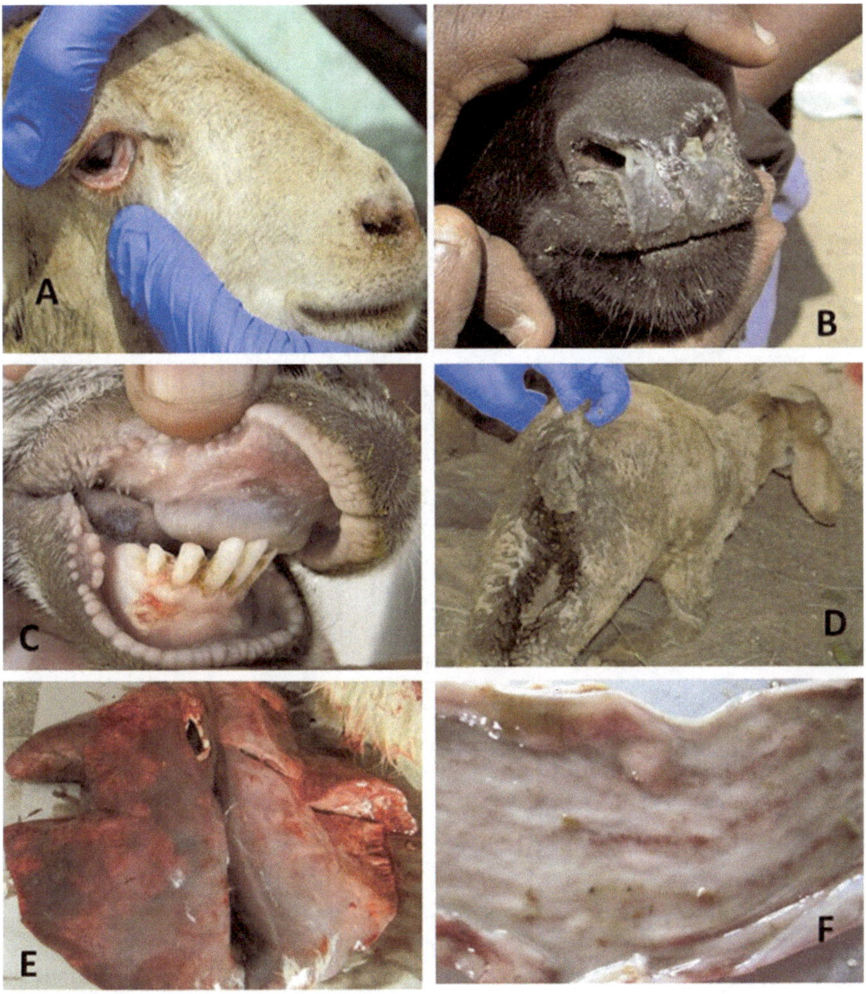

Fig. 13.1 (**a**–**f**) PPRV-infected sheep showing congestion of mucosa in the eye (reddened eye with mild conjunctivitis) (**a**); mucopurulent discharges from nostrils (**b**); ulcers (diphtheritic–plaques) and necrotic lesions in the gum (**c**); severe diarrhoea with solid hind quarters (Frank diarrhoea) (**d**); post-mortem lesions of PPR-affected sheep's lungs showing congestion and consolidation of lobes (**e**); and colon showing discontinuous streaks of congestion and haemorrhages on the mucosal folds (**f**)

primary diagnostics. For large-scale screening of samples, ELISA has widely been used for monitoring and surveillance of disease or even for diagnosis. Advanced understanding on the biology of the virus and molecular biological techniques leads to the development of rapid, highly specific and sensitive molecular assays (RT-PCR, real-time RT-PCR, and loop-mediated isothermal amplification (LAMP) assays) for the detection of the viral genome. Further, point-of-care diagnostics or on-the-field diagnostics tool, in the less equipped laboratory settings including lateral flow assay, has also been developed, but its usage is not up to the expected levels (Raj et al. 2008). Further, recombinant DNA technology made the production of safe recombinant viral proteins easy, as a better alternative to live antigen, and its application as the diagnostic antigen in the immunoassay for disease diagnosis. Recently, various modern diagnostic approaches have been developed by different researchers for detection of PPRV antibodies in serum and PPRV antigen/nucleic acid in the secretion, blood and tissue samples including milk (Clarke et al. 2018) and are summarised in Table 13.1. Parida et al. (2019) reported that nasal swabs are the most suitable to sample when considering the molecular diagnosis. These diagnostics can be efficiently applied at different stages in the PPR control and eradication phases based on the resource availability and the number of samples to be screened (Libeau 2015; Santhamani et al. 2016).

The differential diagnoses to be carried out with other diseases having similar clinical signs are contagious caprine pleuropneumonia (CCPP), pasteurellosis, bluetongue, contagious ecthyma (ORF), foot and mouth disease (FMD), etc. Sometimes, coexisting/concurrent or mixed infection with other viruses causing diseases has also been described (Balamurugan 2017; Kumar et al. 2016; Maan et al. 2018; Adedeji et al. 2019; Malik et al. 2011). Hence, identification of epizootiological and economically significant sets of these viruses or mixed infections could also support in establishing better guidelines for animal trade that could transmit further infection and epizootics (Kumar et al. 2016).

13.10 Socioeconomic Impact of PPR

Sheep and goat rearing provide social, financial and economic security to landless, marginal and small farmers in developing and underdeveloped countries. PPR in sheep and goats devastates the security fabric of the farmers who are already in distress due to various socio-economic-climate-associated risk factors in these regions. Epizootics of PPR can cause 50–80% mortality in naive small ruminants. The disease primarily causes mortality resulting in direct production loss. Further, live weight loss, milk loss and few abortion cases are also visible in the affected flocks. To control the disease, the farmers incur expenditure for treatment and labour cost to nurse the infected animals. In a few cases, distress sale is also pronounced, especially when a larger number of animals die on the farm. Further, at the macro level, the ripple effect on the entire small ruminant value chain and the spillover effects on various livestock interlinked sectors of the economy are evident if PPR outbreaks occur in wider geographical locations in large numbers. Some impacts are direct (production loss), and some are indirect (treatment cost, distress sale, market restrictions, etc.). Some impacts are tangible and easy to measure, but many are intangible and difficult to measure. The tangible impact variables include mortality, reduction in meat and milk, foregone revenue, etc. whereas intangibles include the reduction in functionality like asset building, diversification and income smoothing (de Haan et al. 2015). The consequent impacts of PPR on food access, consumption change, income and expenditure adjustment among the PPR-affected

Table 13.1 Diagnostic techniques for PPR diagnosis

Diagnostic techniques	Features
Gold standard test virus isolation	Primary bovine and sheep cells, Vero cell line, marmoset B-lymphoblastoid-B95 a, VeroNectin-4 (Fakri et al. 2016)
Virus or serum neutralisation test	Using cell culture facilities
Agar gel immunodiffusion test/ counter-immunoelectrophoresis	Tentative diagnosis
Haemagglutination	Simple, inexpensive and provides results within a few hours
Competitive ELISA/blocking ELISA/epitope-blocking ELISA	Neutralising monoclonal antibodies—MAb—H protein/N protein (Singh et al. 2004c; Bodjo et al. 2018)
Immuno-capture ELISA/ sandwich ELISA	Rapid differential identification and MAb—N protein (Singh et al. 2004b)
Indirect ELISA/combined indirect ELISA	PPRV antigen and polyclonal antibodies/G protein (Yousuf et al. 2015)
Simple and aqueous-phase ELISA (SNAP-ELISA)	RT-PCR/ELISA system
Immunohistochemical detection	Detection of antigens within tissues
Immunoperoxidase monolayer assay (IPMA)	IPMA an alternative method of VNT using BHK-21 cell line stably expressing the goat SLAM (Zhang et al. 2016)
Dot-ELISA	Using anti-M protein MAb/anti-N protein MAb
Immunofiltration/antigen-competition ELISA/lateral flow assay	Detection of antigen/antibodies in the clinical samples (Raj et al. 2008)
Recombinant antigen-based ELISA	Using N/H/F antigen (Balamurugan 2017; Basagoudanavar et al. 2018)
Nucleic acid hybridisation and RT-PCR	F gene, N gene or N and M gene in either one- or two-step procedure (Forsyth and Barrett 1995; Couacy-Hymann et al. 2002; Balamurugan et al. 2006)
PCR-ELISA	RT-PCR with ELISA (Saravanan et al. 2004)
Real-time RT-PCR	N or M gene-based SYBR Green/TaqMan real-time RT-PCR either in simplex or duplex format real-time or one-step multiplex (Bao et al. 2008; Balamurugan et al. 2010a, e; Polci et al. 2015; Settypalli et al. 2016)
Loop-mediated isothermal amplification (LAMP)	N gene-based assay (Li et al. 2010)
Luciferase immunoprecipitation system (LIPS)	Rapid detection and specific differentiation of PPRV antibodies (Berguido et al. 2016)
Reverse transcription recombinase polymerase amplification assays (RT-RPA)	Targeting N gene sequences—conventional and real-time RT-RPA (Zhang et al. 2018; Li et al. 2018)
Real-time and lateral-flow strip RT-RPA assay	Real-time fluorescent detection by targeting the N gene of PPRV (Yang et al. 2017)
Lateral-flow immunoassay strip (LFIAS)	Fast and ultrasensitive test-strip system combining quantum dots (QDs) with a LFIAS using recombinant PPRV N protein (Cheng et al. 2017)

flocks and effect on other social variables like women participation, employment, asset loss and family education also need to be assessed to know the impact holistically. Further, the impact of the PPR in smallholder's perspective in the realm of the significance of small ruminants within agriculture and services is needed to implement appropriate interventions (de Haan et al. 2015). Hence, the disease cost associated with PPR is huge at the farm, regional, national and global level. PPR causes severe socio-economic implications to farmers and is reported to cause an estimated economic loss of USD 1.45–2.1 billion annually (OIE and FAO 2015).

The assessment of the impact of the disease in different countries/geographic regions, production environment and management types is essential for developing an appropriate PPR control plan for a region/country/sub-country level. It would be useful to convince the governments and international organisations to support and provide required funds to control PPR, eventually to plan and implement the global PPR eradication campaign. The estimated total cost for the global eradication of PPR is to the tune of US$3080 million (Jones et al. 2016), which includes costs associated with global, regional and national coordination; institutional development, epidemiology and surveillance; establishing diagnostic laboratories, vaccination implementation, training and research; socio-economics; contingency; and emergency response. In India, the recent estimate by Govindaraj et al. (2016) revealed loss to the tune of Rs. 1611 crore at 10% annual incidence level. The studies on the investment benefits of PPR control and eradication revealed considerable benefits to various stakeholders. Furthermore, eliminating PPR will benefit millions of landless, small and marginal farmers, in the most vulnerable societies in developing and underdeveloped countries in African and Asian continents.

13.11 Prevention and Control

Generally, control and eradication of the disease depend on the efficient and timely diagnosis, surveillance, monitoring and execution of vaccination against the disease in the susceptible populations. The available PPR vaccines in India (Sungri 96 Strain) and other countries in the world (Nigeria 75/1 strain) have been widely used for the prevention and control of PPR in sheep and goats. However, these vaccines do not favour the differentiation of infected and vaccinated animals. PPR has an enormous impact on sheep and goat production and restricts trade in endemic countries. Therefore its prevention and control are of high priority. Implementation of effective prophylactic measures is highly imperative for disease control. The terrestrial spread of PPR along with its epidemiological parameters and ways in which it spreads can be abridged through the facility of diagnostic aids (Taylor 2016). Among the available control measures, vaccination is a recommended tool to contain the disease in endemic settings. Even though isolation or quarantine and restriction on the movement of animals in the affected areas limit the spread of the disease, it is difficult to set up at field level due to concomitant problems in underdeveloped and developing countries in the world. Further, in PPR enzootic countries, hygienic and phytosanitary control measures are problematic to sustain. Hence, mass vaccination is the only option available to control the disease effectively. Besides vaccination, quarantine of newly purchased animals for 2–3 weeks, understanding the health condition and the origin of purchased animals, continued monitoring of animals for illness, efficient cleansing of the contaminated places with disinfectants including equipment and clothing, etc. will pave the way for the control of the disease. There is no specific treatment available against PPR. The treatment regime of affected animals includes the use of broad-spectrum antibiotics plus fluid therapy along with vaccination of the affected flock during the first week of an outbreak in endemic settings (Abubakar et al. 2017).

13.11.1 Vaccination

Control strategies vary from different countries based on the prevalence of the disease. However, the choices are limited in underdeveloped and developing countries, as culling out policy is not feasible, owing to various socio-economic reasons. Vaccination is a recommended option for prevention and control of the disease to avoid an immediate economic loss to the farmers (Singh et al. 2009; Singh 2011; Balamurugan et al. 2016). In many countries, the success of rinderpest eradication provided the confidence to implement a similar programme for PPR. Scientific and technical experts, trained veterinarians, technical and para-veterinary staff for handling vaccines at various stages from production or procurement stages to vaccination in the field are paramount importance for the vaccination programme. Professional commitment from veterinarians and ancillary personnel involved in a mass immunisation programme is crucial to succeeding any vaccination programme (Singh et al. 2009). These prophylactic services are being gradually expanded by involving public-private partnership, especially the participation of non-governmental organisations, cooperatives and private veterinary practitioners in implementing and executing the disease control programme as stated (Singh et al. 2009). Further, setting up a network cum database would be of help in developing a coordinated approach towards effective implementation of the control programme. Some of the countries have already initiated PPR control measures either through their resources or with the help of international agencies to augment the small ruminant production (Singh and Bandyopadhyay 2015). Therefore, the launching of control and eradication programme appears to be a technically feasible, economically viable and practically attainable proposition (Singh et al. 2009).

PPR mass vaccination strategies include implementation of the campaign to achieve 80% flock immunity, by considering the dynamics population of small ruminants with the appearance of 30–40% naïve sheep and goat population every year, farming practices and agroclimatic conditions (Singh 2011; Woma et al. 2016). Thus, exhaustive vaccination of the whole population within a niche area in a specified period followed by vaccination of 3–4-month-old naïve animals (Singh 2011) is imperative to evade a window of susceptibility in newborn animals (Balamurugan et al. 2012d). However, vaccination should also be targeted on the high-risk young ones, migratory or trade market flocks, etc. (Singh 2011); otherwise, rigorous vaccinations can also be done in the identified niche area to make “PPR-free zone” in enzootic settings. The prophylaxis and control measures, including vaccination strategies, planning of control and eradication programme, have been discussed in details (Balamurugan et al. 2016; Balamurugan 2017; Raj and Thangavelu 2015).

13.12 Perspectives

PPR is one of the transboundary animal viral diseases of small ruminants, and its control is essential for poverty alleviation in endemic underdeveloped and developing countries, as the existence of PPR can limit domestic and international trade and production. Disease epizootics affects farmers, the livestock industry and the national economy. However, the economic impacts of PPR are underestimated due to the occurrence of other diseases in mixed forms. Due to dynamic populations of sheep and goats, intensive vaccination needs to be carried out frequently for the control of the disease. As of today, PPR has been brought under control in some developing countries by safe and effective PPR vaccines. However, the rapid spread of disease in wider geographical locations necessitates the studies to be undertaken to understand the effect of agroclimatic changes on the disease occurrence in various regions including the mechanism of spread and transmission dynamics of disease and analysing the association between occurrence and risk factors and the formulation of the suitable models for forewarning and forecasting of the disease. Moreover, adapting vaccination strategies to the population dynamics and the specific features of the local epidemiological settings would effect the optimised allocation of limited resources and increase the likelihood of PPR eradication (Fournié et al. 2018). Also, national and international funding agencies’ sponsored vaccination programmes revealed that final eradication could be attained with remarkable swiftness aimed at the culmination of the virus transmission if the origin of the PPRV infection was known (Taylor 2016). Further, research is needed for development and application of newer or latest or next-generation technologies for vaccines and diagnostics, but rather the cautious application of epidemiology and conventional virology to provide quantitative information to support for efficient control and eradication of the disease promptly (Baron et al. 2017). In the current RP-eradicated scenario, it gives the disease an even higher priority, as PPRV can infect large ruminants too. Further, FAO, OIE and other international agencies’ cooperation, understanding, tools and experiences now on proposal advocate that PPR could be controlled and eliminated far more swiftly than rinderpest.

Acknowledgements All the authors of the manuscript thank and acknowledge their respective universities and institutes.

Conflict of interest: There is no conflict of interest.

References

Abraham G, Sintayehu A, Libeau G, Albina E, Roger F, Laekemariam Y, Abayneh D, Awoke KM (2005) Antibody seroprevalences against Peste des petits ruminants (PPR) virus in camels, cattle, goats and sheep in Ethiopia. Prev Vet Med 70:51–57

Abubakar M, Ali Q, Khan HA (2008) Prevalence and mortality rate of Peste des petits ruminant (PPR): possible association with abortion in goat. Trop Anim Health Prod 40(5):317–321

Abubakar M, Mahapatra M, Muniraju M, Arshed MJ, Khan EH, Banyard AC, Ali Q, Parida S (2017) Serological detection of antibodies to Peste des petits ruminants virus in large ruminants. Transbound Emerg Dis 64(2):513–519

Adedeji AJ, Dashe Y, Akanbi OB, Woma TY, Jambol AR, Adole JA, Bolajoko MB, Chima N, Asala O, Tekki IS, Luka P, Okewole P (2019) Co-infection of Peste des petits ruminants and goatpox in a mixed flock of sheep and goats in Kanam, North Central Nigeria. Vet Med Sci 5(3):412–418. https://doi.org/10.1002/vms3.170

Adombi CM, Waqas A, Dundon WG, Li S, Daojin Y, Kakpo L, Aplogan GL, Diop M, Lo MM, Silber R, Loitsch A, Diallo A (2017) Peste des petits ruminants in Benin: persistence of a single virus genotype in the country for over 42 years. Transbound Emerg Dis 64(4):1037–1044

Ali WH, Osman NA, Asil RM, Mohamed BA, Abdelgadir SO, Mutwakil SM, Mohamed NEB (2019) Serological investigations of Peste des petits ruminants among cattle in Sudan. Trop Anim Health Prod 51(3):655–659

Almeshay MD, Gusbi A, Eldaghayes I, Mansouri R, Bengoumi M, Dayhum AS (2017) An epidemiological study on Peste des petits ruminants in Tripoli Region, Lybia. Vet Ital 53(3):235–242

Altan E, Parida S, Mahapatra M, Turan N, Yilmaz H (2019) Molecular characterization of Peste des petits ruminants viruses in the Marmara Region of Turkey. Transbound Emerg Dis 66(2):865–872

Ansari MY, Singh PK, Rajagopalan D, Shanmugam P, Bellur A, Shaila MS (2019) The large protein 'L' of Peste-des-petits-ruminants virus exhibits RNA triphosphatase activity, the first enzyme in mRNA capping pathway. Virus Genes 55(1):68–75

Baazizi R, Mahapatra M, Clarke BD, Ait-Oudhia K, Khelef D, Parida S (2017) Peste des petits ruminants (PPR): a neglected tropical disease in Maghreb region of North Africa and its threat to Europe. PLoS One 12(4):e0175461

Bailey D, Banyard AC, Dash P, Ozkul A, Barrett T (2005) Full genome sequence of peste des petits ruminants virus, a member of the Morbillivirus genus. Virus Res 110:119–124

Balamurugan V (2017) Peste des petits ruminants. In: Bayry J (ed) Emerging and re-emerging infectious diseases of livestock. Springer International Publishing AG, Cham, pp 55–98. ISBN978-3-3-19-47424-3

Balamurugan V, Govindaraj G, Rahman H (2016) Planning, implementation of Peste des petits ruminants control programme and strategies adopted for disease control in India. British J Virol 3:53–62

Balamurugan V, Hemadri D, Gajendragad MR, Singh RK, Rahman H (2014a) Diagnosis and control of peste des petits ruminants: a comprehensive review. Virus 25:39–56

Balamurugan V, Krishnamoorthy P, Raju DSN, Rajak KK, Bhanuprakash V, Pandey AB, Gajendragad MR, Rahman PK (2014b) Prevalence of *Peste-des-petits-ruminant virus* antibodies in cattle, buffaloes, sheep and goats in India. Indian J Virol 25:85–90

Balamurugan V, Krishnamoorthy P, Veeregowda BM, Sen A, Rajak KK, Bhanuprakash V, Gajendragad MR, Prabhudas K (2012a) Seroprevalence of Peste des petits ruminants in cattle and buffaloes from Southern Peninsular India. Trop Anim Health Pro 44:301–306

Balamurugan V, Rahman H, Munir M (2015) Host susceptibility to PPR virus. In: Munir M (ed) Peste des petits ruminants virus. Springer-Verlag Berlin Heidelberg, Berlin, pp 39–50

Balamurugan V, Saravanan P, Sen A, Rajak KK, Bhanuprakash V, Krishnamoorthy P, Singh RK (2011) Sero-epidemiological study of peste des petits ruminants in sheep and goats in India between 2003 and 2009. Rev Sci Tech 30:889–896

Balamurugan V, Saravanan P, Sen A, Rajak KK, Venkatesan G, Krishnamoorthy P, Bhanuprakash V, Singh RK (2012b) Prevalence of peste des petits ruminants among sheep and goats in India. J Vet Sci 13:279–285

Balamurugan V, Sen A, Saravanan P, Bhanuprakash V, Patra RC, Swarup D, Mahesh Kumar MP, Yadav MP, Singh RK (2008) Potential antiviral effect of aqueous extract of *Acacia arabica* leaves on peste des petits ruminants (PPR) virus replication. Pharm Biol 46:171–179

Balamurugan V, Sen A, Saravanan P, Singh RP, Singh RK, Rasool TJ, Bandyopadhyay SK (2006) One-step multiplex RT-PCR assay for the detection of Peste-des-petits-ruminants virus in clinical samples. Vet Res Commun 30:655–666

Balamurugan V, Sen A, Venkatesan G, Bhanot V, Yadav V, Bhanuprakash V, Singh RK (2012c) Peste des petits ruminants virus detected in tissues from an Asiatic lion (*Panthera Leo Persica*) belongs to Asian lineage IV. J Vet Sci 13:203–206

Balamurugan V, Sen A, Venkatesan G, Rajak KK, Bhanuprakash V, Singh RK (2012d) Study on passive immunity: time of vaccination in kids born to goats vaccinated against peste des petits ruminants. Virol Sin 27:228–233

Balamurugan V, Sen A, Venkatesan G, Yadav V, Bhanot V, Bhanuprakash V, Singh RK (2010a) Application of semi-quantitative M gene-based hydrolysis probe (TaqMan) real-time RT-PCR assay for the detection of Peste des petits ruminants virus in the clinical samples for investigation into clinical prevalence of disease. Transbound Emerg Dis 57:383–395

Balamurugan V, Sen A, Venkatesan G, Yadav V, Bhanot V, Bhanuprakash V, Singh RK (2012e) A rapid and sensitive one step-SYBR green based semi quantitative real time RT-PCR for the detection of peste des petits ruminants virus in the clinical samples. Virol Sin 27:1–9

Balamurugan V, Sen A, Venkatesan G, Yadav V, Bhanot V, Riyesh T, Bhanuprakash V, Singh RK (2010b) Sequence and phylogenetic analyses of the structural genes of virulent isolates and vaccine strains of Peste des petites ruminants virus from India. Transbound Emerg Dis 57:352–364

Banyard AC, Parida S (2015) Molecular epidemiology of PPR. In: Munir M (ed) Peste des petits ruminants virus. Springer-Verlag Berlin Heidelberg, Berlin, pp 69–94

Banyard AC, Parida S, Batten C, Oura C, Kwiatek O, Libeau G (2010) Global distribution of peste des petits ruminants virus and prospects for improved diagnosis and control. J Gen Virol 91:2885–2897

Banyard AC, Wang Z, Parida S (2015) Peste des petits ruminants virus, eastern Asia. Emerg Infect Dis 20:2176–2178

Bao J, Li L, Wang Z, Barrett T, Suo L, Zhao W, Liu Y, Liu C, Li J (2008) Development of one-step real-time RT-PCR assay for detection and quantitation of peste des petits ruminants virus. J Virol Methods 148:232–236

Bao J, Wang Q, Li L, Liu C, Zhang Z, Li J, Wang S, Wu X, Wang Z (2017) Evolutionary dynamics of recent peste des petits ruminants virus epidemic in China during 2013–2014. Virology 510:156–164

Baron J, Bin-Tarif A, Herbert R, Frost L, Taylor G, Baron MD (2015) Early changes in cytokine expression in peste des petits ruminants disease. Vet Res 45:22

Baron MD (2015) The molecular biology of PPRV. In: Munir M (ed) Peste des petits ruminants virus. Springer-Verlag Berlin Heidelberg, Berlin, pp 11–38

Baron MD, Diallo A, Lancelot R, Libeau G (2016) Peste des petits ruminants virus. Adv Virus Res 95:1–42

Baron MD, Diop B, Njeumi F, Willett BJ, Bailey D (2017) Future research to underpin successful peste des petits ruminants virus (PPRV) eradication. J Gen Virol 98(11):2635–2644

Barrett T, Amarel-Doel C, Kitching RP, Gusev A (1993) Use of the polymerase chain reaction in differentiating rinderpest field virus and vaccine virus in the same animals. Rev Sci Tech 12:865–872

Basagoudanavar SH, Hosamani M, Muthuchelvan D, Singh RP, Santhamani R, Sreenivasa BP, Saravanan P, Pandey AB, Singh RK, Venkataramanan R (2018) Baculovirus expression and purification of peste-des-petits-ruminants virus nucleocapsid protein and its application in diagnostic assay. Biologicals 55:38–42. https://doi.org/10.1016/j.biologicals.2018.07.0030

Berguido FJ, Bodjo SC, Loitsch A, Diallo A (2016) Specific detection of peste des petits ruminants virus antibodies in sheep and goat sera by the luciferase immunoprecipitation system. J Virol Methods 227:40–46

Birch J, Juleff N, Heaton MP, Kalbfleisch T, Kijas J, Bailey D (2013) Characterization of ovine Nectin-4, a novel peste des petits ruminants virus receptor. J Virol 87:4756–4761

Bodjo SC, Baziki JD, Nwankpa N, Chitsungo E, Koffi YM, Couacy-Hymann E, Diop M, Gizaw D, Tajelser IBA, Lelenta M, Diallo A, Tounkara K (2018) Development and validation of an epitope-blocking ELISA using an anti-haemagglutinin monoclonal antibody for specific detection of antibodies in sheep and goat sera directed against peste des petits ruminants virus. Arch Virol 163(7):1745–1756. https://doi.org/10.1007/s00705-018-3782-1

Chard LS, Bailey DS, Dash P, Banyard AC, Barrett T (2008) Full genome sequences of two virulent strains of peste-des-petits ruminants virus, the Coted'Ivoire 1989 and Nigeria 1976 strains. Virus Res 136:192–197

Chaudhary K, Chaubey KK, Singh SV, Kumar N (2015) Receptor tyrosine kinase signaling regulates replication of the peste des petits ruminants virus. Acta Virol 59:78–83

Cheng S, Sun J, Yang J, Lv J, Wu F, Lin Y, Liao L, Ye Y, Cao C, Fang L, Hua Q (2017) A new immunoassay of serum antibodies against Peste des petits ruminants virus using quantum dots and a lateral-flow test strip. Anal Bioanal Chem 409(1):133–141

Clarke BD, Islam MR, Yusuf MA, Mahapatra M, Parida S (2018) Molecular detection, isolation and characterization of Peste-des-petits ruminants virus from goat milk from outbreaks in Bangladesh and its implication for eradication strategy. Transbound Emerg Dis 65(6):1597–1604. https://doi.org/10.1111/tbed.12911

Couacy-Hymann E, Bodjo C, Danho T, Libeau G, Diallo A (2007) Evaluation of the virulence of some strains of peste-des-petits-ruminants virus (PPRV) in experimentally infected West African dwarf goats. Vet J 173:178–183

Couacy-Hymann E, Koffi MY, Kouadio VK, Mossoum A, Kouadio L, Kouassi A, Assemian K, Godji PH, Nana P (2019) Experimental infection of cattle with wild type peste-des-petits-ruminants virus—their role in its maintenance and spread. Res Vet Sci 124:118–122

Couacy-Hymann E, Roger F, Hurard C, Guillou JP, Libeau G, Diallo A (2002) Rapid and sensitive detection of peste des petits ruminants virus by a polymerase chain reaction assay. J Virol Methods 100:17–25

de Haan NC, Kimani T, Rushton J, Lubroth J (2015) Why is small ruminant health important–Peste des petits ruminants and its impact on poverty and economics? In: Munir M (ed) Peste des petits ruminants virus. Springer-Verlag Berlin Heidelberg, Berlin, pp 195–226

Delpeut S, Noyce RS, Richardson CD (2014) The tumor-associated marker, PVRL4 (nectin-4), is the epithelial receptor for morbilliviruses. Viruses 6:2268–2286

Devireddy LR, Raghavan R, Ramachandran S, Shaila MS (1999) The fusion protein of peste des petits ruminants virus is a hemolysin. Arch Virol 144:1241–1247

Dhanasekaran S, Biswas M, Vignesh AR, Ramya R, Raj GD, Tirumurugaan KG, Raja A, Kataria RS, Parida S, Elankumaran S (2015) Toll-like receptor responses to Peste des petits ruminants virus in goats and water buffalo. PLoS One 9:e111609

Dhar P, Sreenivasa BP, Barrett T, Corteyn M, Singh RP, Bandyopadhyay SK (2002) Recent epidemiology of peste des petits ruminants virus (PPRV). Vet Microbiol 88:153–159

Diallo A (1990) Morbillivirus group: genome organization and proteins. Vet Microbiol 23:55–163

Diallo A, Minet C, Le Goff C, Berhe G, Albina E, Libeau G, Barrett T (2007) The threat of peste des petits ruminants: progress in vaccine development for disease control. Vaccine 25:5591–5597

Dundon WG, Adombi CM, Kanu S, Loitsch A, Cattoli G, Diallo A (2018) Complete genome sequence of a lineage II Peste des petits ruminants virus from sierra Leone. Genome Announc 6(1):e01417

Fakri F, Elarkam A, Daouam S, Tadlaoui K, Fassi-Fihri O, Richardson CD, Elharrak M (2016) VeroNectin-4 is a highly sensitive cell line that can be used for the isolation and titration of Peste des Petits ruminants virus. J Virol Methods 228:135–139

Forsyth MA, Barrett T (1995) Detection and differentiation of rinderpest and peste des petits ruminants viruses in diagnostic and experimental samples by polymerase chain reaction using P and F gene-specific primers. Virus Res 39:151–163

Fournié G, Waret-Szkuta A, Camacho A, Yigezu LM, Pfeiffer DU, Roger F (2018) A dynamic model of transmission and elimination of peste des petits ruminants in Ethiopia. Proc Natl Acad Sci U S A 115(33):201711646. https://doi.org/10.1073/pnas.1711646115

Furley CW, Taylor WP, Obi TU (1987) An outbreak of *peste des petits ruminants* in a zoological collection. Vet Rec 121:443–447

Gargadennec L, Lalanne A (1942) La peste des petits ruminants. Bull Serve Zootech Epizoot Afr Occid Fr 5:16–21

Gari G, Serda B, Negesa D, Lemma F, Asgedom H (2017) Serological investigation of Peste des petits ruminants in East Shewa and Arsi Zones, Oromia Region, Ethiopia. Vet Med Int 17:9769071

Gibbs EP, Taylor WP, Lawman MJ, Bryant J (1979) Classification of peste des petits ruminants virus as the fourth member of the genus Morbillivirus. Intervirology 11:268–274

Gitonga P (2015) Spatial analysis of risk factors and their effects on peste des petits ruminants control strategies in Kajiado and Marsabit pastoral systems of Kenya. Doctor of Philosophy in Dryland Resource Management. https://doi.org/10.13140/RG.2.1.5043.4164

Govindaraj G, Balamurugan V, Rahman H (2016) Estimation of economic loss of PPR in sheep and goats in India: an annual incidence based analysis. Br J Virol 3(3s):77–85

Govindarajan R, Koteeswaran A, Venugopalan AT, Shyam G, Shaouna S, Shaila MS, Ramachandran S (1997) Isolation of pestes des petits ruminants virus from an outbreak in Indian buffalo *(Bubalus bubalis)*. Vet Rec 141:573–574

Gowane GR, Akram N, Misra SS, Prakash V, Kumar A (2018) Genetic diversity of Cahi DRB and DQB genes of caprine MHC class II in Sirohi goat. J Genet 97(2):483–492

Gowane GR, Akram N, Prince LLL, Prakash V, Kumar A (2017) Ovar-MHC polymorphism in Malpura and Avikalin sheep vaccinated for Peste des petits ruminants (PPR) virus. Anim Biotechnol 28(4):306–314

Gowane GR, Najif Akram SS, Misra VP, Kumar A (2016) Assessment of the antibody response to Peste des petits ruminants (PPR) disease vaccination in a flock of Sirohi goat kids. Small Rumin Res 138:20–24

Haffar A, Libeau G, Moussa A, Cècile M, Diallo A (1999) The matrix protein gene sequence analysis reveals close relationship between PPR virus and dolphin morbillivirus. Virus Res 64:69–75

Huyam AMES, Abdelhamid AMEK, Intisar S, Yahia HA (2014) Seroprevalence and risk factors of Peste des petits ruminants in sheep and goats in Sudan. J Advan Vet Anim Res 1:42–49. https://doi.org/10.5455/javar.2014.a12

Jagtap SP, Rajak KK, Garg UK, Sen A, Bhanuprakash V, Sudhakar SB, Balamurugan V, Patel A, Ahuja A, Singh RK, Vanamayya PR (2012) Effect of immunosuppression on pathogenesis of peste des petits ruminants (PPR) virus infection in goats. Microb Pathog 52:217–226

Jaisree S, Aravindhbabu RP, Roy P, Jayathangaraj MG (2018) Fatal peste des petits ruminants disease in Chowsingha. Transbound Emerg Dis 65(1):e198–e201

Jones BA, Rich KM, Mariner JC, Anderson J, Jeggo M, Thevasagayam S, Cai Y, Andrew RP, Peter R (2016) The economic impact of eradicating PPR: a benefit-cost analysis. PLoS One 11(2):E0149982

Kardjadj M, Kouidri B, Metref D, Luka PD, Ben-Mahdi MH (2015) Abortion and various associated risk factors in small ruminants in Algeria. Prev Vet Med 123:97–101

Khalafallaa AI, Saeed IK, Ali YH, Abdurrahman MB, Kwiatekc O, Libeau G, Obeidaa AA, Abbas Z (2010) An outbreak of peste des petits ruminants (PPR) in camels in the Sudan. Acta Trop 116:161–165

Khan HA, Siddique M, Rahman S, Abubakar M, Ashraf MJ (2008) The detection of antibody against peste des petits ruminants virus in sheep, goats, cattle and buffaloes. Trop Anim Health Pro 40:521–527

Khandelwal N, Kaur G, Chaubey KK, Singh P, Sharma S, Tiwari A, Singh SV, Kumar N (2015) Silver nanoparticles impair Peste des petits ruminants virus replication. Virus Res 190:1–7

Kivaria FM, Kwiatek O, Kapaga AM, Swai ES, Libeau G, Moshy W, Mbyuzi AO, Gladson J (2013) The incursion, persistence and spread of peste des petits ruminants in Tanzania: epidemiological patterns and predictions. Onderstepoort J Vet Res 80:593

Kumar N, Barua S, Riyesh T, Chaubey KK, Rawat KD, Khandelwal N, Mishra AK, Sharma N, Chandel SS, Sharma S, Singh MK, Sharma DK, Singh SV, Tripathi BN (2016) Complexities in isolation and purification of multiple viruses from mixed viral infections: viral interference, persistence and exclusion. PLoS One 11(5):e0156110

Kumar N, Khandelwal N, Kumar R, Chander Y, Rawat KD, Chaubey KK, Sharma S, Singh SV, Riyesh T, Tripathi BN, Barua S (2019) Inhibitor of sarco/endoplasmic reticulum calcium-ATPase impairs multiple steps of paramyxovirus replication. Front Microbiol 10:209

Kumar N, Maherchandani S, Kashyap SK, Singh SV, Sharma S, Chaubey KK, Ly H (2014) Peste des petits ruminants virus infection of small ruminants: a comprehensive review. Viruses 6(6):2287–2327

Kumar P, Kumar R, Sharma AK, Tripathi BN (2002) Pathology of peste des petits ruminants (PPR) in goats and sheep : spontaneous study. Indian J Vet Pathol 26:15–18

Kumar P, Tripathi BN, Sharma AK, Kumar R, Sreenivasa BP, Singh RP, Dhar P, Bandyopadyaya SK (2004) Pathological and immunohistochemical study of experimental Peste des petits ruminants virus infection in goats. J Veterinary Med Ser B 51:1–7

Kwiatek O, Ali YH, Saeed IK, Khalafalla AI, Mohamed OI, Obeida AA, Rahman MBA, Osman HM, Taha KM, Abbas Z, Harrak ME, Lhor Y, Diallo A, Lancelot R, Albina E, Libeau G (2011) Asian lineage of Peste des petits ruminants virus, Africa. Emerg Infect Dis 17:1223–1231

Lefèvre PC, Diallo A (1990) Peste des petits ruminants virus. Rev Sci Tech 9:951–965

Li J, Li L, Wu X, Liu F, Zou Y, Wang Q, Liu C, Bao J, Wang W, Ma W, Lin H, Huang J, Zheng X, Wang Z (2017) Diagnosis of Peste des petits ruminants in wild and domestic animals in Xinjiang, China, 2013–2016. Transbound Emerg Dis 64(6):e43–e47

Li L, Bao J, Wu X, Wang Z, Wang J, Gong M, Liu C, Li J (2010) Rapid detection of peste des petits ruminants virus by a reverse transcription loop-mediated isothermal amplification assay. J Virol Methods 170:37–41

Li Y, Li L, Fan X, Zou Y, Zhang Y, Wang Q, Sun C, Pan S, Wu X, Wang Z (2018) Development of real-time reverse transcription recombinase polymerase amplification (RPA) for rapid detection of peste des petits ruminants virus in clinical samples and its comparison with real-time PCR test. Sci Rep 8(1):17760

Libeau G (2015) Current advances in serological diagnosis of PPRV. In: Munir M (ed) Peste des petits ruminants virus. Springer-Verlag Berlin Heidelberg, Berlin, pp 133–154

Liu F (2018) Letter to the editor concerning "First report of peste des petits ruminants virus lineage II in Hydropotes inermis, China" by Zhou et al. Transbound Emerg Dis; 2017: https://doi.org/10.1111/tbed.12683. Transbound Emerg Dis 65(4):1125. https://doi.org/10.1111/tbed.12840

Liu F, Wu X, Zou Y, Li L, Liu S, Chi T, Wang Z (2015) Small interfering RNAs targeting peste des petits ruminants virus M mRNA increase virus-mediated fusogenicity and inhibit viral replication in vitro. Antivir Res 123:22–26

Liu W, Wu X, Wang Z, Bao J, Li L, Zhao Y, Li J (2014) Virus excretion and antibody dynamics in goats inoculated with a field isolate of peste des petits ruminants virus. Transbound Emerg Dis 60(suppl 2):63–68

Ma X, Yang X, Nian X, Zhang Z, Dou Y, Zhang X, Luo X, Su J, Zhu Q, Cai X (2015) Identification of amino-acid residues in the V protein of peste des petits ruminants essential for interference and suppression of STAT-mediated interferon signaling. Virology 483:54–63

Ma XX, Wang YN, Cao XA, Li XR, Liu YS, Zhou JH, Cai XP (2018) The effects of codon usage on the formation of secondary structures of nucleocapsid protein of peste des petits ruminants virus. Genes Genomics 40(9):905–912

Maan S, Kumar A, Gupta AK, Dalal A, Chaudhary D, Gupta TK, Bansal N, Kumar V, Batra K, Sindhu N, Kumar A, Mahajan NK, Maan NS, Mertens PPC (2018) Concurrent infection of bluetongue and Peste-des-petits-ruminants virus in small ruminants in Haryana state of India. Transbound Emerg Dis 65(1):235–239

Mahapatra M, Parida S, Egziabher BG, Diallo A, Barrett T (2003) Sequence analysis of the phosphoprotein gene of peste des petits ruminants (PPR) virus: editing of the gene transcript. Virus Res 96(1–2):85–98

Mahapatra M, Sayalel K, Muniraju M, Eblate E, Fyumagwa R, Shilinde L, Mdaki M, Keyyu J, Parida S, Kock R (2015) Spillover of Peste des petits ruminants virus from domestic to wild ruminants in the Serengeti ecosystem, Tanzania. Emerg Infect Dis 21(12):2230–2234

Malik YS, Singh D, Chandrashekar KM, Shukla S, Sharma K, Vaid N, Chakravarti S (2011) Occurrence of dual infection of peste-des-petits-ruminants and goatpox in indigenous goats of Central India. Transbound Emerg Dis 58(3):268–273

Manjunath S, Kumar GR, Mishra BP, Mishra B, Sahoo AP, Joshi CG, Tiwari AK, Rajak KK, Janga SC (2015) Genomic analysis of host—Peste des petits ruminants vaccine viral transcriptome uncovers transcription factors modulating immune regulatory pathways. Vet Res 46:15

Manjunath S, Mishra BP, Mishra B, Sahoo AP, Tiwari AK, Rajak KK, Muthuchelvan D, Saxena S, Santra L, Sahu AR, Wani SA, Singh RP, Singh YP, Pandey A, Kanchan S, Singh RK, Kumar GR, Janga SC (2017) Comparative and temporal transcriptome analysis of peste des petits ruminants virus infected goat peripheral blood mononuclear cells. Virus Res 229:28–40

Marashi M, Masoudi S, Moghadam MK, Modirrousta H, Marashi M, Parvizifar M, Dargi M, Saljooghian M, Homan F, Hoffmann B, Schulz C, Starick E, Beer M, Fereidouni S (2017) Peste des petits ruminants virus in vulnerable wild small ruminants, Iran, 2014–2016. Emerg Infect Dis 23(4):704–706

Mbyuzi AO, Komba EV, Kimera SI, Kambarage DM (2014) Sero-prevalence and associated risk factors of peste des petits ruminants and contagious caprine pleuro-pneumonia in goats and sheep in the southern zone of Tanzania. Prev Vet Med 116:138–144

Mondal B, Sreenivasa BP, Dhar P, Singh RP, Bandyopadhyay SK (2001) Apoptosis induced by *peste des petits* ruminants virus in goat peripheral blood mononuclear cells. Virus Res 73:113–119

Munir M, Zohari S, Berg M (2013) Molecular biology and pathogenesis of Peste des petits ruminants virus. In: Springer Briefs Anim Sci. Springer Science & Business Media, Berlin, pp 23–32

Muniraju M, Munir M, Parthiban AR, Banyard AC, Bao J, Wang Z, Ayebazibwe C, Ayelet G, El Harrak M, Mahapatra M, Libeau G, Batten C, Parida S (2014) Molecular evolution of peste des petits ruminants virus. Emerg Infect Dis 20:2023–2033

Muthuchelvan D, De A, Debnath B, Choudhary D, Venkatesan G, Rajak KK, Sudhakar SB, Hemadri D, Pandey AB, Parida S (2014) Molecular characterization of peste-des-petits ruminants virus (PPRV) isolated from an outbreak in the Indo-Bangladesh border of Tripura state of North-East India. Vet Microbiol 174(3–4):591–595

Muthuchelvan D, Rajak KK, Ramakrishnan MA, Choudhary D, Bhadouriya S, Saravanan P, Pandey AB, Singh RK (2015) Peste-des-petits-ruminants: an Indian perspective. Adv Anim Vet Sci 3:422–429

Niyokwishimira A, de D Baziki J, Dundon WG, Nwankpa N, Njoroge C, Boussini H, Wamwayi H, Jaw B, Cattoli G, Nkundwanayo C, Ntakirutimana D, Balikowa D, Nyabongo L, Zhang Z, Bodjo SC (2019) Detection and molecular characterization of Peste des petits ruminants virus from outbreaks in Burundi, December 2017–January 2018. Transbound Emerg Dis 66(5):2067–2073. https://doi.org/10.1111/tbed.13255

OIE and FAO (2015). Manual on global strategy for the control and eradication of Pesti des petits ruminants. http://www.fao.org/3/a-i4460e.pdf

OIE—World Organisation for Animal Health (2013) Manual of diagnostic tests and vaccines for terrestrial animals. OIE, Paris, pp 1–14

Omani RN, Gitao GC, Gachohi J, Gathumbi PK, Bwihangane BA, Abbey K, Chemweno VJ (2019) Peste des petits ruminants (PPR) in dromedary camels and small ruminants in Mandera and Wajir counties of Kenya. Adv Virol 2019:4028720

Padhi A, Ma L (2014) Genetic and epidemiological insights into the emergence of peste des petits ruminants virus (PPRV) across Asia and Africa. Sci Rep 4:7040

Parida S, Couacy-Hymann E, Pope RA, Mahapatra M, El Harrak M, Brownlie J, Banyard AC (2015a) Pathology of PPR. In: Munir M (ed) Peste des petits ruminants virus. Springer-Verlag Berlin Heidelberg, Berlin, pp 51–68

Parida S, Muniraju M, Mahapatra M, Muthuchelvan D, Buczkowski H, Banyard AC (2015b) Peste des petits ruminants. Vet Microbiol 181:90–106

Parida S, Selvaraj M, Gubbins S, Pope R, Banyard A, Mahapatra M (2019) Quantifying levels of Peste des petits ruminants (PPR) virus in excretions from experimentally infected goats and its importance for nascent PPR eradication programme. Viruses 11(3):E249

Pawar RM, Raj GD, Kumar TM, Raja A, Balachandran C (2008) Effect of siRNA mediated suppression of signaling lymphocyte activation molecule on replication of peste des petits ruminants virus *in vitro*. Virus Res 136:118–123

Polci A, Cosseddu GM, Ancora M, Pinoni C, El Harrak M, Sebhatu TT, Ghebremeskel E, Sghaier S, Lelli R, Monaco F (2015) Development and preliminary evaluation of a new real-time RT-PCR assay for detection of Peste des petits ruminants virus genome. Transbound Emerg Dis 62:332–338

Pope RA, Parida S, Bailey D, Brownlie J, Barrett T, Banyard AC (2013) Early events following experimental infection with Peste-Des-Petits ruminants virus suggest immune cell targeting. PLoS One 8(2):e55830

Qi X, Wang T, Xue Q, Li Z, Yang B, Wang J (2018) MicroRNA expression profiling of goat peripheral blood mononuclear cells in response to peste des petits ruminants virus infection. Vet Res 49(1):62

Rahaman A, Srinivasan N, Shamala N, Shaila MS (2003) The fusion core complex of the peste des petits ruminants virus is a six-helix bundle assembly. Biochemistry 42(4):922–931

Rahman A-U, Wensman JJ, Abubakar M, Shabbir MZ, Rossiter P (2018) Peste des petits ruminants in wild ungulates. Trop Anim Health Prod 50(8):1815–1819. https://doi.org/10.1007/s11250-018-1623-1626

Rahman UA, Abubakar M, Rasool MH, Manzoor S, Saqalein M, Rizwan M, Munir M, Ali Q, Wensman JJ (2016) Evaluation of risk factors for Peste des petits ruminants virus in sheep and goats at the wildlife-livestock interface in Punjab Province, Pakistan. Biomed Res Int 2016:7826245

Raj GD, Thangavelu A, Munir M (2015) Strategies and future of global eradication of PPRV. In: Munir M (ed) Peste des petits ruminants virus. Springer-Verlag Berlin Heidelberg, Berlin, pp 227–254

Raj GD, Rajanathan TM, Kumar CS, Ramathilagam G, Hiremath G, Shaila MS (2008) Detection of peste des petits ruminants virus antigen using immunofiltration and antigen-competition ELISA methods. Vet Microbiol 129(3–4):246–251

Rajak KK, Sreenivasa BP, Hosamani M, Singh RP, Singh SK, Singh RK, Bandyopadhyay SK (2005) Experimental studies on immunosuppressive effects of peste des petits ruminants (PPR) virus in goats. Comp Immunol Microbiol Infect Dis 28:287–296

Ratta B, Pokhriyal M, Singh SK, Kumar A, Saxena M, Sharma B (2016) Detection of Peste des petits ruminants virus (PPRV) genome from nasal swabs of dogs. Curr Microbiol 73(1):99–103

Rojas JM, Rodríguez-Martín D, Avia M, Martín V, Sevilla N (2019) Peste des petits ruminants virus fusion and hemagglutinin proteins trigger antibody-dependent cell-mediated cytotoxicity in infected cells. Front Immunol 9:3172

Rony MS, Rahman AKMA, Alam MM, Dhand N, Ward MP (2017) Peste des petits ruminants risk factors and space-time clusters in Mymensingh, Bangladesh. Transbound Emerg Dis 64(6):2042–2048

Santhamani R, Singh RP, Njeumi F (2016) Peste des petits ruminants diagnosis and diagnostic tools at a glance: perspectives on global control and eradication. Arch Virol 161(11):2953–2967

Sanz Bernardo B, Goodbourn S, Baron MD (2017) Control of the induction of type I interferon by Peste des petits ruminants virus. PLoS One 12(5):e0177300

Saravanan P, Singh RP, Balamurugan V, Dhar P, Sreenivasa BP, Muthuchelvan D, Sen A, Aleyas AG, Singh RK, Bandyopadhyay SK (2004) Development of a N gene-based PCR-ELISA for detection of Peste-des-petits-ruminants virus in clinical samples. Acta Virol 48(4):249–255

Saravanan P, Balamurugan V, Sen A, Sreenivasa BP, Singh RP, Bandyopadhyay SK, Singh RK (2010) Long term immune response of goats to a Vero cell adapted live attenuated homologous PPR vaccine. Indian Vet J 87:1–3

Schulz C, Fast C, Schlottau K, Hoffmann B, Beer M (2018) Neglected hosts of small ruminant Morbillivirus. Emerg Infect Dis 24(12):2334–2337

Sen A, Saravanan P, Balamurugan V, Bhanuprakash V, Venkatesan G, Sarkar J, Rajak KK, Ahuja A, Yadav V, Sudhakar SB, Parida S, Singh RK (2014) Detection of subclinical peste des petits ruminants virus infection in experimental cattle. Virus 25:488–492

Seth S, Shaila MS (2001) The hemagglutinin-neuraminidase protein of peste des petits ruminants virus is biologically active when transiently expressed in mammalian cells. Virus Res 75:169–177

Settypalli TB, Lamien CE, Spergser J, Lelenta M, Wade A, Gelaye E, Loitsch A, Minoungou G, Thiaucourt F, Diallo A (2016) One-step multiplex RT-qPCR assay for the detection of Peste des petits ruminants virus, Capripoxvirus, Pasteurella multocida and mycoplasma capricolum subspecies (ssp.) capripneumoniae. PLoS One 11(4):e0153688

Şevik M, Oz ME (2019) Detection of Peste des petits ruminants virus RNA in Culicoides imicola (Diptera: Ceratopogonidae) in Turkey. Vet Ital 55(2):173–179

Shaila MS, Purushothaman V, Bhavasar D, Venugopal K, Venkatesan RA (1989) Peste des petits ruminants in India. Vet Rec 125:602

Shaila MS, Shamaki D, Forsyth MA, Drallo A, Groatley L, Kitching RP, Barrett T (1996) Geographic distribution and epidemiology of peste des petits ruminants viruses. Virus Res 43:149–153

Shatar M, Khanui B, Purevtseren D, Khishgee B, Loitsch A, Unger H, Settypalli TBK, Cattoli G, Damdinjav B, Dundon WG (2017) First genetic characterization of peste des petits ruminants virus from Mongolia. Arch Virol 162(10):3157–3160

Singh RK, Balamurugan V, Bhanuprakash V, Sen A, Saravanan P, Yadav MP (2009) Control and eradication of peste des petits ruminants in sheep and goats in India: possibility. Vet Ital 45:449–462

Singh RP (2011) Control strategies for peste des petits ruminants in small ruminants of India. Rev Sci Tech 30:879–887

Singh RP, Bandyopadhyay SK (2015) Peste des petits ruminants vaccine and vaccination in India: sharing experience with disease endemic countries. Virus Dis 26(4):215–224

Singh RP, Saravanan P, Sreenivasa BP, Singh RK, Bandyopadhyay SK (2004a) Prevalence and distribution of peste des petits ruminants virus infection in small ruminants in India. Rev Sci Tech 23:807–819

Singh RP, Sreenivasa BP, Dhar P, Bandyopadhyay SK (2004b) A sandwich-ELISA for the diagnosis of Peste des petits ruminants (PPR) infection in small ruminants using anti-nucleocapsid protein monoclonal antibody. Arch Virol 149:2155–2170

Singh RP, Sreenivasa BP, Dhar P, Shah LC, Bandyopadhyay SK (2004c) Development of a monoclonal antibody based competitive-ELISA for detection and titration of antibodies to peste des petits ruminants (PPR) virus. Vet Microbiol 98:3–15

Sreenivasa BP, Singh RP, Mondal B, Dhar P, Bandyopadhyay SK (2006) Marmoset B95a cells: a sensitive system for cultivation of Peste des petits ruminants (PPR) virus. Vet Res Commun 30:103–108

Tatsuo H, Ono N, Yanagi Y (2001) Morbilliviruses use signaling lymphocyte activation molecules (CD150) as cellular receptors. J Virol 75(13):5842–5850

Taylor W (2016) The global eradication of peste des petits ruminants (PPR) within15 years—is this a pipe dream? Trop Anim Health Prod 48(3):559–567

Taylor WP (1984) The distribution and epidemiology of Peste des petits ruminants. Prev Vet Med 2:157–166

Teshale S, Moumin G, Moussa C, Gezahegne M (2018) Seroprevalence and risk factors for peste des petits ruminants in sheep and goats in Djibouti. Rev Sci Tech 37(3):961–969

Tripathi BN, Chhattopadhyay SK, Parihar NS (1996) Clinicopathological studies on an outbreak of peste des petits ruminants in goats and sheep. Indian J Vet Pathol 20:99

Truong T, Boshra H, Embury-Hyatt C, Nfon C, Gerdts V, Tikoo S, Babiuk LA, Kara P, Chetty T, Mather A, Wallace DB, Babiuk S (2015) Peste des petits ruminants virus tissue tropism and pathogenesis in sheep and goats following experimental infection. PLoS One 9:e87145

Wang Q, Ou C, Dou Y, Chen L, Meng X, Liu X, Yu Y, Jiang J, Ma J, Zhang Z, Hu J, Cai X (2017) M protein is sufficient for assembly and release of Peste des petits ruminants virus-like particles. Microb Pathog 107:81–87

Wasee Ullah R, Bin Zahur A, Latif A, Iqbal Dasti J, Irshad H, Afzal M, Rasheed T, Rashid Malik A, Qureshi ZU (2016) Detection of Peste des petits ruminants viral RNA in fecal samples of goats after an outbreak in Punjab Province of Pakistan: a longitudinal study. Biomed Res Int 2016:1486824

Woma TY, Ekong PS, Bwala DG, Ibu JO, Ta'ama L, Dyek DY, Saleh L, Shamaki D, Kalla DJ, Bailey D, Kazeem HM, Quan M (2016) Serosurvey of peste des petits ruminants virus in small ruminants from different agro-ecological zones of Nigeria. Onderstepoort J Vet Res 83(1):1035

Yang B, Qi X, Guo H, Jia P, Chen S, Chen Z, Wang T, Wang J, Xue Q (2018a) Peste des petits ruminants virus enters caprine endometrial epithelial cells via the Caveolae-mediated endocytosis pathway. Front Microbiol 9:210

Yang B, Xue Q, Qi X, Wang X, Jia P, Chen S, Wang T, Xue T, Wang J (2018b) Autophagy enhances the replication of Peste des petits ruminants virus and inhibits caspase-dependent apoptosis in vitro. Virulence 9(1):1176–1194

Yang Y, Qin X, Zhang X, Zhao Z, Zhang W, Zhu X, Cong G, Li Y, Zhang Z (2017) Development of real-time and lateral flow dipstick recombinase polymerase amplification assays for rapid detection of goatpox virus and sheeppox virus. Virol J 14(1):131

Yousuf RW, Sen A, Mondal B, Biswas SK, Chand K, Rajak KK, Gowane GR, Sudhakar SB, Pandey AB, Ramakrishnan MA, Muthuchelvan D (2015) Development of a single-plate combined indirect ELISA (CI-ELISA) for the detection of antibodies against peste-des-petits-ruminants and bluetongue viruses in goats. Small Rumin Res 124:137–139

Yu R, Zhu R, Gao W, Zhang M, Dong S, Chen B, Yu L, Xie C, Jiang F, Li Z (2017) Fine mapping and conservation analysis of linear B-cell epitopes of peste des petits ruminants virus hemagglutinin protein. Vet Microbiol 208:110–117

Yunus M, Shaila MS (2012) Establishment of an in vitro transcription system for Peste des petits ruminant virus. Virol J 9:302

Zahur AB, Irshad H, Ullah A, Afzal M, Latif A, Ullah RW, Farooq U, Samo MH, Jahangir M, Ferrari G, Hussain M, Ahmad MM (2014) Peste des petits ruminants vaccine (Nigerian strain 75/1) confers protection for at least 3 years in sheep and goats. J Biosci Med 2:23–27

Zakian A, Nouri M, Kahroba H, Mohammadian B, Mokhber-Dezfouli MR (2016) The first report of peste des petits ruminants (PPR) in camels (*Camelus dromedarius*) in Iran. Trop Anim Health Prod 48(6):1215–1219

Zhang J, Liu W, Chen W, Li C, Xie M, Bu Z (2016) Development of an immunoperoxidase monolayer assay for the detection of antibodies against Peste des petits ruminants virus based on BHK-21 cell line stably expressing the goat signaling lymphocyte activation molecule. PLoS One 11(10):e0165088

Zhang Y, Wang J, Zhang Z, Mei L, Wang J, Wu S, Lin X (2018) Development of recombinase polymerase amplification assays for the rapid detection of peste des petits ruminants virus. J Virol Methods 254:35–39

Zhou XY, Wang Y, Zhu J, Miao QH, Zhu LQ, Zhan SH, Wang GJ, Liu GQ (2018) First report of peste des petits ruminants virus lineage II in *Hydropotes inermis*, China. Transbound Emerg Dis 65(1):e205–e209

Zhu Z, Li P, Yang F, Cao W, Zhang X, Dang W, Ma X, Tian H, Zhang K, Zhang M, Xue Q, Liu X, Zheng H (2019) Peste des petits ruminants virus nucleocapsid protein inhibits interferon-β production by interacting with IRF3 to block its activation. J Virol 93(16):e00362

Chapter 14
Sapelovirus

Yashpal Singh Malik, Sudipta Bhat, Anastasia N. Vlasova, Fun-In Wang, Nadia Touil, Souvik Ghosh, Kuldeep Dhama, Mahendra Pal Yadav, and Raj Kumar Singh

Abstract *Sapelovirus* (SPV), an emerging virus in the family *Picornaviridae*, is detected in several animal and bird species irrespective of their age. Amid all SPVs, porcine sapeloviruses (PSVs) are more ubiquitously present all over the world in porcine population. These viruses are highly stable in different environmental conditions and spread easily within the susceptible animals mainly through faeco-oral route. Usually, PSVs cause asymptomatic infections but are also clinically associated with encephalomyelitis, respiratory distress, fertility disorders and skin lesions. PSV-associated outbreaks have been reported where death occurs due to polioencephalitis and respiratory paralysis. Till date, PSVs have been detected from several European and Asian countries with moderate-to-high prevalence and clinical course. Viral capsid proteins are immunogenic and mutations in these pro-

Y. S. Malik (✉) · S. Bhat · K. Dhama · R. K. Singh
ICAR-Indian Veterinary Research Institute, Izatnagar, Uttar Pradesh, India

A. N. Vlasova
Food Animal Health Research Program, Department of Veterinary Preventive Medicine, CFAES, Ohio Agricultural Research and Development Center, College of Veterinary Medicine, The Ohio State University, Wooster, OH, USA

F.-I. Wang
School of Veterinary Medicine, National Taiwan University, Taipei, Taiwan

N. Touil
Faculté de Médecine et de Pharmacie, Centre de Génomique des Pathologies Humaines (GENOPATH), Université Mohamed V, Rabat, Morocco

Laboratoire de Biosécuritéet de Recherche, HôpitalMilitaired'Instruction Med V de Rabat, Rabat, Morocco

S. Ghosh
Department of Biomedical Sciences, One Health Center for Zoonoses and Tropical Veterinary Medicine, Ross University School of Veterinary Medicine, Basseterre, St. Kitts and Nevis, West Indies

M. P. Yadav
ICAR-Indian Veterinary Research Institute, Izatnagar, Uttar Pradesh, India

Sardar Vallabhbhai Patel University of Agriculture & Technology, Meerut, Uttar Pradesh, India

Y. S. Malik et al. (eds.), *Emerging and Transboundary Animal Viruses*, Livestock Diseases and Management, https://doi.org/10.1007/978-981-15-0402-0_14

teins are responsible for the diverse nature of the viruses. Further, genomic analysis shows the varied evolutionary patterns and the presence of recombination within PSV strains. These viruses also exist as concurrent infections with several enteric bacterial, viral and parasitic pathogens. Classical to modern biotechnological assays are in use to detect PSV involving virus isolation in cell culture, immunohistochemistry, conventional nucleic acid amplification techniques, quantitative real-time amplification assays and isothermal amplification molecular techniques. Till date, there is no vaccine available against PSVs.

Keywords Sapelovirus · Picornaviridae · Porcine · Encephalomyelitis · Mutation · Diversity · Recombination · Diagnosis · Epidemiology

14.1 Preamble

Sapeloviruses, formerly known as porcine enterovirus (PEV)-8, are ubiquitous and most commonly responsible for causing asymptomatic infection of the gastrointestinal tract of animals. The genetic analysis revealed unique markers based on which PEV-8 viruses were reclassified under a new genus named *Sapelovirus* (Krumbholz et al. 2002). As on date, these viruses have been found to infect porcine, avian and simian species. Porcine sapelovirus (PSV)-associated clinical symptoms include pneumonia, reproductive defects, skin lesions, enteritis and encephalomyelitis. As the information available on avian sapelovirus (ASV) and simian sapelovirus (SSV) is limited, this chapter comprehends the information on PSV only.

14.2 Virus Characteristics

PSVs belong to the genus *Sapelovirus* under the family *Picornaviridae*. The size of virion particles is small (~35 nm) and they are non-enveloped possessing icosahedral symmetry with a distinct surface morphology visible under the electron microscope (Bai et al. 2018). Mature infectious virions are stable in the environment and could resist acidic pH. Furthermore, virions are not destabilised at elevated temperature (60 °C for 10 min), lipid solvents and some disinfectants. Sodium chlorite or 70% ethanol is effective to inactivate PSV (Horak et al. 2016).The *Sapelovirus* exhibits buoyant density in CsCl of 1.32–1.34 g/cm^3.

14.3 Virus Taxonomy

The current *Sapelovirus* taxonomy is the result of efforts from the picorna study group and the International Committee on Taxonomy of Viruses (ICTV) Executive Committee (www.picornastudygroup.com) that are contributing their expertise to provide a classification scheme that is supported by verifiable data and expert consensus. The genus *Sapelovirus* contains three species, namely *Avian sapelovirus* (ASV), *Sapelovirus A* (SV-A, porcine) and *Sapelovirus B* (SV-B, simian) (ICTV 10th report, https://talk.ictvonline.org/ictv-reports/ictv_online_report/positive-sense-rna-viruses/picornavirales/w/picornaviridae/701/genus-sapelovirus). ASV and SV-A consist of only a single antigenically heterogenous serotype, respectively, known as ASV-1 and PSV-1. However, SSVs are highly divergent and are classified into three genetically defined *Sapelovirus* types as SSV-1 to SSV-3. PSV (strain csh) was first isolated from China in 2011 (Lan et al. 2011). As on date, 50 complete genomic sequences of PSV strains have been published from China, South Korea, Japan, Germany, France, India and the United Kingdom (Lan et al. 2011; Chen et al. 2012; Schock et al. 2014; Son et al. 2014; Ray et al. 2018; Piorkowski et al. 2018; Bai et al. 2018). Different virus strains exhibit affinity towards different target organs, as the PSV-csh strain (China, Shanghai) and PSV-G5 (the United Kingdom) affect mostly nervous system (Lan et al. 2011; Schock et al. 2014), whereas strains like KS0515, KS04105 and KS0552179 and Chinese PSV strain YC2011 are associated with diarrhoea. Notably, PSV-csh is a diarrhoeagenic strain that leads to respiratory distress before the onset of polioencephalomyelitis in infected animals (Lan et al. 2011).

14.4 Viral Genome

PSVs are positive-sense RNA viruses containing a genome-linked viral protein (VPg) at the 5′-terminus and a 3′-polyA tail. The complete genome size of the virus is 7.5–8.3 kb, coding a single polyprotein which is further cleaved post-translationally by virus-encoded proteinases into 12 individual proteins, viz. a leader protein (L), four structural proteins (VP1-4) and seven non-structural proteins (2A-2C, 3A-3D). The deduced polyproteins of the *Sapelovirus* range from 2322 to 2521 amino acids (aa). The layout of a complete genome of PSV (strain V13) with the nucleotide length of each protein is depicted in Fig. 14.1. PSVs have unique L and 2A protein (Krumbholz et al. 2002). Though the exact function of these two proteins is not known yet, both are presumed to be a protease. All three *Sapeloviruses* contain Internal Ribosomal Entry Site (IRES) type IV. ASVs have the longer L protein (451 aa) compared to PSV (84 aa) and SSV (88 aa). Regarding 2A protein, ASVs have very small protein (12 aa) whereas PSV and SSV have 226 aa and 302 aa, respectively. A species demarcation criteria is followed for the genus *Sapelovirus* that is founded on above 70% aa identity in the polyprotein, greater than 64% aa identity in P1 (VP4, VP2, VP3, VP1), greater than 70% aa identity in 2C + 3CD, a similar genome base composition which varies by no more than 1% and a common genome organisation (ICTV 10th report, 2018).

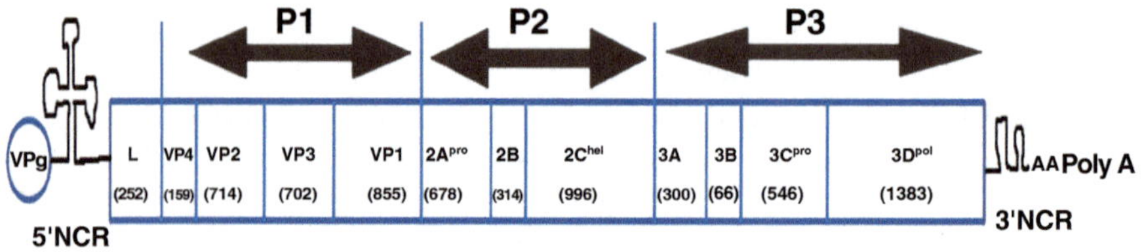

Fig. 14.1 A schematic layout of the porcine Sapelovirus genome (nucleotide lengths in base pair of different genes are given as per the first isolated PSV strain V13, NC_003987). *VPg* genome-linked viral protein, *NCR* noncoding RNA, *L* leader protein, *VP* viral protein, *P1* structural protein, *P2–P3* non-structural protein, *pro* protease, *hel* helicase and *pol* polymerase are depicted in schematic diagram

14.5 Epidemiology

14.5.1 Species Affected

Sapeloviruses have been isolated from pigs, wild boars, monkeys and ducks whereas sapelovirus-like-specific genome has been detected in a variety of species like bats (several species), cats, cattle, dogs, pigeons, quails, and sea lions (Li et al. 2011). Notably, these viruses to date are not known to cause any zoonotic transmission to human.

14.5.2 Geographic Distribution and Strain Variability

PSVs are being detected from different parts of the world in both healthy and diarrhoeic piglets and regarded as emerging enteric pathogens in porcine species. The countries that confirmed the presence of PSVs are shown in Fig. 14.2 (Forman et al. 1982; La Rosa et al. 2006; Buitrago et al. 2010; Lan et al. 2011; Shan et al. 2011; Cano-Gomez et al. 2013; Schock et al. 2014; Donin et al. 2014, 2015). The most immunodominant protein VP1 in the capsid of PSVs is used for estimating the divergence among the strains. Though only a single type of PSV-1 is present, fewer similarities are observed between PSV strains from different geographical regions. Phylogenetic analysis based on VP1 gene sequences shows that PSVs are distributed in different clusters depending upon the country of origin. Both full-length (855 bp) and partial-length (544 bp) VP1 genes have been used for the analysis of PSV diversity. The analysis based on VP1 sequences by SplitsTree4 software segregates three species within the genus *Sapelovirus* (Fig. 14.3) in three different clusters. Variations within PSV isolates are quite evident that indicates that several PSVs are circulating all over the world, and they are different from each other based on VP1 sequence. Studies from China reveal that recombinations occur between PSV strains, which further contribute to the antigenic diversity among PSVs. Recently, identification and genomic characterisation of *Sapelovirus* have been documented from France and India (Ray et al. 2018; Piorkowski et al. 2018;

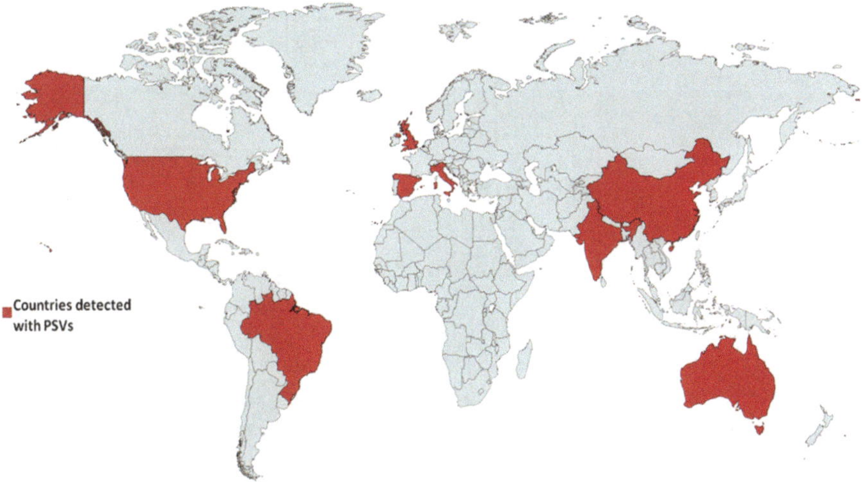

Fig. 14.2 Distribution of PSV throughout the world. The colour-shaded portions are where PSV incidences have been reported

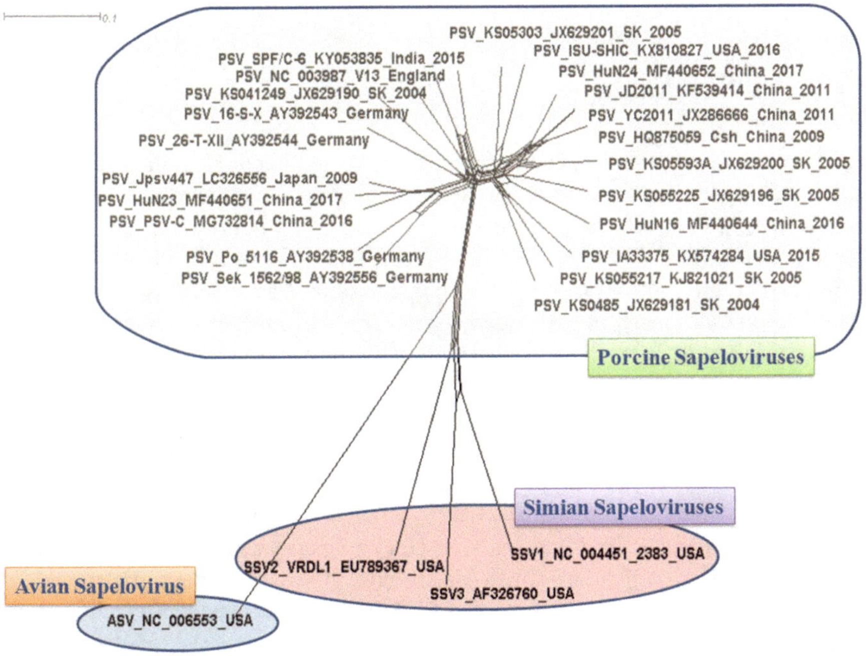

Fig. 14.3 Unrooted maximum-likelihood phylogenetic tree based on partial VP1 (544 bp) nucleotide sequences of the representative members of *Porcine Sapelovirus* PSV, *Simian Sapelovirus* SSV and *Avian Sapelovirus* ASV from China, Japan, England, India, the United States, Germany and SK (South Korea). Tree was developed using SplitsTree4 (Huson and Bryant 2006)

Kumari et al. 2019). A Chinese study reported the recombination hotspot near 3′ of the VP1 gene and showed six interclades within PSV recombinant strains (Yang et al. 2017).

14.6 Transmission and Pathogenesis

The most common route of transmission of the virus is faecal-oral; however, due to its high stability the infectious virions spread through fomites also (Huang et al. 1980; Lan et al. 2011). Several intestinal epithelial cell lines are commonly used for the pathogenicity and virulence studies of PSVs (Lan et al. 2013).

After ingestion through the oral route, PSVs primarily replicate in the intestinal tract. Receptor used by the PSV is unknown, but one recent study shows that a 2,3-linked terminal sialic acid (SA) on the cell surface GD1a ganglioside could be used for PSV binding and infection as a receptor (Kim et al. 2016). Experimental data shows that PSV utilises caveolae-dependent endocytosis pathway for PK-15 cell internalisation and depends upon lower pH, dynamin, Rab7 and Rab11 (Zhao et al. 2019). PSV strains are known to cause viremia through which they access the central nervous system (CNS). The incubation periods vary from 5 to 14 days.

14.7 Clinical Signs

PSVs can cause both symptomatic and asymptomatic infections in the natural field cases and experimentally infected piglets. Commonly, co-infections of PSV with other enteric pathogens like porcine kobuvirus (PKV), porcine enterovirus (PEV), porcine teschovirus (PTV), group A-C rotaviruses (RVA-C), porcine sapovirus (PSaV), porcine norovirus (PNoV), transmissible gastroenteritis coronavirus (TGEV), porcine epidemic diarrhoea coronavirus (PEDV), *Escherichia coli* and Salmonella spp., *Lawsonia intracellularis*, *Brachyspira hyodysenteriae*, *Brachyspira pilosicoli*, *Brachyspira intermedia* and *Clostridium perfringens*. Symptomatic manifestations are mainly neurological, which include spinal cord damage, ataxia, mental dullness, paresis, paralysis and a decreased response to environmental stimuli. Polioencephalitis in grower pigs is characterised by ataxia and paraparesis. Experimental infection of polioencephalitis has been documented in 50–60-day-old grower pigs. A novel PSV isolate from the United States was found to be associated with typical polioencephalomyelitis (Arruda et al. 2016). Gastroenteritis and respiratory distress may also be seen with PSV-induced polioencephalomyelitis (Lan et al. 2011). Intravaginal inoculation at day 15 of gestation causes early embryonic death and foetal resorption, but when inoculated at day 30, it results in a significant increase in foetal death (Huang et al. 1980). PSV is also detected from intestinal contents of stillbirth animals from Indian pig population, which suggests the transplacental transmission of the virus (Kumari et al. 2019).

14.8 Morbidity and Mortality

No specific data is available on the mortality and morbidity associated with PSV infection. Experimental infection through intravaginal and intrauterine route in gilts with PSV on day 30 of gestation results in a 94.4% foetal mortality (Huang et al. 1980). An outbreak reported in the United States in 2016 stated 20% morbidity and 30% case fatality rate (Arruda et al. 2016).

14.8.1 Post-mortem and Histopathology

Lesions are mainly seen in the central nervous system (CNS), consistent with the other neurotropic viral infection, characterised as subacute, multifocal and non-suppurative polioencephalomyelitis, accompanied with punctate haemorrhage and hyperaemia in the dura mater (Lan et al. 2011; Schock et al. 2014). Neuronal vacuolisation and perivascular cuffing are also commonly observed (Lan et al. 2011). Congestion is the most common gross lesion seen in the small intestine along with pronounced loss of villi with haemorrhage in the lamina propria histologically (Lan et al. 2011). In the case of clinical pneumonia, consolidation and multifocal haemorrhage are commonly seen in the lung lobes, where erythrocytes infiltrate throughout the interstitium and alveoli with alveolar wall thinning (Lan et al. 2011). Pathological lesions from naturally infected Indian pig show severe clouding, thickening and congestion of meninges, pleural thickening, frothy exudates in the trachea, congested mesenteric lymph nodes, yellowish discolouration of liver, intestinal (ileum) corrugation and thickening of the intestinal mucosal fold (Kumari et al. 2019)

14.9 Diagnosis

14.9.1 Clinical History

A neurological disorder characterised by ataxia and limb paralysis, with or without other clinical symptoms (diarrhoea or pneumonia), is suggestive of PSV infection (Lan et al. 2011). The PSV-induced reproductive disorder can be suspected in litters with a few to several stillborn or mummified foetuses (Huang et al. 1980).

14.9.2 Samples

Samples from live animal include faeces and blood (serum). In the case of the faecal sample, direct collection from the anus is preferred to avoid sample contamination. Spinal cord and brain are the most common preferred samples from the dead animal for PSV-induced CNS infection diagnosis. Other than these, liver, spleen, trachea, lung and intestine samples are also useful in diagnosing PSV infections. Though PSV has not been successfully isolated from the tissues of stillborn or mummified foetuses, molecular detection has been reported from stillbirth cases (Huang et al. 1980; Kumari et al. 2019).

14.9.3 Detection of Virus, Nucleic Acid or Antigens

Routine laboratory diagnosis of PSVs mainly includes cell culture isolation and virus genome characterisation. Pig kidney cell-derived cell lines including EFN, PK-15, IBRS-2 and LLC-PK are susceptible to PSV with a cytopathic effect (CPE) in these cells (Sozzi et al. 2010; Lan et al. 2011). Additionally, monkey kidney (Vero) and baby hamster kidney fibroblasts (BHK-21) are also susceptible to PSVs where a CPE characterised by cell shrinkage, rounding and detachment is commonly seen after 3–5 passages. Recently, a human hepatocarcinoma cell line, PLC/PRF/5, is normally used to isolate the hepatitis E virus, from swine faecal specimens serendipitously isolated from Japanese PSV strain (Bai et al. 2018). Cultured PSVs are generally identified by using virus neutralisation (VN) (Sozzi et al. 2010) and immunofluorescence antibody (IFA) assays (Son et al. 2014). However, these methods are costly, less specific, labour intensive and time consuming. Several molecular methods like reverse transcription PCR (RT-PCR), nested reverse transcription PCR (nRT-PCR) and real-time quantitative PCR have been developed for PSV detection (Zell et al. 2000; Palmquist et al. 2002; Krumbholz et al. 2003). Table 14.1 depicts the list of assays with target genes and primer sequences used by researchers in detecting PSVs. These molecular methods are specific, sensitive and rapid and do not cross-react with other common pathogens of porcine diseases including pseudorabies virus, porcine reproductive and respiratory syndrome, porcine parvovirus, porcine coronavirus, porcine reovirus or picorna-like virus (Palmquist et al. 2002). Real-time PCR assays developed for the detection of PSVs from field samples could detect a range of 10^2–10^3 copies of DNA (Chen et al. 2014; Kumari et al. 2019). Both these real-time PCR assays utilised the TaqMan chemistry, which is quite specific. Another sensitive diagnostic available for PSV is reverse transcription loop-mediated isothermal amplification (RT-LAMP) that was developed by Wang et al. (2014) with a detection limit of ten copies of DNA. All these easy, rapid and sensitive diagnostics will help in better understanding of the epidemiology of PSV and find a better way for prevention.

Table 14.1 RT-PCR assays used for detection of PSVs

Assays	Primer sequence (5′–3′)	Targeted gene	References
RT-PCR	FP-CCCTGGGACGAAAGAGCCTG	5′UTR	Zell et al. (2000)
	RP-CCTTTAAGTAAGTAGTAAAGGG		
nRT-PCR	FP-CCAAGATTAGAAGTTGATTTG		
	RP-GGGTAGCCTGCTGATGTAGTC		
Duplex PCR for PTV and PSV	FP-GTGGCGACAGGGTACAGAAGAG	5′UTR	Palmquist et al. (2002)
	RP-GGCCAGCCGCGACCCTGTCAG		
RT-PCR (genotyping)	FP-AGGATGTGGTGCAAGCAAGCAT	VP1	Son et al. (2014)
	RP-AGGCAGCACCGTTCTGGTCAA		
RT-LAMP	FP-CCATACCCTACCCTCCCTTC	5′UTR	Wang et al. (2014)
	RP-GCCCATAGTTCACTGCCTAC		
Real-time RT-PCR	FP-GGCAGTAGCGTGGCGAGC	5′UTR	Chen et al. (2014)
	RP-CTACTCTCCTGTAACCAGT		
	Minor groove-binding Taqman probe—CGATAGCCATGTTAGTG		
Real-time RT-PCR	FP-GGAAACCTGGACTGGGYCT	5′UTR	Kumari et al. (2018)
	RP-ACACGGGCTCTCTGTTTCTT		
	Taqman probe—CCAGCCGCGACCCTATCAGG		

14.10 Immunity

There is little information available on the immune response to PSVs. The humoral response is primarily based on the IgA. A study done on porcine intestinal epithelial cells (IPEC-J2) reveals the role of PSV infection in changing the innate immune response pathway (Lan et al. 2013). The study shows both upregulation (e.g. TLR3) and downregulations (e.g. TLR4) of several innate immune system molecules (Lan et al. 2013). Maternal antibodies proved not to be protective in the case of transplacental infection (Huang et al. 1980). Seropositive gilts when inoculated intrauterinely and intravaginally resulted in embryonic and foetal infection (Huang et al. 1980). Though maternal antibody in the colostrums is believed to be protective infections are found in post-weaning animals (Schock et al. 2014). As of now, there is no vaccine available for PSVs.

14.11 Conclusion and Prospects

Although PSV is widely distributed all over the world among the porcine population, detailed information is lacking about their evolutionary pattern and mechanisms driving their divergence. However, precise data about the prevalence of PSV infections and the genetic diversity of this virus in pigs and wild boars have been reported only in a limited number of countries. Moreover, the association of co-infections with other enteric pathogens is exactly not known which raises ambiguity in the pathogenic potential of PSVs. Several recent findings on recombination events between PSV strains emphasise to study genetic variability among PSVs.

Acknowledgements All the authors of the manuscript thank and acknowledge their respective universities and institutes.

Conflict of interest: There is no conflict of interest.

References

Arruda PH, Arruda BL, Schwartz KJ, Vannucci F, Resende T, Rovira A, Sundberg P, Nietfeld J, Hause BM (2016) Detection of a novel sapelovirus in central nervous tissue of pigs with polioencephalomyelitis in the USA. Transbound Emerg Dis 64(2):311–315

Bai H, Liu J, Fang L, Kataoka M, Takeda N, Wakita T, Li TC (2018) Characterization of porcine sapelovirus isolated from Japanese swine with PLC/PRF/5 cells. Transbound Emerg Dis 65(3):727–734

Buitrago D, Cano-Gomez C, Aguero M, Fernandez-Pacheco P, Gomez-Tejedor C, Jimenez-Clavero MA (2010) A survey of porcine picornaviruses and adenoviruses in fecal samples in Spain. J Vet Diagn Investig 22(5):763–766

Cano-Gomez C, Garcia-Casado MA, Soriguer R, Palero F, Jimenez-Clavero MA (2013) Teschoviruses and sapeloviruses in faecal samples from wild boar in Spain. Vet Microbiol 165(1-2):115–122

Chen J, Chen F, Zhou Q, Li W, Chen Y, Song Y, Zhang X, Xue C, Bi Y, Cao Y (2014) Development of a minor groove binder assay for real-time PCR detection of porcine Sapelovirus. J Virol Methods 198:69–74

Chen JW, Chen F, Zhou QF, Li W, Song Y, Pan Y, Zhang X, Xue C, Bi Y, Cao Y (2012) Complete genome sequence of a novel porcine sapelovirus strain YC2011 isolated from piglets with diarrhea. J Virol 86(19):10898–10898

Donin DG, Leme RD, Alfieri AF, Alberton GC, Alfieri AA (2014) First report of porcine teschovirus (PTV), Porcine sapelovirus (PSV) and Enterovirus G (EV-G) in pig herds of Brazil. Trop Anim Health Prod 46(3):523–528

Donin DG, Leme RD, Alfieri AF, Alberton GC, Alfieri AA (2015) Molecular survey of porcine teschovirus, porcine sapelovirus, and enterovirus G in captive wild boars (*Sus scrofa scrofa*) of Parana state, Brazil. Pesq Vet Bras 35(5):403–408

Forman AJ, Pass DA, Connaughton ID (1982) The characterisation and pathogenicity of porcine enteroviruses isolated in Victoria. Aust Vet J 58(4):136–142

Horak S, Killoran K, Leedom Larson KR (2016) Porcine sapelovirus. Swine Health Information Center and Center for Food Security and Public Health. http://www.cfsph.iastate.edu/pdf/shic-factsheet-porcine-sapelovirus

Huang J, Gentry RF, Zarkower A (1980) Experimental infection of pregnant sows with porcine enteroviruses. Am J Vet Res 41(4):469–473

Huson DH, Bryant D (2006) Application of phylogenetic networks in evolutionary studies. Mol Biol Evol 23(2):254–267

Kim DS, Son KY, Koo KM, Kim JY, Alfajaro MM, Park JG, Hosmillo M, Soliman M, Baek YB, Cho EH, Lee JH (2016) Porcine sapelovirus uses α2, 3-linked sialic acid on GD1a ganglioside as a receptor. J Virol 90(8):4067–4077

Krumbholz A, Dauber M, Henke A, Birch-Hirschfeld E, Knowles NJ, Stelzner A, Zell R (2002) Sequencing of porcine enterovirus groups II and III reveals unique features of both virus groups. J Virol 76(11):5813–5821

Krumbholz A, Wurm R, Scheck O, Birch-Hirschfeld E, Egerer R, Henke A, Wutzler P, Zell R (2003) Detection of porcine teschoviruses and enteroviruses by LightCycler real-time PCR. J Virol Methods 113(1):51–63

Kumari S, Ray PK, Singh R, Desingu PA, Sharma GT, Saikumar G (2018) Development of a Taqman-based real-time PCR assay for detection of porcine sapelovirus infection in pigs. Anim Biotechnol:1–4. https://doi.org/10.1080/10495398.2018.1549561. (Epub ahead of print)

Kumari S, Ray PK, Singh R, Desingu PA, Varshney R, Saikumar G (2019) Pathological and molecular investigation of porcine sapelovirus infection in naturally affected Indian pigs. Microb Pathog 127:320–325

La Rosa G, Muscillo M, Di Grazia A, Fontana S, Iaconelli M, Tollis M (2006) Validation of RT-PCR assays for molecular characterization of porcine teschoviruses and enteroviruses. J Vet Med B Infect Dis Vet Public Health 53(6):257–265

Lan D, Tang C, Yue H, Sun H, Cui L, Hua X, Li J (2013) Microarray analysis of differentially expressed transcripts in porcine intestinal epithelial cells (IPEC-J2) infected with porcine sapelovirus as a model to study innate immune responses to enteric viruses. Arch Virol 158(7):1467–1475

Lan DL, Ji WH, Yang SX, Cui L, Yang Z, Yuan C, Hua X (2011) Isolation and characterization of the first Chinese porcine sapelovirus strain. Arch Virol 156(9):1567–1574

Li L, Shan T, Wang C, Côté C, Kolman J, Onions D, Gulland FM, Delwart E (2011) The fecal viral flora of California sea lions. J Virol 85(19):9909–9917

Palmquist JM, Munir S, Taku A, Kapur V, Goyal SM (2002) Detection of porcine teschovirus and enterovirus type II by reverse transcription-polymerase chain reaction. J Vet Diagn Investig 14(6):476–480

Piorkowski G, Capai L, Falchi A, Casabianca F, Maestrini O, Gallian P, Barthélémy K, Py O, Charrel R, De Lamballerie X (2018) First identification and genomic characterization of a porcine sapelovirus from Corsica, France, 2017. Microbiol Resour Announc 7(11):e01049–e01018

Ray PK, Desingu PA, Kumari S, John JK, Sethi M, Sharma GK, Pattnaik B, Singh RK, Saikumar G (2018) Porcine sapelovirus among diarrhoeic piglets in India. Transbound Emerg Dis 65(1):261–263

Schock A, Gurrala R, Fuller H, Foyle L, Dauber M, Martelli F, Scholes S, Roberts L, Steinbach F, Dastjerdi A (2014) Investigation into an outbreak of encephalomyelitis caused by a neuroinvasive porcine sapelovirus in the United Kingdom. Vet Microbiol 172(3–4):381–389

Shan TL, Li LL, Simmonds P, Wang CL, Moeser A, Delwart E (2011) The fecal virome of pigs on a high-density farm. J Virol 85(22):11697–11708

Son KY, Kim DS, Kwon J, Choi JS, Kang MI, Belsham GJ, Cho KO (2014) Full-length genomic analysis of Korean porcine sapelovirus strains. PLoS One 9(9):11

Sozzi E, Barbieri I, Lavazza A, Lelli D, Moreno A, Canelli E, Bugnetti M, Cordioli P (2010) Molecular characterization and phylogenetic analysis of VP1 of porcine enteric picornaviruses isolates in Italy. Transbound Emerg Dis 57(6):434–442

Wang CY, Yu DY, Cui L, Hua X, Yuan C, Sun H, Liu Y (2014) Rapid and real-time detection of Porcine Sapelovirus by reverse transcription loop-mediated isothermal amplification assay. J Virol Methods 203:5–8

Yang T, Yu X, Yan M, Luo B, Li R, Qu T, Luo Z, Ge M, Zhao D (2017) Molecular characterization of Porcine sapelovirus in Hunan, China. J Gen Virol 98(11):2738–2747

Zell R, Krumbholz A, Henke A, Birch-Hirschfeld E, Stelzner A, Doherty M, Hoey E, Dauber M, Prager D, Wurm R (2000) Detection of porcine enteroviruses by nRT-PCR: differentiation of CPE groups I-III with specific primer sets. J Virol Methods 88(2):205–218

Zhao T, Cui L, Yu X, Zhang Z, Shen X, Hua X (2019) Porcine Sapelovirus enters PK-15 cells via caveolae-dependent endocytosis and requires Rab7 and Rab11. Virology 529:160–168. https://doi.org/10.1016/j.virol.2019.01.009

Chapter 15
Hepatitis E Viruses

Harsh Kumar, Nassim Kamar, Gheyath K. Nasrallah, and Dinesh Kumar

Abstract Hepatitis E virus (HEV), member of genus *Orthohepevirus* in the family *Hepeviridae*, is the leading cause of hepatitis E. More than 20 million cases of hepatitis E virus (HEV) infection have been reported globally, out of which 3.3 million suffered from symptomatic illness and 44,000 deaths were recorded. Hepatitis E is considered highly significant on account of its predominance in both developed and developing nations due to poor sanitation and low-grade drinking water. In India and Africa, the major cause of hepatitis E was water contaminated with faecal matter. Reports from East and Southeast Asia signified the role of animal reservoir, especially pigs, for HEV. In Western Asia, the major cases of hepatitis E were associated with blood transfusion. However, in Europe and South and North America, the significant causes of hepatitis E were associated with the consumption of pork products. Till date, major rely is on the enzyme immunoassays involving antibodies for detection and diagnosis purpose. Still, there is a need for the improvement and development of new approaches and methods for diagnosing hepatitis E virus as well as developing HEV vaccines.

Keywords HEV · History · Genome and classification · Pathogenesis · Transmission and epidemiology · Diagnosis · Vaccines · Prevention · Zoonoses

H. Kumar · D. Kumar (✉)
School of Bioengineering & Food Technology, Shoolini University of Biotechnology and Management Sciences, Solan, Himachal Pradesh, India
e-mail: dineshkumar@shooliniuniversity.com

N. Kamar
Department of Nephrology, Dialysis an Organ Transplantation, CHU Rangueil, INSERM U1043, IFR–BMT, University Paul Sabatier, Toulouse, France
e-mail: kamar.n@chu-toulouse.fr

G. K. Nasrallah
Department of Biomedical Science, Biomedical Research Center, College of Health Sciences, QA Health, Qatar University, Doha, Qatar
e-mail: gheyath.nasrallah@qu.edu.qa

Y. S. Malik et al. (eds.), *Emerging and Transboundary Animal Viruses*, Livestock Diseases and Management, https://doi.org/10.1007/978-981-15-0402-0_15

15.1 Overview

Hepatitis E virus (HEV) infection has been estimated to infect >20 million people globally, out of which 3.3 million suffered from symptomatic illness and annually 44,000 mortality cases were recorded in 2015 (Rein et al. 2012; WHO 2017). Being self-limited HEV infection gets resolved spontaneously while few patients experience symptoms like abdominal pain, anorexia, fever, jaundice and severe icteric viral hepatitis (Aggarwal 2011; Mirazo et al. 2014). On the other hand, old, immunosuppressed, HIV/AIDS-positive patients with cancer or receiving organ transplants are highly susceptible to HEV. Patients with chronic hepatitis E require an extended stay at the hospital and show a high mortality rate. During pregnancy, this infection makes the situation critical for both baby and mother (Navaneethan et al. 2008; Fujiwara et al. 2014; Poovorawan et al. 2014).

Four main genotypes of HEV (belonging to the same serotype) have been identified to cause hepatitis E in humans (Emerson and Purcell 2003). Genotypes 1 and 2 of HEV are dominantly found in humans during epidemics in low-income countries like those in Africa and some parts of Asia (Teshale et al. 2010b). In industrialised and developing countries, genotypes 3 and 4 are usually found in sporadic cases of autochthonous hepatitis E (Dalton et al. 2008). Various animals like chicken, deer, rabbits, rats, pigs and wild boars have been reported to be infected by HEV (Intharasongkroh et al. 2017). Similarly, HEV human isolates have shown similarity with swine strain, especially genotype 3 (Suwannakarn et al. 2010; Temmam et al. 2013). In developed countries like France and Japan, the zoonotic potential of HEV infecting humans was found to be associated with ingestion of contaminated meat and pork products, and uncooked sausages (Intharasongkroh et al. 2017).

15.2 History of Hepatitis E

In 1980, serological tests were developed for detecting hepatitis A and B virus. However, samples of 1955–1956 waterborne hepatitis outbreak in New Delhi, after analysis with serological tests, were found to be associated with enteric non-A and -B hepatitis (Viswanathan 1957). A similar type of investigation was reported with a hepatitis outbreak which occurred in Kashmir in 1978–1979 (Khuroo 1980). The result of both these outbreaks showed the presence of HEV as an aetiology mediator. This was the first investigation where the waterborne epidemic was associated with HEV (Wong et al. 1980; Teshale and Hu 2011). In 1983, a Soviet doctor, Mikhail Balayan, created bizarre in the medical field when non-A and -B hepatitis cases were investigated by him in Soviet soldiers during Afghanistan war (Balayan et al. 1983). He also continued his investigation in Moscow, where he prepared yoghurt contaminated with faecal samples of nine soldiers and ingested it during his return to Moscow and waited for the phenotypic symptoms. In this investigation, he experienced the symptoms of severe hepatitis after 35 days. He analysed his faeces and observed new strains of the virus that are also affecting the liver of laboratory animals. This was the time when genotype 1 was first detected (Teshale and Hu 2011; Pinto et al. 2017).

Other cases of hepatitis E infections occurred in Costa Rica in 1975 (Khuroo 1980; Khuroo et al. 2016a). In 1988, an outbreak in Somalia, Africa, was reported to expand to 11,000 people. In Latin America, the first detection of HEV with detection of genotype 2 was reported in Mexico, during an outbreak between 1986 and 1987 (Rendon et al. 2016; Pinto et al. 2017). On the other hand, the largest epidemic was reported in China between 1986 and 1988 and 120,000 people were confirmed with HEV caused by genotype 1 (Aye et al. 1992).

15.3 Virus Classification and Genome

According to taxonomic scheme the family *Hepeviridae* is divided into two genera: *Piscihepevirus* (cutthroat trout virus) and *Orthohepevirus* (mammalian and avian strains) (ICTV 2019). The genus *Orthohepevirus* has been divided into four species, *Orthohepevirus A*, *Orthohepevirus B* (infecting birds), *Orthohepevirus C* (infecting rodents, soricomorphs and carnivores) and *Orthohepevirus D* (infecting bats) (ICTV 2019). The species, *Orthohepevirus A*, includes seven genotypes that infect humans (HEV1, 2, 3, 4 and 7), pigs (HEV3 and 4), rabbits (HEV3), wild boar (HEV3, 4, 5 and 6), mongooses (HEV3), deer (HEV3), yaks (HEV4) and camels (HEV7) (ICTV 2019). HEV is a non-enveloped virus, with 27–34 nm icosahedral capsid. This virus has a positive-sense, single-stranded, 7.2 kb RNA genome capped and polyadenylated at the 5′ and 3′ ends, respectively (Reyes et al. 1990; Tam et al. 1991). The three open reading frames (ORFs) of the HEV genome are shown in Fig. 15.1. ORF1 translates for functional domains (1693 amino acids) of a non-structural protein (Koonin et al. 1992). This functional domain contains a region of a cysteine protease, methyltransferase, RNA helicase as well as RNA-dependent RNA polymerase. ORF2 translates for capsid protein (660 amino acids) of the virus, playing a crucial

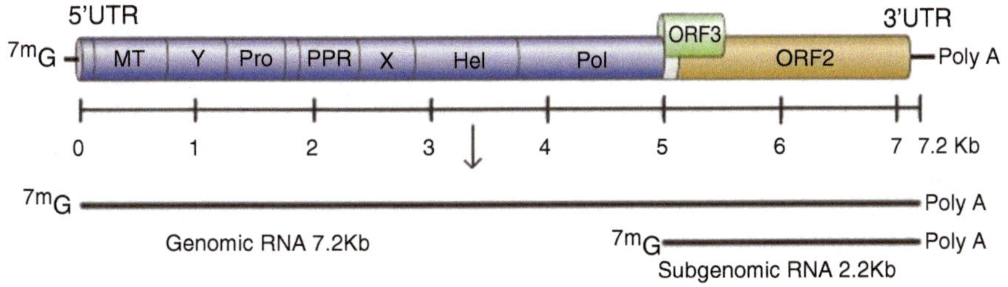

Fig. 15.1 Hepatitis E virus genome. The 5′ end of the RNA genome is capped with a 7-methylguanosine (7 mG) and the 3′ end is polyadenylated (poly(A)). Open reading frame 1 (ORF1) encodes non-structural proteins, including *MT* methyltransferase, *Pro* cysteine protease, *Hel* RNA helicase, *Pol* RNA-dependent RNA polymerase as well as Y domain, *PPR* proline-rich hinge domain, *X* macrodomain (adopted from Kamar et al. 2014)

role during assembly of the virus, interaction with target cells and immunogenicity (Li et al. 1997; He et al. 2008; Kalia et al. 2009; Xing et al. 2011). The ORF2 protein encodes for three linear domains: centre domain (M), shell domain (S) and protruding domain (P). The HEV dominant-neutralising epitopes are located within the P domain (aa 455–602) (Meng et al. 2001; Guu et al. 2009; Yamashita et al. 2009; Xing et al. 2010; Tang et al. 2011). ORF3 overlapping with ORF2 encodes for a minor protein (113–114 amino acids long) which is associated with the morphology of virion and its release (Graff et al. 2006; Yamada et al. 2009a; Emerson et al. 2010). The HEV genomes of a few topographically distinctive isolates demonstrate high similarity with the consensus sequence of HEV domains (Arankalle et al. 1999). Four phylogenetically similar genotypes have been annotated, which are differentiated based on geographical regions, as shown in Fig. 15.2. In African and Asian regions, HEV strains belonging to genotype 1 are more prevalent. Genotype 2 was identified in sole Mexican HEV strain, and rare variants of genotype 2 were also isolated during endemic conditions in Africa. In industrialised nations, genotype 3 with swine and human HEV strains are common, whereas in East Asia especially China, Taiwan and Japan, genotype 4 represents swine HEV and human strains (Chandra et al. 2008). Avian HEV has been reported to show similarity with genotype 5, but this has not yet been confirmed (Haqshenas et al. 2001; Huang et al. 2004).

HEV-5 and -6 genotypes were reported to generally infect wild boar (Al-Sadeq et al. 2018). In the United Arab Emirates, a new genotype of HEV has been identified and named as HEV-7 (Woo et al. 2014). HEV-7 primarily infects dromedary camels, as it was first found in camel faeces. However, the same viral sequence was isolated from the faeces of hepatitis patient, suggesting that the virus can infect both humans and camels (Rasche et al. 2016). Similarly, HEV-8 was identified in Bactrian camels in China, but until now there is no report on the detection of HEV-8 viral sequences in humans (Woo et al. 2016).

15.4 Host-Virus Interactions

It is believed that the severity of HEV infection depends on the host's immune system status. However, viral properties may also contribute significantly to the pathogenesis of HEV. Therefore, HEV genotype is also considered as the main player in the severity of the disease. For instance, genotype 4 showed more severe symptoms of viral hepatitis than genotype 3 in infected patients. However, genotype 1 is considered as the most severe genotypes and can lead to fulminant hepatic failure and serious placental diseases (Al-sadeq et al. 2017, 2018). The capsid protein of HEV binds to a cellular receptor on the host for viral entry and initiates the replication. For instance, experiment on ORF2 peptide binding revealed that ORF2 C-terminal region might mediate HEV entry by binding to heat-shock cognate protein (HSC70) that is found on the surface of the host cell receptor (Zhou and Emerson 2006). Furthermore, various heparan sulphate proteoglycans (HSPGs) have been discovered which are present on the surface of the host cells, and may

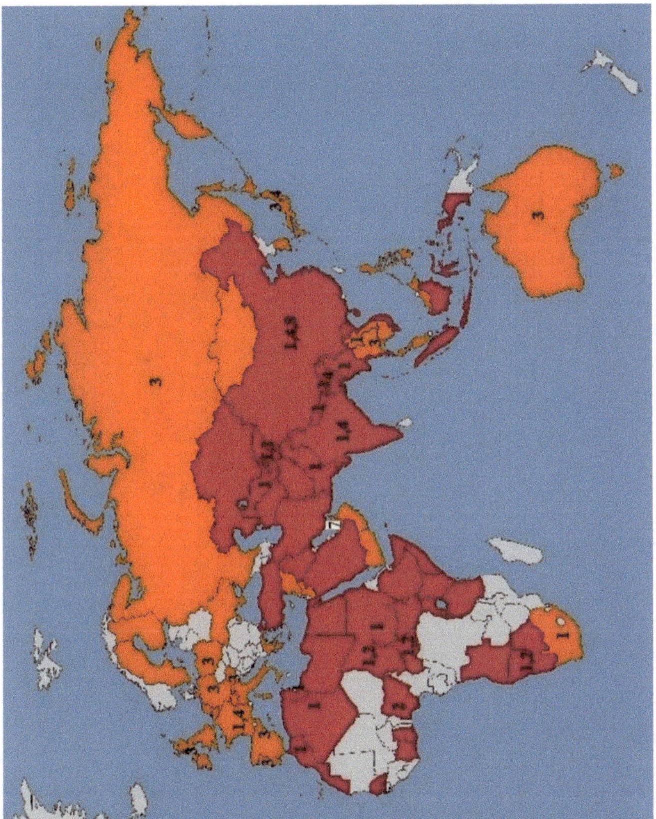

Fig. 15.2 HEV epidemiology and genotype distribution (adopted from Al-sadeq et al. 2017)

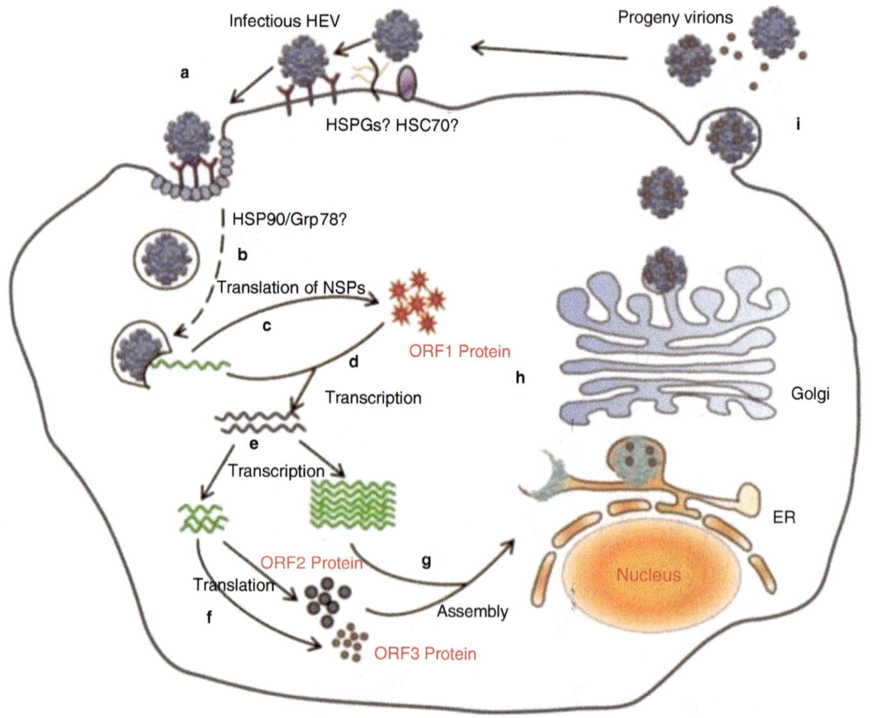

Fig. 15.3 Proposed pathogenesis of HEV. (Step **a**) HEV attaches to the cell surface via HSPGs, HSC70 or other putative attachment receptor(s) and then enters the cell via an unknown specific cellular receptor. (Step **b**) The HEV virion penetrates the membrane and enters the cells. HSP90 and Grp78 may be involved in this transport. The virion then uncoats and releases the positive-sense genomic RNA into the cytoplasm of the cell. (Step **c**) The positive-sense genomic viral RNA serves as the template to translate the ORF1 non-structural polyprotein in the cytoplasm. (Step **d**) The viral RdRp synthesises an intermediate, replicative negative-sense RNA from the positive-sense genomic RNA that (step **e**) serves as the template for the production of positive-sense, progeny viral genomes. (Step **f**) The ORF2 and ORF3 proteins are translated from the subgenomic, positive-stranded RNA. (Step **g**) The ORF2 capsid protein packages the viral genomic RNA and assembles new virions. (Step **h**) The nascent virions are transported to the cell membrane. The ORF3 protein facilitates the trafficking of the virion. (Step **i**) The nascent virions are released from the infected cells (adopted from Cao and Meng 2012)

also act as attachment receptors generally present on the surface of the host cells (Kalia et al. 2009) (Fig. 15.3). That is, a truncated dimer (23 nm size) of HEV capsid protein (HEV 239) was able to bind to the HSPGs (He et al. 2008).

In another study, the same truncated protein was also found to interact with a cellular chaperone protein known as Grp78. In summary, in addition to other reports, the studies above demonstrated that the truncated capsid protein HEV 239 plays an essential role in HEV interaction with the host cell and the life cycle of HEV (Yu et al. 2011). The RNA of HEV gets uncoated via an unknown mechanism after entry of the virus into permissive cells. Geldanamycin, an inhibitor of HSP90, obstructs the intracellular movement of HEV239 capsid protein but is not having any effect on the binding as well as the entry of capsid protein of truncated HEV239. Thus, it confirms that HSP90 plays a critical role in the intracellular transport of the virus (Zheng et al. 2010).

After viral genome uncoating and its release into the cells, 7-methylguanosine of the 5′-noncoding regions (NCR) present in the HEV genome directs ribosomal subunit (the 40S) to start cap-dependent translation of viral genome. The viral RNA-dependent RNA polymerase (RdRp) produces negative-sense, replicative RNA intermediate. This intermediate acts as a template for creating a positive-sense viral genome. It has been found that the HEV RdRp activity is important in HEV replication system. That is, the intermediate viral RNA (negative-sense) was detected within HEV-infected animal tissues (Nanda et al. 1994; Meng et al. 1998; Agrawal et al. 2001; Williams et al. 2001; Graff et al. 2005a). Furthermore binding of *cis*-reactive elements (CRE) with 3′ NCR of HEV genome along with RdRp reveals the importance of this interaction during HEV replication (Agrawal et al. 2001; Emerson et al. 2001; Graff et al. 2005a). The importance of CRE in the junction region (JR) of the HEV genome and its involvement in subgenomic RNA synthesis is well documented (Graff et al. 2005b; Cao et al. 2010). The subgenomic RNA is translated and encodes for small ORF3 protein and capsid protein of HEV.

Finally, both assembly and release mechanism of HEV is not well understood. However, ORF2 protein performs a leading role in the assembly and packaging of progeny virions. The binding domain of 76 nt present at 5′ ends of the HEV genome was recognised to interact with ORF2 protein and shows the importance of this region in the packaging of the HEV genome (Surjit et al. 2004). On the other hand, ORF3 protein is also considered to perform a similar function (Yamada et al. 2009b; Emerson et al. 2010). The ORF3 protein contains motifs that play a central role during discharge of membrane-bound HEV particles from infected cells (Nagashima et al. 2012).

15.5 Transmission and Epidemiology of HEV

Generally, the transmission of HEV takes place via the faecal-oral route (Khuroo 1991). In developing countries, severe hepatitis E outbreaks are associated with the supply of faecal contaminated water (Khuroo et al. 2016b). Moreover, animal- and human-related HEV isolates have been reported from raw water and sewage samples (Vaidya et al. 2003; Ippagunta et al. 2007; Ishida et al. 2012; Masclaux et al. 2013). Open defecation in fields and yards is one of the contributors of faecal contamination of crops, groundwater and waterways (Gurav et al. 2007). For example, In India, about 300 million people defecate openly and contaminate water sources like rivers, wells and streams (Khuroo and Khuroo 2015). Besides this, there are situations where piped water supply also gets contaminated with faecal material. In India, cities and towns get their water supplies through pipes that are either in parallel or cross over along the sewage drains (common practices). Due to the formation of cracks and holes in old water pipes, the sewage contaminates the piped water supply with faecal contamination during scheduled intermittent water supply. Therefore, in India, most of the hepatitis E epidemic cases have been associated with this phenomenon only (Khuroo et al. 2016a, b).

HEV-3 and -4 are zoonotic and found in native pigs, sika deer and wild boar (Pavio et al. 2015). There are three modes of infections of these phenotypes, i.e. direct interaction with the contaminated environment; infected animals; and zoonotic food-borne consumption (Yugo and Meng 2013). Out of above reservoirs transmission of HEV-3 and -4 via food-borne zoonotic has been extensively studied (Izopet and Kamar 2008; Yugo and Meng 2013). Outbreaks of hepatitis E and indigenous cases occur due to the consumption of parboiled flesh of domestic pigs, sika deer and wild boar (Khuroo and Khuroo 2008; Miyashita et al. 2012). In Europe, a common pathway leading to widespread HEV is due to consumption of Corsican figatelli sausage and eating of raw livers of pork purchased from the supermarket (Khuroo et al. 2016a, b).

Although the virus is predominantly transmitted through the faecal-oral route, other transmission routes have been identified recently, including vertical transmission (during pregnancy) and through blood transfusion. In 2004, HEV has been recognised as a transfusion-transmissible infectious agent that may pose a threat to the blood bank supply. Interestingly, although limited data is available concerning the transmission of HEV through blood, transfusion-transmitted infections of HEV have been documented in several countries around the world such as Japan, the United Kingdom, France, Denmark and Saudi Arabia (Al-sadeq et al. 2017). Therefore, blood transfusion has been considered as a potential cause of acute or chronic life-threatening HEV infections, especially in immune-compromised blood recipients. Despite that, there are not enough available studies yet to evaluate whether HEV is a considerable emergent blood pathogen for the blood supply in many countries in the world. Vertically transmitted HEV infection is known to cause acute hepatitis in newborn babies and had evidence of intrauterine HEV infection (Khuroo et al. 2009). HEV contamination (mother-to-foetus transmission) in neonates is a self-restricting illness and did not cause unending viremia or delayed clinical course.

Table 15.1 Hepatitis E cases in India from 2009 to 2018 based on hospital findings

Year	No. of cases reported	No. of deaths	Suspected cases	No. of deaths
2009[a]	0	0	152	3
2010	153	0	1102	3
2011	2750	4	2246	15
2012	3526	8	1641	6
2013	1153	1	400	0
2014	1097	12	746	0
2015	4460	26	145	0
2016	13,997	112	90	0
2017	1611	1	120	1
2018	353	0	255	0

Compiled from source: http://www.idsp.nic.in

[a]Data has been included from 25th week

As per the Integrated Disease Surveillance Programme (IDSP), India, report from 2009 to 2018, an aggregate of 29,100 cases of hepatitis E were reported and out of these 164 deaths occurred. Similarly, 6897 cases of suspected hepatitis have additionally been reported from India with 28 death cases, as shown in Table 15.1 (IDSP 2019). In all the cases, polluted drinking water with faecal containment was the primary reason for flare-ups, and anti-HEV IgM was well established in patients' blood serum. Various studies in the South and Southeast Asian countries confirmed the existence of genotypes 3 and 4 isolated from animals and animal food products (Kumar et al. 2019). However, in Western Asian countries, blood-borne hepatitis E was the main cause of infection (Hesamizadeh et al. 2016; Parsa et al. 2016; Nasrallah et al. 2017). In Europe, the number of hepatitis E cases has exponentially increased from 514 to 5617 in 10 years (2005–2015) (Aspinall et al. 2017). HEV genotype 3 is associated with indigenous cases, and in some cases the same genotype was found in European pork products. On the other hand, travel cases were related to genotype 1 or 4 (Aspinall et al. 2017).

In Chad, Kenya, Somalia, Sudan and Uganda (parts of Africa), an outbreak of hepatitis E infection has been recorded among the people living in shelters and refugee camps due to limited access to clean water and proper sanitary conditions. These peoples were highly vulnerable to HEV infection because malnutrition and crowded living exposed them to higher risk and infection due to a suppressed immune system (Teshale et al. 2010a).

Genotypes 1–3 of HEV have been documented in humans and animals in Argentina (HEV-3), Brazil (HEV-3), Colombia (HEV-3), Mexico (HEV-2, only humans), the United States (HEV-3), Uruguay (HEV-1 and HEV-3) and Venezuela (HEV-1 and HEV-3) (Melgaço et al. 2018). In Canada genotype 3 has been reported in the pork meat marketed in the retail shops and overall HEV rate was found to be 47% in pork pâtés and 10.5% in raw pork liver (Mykytczuk et al. 2017).

15.6 HEV and Zoonoses

Researchers have also started investigating different animal species for HEV markers, as it is known that HEV infection follows a symptomatic route. Till now, only HEV-3 and -4 have been identified in pigs globally. In the same way, both HEV-3 and -4 were discovered in boars and pigs in Japan (Takahashi et al. 2004, 2014). In addition to this, reports published from China and Japan revealed the presence of both HEV-3 and -4, but on the contrary the published report from the United States showed the presence of HEV-3 only in domestic and wild pigs (Pauli et al. 2015).

Moreover, HEV was detected in pork liver as well as other products which were commercially available for consumption in the United Kingdom, the United States, Germany and Japan (Pauli et al. 2015). HEV sample collected from rabbits of different countries like China, France and the United States, on phylogenetic analysis, showed high similarity with HEV-3. In addition, the phylogenetic analysis also revealed that rabbit HEV (rbHEV) has the potential to infect humans also (Izopet et al. 2012).

The HEV-like sequence obtained from rats of Germany phylogenetically was found to be a distinct genotype from HEV-1 to -4. Hence it was categorised into a separate group (Johne et al. 2010). Further, research conducted on the rats of the United States showed comparable HEV sequences (Purcell et al. 2011). Phylogenetical evaluation of different rat isolates confirmed the high variability of HEV sequence in contrast to HEV-1 to -4 (Mulyanto et al. 2013). These results prompted rat HEV to be categorised in *Orthohepevirus* or *Rocavirus* genus (Johne et al. 2014; Smith et al. 2014).

Interestingly, the experiment conducted on the horses of Egypt showed 13% of horses with antibodies against HEV and, moreover, three sequences showed high proximity with HEV-1 sequences (i.e. Egyptian human genotype) (Saad et al. 2007). In Australia and the United States, the hepatitis-splenomegaly syndrome has been recorded in chicken, and serological and genetic analysis showed the relation of avian HEV (aHEV) with human HEV (Payne et al. 1999; Haqshenas et al. 2001). Apart from this, HEV found in bats of Japan on BLASTn showed high similarity with German strain. Although the researcher reported the geographical distribution of HEV of bats with the different bat species, it failed to interlink the presence of a similar strain of HEV in Japan and Germany (Kobayashi et al. 2018).

In Japan, serum samples were also collected from mongooses to examine the presence of HEV antibodies with the help of ELISA and RT-PCR. The ELISA result revealed that only 8.3% of mongooses possessed antibodies against HEV, and RNA RT-PCR was not conducted. This study further proclaimed that mongooses should not be considered as the main HEV zoonotic reservoir (Li et al. 2006). But later study found the existence of HEV genotype 3 in wild mongoose and mongoose-derived HEV sequences were very similar to pigs (Nidaira et al. 2012).

15.7 Diagnosis of HEV

Various approaches, including demonstrative and diagnostic procedures, were applied for confirming and collecting information on HEV. Elevated values of liver parameters at routine intervals, including ALT (alanine aminotransferase), AST (aspartate aminotransferase), bilirubin, gamma-glutamyl transpeptidase and soluble phosphate, are some of non-specific markers of liver damage by HEV (Aggarwal and Jameel 2011). Immune electron microscopy (IEM) identifies HEV infection like particles in faecal samples, and these particles were precipitated with a native antibody to recover HEV from sera (Dienstag et al. 1976). HEV antigen can also be detected in liver biopsy tissue utilising immune fluorescence microscopy where HEV particles absorbed with fluorescent-labelled anti-HEV IgG antibody and complex are observed under a microscope equipped with the epifluorescent gadget (Krawczynski and Bradley 1989). The various approaches used for identification of HEV are shown in Fig. 15.4.

Patients' initial diagnosis, timely collection of faeces or serum samples, and suitable processing determine the precision of methods used for HEV RNA detection. In acute hepatitis E, the HEV RNA is usually detected during weeks 2–6 of illness

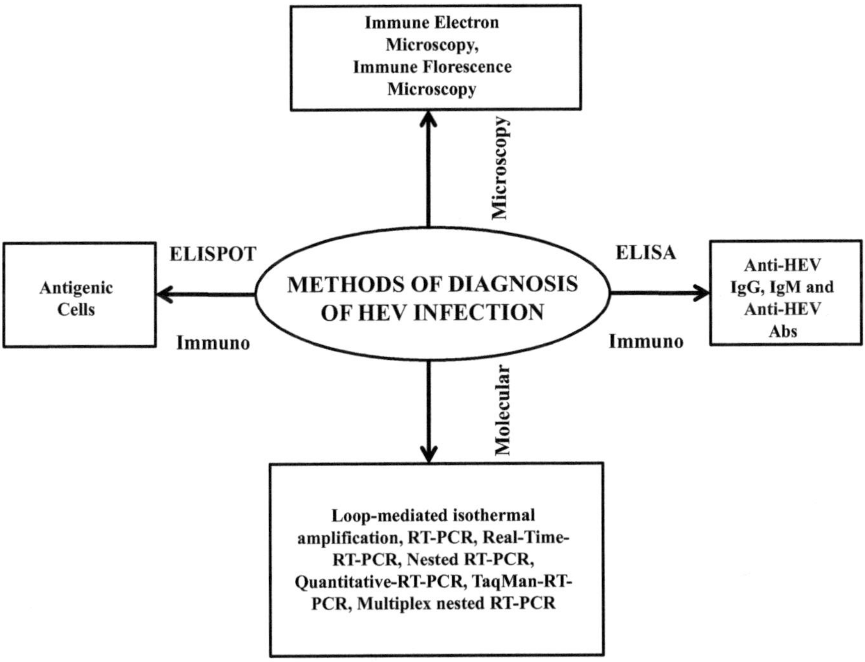

Fig. 15.4 Various approaches used for the diagnosis of hepatitis E virus (adapted from Kumar et al. 2019)

in serum and 3–8 weeks in stool (Al-sadeq et al. 2017, 2018). The detection of viral RNA in biological samples is the gold standard for the identification of severe HEV hepatitis as nucleic acid amplification techniques (NATs) can precisely recognise active infection (Huang et al. 2010). Generally, NAT-based identifications are not available at common diagnostic research centres, and it requires specialised trained human resources for analysis of samples (Mirazo et al. 2014). Over time, NAT tests based on reverse transcription followed by PCR, reverse transcription loop-mediated isothermal amplification and real-time PCR are currently used for detection of HEV RNA in stool and serum samples. NAT assays have been designed and optimised broadly to detect four HEV genotypes of human (Jothikumar et al. 2006; Gyarmati et al. 2007; Lan et al. 2009).

The in-house tests used for HEV nucleic acid detection show high variability in in vitro conditions (Baylis et al. 2011). Multiplex RT-PCR (single step) was used for detecting non-hepatitis A–C and HEV RNA present in the sera of infected patients. In this method, the consensus regions of all the viral genomes were used as a target sequence for amplification, and this method is considered rapid, sensitive and reproducible (Irshad et al. 2013).

A very little information was available regarding immunological aspects of the HEV infection until responses and patterns of antibodies acting against HEV were recognised and studied. The sensitivity and specificity of various commercial ELISA kits used for diagnosis of HEV infection are shown in Table 15.2. Anti-HEV

Table 15.2 Various types of commercial assays based on ELISA used for detection of HEV infection

Commercial assay	Manufacturer and location	Antigens used in HEV assay	Antibody detected	Sensitivity (%)	Specificity (%)	References
Wantai	Wantai Biological Pharmacy Enterprise Co., Beijing, China	IgM: micro-wells are pre-coated with anti-μ chain	IgM	97.7; 75; 65.4; 87.3; 96.7	99.6; >99; NS; 100; 96.7	Abravanel et al. (2013), Pas et al. (2013), Avellon et al. (2015), Zhou et al. (2008), Abravanel et al. (2015)
		IgG: recombinant HEV antigens	IgG	93.2; 72.5	97.8; NS	
Mikrogen	Mikrogen GmbH Neuried, Germany	Recombinant HEV-ORF2 from genotypes 1 and 3	IgM	38; 74; 75; 92	99; 99; NS; 95.6	Norder et al. (2016), Pas et al. (2013), Avellon et al. (2015), Drobeniuc et al. (2010)
			IgG	62; 72.5	99; NS	
Euroimmun	Euroimmun Medizinische Labordiagnostika AG, Lübeck, Germany	Recombinant antigens of HEV from genotypes 1 and 3	IgM	24; 61.5	100; NS	Drobeniuc et al. (2010), Avellon et al. (2015), Norder et al. (2016)
			IgG	42; 57.5	99; NS	
MP Diagnostics	MP Biomedicals, Singapore (Formerly Genlabs)	ORF2 antigen	IgM	74; 88; 59.6; 67.3; 72; 80; 72.5	84; 99.5; NS; 89.1; 93; 86.1; 93	Pas et al. (2013), Legrand-Abravanel et al. (2009), Avellon et al. (2015), Zhou et al. (2008), Abravanel et al. (2015), Wu et al. (2014), Khudyakov and Kamili (2011)
		3 ORF2 recombinant antigen	IgG	70; 73.3	NS; 65.3	
DSI	DSI S.R.l., Saronno, Italy	Recombinant ORF2 and ORF3 peptides from genotypes 1, 2 and 3	IgM	63; 71; 80.8	99; 90; NS	Norder et al. (2016), Pas et al. (2013), Avellon et al. (2015)
			IgG	72; 75	99; NS	
DS-EIA	RPC Diagnostic Systems, Nizhny Novgorod, Russia	ORF2 and ORF3 peptide from genotypes 1 and 3	IgM	71; 98	90; 95.6	Pas et al. (2013), Khudyakov and Kamili (2011)
			IgG	71	90	
DiaPro	Diagnostic Bioprobes S.r.l., Milan, Italy	Four synthetic peptides with conservative epitopes of ORF2 and ORF3 from genotypes 1, 2, 3 and 4	IgM	72; 81; 59.6	100; 98; NS	Norder et al. (2016), Pas et al. (2013), Avellon et al. (2015)
			IgG	98; 77.5	96; NS	

Axiom	Axiom Diagnostics, Worms, Germany	Recombinant carboxy-terminal of ORF2 from genotype 1	IgM	29	99	Norder et al. (2016)
			IgG	95	98	
Adaltis EIAgen	Adaltis S.r.l., Guidonia Montecelio, Italy	Synthetic immunodominant determinants encoded by ORF2 and ORF3	IgM	97.7; 90; 80	100; 100; 87.4	Abravanel et al. (2013), Legrand-Abravanel et al. (2009), Wu et al. (2014)
			IgG	81.8; 80	89.5; 62.9	
Fortress Diagnostics	Fortress Diagnostics, Ulster, UK	IgM: micro-wells are pre-coated with anti-μ chain	IgM and IgG	92; 100; 95	88; 86.2; 97	Galiana et al. (2008), Schnegg et al. (2013), Yan et al. (2008), Bendall et al. (2010)
		IgG: HEV recombinant antigen	IgG	99.5; 98	99.6; NS	
International Immuno-Diagnostics	International Immuno-Diagnostics, Foster City, USA	NM	IgM	82.4	91.7	Khudyakov and Kamili (2011)

NM not mentioned, *NS* not specified
Adapted from Al-Sadeq et al. (2018)

IgM appeared during severe HEV infection and was detectable after 4 days when symptoms of jaundice first appeared and held on for up to 5 months (Favorov et al. 1992). Generally, 90% of patients suffering from severe hepatitis E infection show detectable anti-HEV IgM after 14 days of infection, and in contrast to anti-HEV IgG antibodies are noticeable soon after the occurrence of anti-HEV IgM (Favorov et al. 1992). The way these two classes of antibodies develop at the same time in severe HEV infection makes it hard to diagnose the infection. Therefore, common immunoassays for anti-HEV IgM are only effective at 90–97% level, showing false-positive effects up to 10% and some of these results also show false-negative results with genotype 1 strain-infected patients (Herremans et al. 2007; Legrand-Abravanel et al. 2009; Drobeniuc et al. 2010). In commercial HEV serological test, ORF2, ORF3 antigens and immunodominant peptides are commonly utilised for recognising antibodies (IgA, IgM and IgG) effective against HEV (Takahashi et al. 2005; Zhang et al. 2009b). Sometimes the cross-serological reactivity of different viruses with HEV has also been linked with hepatitis infection (Hyams et al. 2014). The condition of the patient during serological testing is another limiting factor for HEV infection detection as sometimes during infection the seroconversion to HEV antibodies gets delayed or stopped. In such cases, quantiferon assay provides rapid results against HEV infection (Brown et al. 2016). In this assay, 616 overlapping peptides spanning open reading frames [ORFs 1–3] (i.e. genotype 3a peptide library) were used in IFN-γ (interferon-gamma) T-cell ELISpot assay.

15.8 Prevention and Vaccines

Providing clean water and improved sanitary conditions are the two most important preventive measures during HEV outbreaks. Implementation of these strategies on time is also quite challenging, especially when an epidemic occurs (Teshale and Hu 2011). Therefore, there is a need to develop a valid vaccine for regulating and reducing the chances of hepatitis E infection. Two hepatitis E immunisations have been used in clinical preliminaries, and the first of these, rHEV5, was produced by GlaxoSmithKline (GSK, Rixensart, Belgium) (Li et al. 2015). The trails of the rHEV5 were conducted on Nepalese army persons, but the vaccine has not shown any protection after one dose. However, vaccination with two doses may provide protection, but research failed to provide firm results (Shrestha et al. 2007).

On the other hand, HEV 239 (exchange name Hecolin®) was prepared by Innovax (Xiamen, China) (Zhang et al. 2009a; Zhu et al. 2010; Wu et al. 2012). HEV 239 was authorised in China in 2012, and Hecolin® is available in the Chinese market for use in individuals of more than 16 years of age. However, uses in other countries are still under approval (Park 2012; Riedmann 2012). This recombinant vaccine is highly effective as antigenically conserved nature of HEV exhibits only single serotype, which is found to be protective for all four HEV genotypes (HEV-1–4) (Melgaço et al. 2018).

Chinese Center for Disease Control and Prevention is using Internet framework to gather post-showcasing unfriendly responses without any worries with the immunisation. Considering that only China has experience with vaccination against hepatitis E, and HEV infection has remained a health problem, the cost-effectiveness of the HEV vaccine has been debated a lot. The Hecolin HEV 239 vaccine from Xiamen Innovax Biotech costs around USD 17.60–41.70 per dose, which is less expensive than hepatitis A vaccine (USD 23.21 per dose) (Riedmann 2012; Zhao et al. 2016). As immunisation can reduce the cost of hospitalisation and treatment, the launch of the hepatitis E vaccine could be a cost-effective intervention in the vaccine market. Still, the information is lacking in some areas, such as paediatric subjects (<16 years old), the elderly persons (>65 years old), pregnant ladies, people with hidden liver sicknesses or immunosuppressed individuals. These groups face a more prominent danger of HEV infection and are in high need of this vaccine (Li et al. 2015).

15.9 Conclusions and Prospects

Worldwide, HEV is considered to be the chief cause of non-A and -B enterically transmitted severe viral hepatitis. In the developing countries, the major reason for the widespread occurrence of this disease is poor sanitation conditions. However, in developed nations, zoonotic and blood transfusions are the main reasons for HEV infections. In epidemic conditions, the first step for preventing HEV infections is to have good-quality drinking water and proper sanitary conditions for human waste disposal. Different serological and molecular detection methods have been developed for robust detection of HEV infection by detecting HEV antigens, HEV antibody or RNA in patient's sera. Each technique has advantages and disadvantages compared to other approaches used in HEV detection. Detection of HEV RNA remains the standard gold method, and considerable improvements in vaccination and treatment have been achieved, but still limitations are there that bound the success of complete elimination of HEV.

Acknowledgements All the authors of the manuscript thank and acknowledge their respective universities and institutes.

Conflict of interest: There is no conflict of interest.

References

Abravanel F, Chapuy-Regaud S, Lhomme S et al (2013) Performance of anti-HEV assays for diagnosing acute hepatitis E in immunocompromised patients. J Clin Virol 58:624–628

Abravanel F, Lhomme S, Chapuy-Regaud S et al (2015) Performance of a new rapid test for detecting anti-hepatitis E virus immunoglobulin M in immunocompetent and immunocompromised patients. J Clin Virol 70:101–104

Agrawal S, Gupta D, Panda SK (2001) The 3′ end of hepatitis E virus (HEV) genome binds specifically to the viral RNA-dependent RNA polymerase (RdRp). Virology 282:87–101
Aggarwal R (2011) Clinical presentation of hepatitis E. Virus Res 161:15–22
Aggarwal R, Jameel S (2011) Hepatitis E. Hepatology 54:2218–2226
Al-Sadeq DW, Majdalawieh AF, Nasrallah GK (2017) Seroprevalence and incidence of hepatitis E virus among blood donors: a review. Rev Med Virol 2017:e1937
Al-Sadeq DW, Majdalawieh AF, Mesleh AG et al (2018) Laboratory challenges in the diagnosis of hepatitis E virus. J Med Microbiol 67:466–480
Arankalle VA, Paranjape S, Emerson SU et al (1999) Phylogenetic analysis of hepatitis E virus isolates from India (1976–1993). J Gen Virol 80:1691–1700
Aspinall EJ, Couturier E, Faber M et al (2017) Hepatitis E virus infection in Europe: surveillance and descriptive epidemiology of confirmed cases, 2005 to 2015. Euro Surveill 22:30561
Avellon A, Morago L, Garcia-Galera del Carmen M et al (2015) Comparative sensitivity of commercial tests for hepatitis E genotype 3 virus antibody detection. J Med Virol 87:1934–1939
Aye TT, Uchida T, Ma XZ et al (1992) Complete nucleotide sequence of a hepatitis E virus isolated from the Xinjiang epidemic (1986–1988) of China. Nucleic Acids Res 20:3512
Balayan MS, Andjaparidze AG, Savinskaya SS et al (1983) Evidence for a virus in non-A, non-B hepatitis transmitted via the fecal-oral route. Intervirology 20:23–31
Baylis SA, Hanschmann KM, Blümel J et al (2011) Standardization of hepatitis E virus (HEV) nucleic acid amplification technique (NAT)-based assays: an initial study to evaluate a panel of HEV strains and investigate laboratory. J Clin Microbiol 49:1234–1239
Bendall R, Ellis V, Ijaz S et al (2010) A comparison of two commercially available anti-HEV IgG kits and a re-evaluation of anti-HEV IgG seroprevalence data in developed countries. J Med Virol 82:799–805
Brown A, Halliday JS, Swadling L et al (2016) Characterization of the specificity, functionality, and durability of host T-cell responses against the full-length hepatitis E virus. Hepatology 64:1934–1950
Cao D, Huang YW, Meng XJ (2010) The nucleotides on the stem–loop RNA structure in the junction region of the hepatitis E virus genome are critical for virus replication. J Virol 84:13040–13044
Cao D, Meng XJ (2012) Molecular biology and replication of hepatitis E virus. Emerg Microbes Infect 1:e17
Chandra V, Taneja S, Kalia M et al (2008) Molecular biology and pathogenesis of hepatitis E virus. J Biosci 33:451–464
Dalton HR, Bendall R, Ijaz S et al (2008) Hepatitis E: an emerging infection in developed countries. Lancet Infect Dis 8:698–709
Dienstag JL, Alling DW, Purcell RH (1976) Quantitation of antibody to hepatitis A antigen by immune electron microscopy. Infect Immun 13:1209–1213
Drobeniuc J, Meng J, Reuter G et al (2010) Serologic assays specific to immunoglobulin M antibodies against hepatitis E virus: pangenotypic evaluation of performances. Clin Infect Dis 51:e24–e27
Emerson SU, Nguyen HT, Torian U et al (2010) Release of genotype 1 hepatitis E virus from cultured hepatoma and polarized intestinal cells depends on open reading frame 3 protein and requires an intact PXXP motif. J Virol 84:9059–9069
Emerson SU, Purcell RH (2003) Hepatitis E virus. Rev Med Virol 13:145–154
Emerson SU, Zhang M, Meng XJ et al (2001) Recombinant hepatitis E virus genomes infectious for primates: importance of capping and discovery of a cis-reactive element. Proc Natl Acad Sci U S A 98:15270–15275
Favorov MO, Fields HA, Purdy MA et al (1992) Serologic identification of hepatitis E virus infections in epidemic and endemic settings. J Med Virol 36:246–250
Fujiwara S, Yokokawa Y, Morino K et al (2014) Chronic hepatitis E: a review of the literature. J Viral Hepat 21:78–89

Galiana C, Fern_andez-Barredo S, García A et al (2008) Occupational exposure to hepatitis E virus (HEV) in swine workers. Am J Trop Med Hyg 78:1012–1015

Graff J, Nguyen H, Kasorndorkbua C et al (2005a) In vitro and in vivo mutational analysis of the 39-terminal regions of hepatitis E virus genomes and replicons. J Virol 79:1017–1026

Graff J, Nguyen H, Yu C et al (2005b) The open reading frame 3 gene of hepatitis E virus contains a cis-reactive element and encodes a protein required for infection of macaques. J Virol 79:6680–6689

Graff J, Torian U, Nguyen H et al (2006) A bicistronic subgenomic mRNA encodes both the ORF2 and ORF3 proteins of hepatitis E virus. J Virol 80:5919–5926

Gurav YK, Kakade SV, Kakade RV et al (2007) A study of hepatitis E outbreak in rural area of Western Maharashtra. Indian J Community Med 32:182–184

Guu TS, Liu Z, Ye Q et al (2009) Structure of the hepatitis E virus-like particle suggests mechanisms for virus assembly and receptor binding. Proc Natl Acad Sci U S A 106:12992–12997

Gyarmati P, Mohammed N, Norder H et al (2007) Universal detection of hepatitis E virus by two real-time PCR assays: TaqMan and primer-probe energy transfer. J Virol Methods 146:226–235

Haqshenas G, Shivaprasad HL, Woolcock PR et al (2001) Genetic identification and characterization of a novel virus related to human hepatitis E virus from chickens with hepatitis-splenomegaly syndrome in the United States. J Gen Virol 82:2449–2462

Herremans M, Bakker J, Duizer E et al (2007) Use of serological assays for diagnosis of hepatitis E virus genotype 1 and 3 infections in a setting of low endemicity. Clin Vaccine Immunol 14:562–568

Hesamizadeh K, Sharafi H, Keyvani H et al (2016) Hepatitis A virus and hepatitis E virus seroprevalence among blood donors in Tehran, Iran. Hepat Mon 16:e32215

He S, Miao J, Zheng Z et al (2008) Putative receptor-binding sites of hepatitis E virus. J Gen Virol 89:245–249

Huang FF, Sun ZF, Emerson SU et al (2004) Determination and analysis of the complete genomic sequence of avian hepatitis E virus (avian HEV) and attempts to infect rhesus monkeys with avian HEV. J Gen Virol 85:1609–1618

Huang S, Zhang X, Jiang H et al (2010) Profile of acute infectious markers in sporadic hepatitis E. PLoS One 5:e13560

Hyams C, Mabayoje DA, Copping R et al (2014) Serological cross reactivity to CMV and EBV causes problems in the diagnosis of acute hepatitis E virus infection. J Med Virol 86:478–483

ICTV (2019) International committee on taxonomy of viruses. https://talk.ictvonline.org/ictv-reports/ictv_online_report/positive-sense-rna-viruses/w/hepeviridae. Accessed 25 Mar 2019

IDSP (2019). Weekly outbreaks. http://www.idsp.nic.in. Accessed 1 Mar 2019

Intharasongkroh D, Sa-nguanmoo P, Tuanthap S et al (2017) Hepatitis E virus in pork and variety meats sold in fresh markets. Food Environ Virol 9:45–53

Ippagunta SK, Naik S, Sharma B et al (2007) Presence of hepatitis E virus in sewage in Northern India: frequency and seasonal pattern. J Med Virol 79:1827–1831

Irshad M, Ansari MA, Irshad K et al (2013) Novel single-step multiplex real-time polymerase chain reaction assay for simultaneous quantification of hepatitis virus A, B, C, and E in serum. J Gastroenterol Hepatol 28:1869–1876

Ishida S, Yoshizumi S, Ikeda T et al (2012) Detection and molecular characterization of hepatitis E virus in clinical, environmental and putative animal sources. Arch Virol 157:2363–2368

Izopet J, Dubois M, Bertagnoli S et al (2012) Hepatitis E virus strains in rabbits and evidence of a closely related strain in humans, France. Emerg Infect Dis 18:1274–1281

Izopet J, Kamar N (2008) Hepatitis E: from zoonotic transmission to chronic infection in immunosuppressed patients. Med Sci 24:1023–1025

Johne R, Dremsek P, Reetz J et al (2014) Heperviridae: an expending family of vertebrate viruses. Infect Genet Evol 27:212–229

Johne R, Plenge-Bonig A, Hess M et al (2010) Detection of novel hepatitis E-like virus in faeces of wild rats using a nested broad-spectrum RT-PCR. J Gen Virol 91:750–758

Jothikumar N, Cromeans TL, Robertson BH et al (2006) A broadly reactive one-step real-time RT-PCR assay for rapid and sensitive detection of hepatitis E virus. J Virol Methods 131:65–71

Kalia M, Chandra V, Rahman SA et al (2009) Heparan sulfate proteoglycans are required for cellular binding of the hepatitis E virus ORF2 capsid protein and for viral infection. J Virol 83:12714–12724

Kamar N, Dalton HR, Abravanel F et al (2014) Hepatitis E virus infection. Clin Microbiol Rev 27:116–138

Khudyakov Y, Kamili S (2011) Serological diagnostics of hepatitis E virus infection. Virus Res 161:94–92

Khuroo MS (1980) Study of an epidemic of non-A, non-B hepatitis. Possibility of another human hepatitis virus distinct from post-transfusion non-A, non-B type. Am J Med 68:818–824

Khuroo MS (1991) Hepatitis E: the enterically transmitted non-A, non-B hepatitis. Indian J Gastroenterol 10:96–100

Khuroo MS, Kamili S, Khuroo MS (2009) Clinical course and duration of viremia in vertically transmitted hepatitis E virus (HEV) infection in babies born to HEV-infected mothers. J Viral Hepat 16:519–523

Khuroo MS, Khuroo MS (2008) Hepatitis E virus. Curr Opin Infect Dis 21:539–543

Khuroo MS, Khuroo MS (2015) Sanitation and sewage disposal in India. JK-Practitioner 20:43–46. http://medind.nic.in/jab/t15/i1/jabt15i1p43.pdf. Accessed 1 Mar 2019

Khuroo MS, Khuroo MS, Khuroo NS (2016a) Hepatitis E: discovery, global impact, control and cure. World J Gastroenterol 22:7030–7045

Khuroo MS, Khuroo MS, Khuroo NS (2016b) Transmission of hepatitis E virus in developing countries. Viruses 8:E253

Kobayashi T, Murakami S, Yamamoto T et al (2018) Detection of bat hepatitis E virus RNA in microbats in Japan. Virus Genes 54:599–602

Koonin EV, Gorbalenya AE, Purdy MA et al (1992) Computer-assisted assignment of functional domains in the nonstructural polyprotein of hepatitis E virus: delineation of an additional group of positive-strand RNA plant and animal viruses. Proc Natl Acad Sci U S A 89:8259–8263

Krawczynski K, Bradley DW (1989) Enterically transmitted non-A, non-B hepatitis: identification of virus-associated antigen in experimentally infected cynomolgus macaques. J Infect Dis 159:1042–1049

Kumar H, Kamar N, Kumar D (2019) Hepatitis E: current status in India and other Asian countries. J Pure Appl Microbiol 13:141–159

Lan X, Yang B, Li BY et al (2009) Reverse transcription-loop-mediated isothermal amplification assay for rapid detection of hepatitis E virus. J Clin Microbiol 47:2304–2306

Legrand-Abravanel F, Thevenet I, Mansuy JM et al (2009) Good performance of immunoglobulin M assays in diagnosing genotype 3 hepatitis E virus infections. Clin Vaccine Immunol 16:772–774

Li SW, Zhao Q, Wu T et al (2015) The development of a recombinant hepatitis E vaccine HEV 239. Hum Vaccin Immunother 11:908–914

Li T-C, Saito M, Ogura G et al (2006) Serologic evidence for hepatitis E virus infection in mongoose. Am J Trop Med Hyg 74:932–936

Li TC, Yamakawa Y, Suzuki K et al (1997) Expression and self-assembly of empty virus-like particles of hepatitis E virus. J Virol 71:7207–7213

Masclaux FG, Hotz P, Friedli D et al (2013) High occurrence of hepatitis E virus in samples from wastewater treatment plants in Switzerland and comparison with other enteric viruses. Water Res 47:5101–5109

Melgaço JG, Gardinali NR, de VDM M et al (2018) Hepatitis E: update on prevention and control. Biomed Res Int 2018:5769201

Meng J, Dai X, Chang JC et al (2001) Identification and characterization of the neutralization epitope (s) of the hepatitis E virus. Virology 288:203–211

Meng XJ, Halbur PG, Haynes JS et al (1998) Experimental infection of pigs with the newly identified swine hepatitis E virus (swine HEV), but not with human strains of HEV. Arch Virol 143:1405–1415

Mirazo S, Ramos N, Mainardi V et al (2014) Transmission, diagnosis, and management of hepatitis E: an update. Hepat Med 6:45–59

Miyashita K, Kang JH, Saga A et al (2012) Three cases of acute or fulminant hepatitis E caused by ingestion of pork meat and entrails in Hokkaido, Japan: zoonotic food-borne transmission of hepatitis E virus and public health concerns. Hepatol Res 42:870–878

Mulyanto, Depamede SN, Sriasih M et al (2013) Frequent detection and characterization of hepatitis E virus variants in wild rats (*Rattus rattus*) in Indonesia. Arch Virol 158:87–96

Mykytczuk O, Harlow J, Bidawid S et al (2017) Prevalence and molecular characterization of hepatitis E virus in retail pork products marketed in Canada. Food Environ Virol 9:208–218

Nagashima S, Takahashi M, Jirintai S et al (2012) Tsg101 and the vacuolar protein sorting pathway are required for release of hepatitis E virions. J Gen Virol 92:2838–2248

Nanda SK, Panda SK, Durgapal H et al (1994) Detection of the negative strand of hepatitis E virus RNA in the livers of experimentally infected rhesus monkeys: evidence for viral replication. J Med Virol 42:237–240

Nasrallah GK, Al-Absi ES, Ghandour R et al (2017) Seroprevalence of hepatitis E virus among blood donors in Qatar (2013–2016). Transfusion 57:1801–1807

Navaneethan U, Al Mohajer M, Shata MT (2008) Hepatitis E and pregnancy: understanding the pathogenesis. Liver Int 28:1190–1199

Nidaira M, Takahashi K, Go O et al (2012) Detection and phylogenetic analysis of hepatitis E viruses from mongooses in Okinawa, Japan. J Vet Med Sci 74:1665–1668

Norder H, Karlsson M, Mellgren Å et al (2016) Diagnostic performance of five assays for anti-hepatitis E virus IgG and IgM in a large cohort study. J Clin Microbiol 54:549–555

Park SB (2012) Hepatitis E vaccine debuts. Nature 491:21–22

Parsa R, Adibzadeh S, Behbahani AB et al (2016) Detection of hepatitis E virus genotype 1 among blood donors from Southwest of Iran. Hepat Mon 16:e34202

Pas SD, Streefkerk RH, Pronk M et al (2013) Diagnostic performance of selected commercial HEV IgM and IgG ELISAs for immunocompromised and immunocompetent patients. J Clin Virol 58:629–634

Pauli G, Aepfelbacher M, Bauerfeind U et al (2015) Hepatitis E virus. Transfus Med Hemother 42:247–265

Pavio N, Meng XJ, Doceul V (2015) Zoonotic origin of hepatitis E. Curr Opin Virol 10:34–41

Payne CJ, Ellis TM, Plant SL et al (1999) Sequence data suggests big liver and spleen disease virus (BLSV) is genetically related to hepatitis E virus. Vet Microbiol 68:119–125

Pinto MA, de Oliveira JM, González J (2017) Hepatitis A and E in South America: new challenges toward prevention and control. In: Human virology in Latin America. Springer, Cham, pp 119–138

Poovorawan K, Jitmitrapab S, Treeprasertsuk S et al (2014) Risk factors and molecular characterization of acute sporadic symptomatic hepatitis E virus infection in Thailand. Asian Pac J Trop Med 7:709–714

Purcell RH, Engle RE, Rood MP et al (2011) Hepatitis E virus in rats, Los Angeles, California, USA. Emerg Infect Dis 17:2216–2222

Rasche A, Saqib M, Liljander AM et al (2016) Hepatitis E virus infection in dromedaries, North and East Africa, United Arab Emirates, and Pakistan, 1983–2015. Emerg Infect Dis 22:1249–1252

Rein DB, Stevens GA, Besides TJ (2012) The global burden of hepatitis E virus genotypes 1 and 2 in 2005. Hepatology 55:988–997

Rendon J, Hoyos MC, Di Filippo D et al (2016) Hepatitis E virus genotype 3 in Colombia: survey in patients with clinical diagnosis of viral hepatitis. PLoS One 11:e0148417

Riedmann EM (2012) Chinese biotech partnership brings first hepatitis E vaccine to the market. Hum Vaccin Immunother 8:1743–1744

Reyes GR, Purdy MA, Kim JP et al (1990) Isolation of a cDNA from the virus responsible for enterically transmitted non-A, non-B hepatitis. Science 247:1335–1339

Saad MD, Hussein HA, Bashandy MM et al (2007) Hepatitis E virus infection in work horses in Egypt. Infect Genet Evol 7:368–373

Schnegg A, Bürgisser P, André C et al (2013) An analysis of the benefit of using HEV genotype 3 antigens in detecting anti-HEV IgG in a European population. PLoS One 8:e62980

Surjit M, Jameel S, Lal SK (2004) The ORF2 protein of hepatitis E virus binds the 5′ region of viral RNA. J Virol 78:320–328

Suwannakarn K, Tongmee C, Theamboonlers A et al (2010) Swine as the possible source of hepatitis E virus transmission to humans in Thailand. Arch Virol 155:1697–1699

Shrestha MP, Scott RM, Joshi DM et al (2007) Safety and efficacy of a recombinant hepatitis E vaccine. N Engl J Med 356:895–903

Smith DB, Simmonds P, Jameel S et al (2014) Consensus proposals for classification of the family Hepeviridae. J Gen Virol 95:2223–2232

Takahashi K, Kitajima N, Abe N et al (2004) Complete or near-complete nucleotide sequences of hepatitis E virus genome recovered from a wild boar, a deer, and four patients who ate the deer. Virology 330:501–505

Takahashi M, Kusakai S, Mizuo H et al (2005) Simultaneous detection of immunoglobulin A (IgA) and IgM antibodies against hepatitis E virus (HEV) is highly specific for diagnosis of acute HEV infection. J Clin Microbiol 43:49–56

Takahashi M, Nishizawa T, Nagashima S et al (2014) Molecular characterization of a novel hepatitis E virus (HEV) strain obtained from wild boar in Japan that is highly divergent from previously recognized HEV strains. Virus Res 180:59–69

Tam AW, Smith MM, Guerra ME et al (1991) Hepatitis E virus (HEV): molecular cloning and sequencing of the full-length viral genome. Virology 185:120–131

Tang X, Yang C, Gu Y et al (2011) Structural basis for the neutralization and genotype specificity of hepatitis E virus. Proc Natl Acad Sci U S A 108:10266–10271

Temmam S, Besnard L, Andriamandimby SF et al (2013) High prevalence of hepatitis E in humans and pigs and evidence of genotype-3 virus in swine, Madagascar. Am J Trop Med Hyg 88:329–338

Teshale EH, Hu DJ (2011) Hepatitis E: epidemiology and prevention. World J Hepatol 3:285–291

Teshale EH, Howard CM, Grytdal SP et al (2010a) Hepatitis E epidemic, Uganda. Emerg Infect Dis 16:126–129

Teshale EH, Hu DJ, Holmberg SD (2010b) The two faces of hepatitis E virus. Clin Infect Dis 51:328–334

Vaidya SR, Tilekar BN, Walimbe AM et al (2003) Increased risk of hepatitis E in sewage workers from India. J Occup Environ Med 45:1167–1170

Viswanathan R (1957) Infectious hepatitis in Delhi (1955–56): a critical study-epidemiology 1957. Natl Med J India 26:362–377

WHO (2017) The global hepatitis report, 2017. https://apps.who.int/iris/bitstream/handle/10665/255016/9789241565455-eng.pdf;jsessionid=09784545BEEE72E45785F98132749D65?sequence=1. Accessed 1 Mar 2019

Williams TP, Kasorndorkbua C, Halbur PG et al (2001) Evidence of extrahepatic sites of replication of the hepatitis E virus in a swine model. J Clin Microbiol 39:3040–3046

Wong DC, Purcell RH, Sreenivasan MA et al (1980) Epidemic and endemic hepatitis in India: evidence for a non-A, non-B hepatitis virus aetiology. Lancet 2:876–879

Woo PC, Lau SK, Teng JL et al (2014) New hepatitis E virus genotype in camels, the Middle East. Emerg Infect Dis 20:1044–1048

Woo PC, Lau SK, Teng JL et al (2016) New hepatitis E virus genotype in Bactrian camels, Xinjiang, China, 2013. Emerg Infect Dis 22:2219–2221

Wu T, Li SW, Zhang J et al (2012) Hepatitis E vaccine development: a 14year odyssey. Hum Vaccin Immunother 8:823–827

Wu WC, Su CW, Yang JY et al (2014) Application of serologic assays for diagnosing acute hepatitis E in national surveillance of a nonendemic area. J Med Virol 86:720–728
Xing L, Li TC, Mayazaki N et al (2010) Structure of hepatitis E virion-sized particle reveals an RNA-dependent viral assembly pathway. J Biol Chem 285:33175–33183
Xing L, Wang JC, Li TC et al (2011) Spatial configuration of hepatitis E virus antigenic domain. J Virol 85:1117–11124
Yamada K, Takahashi M, Hoshino Y et al (2009a) ORF3 protein of hepatitis E virus is essential for virion release from infected cells. J Gen Virol 90:1880–1891
Yamada K, Takahashi M, Hoshino Y et al (2009b) Construction of an infectious cDNA clone of hepatitis E virus strain JE03-1760F that can propagate efficiently in cultured cells. J Gen Virol 90:457–462
Yamashita T, Mori Y, Miyazaki N et al (2009) Biological and immunological characteristics of hepatitis E virus-like particles based on the crystal structure. Proc Natl Acad Sci U S A 106:12986–12991
Yan Q, Du HL, Wang YB et al (2008) Comparison of two diagnostic reagents to detect anti-hepatitis E virus IgG antibodies. Chin J Zoon 24:1087–1089
Yugo DM, Meng XJ (2013) Hepatitis E virus: foodborne, waterborne and zoonotic transmission. Int J Environ Res Public Health 10:4507–4533
Yu H, Li S, Yang C et al (2011) Homology model and potential virus-capsid binding site of a putative HEV receptor Grp78. J Mol Model 17:987–995
Zhang J, Liu CB, Li RC et al (2009a) Randomized-controlled phase II clinical trial of a bacterially expressed recombinant hepatitis E vaccine. Vaccine 27:1869–1874
Zhang S, Tian D, Zhang Z et al (2009b) Clinical significance of anti-HEV IgA in diagnosis of acute genotype 4 hepatitis E virus infection negative for anti-HEV IgM. Dig Dis Sci 54:2512–2518
Zhao Y, Zhang X, Zhu F et al (2016) A preliminary cost-effectiveness analysis of hepatitis E vaccination among pregnant women in epidemic regions. Hum Vaccin Immunother 12:2003–2009
Zheng ZZ, Miao J, Zhao M et al (2010) Role of heat-shock protein 90 in hepatitis E virus capsid trafficking. J Gen Virol 91:1728–1736
Zhou H, Jiang CW, Li LP et al (2008) Comparison of the reliability of two ELISA kits for detecting IgM antibody against hepatitis E virus. Chin J Prev Med 42:667–671
Zhou Y, Emerson SU (2006) P.302 Heat shock cognate protein 70 may mediate the entry of hepatitis E virus into host cells. J Clin Virol 36:S155
Zhu FC, Zhang J, Zhang XF et al (2010) Efficacy and safety of a recombinant hepatitis E vaccine in healthy adults: a large-scale, randomised, double-blind placebo-controlled, phase 3 trial. Lancet 376:895–902